D0895519

In a very short time, lasers have advanced from laboratory curiosities to increasingly useful, commercially available tools for material processing, precision measurements, surgery, communication, and even entertainment.

This book provides the background in theoretical physics necessary to understand engineering applications of lasers and optics. It summarizes relevant theories of geometrical optics, physical optics, quantum optics, and laser physics, connecting these theories to applications in such areas as fluid mechanics, combustion, surface analysis, material processing, and laser machining. Discussions of advanced topics such as laser Doppler velocimetry, laser-induced fluorescence, and holography are simplified, yet are sufficiently detailed to enable the reader to evaluate existing systems and design new ones.

The book includes numerous examples and homework problems to illustrate important points. A unique feature is the advanced research problems in each chapter that simulate real-world research and encourage independent reading and analysis.

Engineers and students will find this book a thorough and easy-to-understand source of information.

Introduction to
Optics and Lasers in Engineering

Introduction to Optics and Lasers in Engineering

Gabriel Laufer
University of Virginia

CAMBRIDGE
UNIVERSITY PRESS

Published by the Press Syndicate of the University of Cambridge
The Pitt Building, Trumpington Street, Cambridge CB2 1RP
40 West 20th Street, New York, NY 10011-4211, USA
10 Stamford Road, Oakleigh, Melbourne 3166, Australia

First published 1996

Printed in the United States of America

Library of Congress Cataloging-in-Publication Data

Laufer, Gabriel

Introduction to optics and lasers in engineering / Gabriel Laufer.

p. cm.

Includes bibliographical references and index.

ISBN 0-521-45233-3 (hc)

1. Lasers in engineering. 2. Optics. I. Title.
TA367.5.L39 1996
621.36′6 - dc20 95-44046
CIP

A catalog record for this book is available from the British Library.

ISBN 0-521-45233-3 Hardback

In memory of my dear parents,
Dr. Dov and Shoshana Laufer

Contents

Preface

Over the last two decades, such new technologies as laser-aided material processing, optical communication, holography, and optical measurement techniques have advanced from laboratory curiosities to commercially available engineering tools. Faced with these developments, engineers must now apply in their practice many concepts of physics that in the past were considered outside the boundaries of classical engineering. Advanced levels of electromagnetic theory and of quantum mechanics are now required to answer fundamental questions about interference, diffraction, or polarization of light and their applications. For example: How does radiation interact with gases, liquids, and solids? How does one obtain optical gain? When should a laser be used, and when would an ordinary light source suffice? How can a laser beam be produced and focused? Answers to these questions are essential for selecting an optical technique for measuring the properties of gas, liquid, or solid phases, or when designing a laser system for material processing, surgery, communication, or entertainment. Although most engineering students attend two or three undergraduate courses in physics, they seldom acquire the proficiency required to fully understand the intricacies of modern optics or laser applications. Similarly, midlevel engineers who obtained their formal education before many new techniques were developed may find an increasing gap between the knowledge acquired in undergraduate physics classes and the requirements of professional practice. Because most applications of engineering interest can be analyzed using relatively simple models of the physics of light and lasers, a single course and a single textbook depicting these models and the engineering applications can provide sufficient background for most engineers. Unfortunately, the choice of such textbooks is very limited.

This book was developed to bridge a gap between abstract concepts of physics and practical engineering problems. It summarizes the relevant theories of geometrical optics, physical optics, quantum optics, and laser physics, and connects them to numerous engineering applications. This text navigates the fine line separating extensive descriptions of theoretical physics and the approximate solutions needed for most applications. Relevant theories are described at an advanced level while maintaining a relatively simple mathematical presentation. In addition, an effort was made at each step to elucidate the physical meaning of the mathematical results and the engineering implications. The book has been written primarily for use in graduate courses in mechanical, aerospace, chemical, and nuclear engineering programs. Nevertheless, practicing scientists and engineers in these and other disciplines who are interested in this rapidly developing field should also find the book useful. Owing to the diversity of tasks that these engineering disciplines must face, the scope of the book is broader than that of most engineering optics books. However, with the broad background so provided, it is expected that the reader may successfully follow the advanced research and technical literature in this field.

This book is an outcome of a graduate-level course on laser applications in engineering that I have taught since 1980 and continuously updated. The content, examples, and problems have evolved significantly, with more technical papers reviewed and added to the course, new applications included, many more homework problems developed and solved, and new classroom demonstrations tested. Some recent technical papers became the subject of in-class examples and of homework assignments. Through interaction with students, the methods of presentation were modified and new topics were included to accommodate the needs and the backgrounds of engineering students. I believe that, after a decade of refinement, the contents and format of this text reflect the interests and needs of a large audience.

Although an important objective of the book is to introduce as many applications of engineering interest as possible, it must follow a sequence dictated by the gradually increasing complexity of the physical theories of optics and lasers. Therefore, the organization of the book follows the logical sequence of these theories, which can be effectively divided into four parts: geometrical optics (covered by Chapter 2); physical optics (Chapters 3, 4, 5, 6, and 7); quantum optics (Chapters 8, 9, 10, and 11); and laser physics (Chapters 12 and 13). Accordingly, the discussion of technical applications had to fit into this general structure. Hence, applications that can be explained by the simpler models of geometrical optics are discussed in Chapter 2, whereas discussion of applications requiring more advanced theories such as quantum optics are discussed in later chapters. This is in contrast to most monographs, where applications to one field of engineering are discussed separately from applications to another field. Some of the most important applications are discussed in detail in dedicated sections. Less notable applications are introduced through sample problems or as homework assignments, an attractive approach to engineering students who prefer to solve problems of practical relevance.

The chapter on geometrical optics has two objectives: to demonstrate how simple models can often be used to analyze advanced technical applications, and to introduce the concept of ray transfer matrices for the analysis of multi-element optical systems. Since most students are expected to be versed in geometrical optics, simple concepts such as the laws of refraction and reflection or the lens equation are mentioned only in passing. Instead, more involved problems associated with these laws are discussed – for example, propagation through media with continuously varying index of refraction, Fresnel lenses, propagation through multi-element optical systems, and the shadowgraph and schlieren techniques for flow visualization.

The chapters covering physical optics consist of a revision of topics included in undergraduate physics classes. They are distinguishable from similar discussion in standard texts in that they are coupled with related discussions of engineering applications. Chapter 3, on the electromagnetic theory and Maxwell's equations, adds completeness to the book, provides an easy reference, and prevents the use of inconsistent terminology and nomenclature. It is anticipated that most instructors using this book as a text will want to review these topics. The wave equations are derived from Maxwell's equations in Chapter 4, and are solved in one dimension. These results are used in subsequent chapters to explain such phenomena as polarization (Chapter 5), interference (Chapter 6), and diffraction (Chapter 7). Also described are optical devices that rely on these phenomena: polarizers, diffraction gratings, quarter- and half-wave plates, Brewster windows, liquid crystal displays, and more. Examples of technical applications discussed in this part of the book include laser Doppler velocimetry, ellipsometry, interferometry, moiré techniques, and speckle pattern interferometry.

The third part of the book, on quantum optics, provides the link between quantum mechanics and applied optics. Although solutions to problems that involve atomic or molecular systems require abstract mathematical techniques such as operator algebra, most microscopic systems of engineering interest (e.g., the N_2, O_2, H_2O, and CO_2 molecules) have been studied in great detail, and their critical parameters – absorption and emission wavelengths, transition probabilities, and energies of most levels – are tabulated in the literature to a high degree of accuracy. Consequently, most engineers rarely need to derive or measure any of these parameters. They need, rather, to understand the fundamental principles of quantum mechanics, and must acquire the ability to identify and locate the necessary parameters. Therefore, the discussion in this part of the book skips the elaborate mathematical techniques of quantum mechanics while highlighting its physical interpretation. The postulates of quantum mechanics are introduced in Chapter 8, along with the historical background leading to them. Using these postulates, such fundamental observations as the quantization of energy, the probabilistic nature of microscopic systems, and the duality in the representation of waves and matter are presented. This background is sufficient to discuss the structure of single atoms, diatomic molecules, rotational, vibrational, and electronic levels, and the selection rules for single-photon transitions (Chapter 9). The three modes of radiative interaction –

absorption, spontaneous emission, and stimulated emission – are introduced in the following chapter, together with the concepts of line broadening and predissociation. The statistical models for systems in thermodynamic equilibrium are also introduced briefly in Chapter 10, and include (without proof) the equations of blackbody radiation and Boltzmann's distribution. These models allow the reader to connect the behavior of a single molecule or a single atom to the behavior of practical systems containing large numbers of molecules interacting with one another and with a large number of photons. Spectroscopic techniques, which use the effects of blackbody radiation or the equilibrium population distribution for measurements of temperature and species density, are discussed in Chapter 11. Examples include applications of laser-induced fluorescence techniques for gas diagnostics, thermographic phosphor and pressure paint techniques for surface analysis, and more.

The objective of the fourth part of the text is to introduce the fundamental principles of laser oscillators (Chapter 12) and propagation of laser beams (Chapter 13). Just as in the previous chapters, these discussions are coupled with a description of relevant devices and engineering applications. To keep the explanations at a fundamental level, most of the cases described in this section are idealized (e.g., the small-signal gain or the Gaussian distribution of a laser beam); more complex problems (e.g., high-order transverse laser modes) are mentioned only briefly. Nevertheless, even with such a simplified approach, techniques for pulsing laser beams such as *Q*-switching and mode-locking can be described, along with requirements for successful operation and estimates of pulse duration and pulse power. Similarly good estimates of the geometrical characteristics of a laser beam – for example, the variation of its diameter along the propagation path or its divergence angle – can be obtained from such fundamental analysis. This part is also used to present criteria for successful application of lasers for material processing (e.g., cutting or heat treatment) or for the delivery of laser beams over a long distance (e.g., space communications or distance measurement).

The examples and homework problems in the text were often derived from problems encountered in engineering practice or research. Since the objective of most graduate programs is to prepare students for independent research, I included in each chapter a few research problems, which are a unique feature of the book. To solve them, students must obtain, read independently, analyze, and critically evaluate papers describing various techniques that were not discussed in the book. These problems are intended to simulate real-life research, where earlier results are coupled with existing knowledge to produce new results. All problems in this book – including the research problems – have been solved by my students in previous graduate courses. These students are too numerous to be named individually, but I would like to thank and acknowledge each one of them for their critical review of these problems and for their many helpful suggestions. A special word of appreciation is due to the editors of this book, Florence Padgett and Matt Darnell. Their meticulous work behind the scenes has added much quality to the technical side of the book, its consistency, and appearance. Finally, I would like to thank my dear wife Liora and my children Aharon, Tammar, and Dan. While theirs is not a technical contribution, their love and support were just as essential.

Introduction

I see trees of green, red roses too,
I see them bloom for me and you,
and I think to myself
What a wonderful world.

I see skies of blue and clouds of white,
the bright blessed day, the dark sacred night,
and I think to myself
What a wonderful world.

The colors of the rainbow, so pretty in the sky
are also on the faces of the people goin' by . . . . *

Explanation of the various effects of *light* is a very elusive task. Although light has captured the imagination of human beings since the dawn of civilization, science has yet to deliver a single, comprehensive explanation of all its effects. The advanced theories that now exist create many new questions along with new answers. Part of the confusion can be blamed on our tendency to explain physical phenomena using the perception of our senses. Unfortunately, our senses do not tell the full story. Although we can see light, and even distinguish among some of its colors, we cannot see most of the radiation emitted by the sun. Even our ability to visually determine the brightness of light sources is limited by the rapid saturation of the eye retina. Our senses tell us that light propagates in straight lines, yet careful experiments have demonstrated that the trajectories of light can be bent by gravitation. We cannot even capture and store light. We may observe that light propagates from a source until it is trapped by a target, but it does not seem to require any medium to carry it. We can demonstrate that the behavior of light is wavelike, but we also have sufficient evidence to indicate that it is made of particles that cannot be split. Although we know that these particles of light have momentum, they do not have any mass when at rest. Furthermore, although these momentum-carrying particles travel faster than any other particle, some cannot penetrate glass whereas others can penetrate metals.

To most people, light is simply what makes vision possible. However, on a bright day we can sense sunlight by its warmth on our skin even when our eyes are closed. We also know that the visual capabilities of animals and insects are different from ours. Therefore, to them light is not the same as it is to us. Evidently, the perception afforded by our eyes is insufficient to understand the nature of light. Contributions by celebrated physicists such as Archimedes, Newton, Maxwell, and Einstein helped us to develop theories that explain most of our observations. Using these theories, we have learned how to construct many of our modern devices. Theories formulated by Archimedes are still used to calculate simple focusing mirrors, and theories derived by Einstein were required before the first laser could be constructed.

Despite the advances of modern technology, many of the mysteries of light remain unresolved. Historically, new interpretations of the physical nature of light not only helped us to augment our understanding but also led to new technical applications. The most dramatic leap occurred at the turn of this century, when the theories of relativity and quantum mechanics emerged. Many of the applications discussed in this book could not be developed before these theories were understood. Today it is hard to imagine that many of the interpretations of the nature of light were not known even 100 years ago. However, despite the apparent progress, we still do not have one universal description of the nature of light. Instead, we have created several theories, each explaining only some of the effects while failing to describe others. As with many other engineering theories, the strength of this approach is that we can use these theories to design and build applicable devices, which operate as predicted as long as their range of application does not exceed the assumptions made by the theory. Thus, lenses and imaging devices may be designed by relatively simple geometrical optics techniques. If an evaluation of the chromatic correction for a lens is required, some aspects of physical optics must be introduced. To estimate the attenuation of imaged radiation due to absorption by some of the optical elements (e.g. filters), or to evaluate the emission by some light sources, quantum mechanical concepts may need to be used.

Any design in which optics or lasers are an integral part requires consideration of some of the theories of optics and radiation. Although these theories were traditionally part of other disciplines, the advent of modern applications has eliminated this artificial distinction. This book is an attempt to bridge the gap that may still exist between purely physical concepts and engineering applications. For simplicity, the text is structured along the lines of major theories of light and of radiation. The simplest theory, geometrical optics, is thus presented first, followed by physical optics and then theories about the interaction of radiation with matter.

1 Radiometry

1.1 Introduction

In the introduction we saw that explaining the concept of light may require more than just one theory. However, before we begin our journey through the disciplines of optics, we must identify the physical parameters needed to quantify the phenomena that are associated with light. But even before that, we should recognize that the phenomenon we call light is only a part of the broader phenomenon of radiation. If we consider radiation to be

the emission and/or propagation of energy through space in the form of electromagnetic waves or indivisible energy quanta,

then light may be defined (American National Standard 1986) as

the part of radiation that is spectrally detectable by the eye.

(This definition is sometimes extended to include ultraviolet and infrared radiation.) Note that, by the present definition, radiation that is spectrally detectable to the eye will be called light even if it is too faint to be seen. Although these definitions include terms (such as electromagnetic waves and spectra) that are yet to be explained, it is evident that light represents a subcategory of radiation. Thus, once the physical quantities that specify radiation are defined they may also be used for the quantification of light. Furthermore, the measurement of radiation – or *radiometry* – does not depend on how we define what is detectable by the eye. Therefore, radiometry is more objective than *photometry,* which is the process of measuring light. The quantities obtained by radiometry are sufficient for most engineering applications. Photometry is used primarily in technology associated with vision, which includes specifications of computer monitors, photographic films, color mixing, and lighting. These topics are beyond the scope of this book. (See American National Standards 1986 for nomenclature and definitions of photometry terms.) In the sections that follow we will discuss the radiometry terms to be used in this book.

The definition of radiation (see e.g. Edwards 1989 and Roberts 1991) suggests that radiometry must include: measurements of the energy that is transferred; the properties of the waves that carry it, such as wavelength or frequency; and the properties of the energy quanta, which are called photons. Since the properties of photons can be specified using properties of the electromagnetic waves, the list of parameters required for radiometry can be somewhat reduced. Using the American National Standard (1986), we will define parameters that specify the rate of transfer of the radiative energy, and will separately describe the characteristics of the electromagnetic wave that carries that energy.

1.2 Energy Transfer by Radiation

The primary effect of radiation is to transfer energy from a source to a target. However, since radiation can neither be stored nor brought to rest, a more natural parameter to consider is the *radiant flux* Φ, which is the time rate of flow of radiant energy. Using SI units, this flux is given in watts. As an example, the flux of the sun in Figure 1.1(a) is the rate at which all the emitted energy crosses an imaginary envelope marked by the dashed circle. Clearly, that flux is independent of the radius of the imaginary envelope. Of course, we do not expect to capture all the sun's energy. Therefore, an alternative measure of the radiation emitted by the sun will be the exitance $M(x)$ of a point x on the sun; see Figure 1.1(b). *Exitance* is the total radiant flux $\Phi(x)$ leaving an area element dA_S at point x on the source:

$$M(x) = \frac{d\Phi(x)}{dA_S} \text{ W/m}^2. \tag{1.1}$$

Previous definitions distinguished between emittance, or radiation emitted *by* the source, and exitance, which includes all radiation (e.g. reflection and transmission) emerging from a surface. The present American National Standard defines only the term exitance. For measurements of the exitance, all rays that emerge from a surface must be included. Technically this task may prove to be impossible. However, the exitance can sometimes be inferred from other measurements such as the source color or temperature.

After emerging from the source, radiation may propagate indefinitely until it encounters a scattering source or a target. There is very little interest in quantifying radiation along its path, but quantifying it at the target is an essential part of the design of any optical device. Therefore, parameters that are specific to the point of incidence must be defined. For example, the design of a planar solar collector requires that the power that falls on an area element from all possible directions be specified. This parameter, the irradiance, is similar to the exitance, except that its measurement includes radiation that is approaching the surface from all possible directions. Therefore, the *irradiance* $E(x)$ is defined as the total radiation energy from all possible directions that falls per unit time on a unit area A_T at point x:

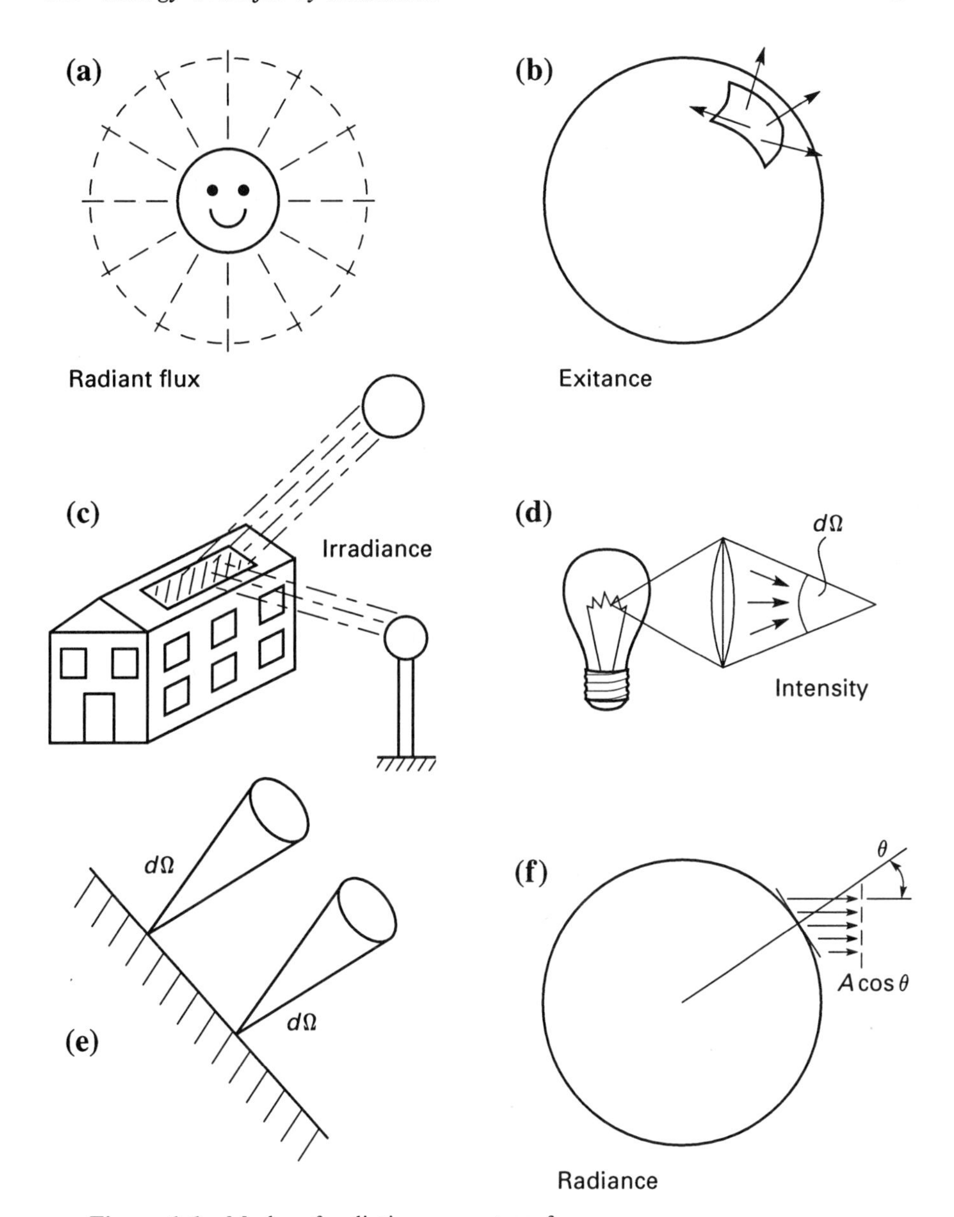

Figure 1.1 Modes of radiative energy transfer.

$$E(x) = \frac{d\Phi(x)}{dA_T} \text{ W/m}^2. \tag{1.2}$$

When more than one source is present, as in Figure 1.1(c), evaluation of the irradiance requires that radiation from all sources be included.

The design of telescopes or lenses requires that the power captured by these devices be specified. When the source is near the imaging device, as in Figure 1.1(d), the captured radiative power increases with the collection solid angle Ω. To determine the power collected by this lens, the intensity $I(\theta)$ must be specified. The *intensity* is defined as the total radiation emitted by the source, per unit time, along a line in the θ direction and within a solid angle $d\Omega$:

$$I(\theta) = \frac{d\Phi(\theta)}{d\Omega} \text{ W/sr.} \tag{1.3}$$

If uninterrupted, all the radiation power that is included within a solid angle Ω will be captured by an optical device with the same or larger collection angle. Note that mathematically a solid angle must have an apex and so the definition of intensity may strictly apply only to point sources. However, in reality a lens (or a telescope) can image a finite area of the source. Therefore, if the size of the imaged source is negligible relative to the distance, the intensity may include all the small cones that are oriented in the θ direction as in Figure 1.1(e) and that have a solid angle of $d\Omega$.

When the distance between an imaging device and a source increases, so that the collection solid angle Ω is approaching the infinitesimal limit $d\Omega$, all rays that are collected by the lens are almost parallel to each other. This is always true in astronomical applications. If the plane of the lens is perpendicular to these rays then the total radiation power captured by the lens can be defined by the radiance $L(x,\theta)$. The radiance is defined as the portion of the radiative flux that is contained within a cone with a solid angle $d\Omega$ pointing in a direction θ and crossing an area element that is projected normal to the direction of propagation; see Figure 1.1(f). Mathematically, this may be written as

$$L(x,\theta) = \frac{\partial^2\Phi(x,\theta)}{\partial\Omega\,\partial A(x)\cos\theta} = \frac{\partial I(x,\theta)}{\partial A(x)\cos\theta} \text{ W/sr-m}^2. \tag{1.4}$$

This definition does not distinguish between the radiation leaving a source or radiation incident upon a target. Therefore, the term ∂A in (1.4) represents an area element on an imaginary envelope either surrounding the source or attached to it. With this definition, the target area (e.g. collection lens) is normal to the incident rays, and the radiance specifies the power per unit area that falls on that target or lens.

1.3 Spectral Parameters of Radiation

In the previous section we discussed the terminology and units that are needed to describe the energy transferred by radiation. The wave or corpuscular nature of light were not part of these definitions because all the energy carried by the radiation was considered. However, in many applications the composition of the radiation must be defined as well. If we assume that radiation consists of electromagnetic waves that can be described by the sinusoidal waves shown in Figure 1.2, then the wavelength λ can be used to uniquely specify a certain wave. Although wavelength could be defined as the distance traveled by a wave between two points with the same phase, such a definition would not be complete because the wavelength can vary while the wave is traveling through media other than free space. Therefore, we define the *wavelength* as

> *the physical distance covered in free space by one cycle of that sinusoidal wave.*

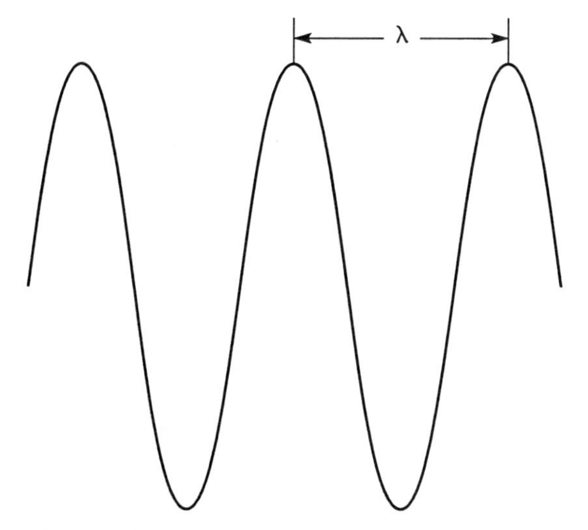

Figure 1.2 Wavelength of an electromagnetic wave.

This definition specifies explicitly that the measurement of the wavelength must be made in free space. Although wavelength measurements obtained in atmospheric air are sufficiently accurate for most applications, some tables list the results of wavelength measurements in vacuum (see Weast and Astle 1980). Included in these tables are the wavelengths of the radiation emitted by such sources as cesium, sodium, and mercury lamps. These lamps emit several monochromatic lines that can be used for the calibration of detectors or spectrum analyzing devices. Owing to their high efficiency, sodium and mercury lamps are also used for highway and street lighting.

The range of wavelengths that make up the radiative spectrum is unlimited. However, this infinitely wide spectrum can be naturally divided into sections in which the electromagnetic waves have comparable wavelength and all share some characteristic behavior or applications. Figure 1.3 presents the division of the radiative spectrum into such groups. The group of the shortest of wavelengths belongs to γ-rays, where $\lambda \approx 10^{-10}$ m. The longest waves on this chart belong to radio and microwave radiation, where λ ranges from 1 mm to hundreds of meters. The visible spectrum, marked by the shaded area, is seen to occupy a very narrow range.

Because of the large variation in wavelengths, the *units* for λ depend on where along the λ axis in Figure 1.3 the measurement is made. In this text we will be concerned primarily with the visible spectrum, where λ is specified in nanometers: 1 nm $= 10^{-9}$ m. For example, the nominal wavelength of a red He-Ne laser beam is $\lambda_{\text{He-Ne}} = 632.8$ nm. In textbooks the visible spectrum is sometimes specified in units of Angstroms (1 Å $= 10^{-10}$ m). The infrared spectrum is usually specified in units of micrometers (or *microns*), where 1 μm $=$ 10^{-6} m. For example, the nominal wavelength of a CO_2 laser beam is $\lambda_{CO_2} =$ 10.6 μm. Microwaves and radio waves are normally specified in millimeters and meters, respectively.

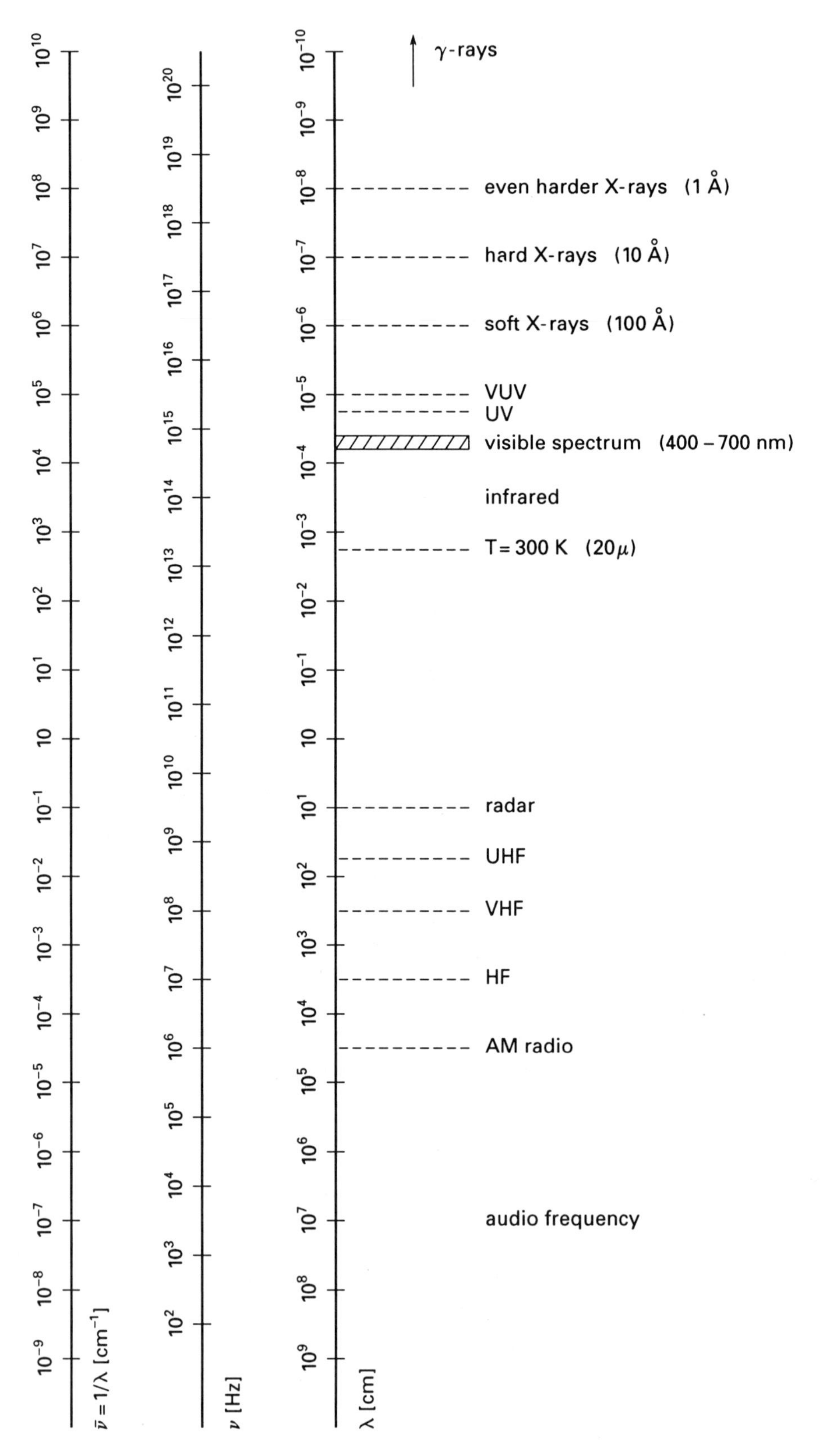

Figure 1.3 Division of the radiative spectrum into its spectral groups.

An alternative way to specify the spectral properties of radiation is by its frequency ν. Since the frequency and wavelength of radiation are related by the speed of light, specifying either the wavelength or the frequency is usually sufficient. The frequency of electromagnetic waves is defined in units of hertz [Hz] as the number of cycles completed by the wave in a second. Unlike the wavelength, the frequency of an electromagnetic wave is independent of the medium in which it travels. This, of course, presents an advantage because the same parameter applies to all media.

The frequency of electromagnetic waves in the visible spectrum approaches 10^{15} Hz. Since the frequency varies inversely with wavelength, the higher frequencies belong to the shorter wavelengths and the low frequencies represent radio waves and microwaves.

In most engineering applications, five significant figures are required to accurately specify the frequency. Since the frequencies of the visible, infrared, and ultraviolet spectra carry a high exponent, a new unit in which this high exponent is eliminated would be useful. Dividing the radiation frequency [in Hz] by the speed of light in free space ($c_0 = 2.9979 \times 10^{10}$ cm/s) yields a new representation of the frequency, $\bar{\nu}$, in wavenumbers [cm^{-1}]. For example, the frequency of the radiation of a red He–Ne laser beam is approximately $\nu = 4.74 \times 10^{14}$ Hz or $\bar{\nu} = 15{,}802.8\ \text{cm}^{-1}$.

1.4 Spectral Energy Transfer

The terms presented in Section 1.2 are useful for the description of the total energy transfer. In order to evaluate these terms, all the spectral components of the radiation (either emitted or collected) must be included. However, most optical applications are concerned with radiation within a limited spectral range. Thus, application may include considerations of the total energy transferred through an optical filter that is transparent only to ultraviolet radiation. Furthermore, most filters may transmit different portions of the incident radiation at different wavelengths. Alternatively, atmospheric applications may require specification of the solar energy transmitted through the ozone layer. For these applications, new terms must be used. As a rule, all the terms described in Section 1.2 can now be redefined per unit wavelength. Thus the *spectral flux* Φ_λ is defined as the radiant flux per unit wavelength interval at wavelength λ:

$$\Phi_\lambda = \frac{d\Phi}{d\lambda}\ \text{W/nm}. \tag{1.5}$$

The *spectral excitance* $M_\lambda(x)$ and *spectral irradiance* $E_\lambda(x)$ are similarly defined as the exitance and irradiance per unit wavelength at wavelength λ. The *spectral intensity* $I_\lambda(\theta)$ is the intensity per unit wavelength at the wavelength λ, and the *spectral radiance* $L_\lambda(x,\theta)$ is the radiance per unit wavelength at wavelength λ.

Other parameters and units may be used for the description of radiation. However, for most engineering applications, the terms described here will be sufficient. In the next chapters we will have more opportunities to familiarize ourselves with these concepts and their importance.

References

American National Standard (1986), Nomenclature and definitions for illuminating engineering, ANSI/IES RP-16-1986, Illuminating Engineering Society of North America, New York.

Edwards, I. (1989), The nomenclature of radiometry and photometry, *Lasers and Optronics* 8(8): 37–42.

Roberts, D. A. (1991), Radiometry/photometry terms. In *The Photonics Design and Applications Handbook,* 37th ed., Pittsfield, MA: Laurin, pp. H-58–H-61.

Weast, R. C., and Astle, M. J., eds. (1980), *CRC Handbook of Chemistry and Physics,* 60th ed., Boca Raton, FL: CRC Press, pp. E217–E397.

2 Geometrical Optics

2.1 Introduction

The classical description of radiation and optics provides two alternative approaches. In the first and more rigorous approach, radiation is viewed as waves of electric and magnetic fields propagating in space. In the second approach, radiation is modeled by thin rays traveling from a source to a target while neglecting all aspects of its wave nature. Rigorous considerations show that the second approach, *geometrical optics,* is merely a class within the broader picture described by the first approach, which is called *physical optics* or *electromagnetic theory.* Electromagnetic theory is normally used to describe the propagation characteristics of electromagnetic waves. It is a very general theory that can depict most effects associated with the propagation of light. Many effects - such as the interference between several waves and diffraction - can be explained *only* by electromagnetic theory. Electromagnetic theory can also be used to design imaging and illuminating optical devices such as telescopes, microscopes, projectors, and mirrors. However, many of the wave characteristics of radiation are irrelevant for the successful design of these devices; only higher-order corrections require electromagnetic wave considerations. Therefore, in applications where the wave nature of radiation can be neglected, the alternative description of radiation and optics - geometrical optics - can be used. Although the information generated by geometrical optics is less detailed than results of electromagnetic theory, it is far less complex and yet provides a remarkable prediction of the performance of imaging and projecting optical devices.

Methods of geometrical optics are used for the design of devices with an optical aperture that is large relative to the wavelength of the radiation. For example, the wavelength of visible radiation ranges from 420 nm to 680 nm, while the diameter of a slide-projector lens is approximately 3 cm. Therefore, the radiation wavelength can be assumed to approach zero. Although the response of optical media to the color of the incident radiation does affect the results of geometrical optics, the effects associated with wave phenomena (e.g. interference and diffraction) are neglected. By neglecting the wave nature of radiation, its propagation can be described by infinitesimally thin pencil-like rays that can be drawn schematically using relatively simple geometrical rules. The trajectories of these rays can be traced, or computed, for propagation through free space or through dense media, when crossing interfaces of prescribed shapes, or when bouncing off flat and curved surfaces. These rules can be used to simulate mathematically or graphically the focusing properties of lenses or the reflection by mirrors. For manual calculations or graphical tracing, several rays may be used to simulate individual optical elements or systems. However, with the advent of cheap and powerful computers, commercially available *ray-tracing* software may be used to simulate multi-element optical devices. In these programs, the geometrical parameters of each ray (such as its slope and the coordinate of its point of incidence) are calculated at each interface surface separating two media. Hundreds and thousands of rays passing through lenses, mirrors of prescribed curvature, prisms, or masks can be drawn using the graphical capabilities of desktop computers to visualize the imaging or projecting capabilities of a system. With a strike of a key, optical elements can be replaced or modified, the spacing or orientation of any element can be changed, and the effect of these modifications on the performance of a system can be evaluated qualitatively and quantitatively.

Chronologically, the field of geometrical optics was developed before the wave nature of radiation was known. In 1657, Pierre de Fermat (1601–1665) formulated the fundamental principle of geometrical optics. Although the statement of his principle resulted from teleological arguments, it accurately described the principles of reflection that were observed by the ancient Greeks as well as the principle of refraction codified by Snell in 1621. After the development of electromagnetic theory, Fermat's principle was shown mathematically to be a special case representing radiation with negligibly short wavelength (see e.g. Born and Wolf 1975). The logical sequence would thus be to discuss geometrical optics after electromagnetic theory is developed. However, owing to their relative simplicity, we will discuss the theories of geometrical optics first. The results obtained from geometrical optics will then be confirmed in subsequent chapters after the concepts of physical optics are developed.

2.2 Principles of Geometrical Optics

Since we do not yet have the tools of physical optics, we will use a simple mechanical example to illustrate the simplifications required by geometrical optics. We will show how the results of an analogy to wave propagation are compatible with results that can be derived from a fundamental law of geometrical optics – Fermat's principle. Our example incorporates a multiterrain HUMVEE, traveling at a constant velocity $\mathbf{v}_1$ on a paved surface. The front

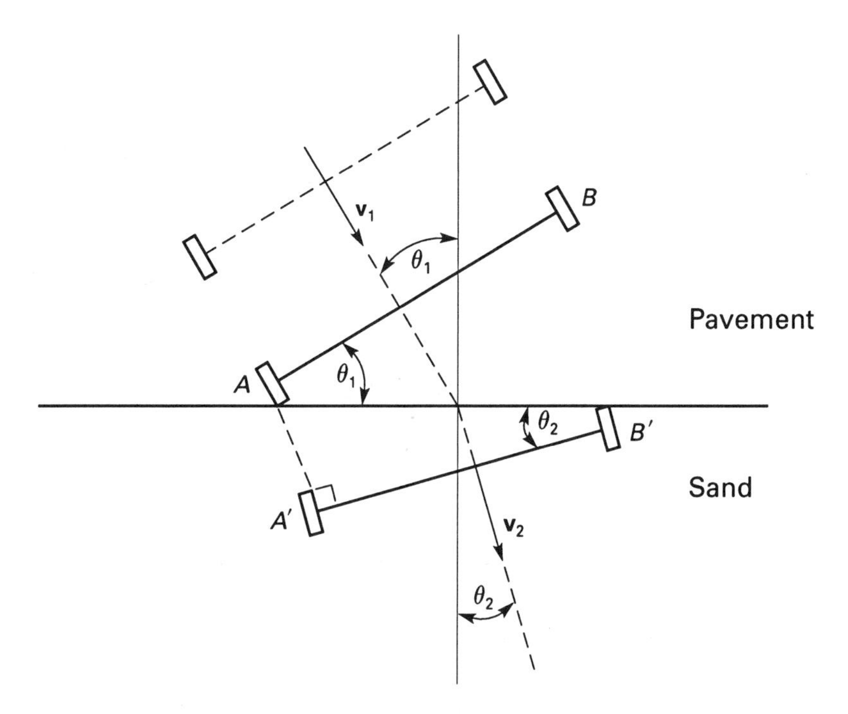

Figure 2.1 Front axle of a car as it crosses from pavement to a sandy area.

axle is free to turn around a pivot symmetrically located between the front wheels. However, initially the two axles are parallel to each other. See Figure 2.1. The vehicle is approaching the edge of the pavement beyond which lies a sandy area. Owing to increased resistance, the velocity of the wheels on the sandy area is reduced to $\mathbf{v}_2$. The angle between the front axle and the edge, just when the left wheel is about to cross it, is θ_1. Assuming that the front axle is free to turn, we would like to find the angle θ_2 between the front axle and the edge of the pavement after both front wheels have crossed into the sandy area.

This mechanical problem approximately simulates some of the characteristics of electromagnetic waves. It was selected for presentation at this stage to avoid the need to define some of the wave characteristics of radiation. Therefore, the method of solution represents well the behavior of electromagnetic waves when crossing from one medium to another while somewhat compromising the accurate description of the mechanical behavior of the car.

The front axle of the car, just when the left wheel touches the edge, is marked in the figure by line *AB*. The velocity vector of the car is collinear with the car centerline and is *normal* (perpendicular) to the front axle. Therefore, the angle θ_1 between the velocity vector and the normal to the pavement edge is equal to the angle between the axle and the edge. After both front wheels have crossed into the sandy terrain, the axle is marked by *A′B′* and the velocity vector $\mathbf{v}_2$ of the axle forms an angle θ_2 with the normal to the interface. In reality, the motion of the vehicle involves both rolling of the wheels and plowing along a curved trajectory. However, to simulate the behavior of electromagnetic waves, assume that the trajectory of wheel A on the sandy terrain can be approximated by a straight line normal to $A'B'$ while the trajectory of wheel B on the pavement is approximated by a straight line normal to AB. Although this approximation is inaccurate for modeling the motion of the car, we will see that it is an accurate representation of the propagation of electromagnetic waves. Since the velocity vector $\mathbf{v}_1$ is parallel to the straight trajectory BB', the time t required by wheel B to travel from B to B' is

$$t = \frac{AB' \sin\theta_1}{\mathbf{v}_1}. \tag{2.1}$$

Meanwhile, the distance traveled by wheel A is

$$\mathbf{v}_2 t = AB' \sin\theta_2. \tag{2.2}$$

One outcome of this approximation is that the length of the front axle changes from $AB' \cos\theta_1$ to $AB' \cos\theta_2$. This is clearly unacceptable as a solution for a mechanical problem, but it is an accurate representation of the behavior of electromagnetic waves. Combining (2.1) and (2.2), we see that

$$\frac{\mathbf{v}_1}{\mathbf{v}_2} = \frac{\sin\theta_1}{\sin\theta_2}. \tag{2.3}$$

This result is similar to Snell's law of refraction. To present this result in the form derived by Snell, consider a single, infinitesimally thin ray traveling from medium 1 toward medium 2 as in Figure 2.2. The ray corresponds to the velocity vector of the car of the previous example. Therefore, consistent with our example, the *incidence angle* θ_1 is defined as the angle between the incident ray and the normal to the interface. The plane defined by the ray and the normal to the interface is the *incidence plane.* After crossing the interface, the ray is said to be *refracted.* The *refraction angle* is defined as the angle between the refracted ray at the normal to the interface. Equation (2.3) can be used to calculate the refraction angle if the incidence angle and the speeds of light c_1 and c_2 (in the first and second media, respectively) are known. However, the ratio

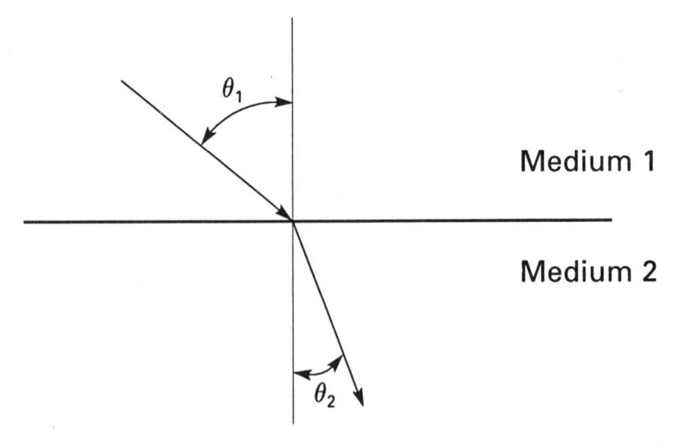

Figure 2.2 Geometry of refraction across the interface between two media.

c_1/c_2 is defined by the optical properties of the two media and is independent of θ_1 or θ_2. Therefore, the velocity ratio can be replaced by a material constant n_{12}. Replacing the velocity ratio $\mathbf{v}_1/\mathbf{v}_2$ in (2.3) with the ratio of the speeds of light, we obtain

$$\frac{c_1}{c_2} = \frac{\sin\theta_1}{\sin\theta_2} = n_{12}. \tag{2.4}$$

The constant n_{12} is unique to each pair of materials. Similarly, the ratio between the speed of light in free space and its speed in any medium is a property of that medium; this ratio is the *index of refraction*. The index of refraction of free space, by this definition, must be unity. The index of refraction of any substance depends on its thermodynamic properties, such as temperature or pressure (for gases), or strain (for solids), as well as on the wavelength of the incident radiation. The wavelength dependence of the index of refraction is called *dispersion*. The relation between the constant n_{12} and the indices of refraction of the two participating media can be found from the following simple ratio:

$$\frac{c_0/c_2}{c_0/c_1} = \frac{c_1}{c_2} = n_{12} = \frac{n_2}{n_1}, \tag{2.5}$$

where c_0 is the speed of light in free space and n_1 and n_2 are the indices of refraction of the two media. Combining (2.4) and (2.5), we find the following general relation between the indices of refraction and the incidence angle θ_i and the refraction angle θ_t:

$$\frac{\sin\theta_1}{\sin\theta_2} = \frac{\sin\theta_i}{\sin\theta_t} = \frac{n_2}{n_1}, \tag{2.6}$$

where θ_i is positive when measured counterclockwise from the normal in medium 1 and θ_t is positive when measured counterclockwise in medium 2. Equation (2.6) is *Snell's law* for refraction. Although the speed of light was used to derive this equation, it does not participate in it directly. Only material properties of the media at both sides of the interface are required to determine the refraction angle for any incidence angle.

The indices of refraction of many substances and their dispersion data are well tabulated. Usually they are obtained from geometrical measurements of the incidence and refraction angles across the interface between air or vacuum and the tested substance. Such goniometric techniques are mostly accurate for the measurement of the index of refraction of sufficiently large samples with flat or geometrically well-defined surfaces. However, even then, accurate goniometric measures may not be possible when the index of refraction is nearly 1. For example, for air at 15°C and pressure of 760 mm of mercury and for radiation at 300 nm, $n-1 = 2.907\times10^{-4}$. For an incidence of 85° on an imaginary vacuum–air interface, the refraction angle will be 84.8°. This difference between the incidence and refraction angles may prove to be too small for accurate measurement. Instead, interferometric techniques are used to measure the indices of refraction of gases (see Chapter 6).

Geological and crystallographic experiments require that the indices of refraction of small fragments of minerals or crystals be measured. These samples are usually small and geometrically irregular, and are therefore not amenable to either goniometric or interferometric measurements. However, when a sample is immersed in a clear liquid with an index of refraction matching that of the sample, the refraction between the liquid and the sample does not involve any redirection of the incident rays. Without visible refraction the sample appears perfectly transparent. The index of refraction of such a sample is determined by dipping it in liquids with various indices of refraction until a solution with a matching index is identified.

2.3 Fermat's Principle

Snell's law is one of the fundamental laws of optics. However, it is part of a more general principle that can be used to derive most equations of geometrical optics. This universal principle describes the propagation of rays through a variety of media and through refractive and reflective interfaces of diverse shapes. Historically, this principle was deduced by Fermat. Three centuries later it was shown mathematically to be part of the even more general field of electromagnetic theory. Here we present this principle without any proof and will consider it, as Fermat did, to be an empirical law – that is, we will postulate that such behavior is expected. By demonstrating that it accurately describes the geometrical parameters of ray propagation for select situations, we will deduce its generality. Fermat's principle states that

of all the paths connecting two points, radiation travels along that path which requires the least time of travel.

The parameter to be minimized by Fermat's principle is the time for passage between the two points. This should not be confused with the minimum distance between these points. Only in special circumstances (e.g., propagation through a homogeneous medium) will the path with the minimum time of passage between two points coincide with the path of minimum distance. Although the time for passage between two points must be minimized and only one minimum is expected to exist, more than one path can meet this minimum. Therefore, several noncoincident rays emerging at one point can be directed (e.g., by a lens) to a second point.

To illustrate the use of Fermat's principle, consider the reflection of a single ray incident upon the mirror aa' in Figure 2.3. In particular, we wish to identify the relation between the incidence angle θ_i and the reflection angle θ_r for the ray that connects points A and B while passing through the mirror aa'. Although there exists a ray that can propagate directly from A to B along the straight line connecting the two points, for reflection we must select only the rays that emerge from A toward the mirror. The path of minimum time of passage between A and B that also bounces off the mirror can be determined by producing an image B' of point B beneath the surface aa'. The location of B'

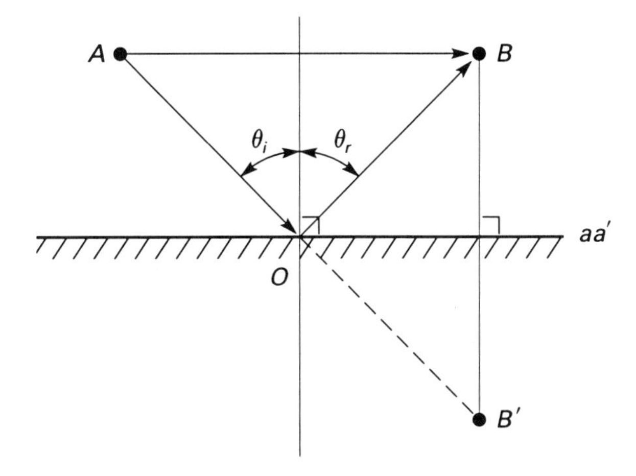

Figure 2.3 Geometry of reflection by a plane surface.

is along the extension to the normal from B to aa' at a distance from aa' that equals the distance between B and aa'. If propagation were to be allowed between A and the imaginary point B' and if the medium beneath aa' were to have the same properties as the medium above aa', then the shortest time for passage between A and B' would be along the straight line intersecting aa' at the point O and forming an angle θ_i with its normal. However, since $OB = OB'$, the times for passage along AB' and along AOB are identical. Since AB' represents the path for minimum time of passage between AB', AOB represents the path for minimum time of passage between A and B for a ray that also bounces off aa'. Thus, using simple geometry, it can be shown that the incidence and reflection angles are related by

$$\theta_i = \theta_r. \tag{2.7}$$

Equation (2.7) is known as the *law of reflection*. This result can also be obtained from (2.6) using $n = -1$; that is, θ_r is measured clockwise from the normal and is therefore the negative of θ_t. Although no such substance is known to exist, representation of the reflection by (2.6) is useful for ray-tracing applications.

Fermat's principle may also be used to describe refraction. If point A is situated in a medium in which the index of refraction is n_1 and point B is situated in a medium in which the index of refraction is n_2, the path of shortest time of passage is not a straight line; see Figure 2.4. Calculus of variations (Hildebrand 1965) must be used to directly determine the path between these points. Alternatively, it can be shown that the path defined by Snell's law (eqn. 2.6) is also the path for minimum time of passage between these points. The path between points A and B that satisfies (2.6) is shown in Figure 2.4 by the solid line. The time for passage along this path is mathematically an extremum. Therefore, when the incidence angle is varied by an infinitesimal amount $d\theta_1$, the time for passage along the varied path $AO'B$ equals the time for passage along the original path. If the section AO of the original path is projected onto section

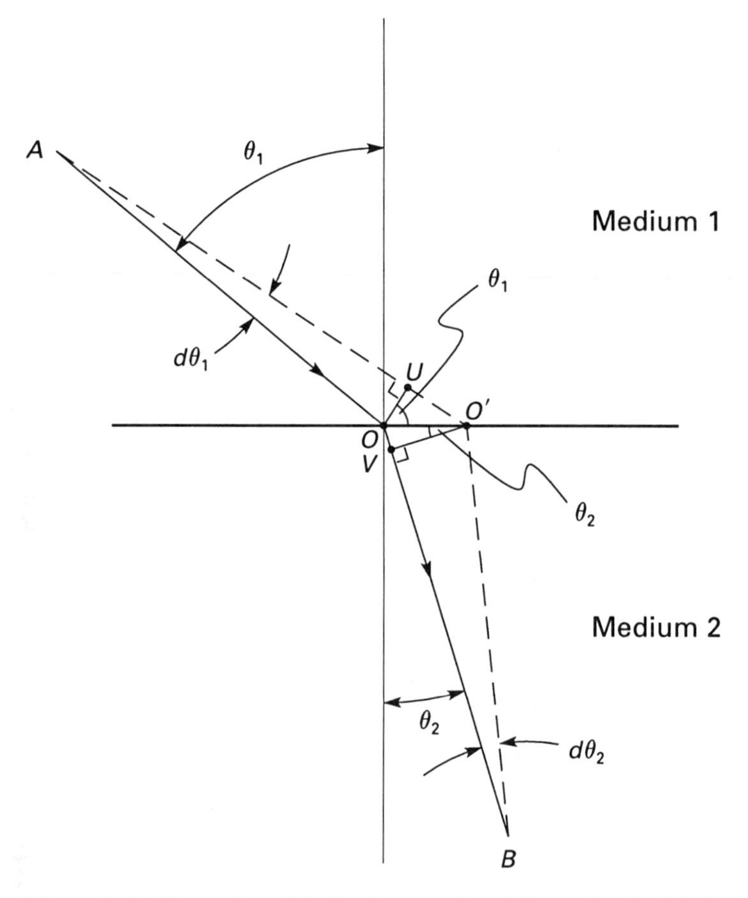

Figure 2.4 Illustration of the implementation of Fermat's principle for refraction.

AO', the segment *UO'* represents the path increment in the first medium. Similarly, section *VO* represents the infinitesimal decline in path length in the second medium. Using θ_1 as the approximate angle between *OU* and the interface and using θ_2 as the approximate angle between *O'V* and the interface, the increase in the time for travel in the first medium, Δt_1, and the decrease in the time for travel in the second medium, Δt_2, may be written as

$$\Delta t_1 = \frac{OO' \sin\theta_1}{c_0/n_1} \quad \text{and} \quad \Delta t_2 = \frac{OO' \sin\theta_2}{c_0/n_2}.$$

If (as expected) the time for passage along *AOB* equals the time for passage along *AO'B*, then $\Delta t_1 = \Delta t_2$ and consequently θ_1 and θ_2 satisfy (2.6). It is left to the reader to show that the time for travel along a path adjacent to *AOB* but to its left also equals the time for passage along *AOB*.

Although Snell's law (eqn. 2.6) can be used almost universally to calculate the refraction between any pair of media, a few exceptions do exist. Before identifying these exceptions we should examine some of the characteristics of refraction. A most notable observation is that refraction occurs only if the incidence is oblique. For normal incidence, the refracted ray remains normal to the interface regardless of the properties of the participating media. This is similar to the propagation of the HUMVEE in our mechanical example. As the front wheels simultaneously cross the interface between the paved and the sandy terrains, the speed is changed abruptly; however, owing to the lack of a mechanical moment, the car does not turn. On the other hand, when traveling along the pavement edge with the right wheels in the sand and the left wheels on the pavement, a moment that turns the car is developed. By analogy one may expect that variations in the index of refraction that are transverse to the incident beam's propagation direction are needed to deflect a ray upon refraction. Equation (2.6) supports this observation. As the incidence angle θ_i increases, the variation in the index of refraction transverse to the ray increases and so does the deflection due to refraction, $\theta_i - \theta_t$ (see Problem 2.5). From this, one may observe that the higher-index-of-refraction medium acts as molasses that "sucks" the rays from the lower-index-of-refraction medium. The largest deflection results when an incident ray is almost tangent to the interface at the side with the lower index of refraction. For this $\theta_i = 90°$ incidence, both the total deflection and the refraction angle θ_t attain their maxima.

The maximum refraction angle is the *critical angle* θ_{crit}. This angle is an important parameter of the refraction configuration. The unique feature of this angle is evident when the incidence is at θ_{crit} and is from the medium with the higher index of refraction, that is, when $n_1 > n_2$. From Snell's law, the refraction angle in this case is $\theta_t = 90°$; that is, the refracted ray is grazing the interface. However, when $\theta_i > \theta_{\text{crit}}$ the refraction angle cannot be obtained from (2.6). Instead, a total reflection by the interface back into the high-index-of-refraction medium takes place with $\theta_r = \theta_i$, where θ_r is the angle between the reflected ray and the normal to the interface. This sudden transition from refraction to reflection is known as *total internal reflection*.

Snell's law successfully describes refraction of rays when the change in the index of refraction across an interface is abrupt and when the incidence angle $\theta_i \leq \theta_{\text{crit}}$. However, when propagating through fluids, rays may be gradually refracted by gradual variations in the index of refraction that follow temperature, density, or composition gradients. Gradual variations in the index of refraction are also induced artificially by doping techniques to produce desired effects in optical fibers or lenses. Unfortunately, (2.6) – which is specific to refraction across interfaces with an abrupt change in the index of refraction – can no longer be used to determine refraction in media with a gradually varying index of refraction. Nevertheless, by dividing the medium into planar layers, each with an incremental change in the index of refraction, we can show that propagation normal to these hypothetical layers results in no deflection of the incident ray. However, rays propagating obliquely through these layers do deflect. To estimate the magnitude of that deflection, we again invoke Fermat's principle. In the presence of a gradient in the index refraction of magnitude

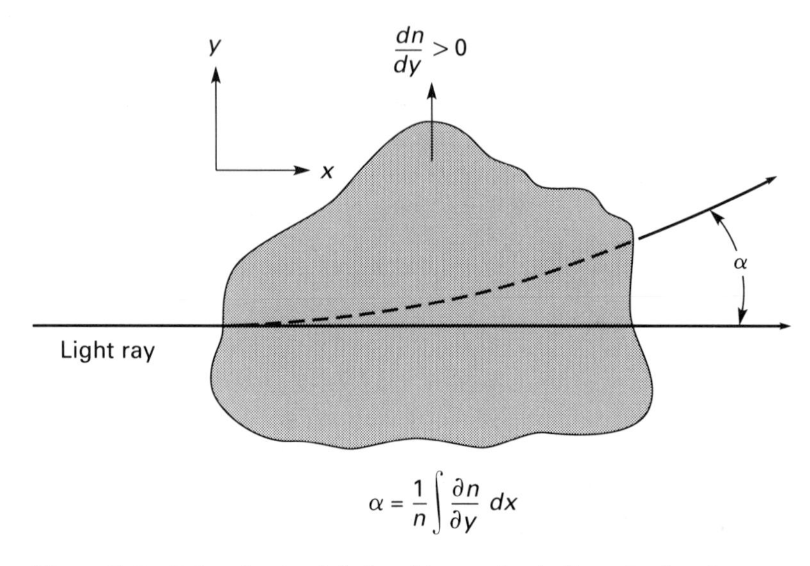

Figure 2.5 Refraction in air induced by varying indices of refraction.

$$\nabla n = \frac{\partial n}{\partial x}\mathbf{i} + \frac{\partial n}{\partial y}\mathbf{j} + \frac{\partial n}{\partial z}\mathbf{k},$$

where **i**, **j**, and **k** are unit vectors in the x, y, and z directions, respectively, the total deflection angle $d\alpha$, following propagation along a distance $d\mathbf{s}$ (Figure 2.5), can be determined from Fermat's principle using calculus of variation. For the simpler two-dimensional case, the deflection of a ray in the (x, y) plane (see Problem 2.3) is

$$d\alpha = \frac{1}{n}\frac{\partial n}{\partial y}\, dx, \tag{2.8}$$

where the positive x direction was arbitrarily selected to coincide with the direction of propagation of the ray. This result is compatible with our empirical observation that only the transverse component of the gradient of the index of refraction (the y component of the gradient in eqn. 2.8) contributes to the deflection by refraction. In the absence of such a component the propagation is without deflection, whereas in its presence the deflection curvature is in the direction of increasing density. In three-dimensional problems, when all three components of the gradient are required to solve $d\alpha$, the partial derivative of n in (2.8) is replaced by the vectorial expression of the gradient. The deflection is then in a plane defined by the ray and the gradient and the curvature is in the direction of increasing density.

The last principle we need to develop before the behavior of various optical components can be analyzed is the principle of reciprocity:

radiation travels from any point A to any point B along the same path that it would travel from point B to point A.

This principle is merely an extension of Fermat's principle, and results from the observation that the *time* for travel from any point A to any point B must be the same as for travel in the reverse direction from B to A. This hypothesis is justified only if the function describing the time for passage between these points has only one minimum. Then, because the time for travel in both directions must meet Fermat's requirement for minimum time of passage, the propagation times in both directions must be identical. The assumption that the time for travel between two points is the same for both forward and reverse directions appears intuitively to be justified, but it is actually a consequence of another assumption: that the speed of light is *isotropic* – that is, the same in all directions. However, this simple assumption was not all that obvious until 1881, when Michelson showed that the speed of light is in fact uniform in all directions.

The principle of reciprocity is obeyed universally by all optical elements that are not subject to intense electric or magnetic fields. Such optical components are called *passive,* and most optical components fall into this category. However, for some applications the radiation propagating in one direction needs to be separated from radiation propagating in the reverse direction. In one such application, the back-reflection of a fraction of an intense laser beam by optical elements must be rejected to prevent it from returning into the laser itself and damaging its mirrors. An active device that is frequently used for this rejection is the *Faraday isolator,* which incorporates an optical crystal subjected to an intense magnetic field. The application of this magnetic field eliminates the symmetry associated with the principle of reciprocity. Propagating in the forward direction through the isolator, a laser beam can pass with only a minor loss, whereas in the reverse direction one of the beam properties – the polarization – is changed and the beam can be turned by 90° into a beam dump where it is absorbed.

The combination of Fermat's principle and the principle of reciprocity – as well as their derivatives, Snell's law and the law of reflection – may now be used to describe the operation of some elementary optical components. Most optical elements are classified into one of two primary groups: reflecting optical elements and refracting elements. Although all refractory optics are partially reflecting and many reflecting elements are partially refracting, their classification is determined by the intended application. However, in the design of precision equipment, both reflection and refraction by each element must be considered. Otherwise, stray rays resulting from undesired reflections or refractions may spoil the characteristics of the system. In these designs each element is included twice: once as a reflecting and once as a refracting component. Optical elements that are not classified as part of either group include apertures, diffraction gratings, polarizers, and acousto- and electro-optical devices. Analysis of these elements requires application of electromagnetic theory. However, even these elements present some reflection and refraction that may need to be considered. For now we will limit our discussion to reflecting and refractory optics.

2.4 Reflecting Optical Elements

This group of optical elements is probably the simplest to describe and one of the most useful. Although full explanation of reflections at an interface is based on electromagnetic theory, the behavior of rays bouncing off that interface is easy to model mathematically. Reflecting optical elements primarily include planar mirrors that are used to redirect and stir rays and curved mirrors that are used to project or collect light. Most reflectors are made of a substrate shaped to certain specifications and covered by fully or partially reflecting coatings. The coating may consist of a thin film of evaporated metal, usually aluminum or gold, or multiple layers of dielectric films that are vacuum deposited in precise thickness, purity, and order. Metallic mirrors are relatively cheap because the coating process is less demanding than that for *dielectric mirrors*. Metallic mirrors are also highly reflecting, with approximately 96% of the incident flux reflected over most of the visible, near-UV, and near-IR spectral range. With the application of a protective overcoat layer, the reflectivity of aluminum can be extended to wavelengths below 200 nm and beyond 10 μm. Thus, these mirrors are used in most optical applications where spectral selectivity is not needed or when the incident flux is low. In high-power applications (e.g., incident continuous irradiance of greater than 100 W/cm^2), the reflection losses by absorption by the metallic film result in the deposition of energy in a layer with a thickness of only a few atoms. Thus, even a small amount of absorbed irradiance is sufficient to evaporate the thin absorption volume, thereby damaging the mirror permanently.

By contrast, dielectric coatings can be designed for extremely efficient reflectivity (in excess of 99%). The small fraction of the radiation not reflected by the dielectric film is mostly transmitted through it. Therefore, the irradiance deposited in the reflecting dielectric layer is a minute fraction of the incident irradiance, and the durability of the mirror in the face of high-power incident radiation is extended. Most mirror manufacturers will specify the damage limits of their products as the allowed incident irradiance for specified wavelength and pulse duration. For incident pulses with durations of approximately 10 ns in the visible spectrum, the irradiance for damage threshold of some dielectric mirrors may exceed 1 MW/cm^2. However, the efficient reflectivity of dielectric mirrors is limited to a relatively narrow spectral range. This spectral selectivity enables their use as optical filters or as *dichroic mirrors* – that is, mirrors that reflect an incident beam of one color but transmit incident beams of different colors. For a premium, most mirror manufacturers will design and manufacture dielectric mirrors that match specific reflection and transmission requirements.

Although the design of a reflective coating requires the application of electromagnetic wave theory, the shape of substrates can be designed using Fermat's principle. Many nonplanar substrates can be described by simple geometrical functions such as the conical sections. Of these, the most common are the spherical and paraboloid mirrors. Paraboloid reflectors (usually with circular symmetry) are used when precise and efficient collection or projection of

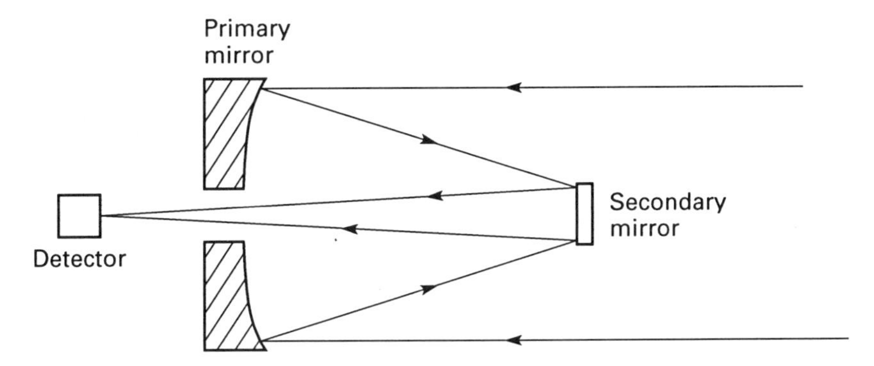

Figure 2.6 Focusing of a collimated beam by a Cassegrainian telescope.

radiation is required. In one such precision application, a parabolic mirror is the primary reflector of the Cassegrainian telescope described schematically in Figure 2.6. Radiation from a distant light source is collected by the parabolic primary mirror and focused onto a secondary mirror. Reflection from the secondary mirror is focused on a detector, located behind an aperture in the primary mirror. The secondary mirror casts a shadow on the primary mirror, so the use of this telescope for the imaging of near objects is limited. However, for collection of radiation from distant objects, the effect of this shadow is only to reduce the total irradiance falling on the detector. To determine the geometrical parameters of the primary mirror and the location of the focal point where the detector should be placed, we consider both the light source and the detector as points. Thus, all rays that emerge at the point source and fall on the pointlike target, after bouncing off the primary and secondary reflectors, traveled along paths of minimum time of passage. Assuming that only one such minimum time can exist, the time for passage for all rays traveling between these two points must be the same. Consequently, if all the paths are contained in the same medium, their lengths must be equal.

To illustrate how light from a distant object is focused into a point, a circularly symmetric paraboloid mirror – illuminated by radiation from a distant point source on the parabola axis – is represented in Figure 2.7 by a section along the diametric plane. The mirror is obtained by revolving the diametric parabola around its axis. For a sufficiently distant object, the incident beam can be regarded as collimated. The directrix aa' of the diametric parabola is perpendicular to the reflector axis as well as to the incident rays. Thus, the distance between the point source and the directrix aa' is constant along all incident rays. Also, for any parabola the distance AB between the focus and any point B on the parabola equals the distance BB' between that point and the directrix. Because the incident rays with equal path length between the point source and the directrix are prevented by the mirror from reaching the directrix, they must converge at A if their path lengths are to remain equal. This argument can be extended to all parabolas obtained by revolving the parabola of Figure 2.7 around the mirror axis. The shape of the substrate of the primary

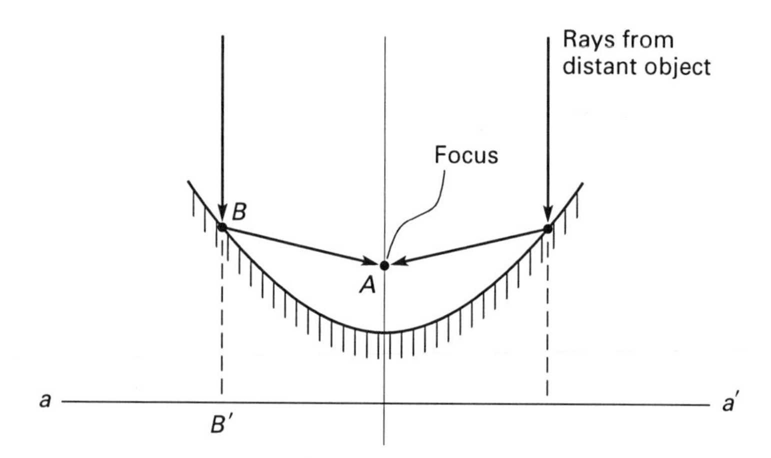

Figure 2.7 Focusing of a collimated beam by a parabolic mirror.

mirror can be designed using the loci of the directrix and the focus and the mirror diameter as free parameters. Although the detector of the Cassegrainian telescope may be placed directly at the focus, a secondary mirror is used either to redirect the beam or to correct for focusing aberrations due either to divergence of the incident beam or to imperfections of the paraboloid mirror.

Using the principle of reciprocity, it can be shown that a parabolic mirror may also be used to collimate radiation emitted by a point source (e.g. a light bulb) located at the focus. Indeed, paraboloid mirrors are frequently used for projection and illumination. Unfortunately, the manufacturing costs of such mirrors are high. Therefore, spherical mirrors are often used as replacements in applications, such as automobile headlights, where projection quality can be compromised while cost containment is essential. However, even when parabolic mirrors are used for projection, the emerging beam is never perfectly collimated. Aberrating effects, such as the finite size of the source and imperfections in the shape of the paraboloid, combine to diverge the beam. Effects (such as diffraction) associated with the wave nature of radiation also contribute to the divergence of the beam.

Some applications require that radiation emitted by linear sources (e.g. a flashlamp) be collimated, or that radiation from a point source (e.g. the sun) be focused on a linear target such as a pipe containing thermal fluid. Using similar arguments, we can show that in such cases cylindrical mirrors with circular or parabolic cross section must be used.

In other applications, radiation emitted by a linear light source must be focused onto a linear target. This can be seen in some lasers where radiation from a flashlamp is used to excite the laser-active medium, which is shaped as a slender rod. For efficient operation, all the radiation emitted by the flashlamp must be focused on the laser medium. By placing the flashlamp in one focus of an elliptic cylinder and the laser rod in the other focus (Figure 2.8), all rays that emerge normal to the lamp axis bounce off the cylinder wall and converge on

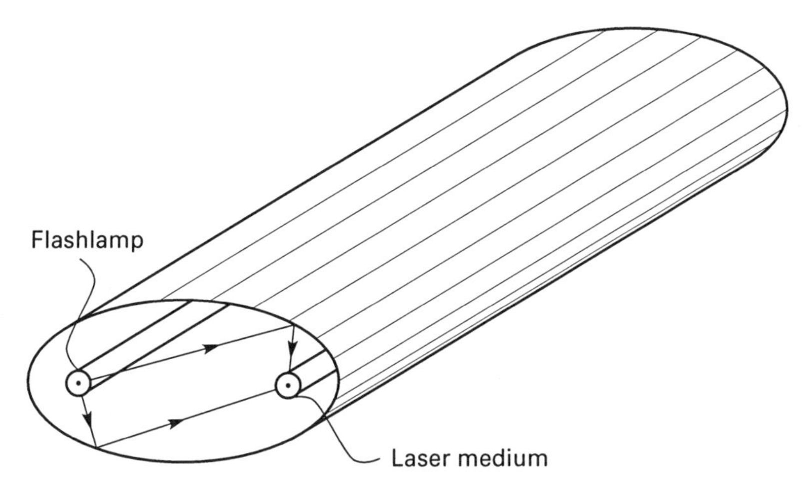

Figure 2.8 Cylindrical mirror with an elliptic cross section. Shown are a linear laser medium at one focus of the ellipse and a linear flashlamp at the other.

the laser rod. It follows from the definition of an ellipse (the locus of all points whose sum of distances from two foci is the constant) that all path lengths of these rays are identical and hence their times for travel, as required by Fermat's principle, are identical as well. Although this is an ideal linear focusing configuration, the development of more efficient laser sources has allowed for the replacement, in most flashlamp laser excitations, of the elliptical cylinders with the more economical circular cylinders.

2.5 Refracting Optical Elements

In ideal reflecting optical elements, the optical properties of the substrate do not influence the optical performance of that component. This of course presents an advantage, because undesired interactions such as absorption or scattering by optical impurities in the substrate are avoided. Furthermore, only one surface, the reflecting surface, must meet optical quality requirements. Therefore, reflectors are preferred in numerous applications, most notably when large elements are required or for outer-space applications where the mass of the element must be low. On the other hand, for imaging of objects within a large field of view, or for compact integrated systems, refracting optical components offer superior performance. Refracting elements are usually designed to transmit most of the incident radiation. Therefore, unlike reflectors, they can be stacked axially in multi-element systems. The minute reflection losses at both interfaces of these components, and the absorption and scattering losses within the refracting medium, are either neglected or evaluated separately. Furthermore, reflection losses can be reduced from a characteristic 4% per surface to only 1% by depositing on each surface a special antireflection coating.

Refracting optical elements primarily include elements with either flat surfaces or curved surfaces. Among the flat-surfaced elements are spacers or retarders with two parallel surfaces, as well as prisms, which may have any number of surfaces ranging from three and up. Lenses are the most recurrent among the curved-surface elements.

All refractory elements are used to redirect the incident radiation. With refractory components we can redirect a collimated beam while maintaining its collimation, or simply shift it transversely while maintaining its original pointing direction. By redirecting individual rays, we can focus incident radiation into a point or an image or diverge incident radiation. In all these functions, redirecting the incident radiation occurs because the speed of light in the refracting element is different (usually slower) from the speed of light in the surrounding medium. Refractory elements with flat surfaces are used when the collimation or divergence characteristics of the incident beam must be preserved, while curved-surface elements are required for focusing, imaging, and diverging beams.

The simplest flat-surface refractory element is a plate. Plates are normally either rectangular or circular and the two surfaces are nominally parallel. The index of refraction of the plate usually exceeds the index of refraction of the surrounding medium. Plates are often included in such optical systems as filters, attenuators, polarizers, and the like. Although refraction by these flat elements is not intended, its effect must be recognized in the design of imaging optics. For normal incidence, no refraction occurs. However, back-reflection of approximately 4% from each surface may occur unless suppressed by antireflection coatings. However, even when suppressed, back-reflection in some laser applications may introduce undesired effects. Therefore, for these applications the plate is replaced by a wedged element. With a typical wedge angle of 5°, reflections can be directed away from the optical axis of the system. For oblique incidence, a complex pattern of reflected and refracted rays is formed even when the incident beam is collimated. Figure 2.9 shows a transmitted ray (solid line) and the reflected and refracted secondary rays (dashed lines) following multiple reflections and refractions. This multitude of reflected and refracted rays may interfere with many imaging or high-power applications, but many of these rays can be eliminated by well-placed masks.

Although multiple reflections and refractions do take place between the plate surfaces, most of the incident radiation is transmitted by the plate after one refraction at each surface. Both the first refraction angle θ_t' and the second refraction angle θ_t'' can be computed from (2.6). However, when the plate is immersed in a single homogeneous medium, $\theta_t'' = \theta_i$. Thus, the ray refracted by the plate remains parallel to the incident ray. The only effect of this refraction is to shift each incident ray parallel to itself. The extent of this parallel shift d increases with θ_i and the thickness of the slab, and can be controlled by rotating the plate.

Prisms present a large and diverse group of refractory elements with a multitude of applications. Almost any imaging or viewing device contains prisms

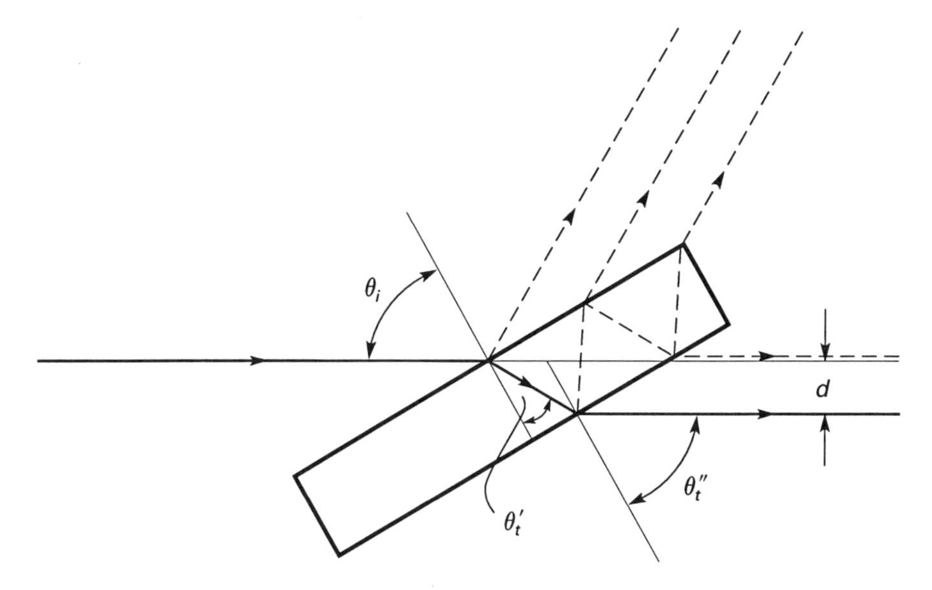

Figure 2.9 Offsetting of a beam trajectory by a plate.

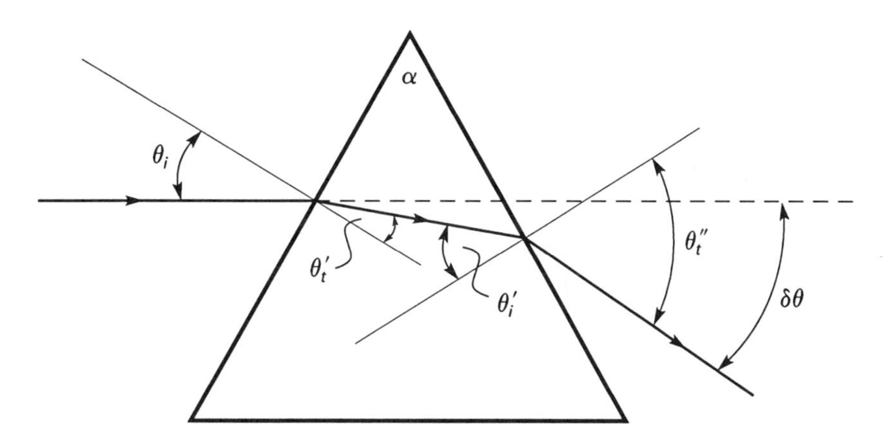

Figure 2.10 Deflection of a ray by a triangular prism.

that are used to reorient or "fold" a beam. Images can be inverted or reverted using prisms, and radiation can be separated into its spectral components. In all these applications, the direction of the incident beam is recomputed after encountering an interface. If refraction is possible (i.e., if $\theta_i < \theta_{\text{crit}}$), the refraction angle is computed using (2.6). Otherwise, when total internal reflection occurs, the reflection angle is determined using (2.7). Figure 2.10 presents refraction through a triangular prism; this is the most fundamental prism. It is left to the reader to show that for small prism angles α (wedges), the *deviation angle* $\delta\theta$ – that is, the angle between the incident and refracted rays – is

$$\delta\theta = (n-1)\alpha.$$

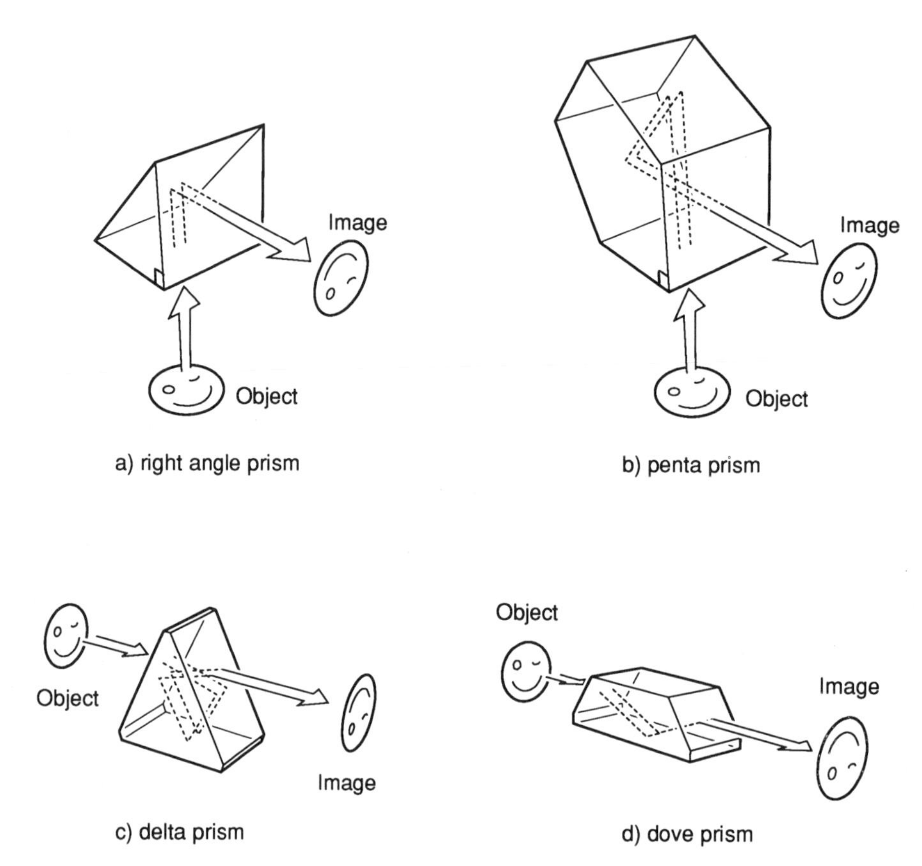

Figure 2.11 Selected prisms for optical applications. (**a**) Right-angle prisms are used like mirrors to turn beams through an angle of 90°. Unlike most mirrors, a prism can sustain high incident beam energy and is therefore useful for many laser applications. (**b**) Penta prisms are used mostly to turn an image through a fixed angle (see Problem 2.6). In typical applications, a penta prism is used as an "optical square." (**c**) Delta prisms and (**d**) dove prisms are used to "fold" beams, particularly in applications where the required optical path length exceeds available space (e.g., telescopes or binoculars).

If the prism medium is dispersive (i.e., if n varies with wavelength) then $\delta\theta$ varies with wavelength as well. Thus, radiation can be separated by a prism into its spectral components. Although the spectral separation increases with the prism angle, at excessively large α the incidence angle at the exit face may exceed θ_{crit} and radiation will not emerge through the second face; see Figure 2.11(a).

When total internal reflection is encountered inside a prism, the reflecting surface acts as a perfect mirror for all wavelengths for which θ_{crit} is exceeded. Images reflected by that surface are reverted. Reflection losses, relative to losses at coated mirrors, are small, while the threshold for damage by high-power beams is high. Thus, large-angle (e.g. 90°) prisms are used as mirrors in high-power applications, whereas multifacet prisms are frequently used to turn (see Problem 2.6), invert, or revert images. Figure 2.11 shows several prisms frequently encountered in optical applications.

The analysis of refracting optical elements with curved surfaces requires that the time of passage of individual rays be compared to the time of passage of other rays; this is how parabolic and elliptical mirrors were analyzed. However, with mirrors all rays were propagated within the same medium, so the comparison of the times of passage could be easily diverted to the comparison between path lengths. This, of course, is not the case for propagation through refracting elements, wherein the speed of light is different from that in the surrounding medium. Thus, instead of comparing geometrical path lengths, we introduce a new parameter. The *optical distance* is defined as

> *the distance L_0 that would be propagated by radiation in free space during the time it would take to propagate through a refracting medium of prescribed length L.*

The optical distance in a refracting medium in which the speed of light is c is

$$L_o = L \cdot \frac{c_0}{c} = L \cdot n. \tag{2.9}$$

Equation (2.9) also implies that the time of propagation is the same for all rays traveling through the same optical distance. The actual geometrical distances and the properties of the refracting media are all accounted for by (2.9). Thus, propagation through an optical slab with an index of refraction of 1.33 and a thickness L lasts just as long as the propagation through a distance of $L_o = 1.33L$ in free space. The concept of optical distance will be used extensively to describe the operation of interferometers and laser systems.

Using this concept of optical distance, we can now attempt to describe the operation of lenses. The purpose of a lens is to collect radiation emitted at a point, called the *object,* and transfer it to another point (real or imaginary), called the *image*. Figure 2.12 conveys this idea schematically. Rays that are emitted by the object point O within a cone with a solid angle Ω_o are to be transmitted – that is, *imaged* – to the image point I within an angle Ω_i. Although the object may emit radiation outside the assigned cone, the optical component that transmits the rays to I is expected to capture only rays that are within Ω_o.

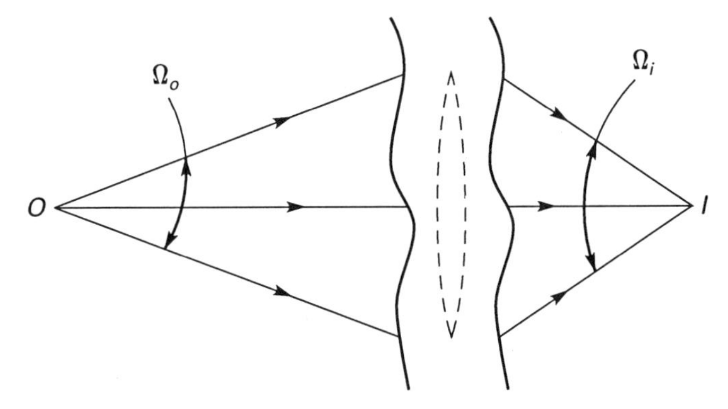

Figure 2.12 Geometry of optically imaging one point source into another.

The actual shape of this imaging optical element is yet to be determined. Meanwhile, it is marked in the figure by the abstract waved solid lines. The object O in Figure 2.12 is shown to the left of the imaging device, while the image is to the right. This is a standard representation in geometrical optics, where propagation is sketched from left to right. The region to the left of an imaging component is the *object plane,* and the region to the right is the *image plane*. According to Fermat's principle, rays emerging at O can intersect at I only if their times of propagation - or their optical distances - are identical. However, the geometrical distance along rays that are on the envelopes of the cones in the object and image planes are longer than the geometrical distances along the cones' axes. To compensate for these differences, the optical distance of the axial ray must be extended relative to the optical distance of the rays on the cones' envelopes. This can be accomplished by simply inserting, between the object and image planes, a medium with a refractive index larger than the surrounding medium and with thickness along the axis of the collection cone that exceeds the thickness at the edges. The surface shapes of lenses are designed using this principle.

In order to calculate the exact shape of both surfaces of a lens, the trajectory of each ray must be computed analytically or graphically, including the refraction at the first interface, the actual optical path within the lens, and the refraction at the exit. By requiring that the optical path along each ray be the same, the actual shapes can be obtained analytically, numerically, or graphically. Lens surfaces computed by this method can be described by fourth-order polynomials. Figure 2.12 and the discussion thereof concern the imaging of linear objects perpendicular to the figure plane or point sources. For imaging a point object into a point image. a lens with circular symmetry of the fourth-order polynomial is required, whereas for linear imaging a cylindrical lens is needed. Unfortunately, the production of fourth-order lenses, either cylindrical or quasispherical, is too costly. Instead, realistic lenses are approximated by spherical surfaces for point imaging or by circular surfaces when cylindrical lenses are needed. Mathematically, circular or spherical surfaces are described by second-order polynomials; that is, the higher-order terms from the polynomials describing an ideal lens surface are rejected.

A typical lens is shown in Figure 2.13. Unless stated otherwise, spherical symmetry is assumed. Often, both faces of a lens are convex spherical surfaces, as in the figure. The radii of the two spherical surfaces need not be identical. Other combinations of spherically symmetric surfaces - such as plano-convex, plano-concave, concave-concave, and concave-convex - are available. The line connecting the centers of the two spherical surfaces is called the *lens axis*. When several lenses are used in an optical system, their axes are lined up to form an *optical axis*. Near the lens axis, the approximation of the fourth-order polynomial by a spherical surface is excellent and so imaging using rays that propagate along or near the axis can be obtained with little or no distortion. However, as the angle between an incident ray and the axis increases, the effect of the higher-order polynomial terms that were neglected becomes appreciable

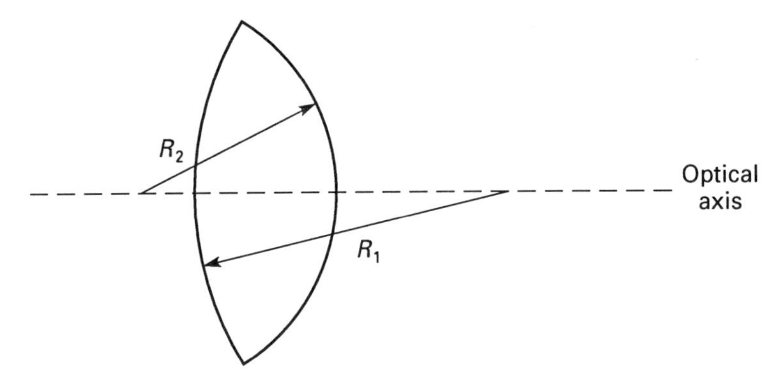

Figure 2.13 Imaging of a point source by a spherical lens.

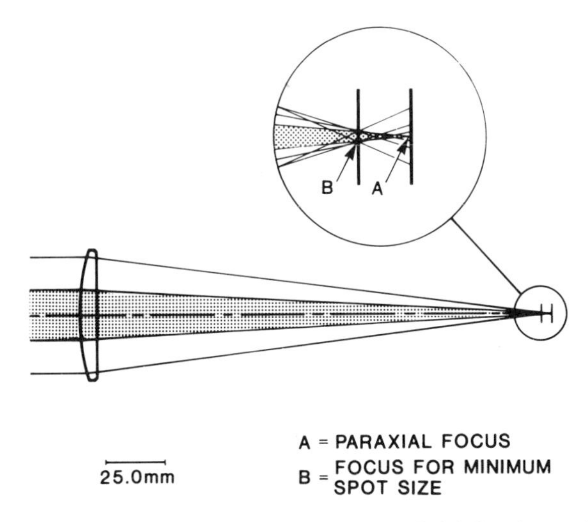

Figure 2.14 Spherical aberration. [Oriel Catalog, vol. 3, 1990, © Oriel Corporation, Stratford, CT]

and distortion of the image increases. Thus, the approximation associated with the spherical surfaces of lenses is called the *paraxial approximation*. The distortion induced by the spherical shape of the lens is called *spherical aberration*.

Figure 2.14 illustrates the effect of spherical aberration. A collimated beam is focused by a plano–convex lens. Rays (close to the axis) for which focusing by the paraxial approximation is achievable are marked by the shaded region; these rays converge into a well-defined point A. However, rays outside that region intersect at a point B ahead of A. Thus, at point A, a sharp point surrounded by a blurred halo is observed. The size of the focal spot may be reduced by placing a screen at point B ahead of point A. The extent of the spherical aberration can be minimized by judicious choice of lenses and their orientation. For example, when focusing a collimated beam, the aberration is minimized by inserting the lens with the convex surface facing the collimated beam. With that orientation, the bending of each ray is divided between the two faces of the lens. With the alternative orientation (i.e., the plane surface facing the collimated beam), all the bending is done at the rear spherical plane. These remarks may appear to violate the principle of reciprocity, but recall that this principle applies only to the reversal of the direction of propagation. Thus, if the converging rays in Figure 2.14 were reversed, they would emerge collimated at the left side of the lens. Turning the lens is not identical to the reversal of the direction of propagation.

An important parameter in the design of a lens is the ratio between the index of refraction of the lens and the index of refraction of the surrounding medium. Because of dispersion, lens properties are not uniform throughout the spectrum. This presents little or no problem for coarse imaging applications. However, in high-resolution photography, dispersion effects may cause images that appear sharp under monochromatic illumination to appear blurred when illuminated by multicolor light; this phenomenon is known as *chromatic aberration*. Designers of camera, microscope, and telescope lenses must consider correction for chromatic aberrations as well as for other aberrations. Chromatic aberration can be corrected by combining two lenses of different shapes and substances into doublets. The lenses are designed so that the chromatic aberration induced by one lens is compensated by the other lens. However, even the most carefully designed lenses are chromatically corrected within a limited spectral range. Lenses that are corrected for the visible spectrum show chromatic aberrations when used to image objects illuminated by IR or UV radiation. Other aberrations and their corrections are discussed by O'Shea (1985).

Properties of lenses are fully defined by the curvatures of both surfaces, the loci of the centers of curvature, and the ratio between the index of refraction of the lens and the index of refraction of the surrounding medium. With this data, a lens grinder can select the material, measure the diameter of the blank, and determine the grinding procedure that will turn that blank into a polished lens. Many ray-tracing programs require similar information to graph rays as they emerge from an object, pass through a lens, and intersect to form an image in the image plane. Effects of spherical and other aberrations related to the shape of the lens can be evaluated by these programs. By including dispersion data, chromatic aberration can be assessed as well. More advanced software contains files of the indices of refraction and dispersion parameters for many lens materials and other media, and so can account for immersion in air, water, or other solutions.

Although detailed representation of all lens parameters is required for its manufacture or for analysis of aberrations and optimization of the imaging, most applications fall within the paraxial approximation. When this approximation is acceptable (i.e., if spherical aberration is negligible) then the equations describing lenses can be significantly simplified, and imaging by a lens can be described by a single parameter – the *focal length*. Figure 2.15 shows two rays propagating from an object located on the axis of a spherical lens at a distance s_o from the center plane of the lens. One ray propagates toward the lens along the axis while the other intersects the lens, at the center plane, at point V

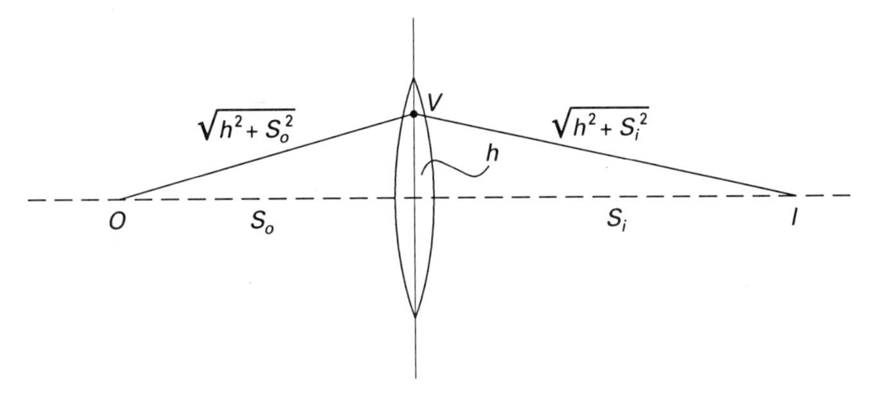

Figure 2.15 Geometry of paraxial imaging by a thin lens.

located at a distance h from the axis. The image I is formed at a distance s_i from the lens by the intersection of ray VI with the axial ray. The paraxial approximation requires that $h \ll s_o$. Therefore, the lengths of rays OV and VI can be approximated by

$$\sqrt{h^2+s_o^2} \approx s_o + \frac{h^2}{2s_o} \quad \text{and} \quad \sqrt{h^2+s_i^2} \approx s_i + \frac{h^2}{2s_i}.$$

Using simple geometrical arguments (see e.g. Feynman, Leighton, and Sands 1963) and also assuming that the lens is thin, the distance s_i between the real image and the lens is related to s_o by

$$\frac{1}{s_o} + \frac{1}{s_i} = \frac{1}{f}, \tag{2.10}$$

where f, the local length of the lens, incorporates the radii of the spherical surfaces, the loci of their center, and the index of refraction enclosed by them. Equation (2.10) is known as the *Gaussian form* of the lens equation. When $s_o \to \infty$, the image would ideally be a point located at a distance f from the lens. This is the *focal point* or focus of the lens. Imaging of a remote object can be used to approximate the focal length of a lens. When a point source is placed at the focus of a lens in the object plane, the image is formed at $s_i = \infty$. This is compatible with the principle of reciprocity, since a remote object forms an image at the focal point of the lens.

When $s_o < f$ – that is, when the object is placed between the focal point and the lens – the distance s_i to the image is negative. A negative value for s_i indicates that the image is located in the object plane of the lens. Since the lens is transmitting, a real image cannot exist in the object plane. Instead, the rays enter the image plane and diverge away from the lens. However, by extending imaginary lines from the diverging rays back into the object plane, an imaginary intersection or *imaginary image* can be formed. The absolute value of s_i is the distance to the imaginary image. Similarly, for lenses that are designed to diverge collimated beams, $f < 0$. The location of an image of a remote object or the focus of a collimated beam is calculated according to (2.10) with $s_i < 0$.

These last results can be generalized to form a sign convention that will be used for the analysis of multicomponent systems. This sign convention, which will be applied to both refracting and reflecting elements, will assist us in determining the location of objects and images relative to those optical elements. In most graphical and mathematical representations, the object plane is assumed to be to the left of the lens and the image plane to the right. Therefore:

1. the distance of the object to the lens, s_o, is positive when the object is located to the left of the lens;
2. the distance of the image to the lens, s_i, is positive when the image is located to the right of the lens;
3. the angle between a ray and the lens axis (or the optical axis) is positive when measured in the clockwise direction;
4. the radius of curvature of a lens or mirror is positive if the center is located to the right of the surface.

To include reflecting surfaces in the generalized analysis, we assume that mirrors behave as refractory elements (i.e., as single-surface lenses) with the exception that the index of refraction in the imaginary domain behind the mirror is $n = -1$.

Using the paraxial approximation, the imaging by a spherical mirror is represented by

$$\frac{1}{s_o} - \frac{1}{s_i} = -\frac{2}{R}. \tag{2.11}$$

Figure 2.16 presents the imaging of a point source, located at O, by a typical spherical mirror. The center of the sphere is at point C, which is at the left side of the mirror. Therefore, the mirror curvature R to be used in (2.11) is negative. When the distance from the object to the mirror $s_o > R$, the image is obtained at $s_i < 0$; that is, the image appears in the object plane. However, unlike an

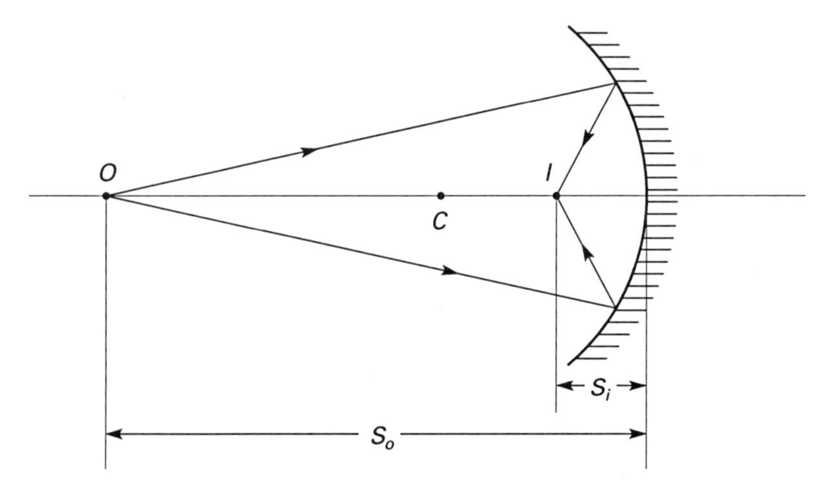

Figure 2.16 Imaging of a point source by a spherical mirror.

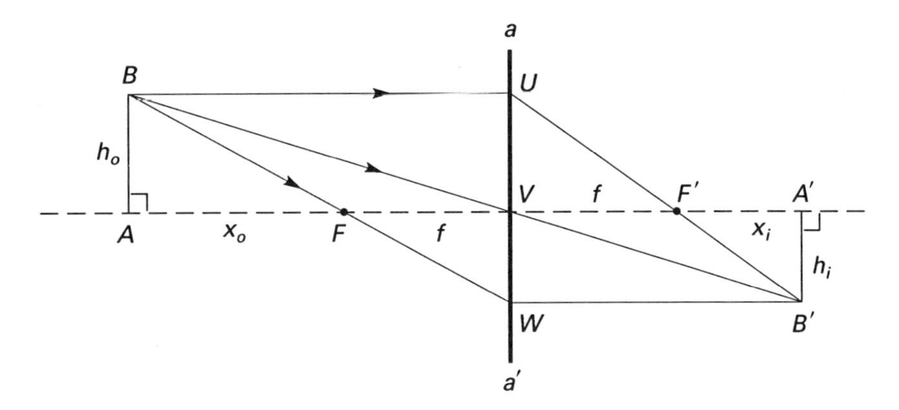

Figure 2.17 Imaging of a finite-sized object by a thin lens.

image created by a lens in the image plane, this image is real: it is formed by the intersection of rays. When $s_o \to \infty$ the image is located at $s_i = R/2$. Thus, for spherical mirrors the focal length is $R/2$.

Our results, describing imaging of a single point source by a lens or a mirror, may be used primarily to determine collection parameters – for example, the amount of radiation collected by a light-gathering lens with a prescribed aperture. However, lenses are often used to image an extended light source (either linear, planar, or volumetric) onto a target such as photographic film, to magnify a microscopic object, or to project images on screens. An *extended source* can be viewed as an array of either continuous or discrete point sources. Evaluation of the imaging of extended sources must include parameters such as the location of the image, its size, and the magnification or demagnification of the image. Effects of aberrations, image resolution, or levels of illumination may also be analyzed theoretically and empirically. Here we develop only the geometrical parameters of the imaging of extended sources.

Figure 2.17 shows a linear extended object imaged by a lens. The object, which is perpendicular to the lens axis, is represented by line AB of length h_o at a distance x_o from the focus F of the lens. The thin lens is marked by the line aa', which is also perpendicular to the axis. Any point on the object can be viewed as a point source emitting radiation in all directions; however, for simplicity we will show the imaging of only two points, A and B, at either end of the object. The images of other points between A and B are expected to be formed in between. The image of point B can be found geometrically or analytically by assuming that, for an ideal lens, all rays that emerge at B should intersect at a single point in the image plane; therefore, the image of B can be located by tracing at least two rays and finding their point of intersection. When aberrations are small (when the paraxial approximation is satisfied), any two rays may be selected: for example, the ray that emerges from B parallel to the lens axis, and the ray emerging from point B toward the focus of the lens in the object plane F. The first ray, parallel to the lens axis, is turned after passing the lens at point U and brought to the image plane through the lens focus F'.

Similarly, the second ray, after passing the lens focus F in the object plane, is turned after passing the lens at point W and propagates in the image plane parallel to the axis. The two rays intersect at point B', which denotes the image of B. The location, along the lens axis, of the image of point A can be calculated using (2.10). Similar ray-tracing analysis for points between A and B produces the full extended image $A'B'$ of the object. Because of the paraxial approximation, the image $A'B'$ is perpendicular to the axis. To obtain the magnification M of the image, we consider the two similar triangles FAB and FVW and the two similar triangles $F'A'B'$ and $F'VU$. From the proportions that can be derived from the similarity of these triangles, we can show that

$$x_o x_i = f^2. \tag{2.12}$$

Equation (2.12) is the Newtonian form of the lens equation. The reader may also show, by substituting $x_o + f$ for s_o and $x_i + f$ for s_i, that Newton's equation is equivalent to the Gaussian form (eqn. 2.10).

Finally, the magnification M is defined as the ratio between the image size h_i and the object size h_o. Using the similarity of triangles FAB and FVW and (2.12) (see Problem 2.8), the magnification is

$$M = \frac{h_i}{h_o} = \frac{s_i}{s_o}. \tag{2.13}$$

The ratio s_i/s_o, which is determined from the Gaussian or the Newtonian lens equation, controls the magnification of the imaging device. Thus, if a large object is to be imaged onto a small-frame film, the film must be placed near the focal plane of the lens while the object must be placed at a sufficient distance at the other side of the lens. Alternatively, when the object is placed at the focus, the image is pushed to infinity. An indirect consequence of (2.13) is that points B, V, and B' lie on a straight line. This follows from the similarity of triangles VAB and $VA'B'$, which leads to the equality $\measuredangle AVB = \measuredangle A'VB'$ that is necessary for these points to be on a straight line. Thus, an image can be constructed graphically by intersecting ray $UF'B'$ with the extension of a ray emerging from B and passing through V.

The magnification (or demagnification) of an image is an important design parameter in most imaging applications. Ultimately, the image size and hence the magnification is determined by the characteristic dimensions of the recording element. However, other parameters such as resolution or the level of detector illumination may dictate other magnification. The imaging of a planar object on the detector array of an electronic camera (Figure 2.18) can be used to illustrate some of these considerations.

With the advent of sensitive – and relatively cheap – electronic cameras, many imaging tasks are now accomplished electronically. These cameras convert the optical input into an electronic signal that can be stored digitally and processed almost instantaneously. The detector array of an electronic camera is divided into rectangular elements, the *pixels*. The number of pixels in a camera is specified by the manufacturer. Clearly, as the number of pixels increases

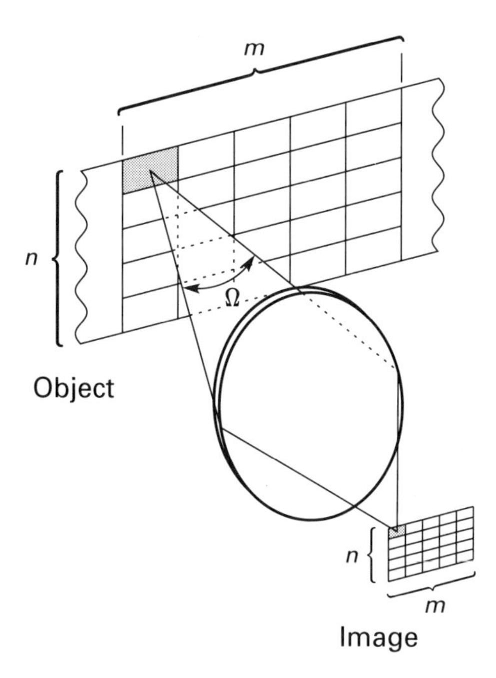

Figure 2.18 Imaging of a two-dimensional object by a lens and recording the image by a two-dimensional electronic detector array.

the recorded image can include finer details of the object, although the electronic storage space required for recording this image increases accordingly. For illustration, the image in Figure 2.18 is divided into $m \times n$ pixels. When the imaging is free of aberrations, each pixel can be correlated with a defined area element in the object plane. Therefore, the object can also be divided into $m \times n$ imaginary squares. The magnification can thus be viewed as the ratio between the dimension of a pixel and the dimension of its projection on the object. If the entire object is to be imaged by a single frame then the magnification needs to be adjusted accordingly. When aberrations are negligible, the *resolution* of the electronic camera – that is, the smallest detail on the object that can be recorded by the camera – is determined by the number of pixel elements. If, on the other hand, the imaging must reveal fine details of the object, then the magnification needs to be adjusted to allow these details to be recorded by several pixels. If the distance between the imaging lens and the detector array is approximately constant, the magnification (or demagnification) is controlled by the distance between the object and the imaging lens. Focusing of the image is then obtained by slightly translating the lens relative to the detector array.

Although the choice of magnification is largely controlled by the geometrical parameters of the detector array and the object and by resolution requirements, other parameters, such as the irradiance falling on each pixel, may need to be considered. If the pixel height and width and its distance to the lens are fixed parameters, then the size of the object element imaged by that pixel increases linearly with s_o and the area imaged by that pixel increases quadratically with s_o. If, in addition, the object is uniformly illuminated, then the total radiant flux available for imaging by the pixel also increases quadratically with s_o. However, the total irradiance eventually falling on a pixel depends also on the radiant power that is intercepted by the lens. This parameter can be best specified by the solid angle Ω, which is the angle subtended by a cone whose apex is at the object element and whose base is the collection lens. If the emission intensity (eqn. 1.3) is uniform within Ω, then the total radiation flux collected by the lens in the θ direction is

$$\Phi(\theta) = \int_\Omega I(\theta)\,d\Omega = I\Omega.$$

To evaluate Ω, consider the cone apex as a point source emitting uniformly in all directions. If the camera lens is located at a distance s_o from the object, then the ratio between (a) the flux $\Phi(\theta)$ captured by the camera and (b) the total flux Φ emitted by the point source into a spherical envelope with radius s_o can be expressed by the ratio between the solid angle Ω and the solid angle 4π subtended by a sphere. This is approximately the ratio between the area of the camera lens and the area of the spherical envelope. For a lens diameter d_L, this ratio is

$$\frac{\Phi(\theta)}{\Phi} = \frac{\Omega}{4\pi} \approx \frac{\pi d_L^2/4}{4\pi s_o^2}$$

or, after reduction,

$$\Omega \approx \frac{\pi}{4}\left[\frac{d_L}{s_o}\right]^2.$$

The ratio d_L/s_o between the lens diameter and the distance to the object is a standard measure of the collection efficiency of lenses. It is usually expressed in terms of the *numerical aperture* (NA) of the lens, which is defined by

$$\mathrm{NA} = n \sin\left[\frac{\phi}{2}\right] \approx \frac{d_L}{2s_o}, \tag{2.14}$$

where ϕ is the cone angle (not to be confused with the three-dimensional solid angle Ω). The last relation expressed in (2.14) is approximately valid if $\phi \ll \pi$ and $n = 1$. When the distance to the object is twice the diameter of the lens, the numerical aperture is said to be 0.25. The collected flux increases quadratically with NA and decreases quadratically with s_o. This decrease in the incident flux on each pixel is exactly offset by the increasing area of the object imaged by the pixel. Thus, although each pixel represents a larger area of the object when s_o is increased, the total radiant energy falling on the pixel remains unchanged. If, on the other hand, the lens is used to image a linear object (such as the scattering from a laser beam), the irradiance per pixel *decreases* linearly with s_o.

The collection efficiency of lenses may also be specified by the ratio f/d_L between the focal length and the lens diameter. This ratio is the *f-number* ($f/\#$).

In some publications this number also represents the ratio between the distance to the object and the lens diameter. In both definitions, a smaller $f/\#$ represents a more efficient collection. In photographic applications, the distance between the lens and the recording medium (which is located near the lens focus) is approximately constant, so the $f/\#$ of a camera lens specifies its aperture. Thus, for a fully open aperture, an efficient (or "fast") camera lens may have $f/\# = 1$. By reducing the shutter aperture the $f/\#$ may be increased to 16.

2.6 Alternative Lenses

The manufacturing of lenses, particularly those with large diameters or short focal lengths, is a fine balancing act between optical requirements, design and manufacturing limitations, and cost. Therefore, alternative means for light gathering and focusing using refractory optical methods are being developed. Two of these, the gradient index (GRIN) lens and the Fresnel lens, have emerged as potentially useful techniques. In both approaches, an expensive large-aperture lens is replaced by a thin, easy-to-manufacture lens.

The Fresnel lens consists of a set of concentric transparent rings each with conical surface. The slopes of the conical surfaces increase progressively for rings away from the center. Figure 2.19 shows a cross section of a Fresnel lens that replicates a plano-convex lens. Two rays passing through two adjacent rings are depicted in the figure. These rays are bent by refraction toward a focal point that is similar to the focus of a regular lens. However, since each ring has a finite width, rays are bent in bundles toward the focus, where they form a somewhat blurred image of the source. The sharp discontinuities between the rings further deteriorate the image quality. On the other hand, since the focusing parameters of this lens are determined only by the slopes of the conical surface and their progression, the lens can be made thin even for large apertures. Fresnel lenses are used primarily for applications in which the quality of the image is a secondary requirement. Thus, Fresnel lenses have originally been installed in lighthouses, where large beams need to be collimated to distances

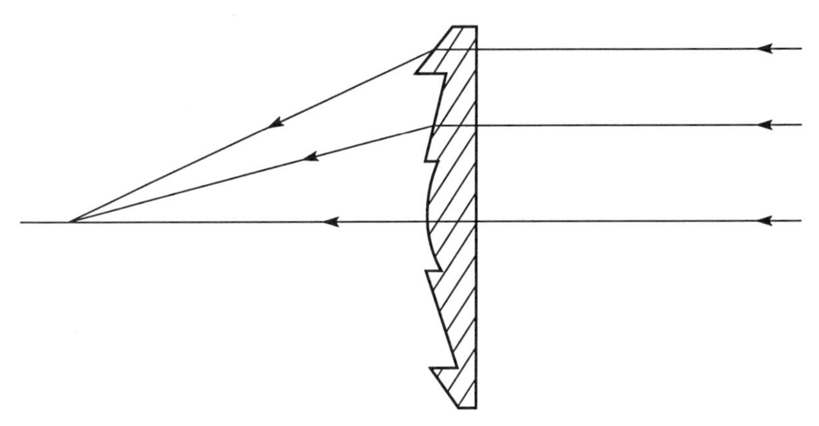

Figure 2.19 Cross section of a Fresnel lens.

of 20–30 km (see Problem 2.9). Flat sheets of Fresnel lenses are used in overhead projectors to gather and collimate light, or for large-field viewing in automobile rear-view mirrors.

To overcome the disadvantage presented by the stepwise focusing properties of the Fresnel lens, new refractory optical materials are being developed. By seeding the optical host material with ion dopants, the index of refraction can be controlled and a gradual variation in the index of refraction can be induced. By inducing index variation in the radial direction in a disk-shaped element, the optical path length for propagation through the center of the medium can be made longer relative to the optical path length at the edges. Recall that, according to Fermat's principle, radiation from one point is focused into another point when all rays propagate through the same optical distance. The classical lens was designed to meet this requirement by varying its thickness in the radial direction, whereas GRIN lenses are expected to meet this requirement by variation of the index of refraction. For a radially varying index, the GRIN lens can be kept flat. Rays passing through the lens propagate in a medium in which the gradient in the index of refraction is transverse to their direction of propagation. Thus, from (2.8), a deflection that depends both on the thickness of the medium and the magnitude of the gradient must take place. Desired focusing parameters can be obtained by designing the thickness of the lens and the variations in the index of refraction. Alternative techniques for GRIN lens manufacturing allow for gradients in the index of refraction in the axial direction. To adjust the optical path length between the object and the image, portions of the substance are ground off.

2.7 Refraction in Gases

The refraction of rays by media in which the index of refraction is gradually varying was discussed in Section 2.3. It was shown that when the gradient in the index of refraction is transverse to the direction of propagation of the ray, a deflection with curvature toward the increasing index of refraction takes place. For many gas mixtures (including air), the index of refraction is a complex function of the gas density and the wavelength of the incident radiation. However, at atmospheric pressure and temperature and around the visible spectrum, the index of refraction increases approximately linearly with density. Thus, when the density of a homogeneous gas is changed – by perturbation of either the temperature or the pressure – the curvature of the refracting ray points toward the increasing density function.

Many atmospheric effects, such as mirages and the apparent break-up or distortion of the circular shape of the sun during sunset or sunrise, are explained by this phenomenon of refraction. Figure 2.20 illustrates a situation in which a mirage is seen only by the child and not by his adult companion. Naturally occurring mirages are usually observed on hot days and over hot surfaces. In the present example the black pavement is heated by absorption of sunlight, and its temperature, together with the temperature of the adjacent boundary

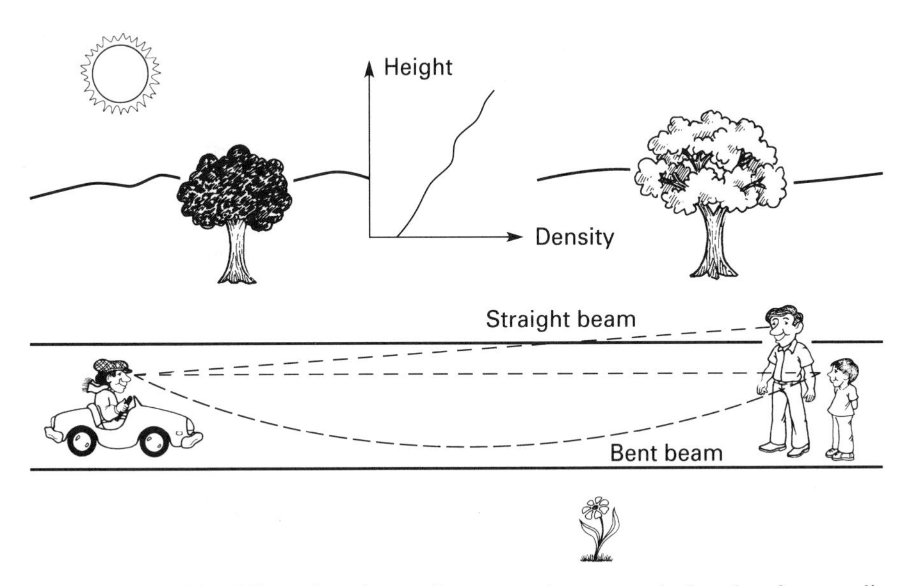

Figure 2.20 Effect of a mirage. Shown are the geometrical paths of one undistorted ray and one ray distorted by refraction in hot air.

layer of air, is raised significantly relative to the ambient temperature. The temperature of the boundary layer is highest at the pavement surface and decreases (see inset) with height above the surface. The density, following ideal gas behavior, increases with height. Although the thickness of the boundary layer depends on several atmospheric parameters such as wind and humidity, its sharp gradients diminish at eye level. Therefore, both the horizontal ray seen by the child and the upward sloping straight ray seen by the adult propagate in straight lines with no apparent refraction. However, a third ray, emerging from the car at a downward slope, is refracted by the density gradient in the inverted boundary layer (eqn. 2.8). If the refraction is sufficient to reverse the downward slope then this ray may be seen by the child, and if the density gradients and propagation distance are sufficient then also by the adult. To an observer, this curved ray and many other rays from the car and its background appear as an image emerging from the road. Since both the car and the driver are also seen directly, the second image is interpreted as a reflection that, owing to the flickering of the image caused by air currents, appears to come from a wet surface or from a body of water. Inverse density gradients, which occur when surfaces are heated relative to their ambient by exceptionally bright sunshine, are required for the observation of a mirage. But even under ideal conditions the total deflection by refraction is small; therefore, the upward slope of the refracted ray is small and the mirage can be detected by observers who are either close to the surface (e.g., the child in Figure 2.20) or at a long distance from the apparent point of reflection.

The refraction by air, which we have shown to cause the mirage, is sensitive to variations in the gas density, which in turn depends on the local temperature and pressure. Therefore, optical applications involving transmission of light through long optical paths in air or through large property gradients must either avoid or discount for distortions induced by refraction. Examples include imaging from space, optical tracking and interception of flight vehicles, and astronomical observations. Distortion by refraction is best avoided by limiting applications to times with optimal atmospheric conditions. On the other hand, the newly developing technology of *aero-optics* is expected to use advanced computational techniques to correct for the distortion by shock waves and temperature gradients of optical transmission through windows of supersonic and hypersonic vehicles.

The refraction of light by gases – which is an impediment to high-resolution imaging through air – is used for visualizing the structure of gas flows in many aerodynamics and combustion testing facilities. Several techniques, all relying on the refraction induced by variations in the index of refraction, have been developed to obtain images of gas density variations and their derivatives. Most notable are the *shadowgraph* and the *schlieren* techniques. In both techniques, the variations in the index of refraction are related to derivatives of the gas density and are optically imaged and recorded. The techniques differ, however, in their complexity and type of information generated.

The shadowgraph technique is one of the simplest flow visualization techniques available. The collimated beam in the diagram of Figure 2.21, which represents a typical shadowgraph system, is passed through a gas flow (e.g., the test section of a wind tunnel) and projected onto a screen. Either a laser or an incandescent lamp can be used as the light source. Lasers can provide the brightness for illumination of large test sections as well as the monochromacity needed for discrimination against the background luminosity that may exist in combusting or hot flows. Incandescent sources offer extra simplicity when the advantages of a laser system are not needed.

In describing the principles of the shadowgraph technique, we may view the collimated beam as a bundle of collimated rays. As long as the gas density

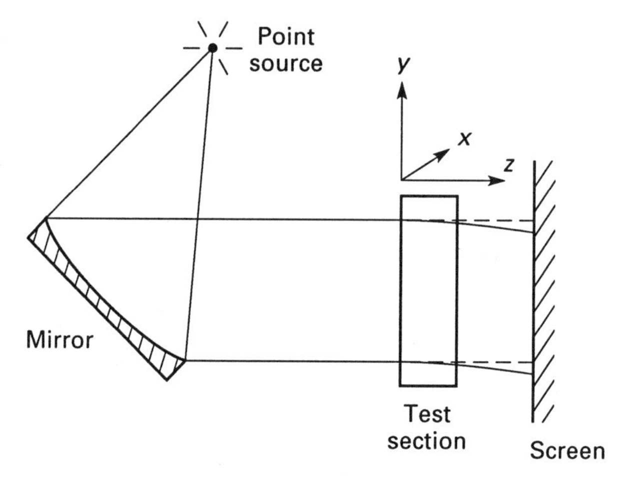

Figure 2.21 Layout of a shadowgraph system.

along the beam path - and, in particular, in the test section - is uniform, all rays within the bundle remain parallel to each other. However, when a density gradient with a component oriented transversely to the beam is developed, each ray is deflected by a total angle $\alpha = \int d\alpha$ that can be computed from (2.8). It is evident that this integration is performed through the entire optical path. Thus, the technique lacks any spatial resolution in the longitudinal direction; that is, the index of refraction or any of its derivatives cannot be determined at any specified longitudinal location. Because the index of refraction is usually proportional to the gas density ρ, α can be represented directly by ρ through substitution in (2.8) and the deflection of any ray can be attributed to transverse density gradients in the test section.

The simplest case of inducing ray deflection is represented by transverse density gradients that are transversely uniform; that is, the second derivative of the density in the transverse direction is zero. In this case, the deflection angle α is uniform for the entire beam and so all the rays of the bundle emerge from the test section still parallel to each other, albeit slightly shifted and tilted. The illumination of the screen by this bundle of parallel rays remains uniform. However, the objective in all flow visualization experiments is to create noticeable variations in the screen illumination - bright or dark regions that can be attributed to flow structures. Such detectable variations require that the derivatives, or gradients, of the screen illumination be nonzero. The simple case of uniform density gradients clearly fails to meet this requirement: when the second transverse derivative of the density in the test section is zero, the derivative of the screen illumination is zero as well. We can anticipate detecting nonzero illumination gradients only when the second transverse density derivative is nonzero. This is demonstrated formally in Liepmann and Roshko (1957), assuming that the irradiance at the screen can be represented by the number of rays per unit area that are incident on the screen. The variation in the irradiance $I(x, y)$ at the screen is related to the second derivative of the density as follows:

$$\frac{\Delta I(x,y)}{I(x,y)} = L \int_{z_1}^{z_2} \left(\frac{\partial^2 \rho}{\partial x^2} + \frac{\partial^2 \rho}{\partial y^2} \right) dz, \tag{2.15}$$

where x and y are the coordinates transverse to the propagation direction z, z_1 and z_2 are the coordinates of the input and output planes of the test section, and L is the distance between the screen and the test section.

The shadowgraph technique is normally used to image gas flows with abrupt density variations, essentially physical discontinuities such as shock waves, detonation waves, or flame fronts (Merzkirch 1987). In these discontinuities, the second derivative of the density transverse to the illuminating beam is large and so their shadowgraph images appear as sharp, high-contrast lines. From these images, the location of the discontinuities or their geometrical features (e.g. slope) can be accurately determined. Since their optical configuration is extremely simple, shadowgraph systems are available in most testing facilities. On the other hand, the direct information produced by the method (i.e., the second derivative of the density distribution) is of secondary importance, either because it represents an integration along the beam path or because other important parameters cannot be easily derived from this measurement. Furthermore, since the method is insensitive to the first derivative of the density distribution, minor density variations that are present in free convection and in most subsonic flows are barely discernible. Visualization of non-uniform flows, with moderately distributed density variations and thus small second derivatives, requires an alternative optical technique that is sensitive to either the density itself or to its first derivative. Interferometry is normally used for visualizing density distributions, whereas schlieren techniques are used for visualizing the distribution of the transverse component of density gradients. Discussion of interferometry requires the use of physical optics and will therefore be deferred to Section 6.4. However, the principles of operation of the schlieren technique can be well understood using geometrical optics.

The schlieren technique and its derivatives were designed to enhance the detection sensitivity of gentle density variations. Schlieren systems are most often used to analyze subsonic flows, heat convection, or combustion. However, they can also be used to image flow discontinuities, just like the shadowgraph. Since the schlieren technique is designed to image the distribution of density gradients tranverse to the incident beam, the irradiance of a schlieren image is expected to change even when the density gradient transverse to the incident beam is uniform. This is in contrast to the shadowgraph technique, where no detectable change is expected in the illumination intensity when the density gradient is uniform.

A typical schlieren set-up is presented in Figure 2.22. It consists of two Galilean telescopes assembled in series. A *Galilean telescope* consists of only two lenses. Thus, the middle lens in the figure is part of the first as well as the second telescope. If the lenses of a Galilean telescope are ideal (i.e., have negligible aberrations) and parallel to each other, then radiation from a point source placed at the focus of the front lens – the objective – is collimated into a beam that is parallel to the optical axis. The second lens, the eye lens, is placed anywhere along the collimated beam and refocuses the beam to form an image of the point source at the focal point behind the lens. By selecting the focal lengths and diameters of both lenses, the telescope can be used for magnification (and reduction) as in Figure 2.22, or for the use intended by Galileo: expansion (or compression) of collimated beams. The magnification is uniquely defined by the ratio of the focal lengths of the two lenses. The image produced by a Galilean telescope can be focused by moving one or both lenses with no effect on the magnification.

The first telescope of the schlieren system consists of lenses L_1 and L_2. The second telescope consists of lenses L_2 and L_3. A light source, which realistically must have a finite size, is shown in the figure by line ab, which is perpendicular to the optical axis. This extended source can be either a linear source or a planar source parallel to the (x, y) plane. Each point on this extended source can be considered as an independent source that forms, past the objective lens L_1, a collimated beam. Therefore, the space between L_1 and L_2, where the test

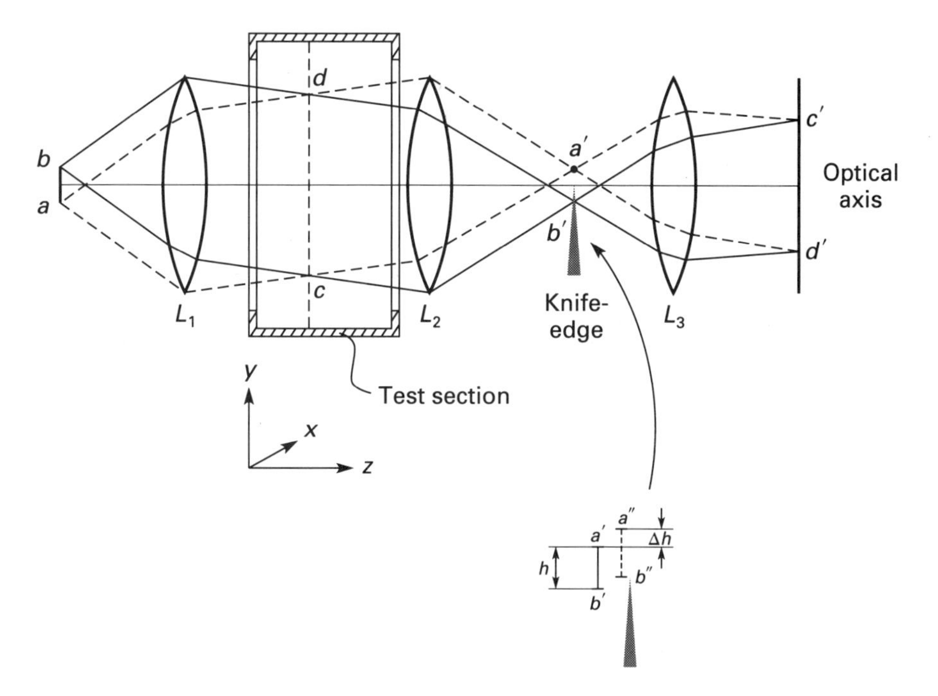

Figure 2.22 Layout of a schlieren system.

section is located, is filled by an infinite number of superimposed collimated beams, each originating from a different point on the extended source. This is illustrated by rays emerging from two points located at the opposite ends of the extended source *ab*. Although each collimated beam consists of parallel rays, only the beam originating at the optical axis is parallel to that axis. The remaining beams, originating at points off the optical axis, although collimated are not parallel to the axis. Instead, a beam originating at point *a* is slightly tilted upward relative to the optical axis while a beam emerging at *b* is tilted downward. Because of these slight tilts the collimated beams, although superimposed, are distinguishable. At the focal plane of L_2 where all rays are focused, the slight tilt of each beam translates into a specified vertical location off the optical axis; the collimated beam that was parallel to the optical axis creates an image at the axis, while the beam originating at point *a* creates an image at point a' above the optical axis; the image of *b* is at b' below that axis. The image plane represents an inverted one-to-one transformation of the source, magnified by the ratio of the focal lengths of the two lenses (see Problem 2.11). Because the image is created by an infinite number of beams, each evenly filling the space between L_1 and L_2, any perturbation within the test section must influence all the image points.

The second telescope of the schlieren system consists of lenses L_2 and L_3, with lens L_2 serving as the eye lens of the first telescope and as the objective lens of the second telescope. Unlike the first telescope, this telescope is designed to re-collimate the rays passing through the image plane $a'b'$ and project them onto a screen behind L_3. This is achieved by matching the distance between L_2 and L_3 with the sum of the focal lengths of these two lenses (see Problem 2.11). Thus, the collimated beams that fill the space between L_1 and L_2 are re-collimated past L_3. Ideally, not only the beams but also their individual orientations are reproduced by this collimation step. With such an ideal re-collimation, any transverse plane (or line) in the test section (e.g. plane *cd*) that is defined by the intersection of two collimated beams is imaged by L_3 to *c'd'*. This imaging transmits along with the beams' orientations all the perturbations induced by density variations and incurred by the beams as they traversed the test section. Although this double telescope system is more complex, the results obtained are essentially similar to those of the shadowgraph. When the density gradients in the test section are uniform, all beams shift uniformly and the screen illumination remains unchanged. Thus, in order to gain sensitivity to the first-order density gradient, a new element must be added to the system; this is the knife-edge placed at the focal plane of L_2. The knife is placed so that half of the image *a'b'* is clipped, thereby reducing the screen illumination. However, when the flow is perturbed (even by a uniform gradient) in the y direction, the image *a'b'* is shifted – either up or down – and the illumination of the screen at *c'd'* increases or decreases accordingly. Thus, intensity measurements can now be used to determine the extent of the density gradient in the y direction. Obviously, density gradients in the x direction (i.e., parallel to the knife-edge) have no effect on the illumination of the screen and therefore remain undetected.

The effect of the knife-edge is best understood by recalling that each point of the image *a'b'* receives light from every point in plane *c*. Thus, although insertion of the knife blocks some of the radiation, all details of the plane *cd* that are required for reconstruction of the image *c'd'* are still transmitted. This is as if the test section were illuminated by part of the light source. For optimal results, the knife is adjusted to clip half of the image *a'b'*, thereby reducing uniformly the illumination at *c'd'* by half. When a uniform density gradient in the y direction is generated inside the test section, all rays deflect evenly by a total angle α that can be calculated using the integration of (2.8). If the gradient of the index of refraction has a component in the positive y direction, the deflection will result in raising the image *a'b'* by an amount

$$\Delta h = f_2 \alpha_y,$$

where α_y is the y component of the deflection angle induced by the flow inhomogeneities and f_2 is the focal length of L_2. Although α_y depends on variation in the index of refraction, it can be related to the density variations in the flow. Thus, the irradiance at a point (x, y) on the screen will increase by the relative amount

$$\frac{\Delta E(x,y)}{E(x,y)} = \frac{\Delta h}{h} = \frac{f_2 \alpha_y}{h} = \frac{1}{h} \int \frac{1}{\rho(x,y,z)} \frac{\partial \rho(x,y,z)}{\partial y} \, dz. \tag{2.16}$$

It is evident from (2.16) that variations in the irradiance of a schlieren image are induced by density gradients in the y direction, but the density may vary in all directions and so the image at $c'd'$ is a result of effects integrated along the beam path while gradients in either the x or z direction are not resolved. Thus, a y component of the density gradient near the entrance window cannot be distinguished from a y component of the density gradient near the exit. This presents a serious limitation on the application of schlieren imaging of highly three-dimensional flows. On the other hand, when most of the structural variations are limited to an (x, y) plane (i.e., when the flow is two-dimensional), quantitative measurements of the density gradients are possible. Several methods have been developed to improve the resolution of the schlieren technique in the z dimension. These include the focused schlieren (Kantrowitz and Trimpi 1950; Weinstein 1993) and holographic methods (Havener and Kirby 1993). Both techniques permit imaging of the total deflection of the rays at a selected plane within the test section. However, even with a spatially resolved technique, the image at the resolved plane results from the cumulative deflection along the path ahead of that plane. An alternative method (Yates 1993) employs computational techniques that generate simulated schlieren images from computed three-dimensional flow fields. By comparing the simulated images with experimentally recorded images, three-dimensional flow characteristics are confirmed.

2.8 Ray Transfer Matrices

In this chapter we have discussed the utility of geometrical optics for the design and evaluation of multi-element optical systems. Until now, most of the examples have dealt with the calculation of systems that include a single element or relatively few elements (e.g., the schlieren system). However, the design of complex imaging systems requires calculation of the propagation of rays through numerous elements, and must also account for the spaces of air or other media between these elements. Even more demanding is the design of laser systems, where the paths of beams bouncing back and forth between two mirrors while passing through various optical elements must be calculated before important stability criteria can be determined. Although these calculations are feasible with the mathematical tools that are already available, the development of a mechanical routine could simplify the development of computational techniques for graphically simulating the optical layout and the propagation of light through it. The objective of such a mechanized calculation method will be to represent each optical element by a matrix operator that transforms the incident (input) ray into an output ray. The terms of the operator matrix will include the numerical parameters of the optical element: the focal length of a lens or a mirror, the optical depth of a spacer, and so forth. The optical layout of a system will then be represented by a series of operators, each one operating on the output of the previous optical element.

Ray transfer matrices were developed to achieve this objective. In its simplest form, a ray transfer matrix is an operator that represents a single optical element. By operating on the parameters of an incident ray, the parameters of the output ray – that is, the ray immediately behind that element – are obtained.

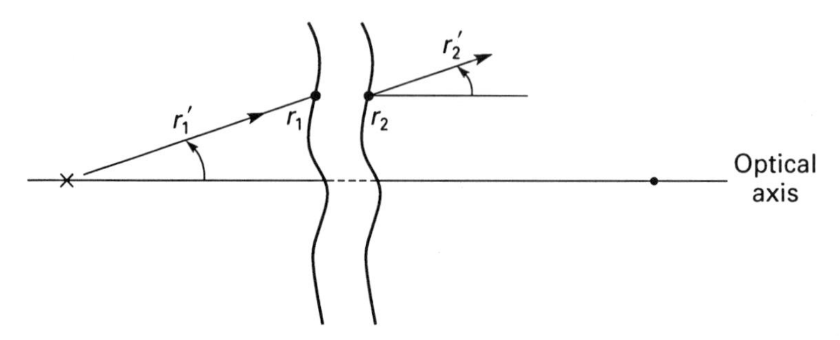

Figure 2.23 Geometrical parameters of a ray entering an optical element, and the parameters of that ray after transformation by that element.

For a lens, the input plane is the front surface while the output plane is the rear surface; the parameters of the lens include its focal length and the indices of refraction at both sides of the lens. Complete geometrical representation of a linear ray entering or leaving an element requires that the two coordinates of its interception with a given plane be specified along with two angles of its slope. Fortunately, most optical systems are either circularly symmetric or can otherwise be reduced to two-dimensional analysis. With that simplification, a ray crossing a specified plane can be represented by only one coordinate (e.g., the distance from the optical axis) and one slope angle. This pair of numbers can be written as a two-component vector. The first component, r_1, is the distance between the point of incidence and the optical axis in the input plane of an arbitrary optical element; the second component of the incidence vector is the slope, r_1' (see Figure 2.23). Similarly, the ray at the exit plane of that element is represented by two components, r_2 and r_2'. For paraxial approximations we have $r' \approx \sin\theta \approx \theta$, where θ is the angle between the ray and the optical axis measured in the counterclockwise direction. Transformation of any two-dimensional vector into any other two-dimensional vector requires a 2×2 transformation matrix. By multiplying the input ray vector by the ray transfer matrix, a new vector representing the exit coordinate and slope of the new ray is obtained. This transformation is shown in the following equation, where the arbitrary element of Figure 2.23 is represented by a general four-element ray transfer matrix:

$$\begin{bmatrix} r_2 \\ r_2' \end{bmatrix} = \begin{bmatrix} A & B \\ C & D \end{bmatrix} \begin{bmatrix} r_1 \\ r_1' \end{bmatrix}. \tag{2.17}$$

The four elements of a ray transfer matrix can be calculated for most optical elements using simple geometrical considerations in which the parameters of the exit ray are expressed by the parameters of the incident ray. To illustrate this procedure, consider the lens in Figure 2.15. When the lens is thin, $r_1 = r_2$. This is the first of the two equations needed to establish the transformation between the input and output rays. To express the slope of the exit ray in terms of the parameters of the incident ray, we observe that the slope of the incident ray in Figure 2.15 is r_1/s_o and the slope of the exit ray is $-r_2/s_i$. Therefore, multiplying both sides of (2.10) by r_1 and using $r_1 = r_2$, we obtain

$$r_2' = r_1' - r_1/f.$$

Thus, the ray transfer matrix of a lens is:

$$\begin{bmatrix} r_2 \\ r_2' \end{bmatrix} = \begin{bmatrix} 1 & 0 \\ -1/f & 1 \end{bmatrix} \begin{bmatrix} r_1 \\ r_1' \end{bmatrix}.$$

The reader can easily confirm that, by the completion of this matrix multiplication, the two equations relating r_1 and r_1' to r_2 and r_2' are reproduced.

Table 2.1 lists the ray transfer matrices of frequently encountered elements. It is left to the reader to verify these matrices. The first element in Table 2.1, the flat medium, is the simplest; it has no effect on the transformation when $r_1' = 0$, but when $r_1' \neq 0$ the coordinate of the exit ray $r_2 \neq r_1$. Modeling an optical system requires inclusion of a ray transfer matrix for every space between elements. Air-spaced gaps may be represented by their length only, but other flat media such as filters or attenuators must be represented by their index of refraction and their thickness. The third element in Table 2.1 represents a medium of length L in which the index of refraction varies quadratically with distance r from the optical axis. Such behavior can occur in laser crystals illuminated by bright flashlamp pulses. The envelope of the crystal is heated by the external flashlamp radiation, while the cooler core of the crystal carries the laser beam. Depending on the lamp irradiance and the beam energy, the crystal temperature and its index of refraction may not be uniform in the radial direction. Note that the path of a ray traveling through this medium is sinusoidal.

Ray transfer matrices of other optical elements can be formed by combining the matrices 4, 5, or 6 of various interfaces. If necessary, the ray transfer matrix of a flat medium with depth d may also be included. Thus, the first element in Table 2.1, the flat medium, is a combination of two dielectric interfaces separated by a spacer with a depth d. Since the interfaces already account for the abrupt change in the index of refraction, the index of refraction to be used here for the flat medium is $n = 1$. The output of the first interface is the input to the spacer and the output from that spacer is the input to the next dielectric interface. The transformation between the input to the first element and the output of the last element can be obtained by repeatedly applying (2.17). The following example illustrates this procedure.

Example 2.1 Derive the ray transfer matrix of a flat plate with index of refraction n_2 when immersed in a homogeneous medium with an index of refraction n_1.

Solution The ray transfer matrix of this plate consists of the combined matrix of two dielectric interfaces, from n_1 to n_2 and from n_2 to n_1, and one spacer with a depth of d. The equivalent transformation of the three matrices is

Table 2.1 *Ray transfer matrices for selected optical elements*

No.	Optical element	Ray transfer matrix
1	Flat medium	
	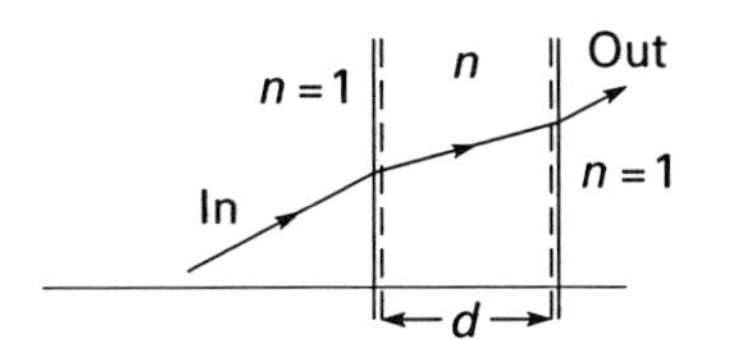	$\begin{bmatrix} 1 & d/n \\ 0 & 1 \end{bmatrix}$
2	Lens	
	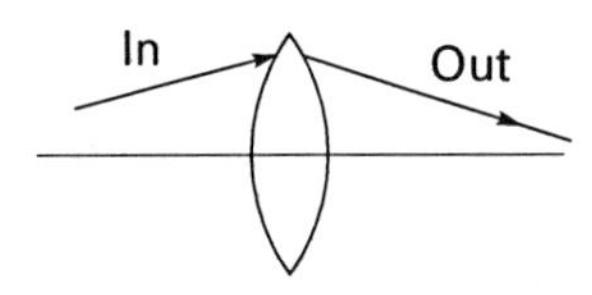	$\begin{bmatrix} 1 & 0 \\ -1/f & 1 \end{bmatrix}$
3	Lenslike medium with quadratic index	
	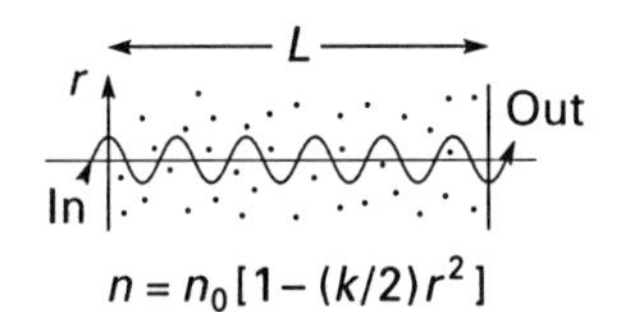	$\begin{bmatrix} \cos(kL) & k\sin(kL) \\ -k\sin(kL) & \cos(kL) \end{bmatrix}$
4	Dielectric interface	
	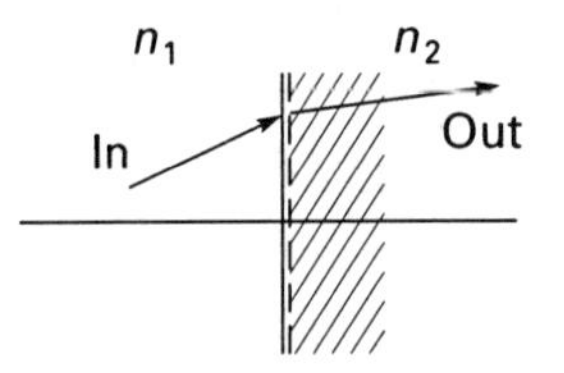	$\begin{bmatrix} 1 & 0 \\ 0 & n_1/n_2 \end{bmatrix}$
5	Spherical dielectric interface	
	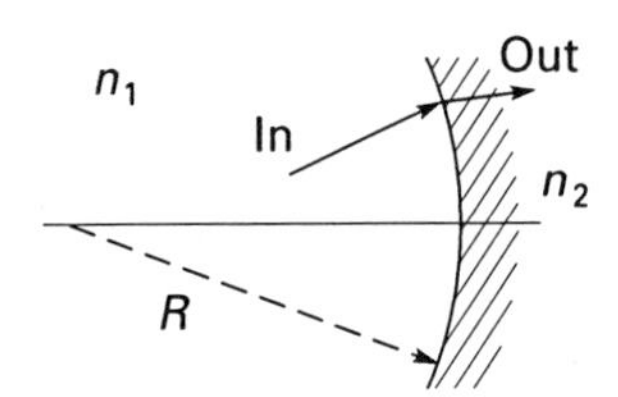	$\begin{bmatrix} 1 & 0 \\ \dfrac{n_1-n_2}{n_2}\dfrac{1}{R} & \dfrac{n_1}{n_2} \end{bmatrix}$
6	Spherical mirror	
	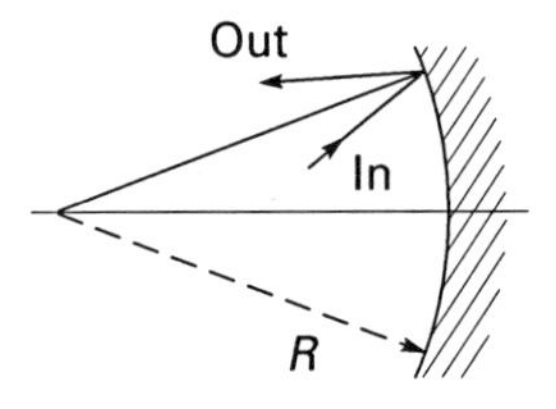	$\begin{bmatrix} 1 & 0 \\ -2/R & 1 \end{bmatrix}$

$$\begin{bmatrix} r_4 \\ r_4' \end{bmatrix} = \begin{bmatrix} 1 & 0 \\ 0 & n_2/n_1 \end{bmatrix} \begin{bmatrix} 1 & d \\ 0 & 1 \end{bmatrix} \begin{bmatrix} 1 & 0 \\ 0 & n_1/n_2 \end{bmatrix} \begin{bmatrix} r_1 \\ r_1' \end{bmatrix} = \begin{bmatrix} 1 & dn_1/n_2 \\ 0 & 1 \end{bmatrix} \begin{bmatrix} r_1 \\ r_1' \end{bmatrix}.$$

Thus, the matrix of the first interface operates on the vector $\{r_1, r_1'\}$ of the incident ray, the matrix of the space d between the interfaces operates on the result of the first transformation, and the matrix of the exit interface is operating on the result of the second transformation. (Here $\{a, b\}$ denotes $[a, b]^{\mathrm{T}}$, where "T" stands for "transpose.") The output ray $\{r_4, r_4'\}$ is then the result of three consecutive transformations, which can be represented by a single ray transfer matrix. Although the elements of this transformation are presented graphically from left to right, the mathematical representation requires that the first element be presented at the extreme right and the last element at the extreme left. ■

A generalized procedure can be developed for the derivation of the ray transfer matrix of a group of optical elements stacked in a series. The output-ray vector $\{r_3, r_3'\}$, following the propagation through two elements, can be related to the input ray by

$$\begin{bmatrix} r_3 \\ r_3' \end{bmatrix} = \begin{bmatrix} A' & B' \\ C' & D' \end{bmatrix} \begin{bmatrix} r_2 \\ r_2' \end{bmatrix} = \begin{bmatrix} A' & B' \\ C' & D' \end{bmatrix} \begin{bmatrix} A & B \\ C & D \end{bmatrix} \begin{bmatrix} r_1 \\ r_1' \end{bmatrix},$$

where the first equation represents the transformation by the second element and the second equation represents the successive transformation by the two elements.

Example 2.2 Find the ray transfer matrix that represents propagation from the object plane of the lens in Figure 2.15 to the image plane. Assume that the focal point in the object plane is located between the object and the lens.

Solution The propagation sequence here includes three components: the free space between the object plane and the lens (element 1 in Table 2.1), the lens itself (element 2 in Table 2.1), and the free space between the lens and the image plane (element 1). Using the procedure of successive transformations and the parameters of Figure 2.15, the input ray vector $\{r_o, r_o'\}$ is transformed into a vector $\{r_i, r_i'\}$ in the image plane by the following transformation:

$$\begin{bmatrix} r_i \\ r_i' \end{bmatrix} = \begin{bmatrix} 1 & s_i \\ 0 & 1 \end{bmatrix} \begin{bmatrix} 1 & 0 \\ -1/f & 1 \end{bmatrix} \begin{bmatrix} 1 & s_o \\ 0 & 1 \end{bmatrix} \begin{bmatrix} r_o \\ r_o' \end{bmatrix}.$$

Again the successive transformation appears with the matrix of the first element – the free space ahead of the lens – at the right end of the sequence while the last element appears at the extreme left. Using the rules for matrix multiplication, this successive transformation can be reduced to the following single ray transfer matrix:

$$\begin{bmatrix} r_i \\ r_i' \end{bmatrix} = \begin{bmatrix} 1 - s_i/f & s_o - s_i s_o/f + s_i \\ -1/f & -s_o/f + 1 \end{bmatrix} \begin{bmatrix} r_o \\ r_o' \end{bmatrix}.$$

This transformation contains, in a condensed form, many of our previously identified results for imaging by a lens. To illustrate this, we complete the last step of the transformation and write explicitly the terms for r_i and r_i' as follows:

$$r_i = \left(1 - \frac{s_i}{f}\right) r_o + \left(s_o - \frac{s_i s_o}{f} + s_i\right) r_o',$$
$$r_i' = -\frac{r_o}{f} + \left(-\frac{s_o}{f} + 1\right) r_o'. \tag{2.18}$$

If $r_o = r_i = 0$ – that is, if the transformation is of a point source on the optical axis into an image point – then the first of equations (2.18) reproduces the Gaussian form of the lens equation (eqn. 2.10). Similarly, for $r_o = 0$ the second of equations (2.18) produces the ratio between the slopes of the object and the image rays in terms of their distances from the lens:

$$\frac{r_i'}{r_o'} = -\frac{s_o}{f} + 1 = -\frac{s_o}{s_i}. \tag{2.19}$$

If a numerical aperture for the image plane is defined similarly to the numerical aperture in the object plane (eqn. 2.14), then the absolute value of the ratio of the slopes in (2.19) is also the ratio of these numerical apertures.

The magnification of this imaging configuration can also be obtained from the ray transfer matrix. Recall from Figure 2.17 that, for an object of finite size, at least one point must be at $r_o \neq 0$. Graphical representation of the imaging of this point requires that several rays be drawn. If one of these rays is used in (2.18) – for example, the ray with $r_o' = 0$ – then the first of these equations reduces to

$$\frac{r_i}{r_o} = 1 - \frac{s_i}{f} = -\frac{s_i}{s_o}.$$

This is identical to the magnification by a lens (eqn. 2.13), with the exception that the negative sign indicates that the image is inverted.

This result could easily be obtained by direct geometrical considerations. However, this example demonstrates the utility of the mathematical technique. Using tabulated mathematical terms and relatively few rules, complex systems can be readily simulated. This method is labor-intensive; however, owing to its simplicity and mechanized structure, it is particularly suitable for computerized calculations. ■

The brief outline in this chapter of the principles of geometrical optics covered topics that are essential for successful application of optical techniques in most fields of engineering. Although a few applications, such as the shadowgraph or the schlieren, could be described using concepts from this chapter exclusively, design and analysis of many engineering devices require techniques from physical optics or quantum mechanics. Therefore, the results of this chapter present a somewhat narrow perspective of the field of optics. However, many specialized topics – such as detailed analysis of the characteristics of complex imaging systems, mentioned here only briefly - can be well analyzed by application of geometrical optics only. For more details see O'Shea (1985).

References

Born, M., and Wolf, E. (1975), *Principles of Optics,* 5th ed., Oxford: Pergamon, pp. 113–23.
Cloupeau, M., and Klarsfeld, S. (1973), Visualization of thermal fields in saturated porous media by the Christiansen effect, *Applied Optics* 12: 198–204.
de Fermat, P. (1891), *Oeuvres de Fermat,* vol. 2, p. 354.
Feynman, R. P., Leighton, R. B., and Sands, M. (1963), *The Feynman Lectures on Physics,* vol. I, Reading, MA: Addison-Wesley, Chapter 27.
Havener, G., and Kirby, D. (1993), Aero-optical phase measurements using Fourier transform holographic interferometry, *AIAA Journal* 31: 426–33.
Hildebrand, F. B. (1965), *Methods of Applied Mathematics,* 2nd ed., Englewood Cliffs, NJ: Prentice-Hall, pp. 119–93.
Kantrowitz, A., and Trimpi, R. L. (1950), A sharp-focusing schlieren system, *Journal of the Aeronautical Sciences* 5: 311–19.
Liepmann, H. W., and Roshko, A. (1957), *Elements of Gas Dynamics,* New York: Wiley, pp. 153–63.
Merzkirch, W. (1987), *Flow Visualization,* 2nd ed., London: Academic Press.
Michelson, A. A. (1881), The relative motion of the earth and the luminiferous ether, *American Journal of Science* 22: 120–9.
Nussenzveig, H. M. (1977), The theory of the rainbow, *Scientific American* 236: 116–27.
O'Shea, D. C. (1985), *Elements of Modern Optical Design,* New York: Wiley, Chapters 5 and 6.
Snell, W. (1637). [The law was never published by Snell; it was first published by Descartes in *Dioptrique Météores,* Leyden.]
Walker, J. D. (1976), Multiple rainbows from single drops of water and other liquids, *American Journal of Physics* 44: 421–33.
Weinstein, L. H. (1993), Large-field high-brightness focusing schlieren system, *AIAA Journal* 31: 1250–5.
Yates, L. A. (1993), Images constructed from computed flow fields, *AIAA Journal* 31: 1877–84.

Homework Problems

Problem 2.1

Visually, the size and distance of an object are determined by the *visual angle* - that is, the angle subtended by the object at the point of observation. Determine how much larger a fish in a pool ($n = 1.33$) would appear to an observer standing outside the pool, relative to its size in air. Assume that the observer is at a distance l_o and the fish is at distance l_i from the water–air interface and that the fish is viewed straight down.

Problem 2.2

Figure 2.24 presents a typical laser Doppler velocimetry (LDV) system. The velocity is measured at the point where the two laser beams intersect. The velocity distribution in a transparent pipe containing fluid with an index of refraction n is obtained by translating the optical system transversely to the pipe axis. Find the relationship between the distance traveled by the optical system and the distance traveled by the intersection point

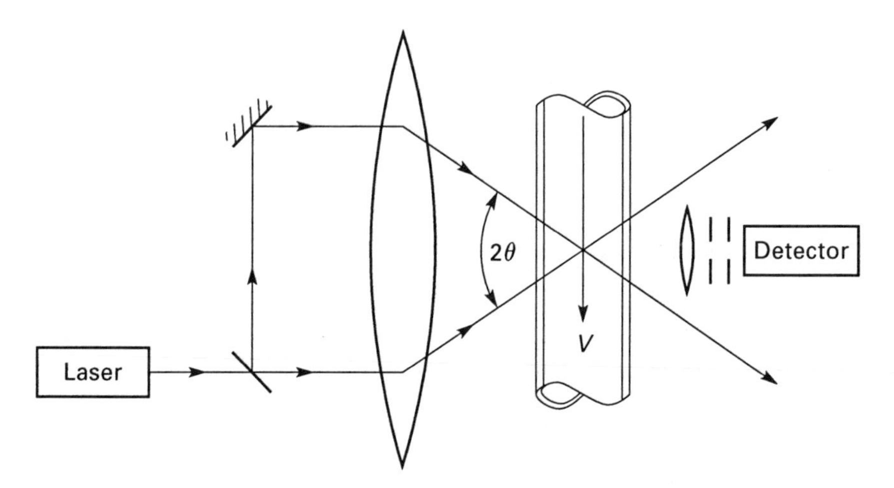

Figure 2.24 Layout of a laser Doppler velocimetry system.

of the beams inside the pipe. You may assume that the walls of the pipe are thin and that their optical effects can be neglected. You may also assume that the plane defined by the laser beams is in the diametrical plane of the pipe.

Problem 2.3

Show by using calculus of variation that the path of minimum time of passage for propagation through a two-dimensional medium with continuously varying index of refraction $n = n(x, y)$ satisfies the condition given by (2.8). Also show that the $\partial n/\partial x$ of the gradient of the index of refraction does not contribute to the deflection angle.

Hint: Without loss of generality, you may assume that the incident ray is pointing in the positive x direction. The distance ds traveled along the curve formed by the ray is $ds = (1+y'^2)^{1/2}\,dx$. Write the integral that describes the time of passage, t. Assume that $n = n(x, y)$. The condition that the integrand $F(x, y, y')$ must meet in order to minimize t is given by Euler's equation:

$$\frac{d}{dx}\left(\frac{\partial F}{\partial y'}\right) - \frac{\partial F}{\partial y} = 0.$$

Problem 2.4

A corner reflector consists of three mutually perpendicular mirrors. Show that, with the exception of a slight parallel shift, an incident ray is reflected back on itself. Note that a similar effect can be achieved by corner cube prisms, where the beam enters through the hypotenuse.

Problem 2.5

Show that the deflection $\theta_i - \theta_t$ due to refraction across an interface separating two media with indices of refraction n_i and n_t increases with the incidence angle. (*Hint:* Find from eqn. 2.6 the value of $d\theta_i/d\theta_t$ for refraction when $n_i < n_t$ and generalize the result for other interfaces.)

Problem 2.6

A top view of a typical penta prism is shown in Figure 2.25. The angle between faces AB and AE is 2α and between faces BC and ED is α. The incidence on face AB is normal.

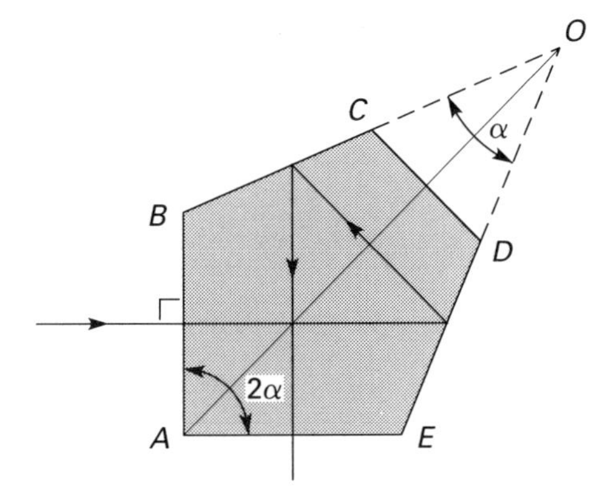

Figure 2.25 Refraction through a penta prism.

(a) If the prism is symmetric around line OA then find the value of α, in terms of θ_{crit}, for which a ray incident on face AB will be reflected by face ED.
(b) Show that, with that angle, that ray will also be reflected by face BC.
(c) Find the angle between the ray and the exit face AE.
(d) Find the angle between the entering ray and the exit ray when the prism is turned around an axis normal to the page. (Note that the deflection angle between the two rays is independent of the turning angle of the prism. Therefore, penta prisms with $\alpha = 45°$ are used as "optical squares" for surveying.)

Problem 2.7

Using the paraxial approximation, prove equation (2.11) (focusing by spherical mirrors).

Problem 2.8

Prove (2.13) by using the similarity of triangles FAB and FVW in Figure 2.17. Also show that, as a result, $\measuredangle AVB = \measuredangle A'VB'$; that is, line BVB' is a straight line.

Problem 2.9

Figure 2.26 (overleaf) illustrates the cross section of the lens in the Block Island lighthouse. The lens consists of concentric crown glass annular rings with spherical convex curvature on one side and a plane surface on the other. The lens simulates a regular plano-convex lens for which

$$\frac{1}{f} = \frac{n-1}{R}.$$

The dimensions (in millimeters) of each of the lens components are given in Table 2.2 (overleaf), where x_i and y_i are the centers of curvature of the ith component, R_i is the radius of curvature of its spherical surface, and r_i is the radius of the annular ring. The lens was designed for a nominal focal length of 920 mm.

(a) Assuming that the focal length of the center ($i = 1$) lens is 920 mm, find its index of refraction.
(b) For a collimated beam incident from the left, find the trajectories of the rays passing through the innermost and outermost circles of each annular

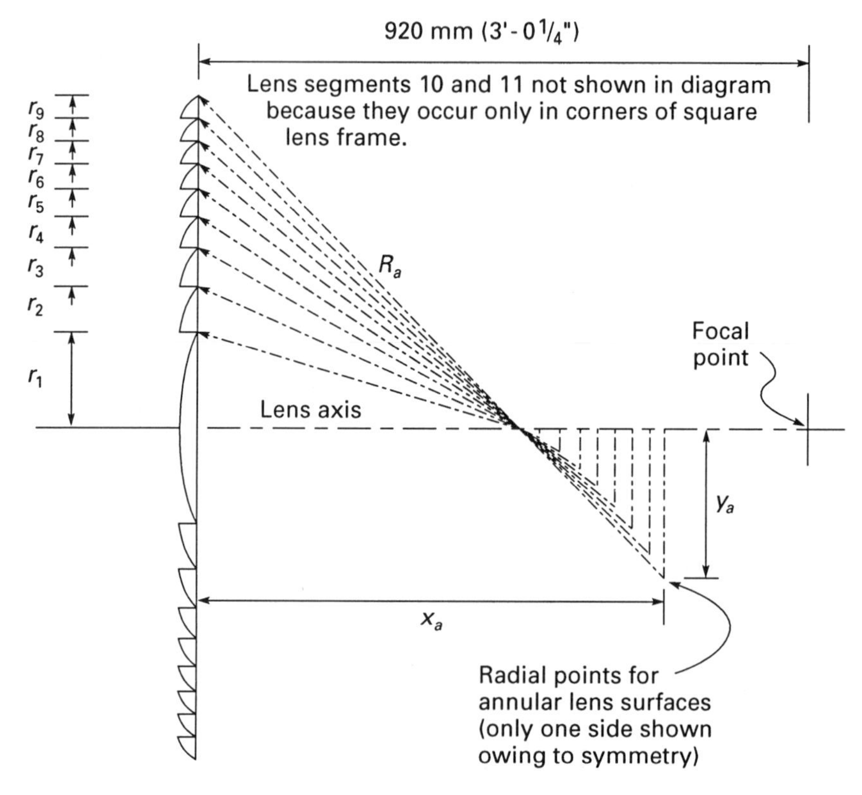

Figure 2.26 Geometry of the cross section of the lens of the Block Island lighthouse. [Reproduced from the Historic American Engineering Record RI-27, National Park Service, Mabel A. Baiges, delineator, 1988]

Table 2.2 *Parameters of the Fresnel lens in the Block Island lighthouse (see Figure 2.26)*

	$i=1$	$i=2$	$i=3$	$i=4$	$i=5$	$i=6$	$i=7$	$i=8$	$i=9$
x_i	454.8	488.6	513.4	540.7	565.3	588	614.4	636.9	660.1
y_i	00.00	13.08	31.72	57.0	84.86	114.9	151.5	189.6	230.2
r_i	140	208.2	262.4	309.2	350.5	387.4	422.2	456.2	490.0
R_i	483.5	543.6	598.6	659.8	719.8	779.5	846.5	911.3	980.3

component of the lens. Determine the coordinates of the points of intersection of each of these rays with the lens axis, and determine the optimal location of the lamp.

Problem 2.10

Scattering by raindrops in the atmosphere is the cause of rainbows. When the diameter of these droplets is much larger than the wavelength of the incident radiation, the scattering can be described by principles of geometrical optics. To illustrate the rainbow effect graphically (or by computer simulation), consider a single spherical drop and represent it by drawing its major circle. Uniform illumination by the sun can be represented by a set of equally spaced parallel rays (see Nussenzveig 1977 and Walker 1976). However, in order to simplify the graphical representation, draw only the rays at the upper hemisphere. For each incident ray, draw the refracted ray assuming that the index of refraction of water is 1.33. Extend each refracted ray until it intercepts the back surface. Using the laws of reflection and refraction, draw the reflected and refracted rays. Note that rays entering the droplet at the top of the upper hemisphere emerge after reflection at the bottom hemisphere where they cluster together. These are the rays that are visible in the rainbow.

(a) Determine the *rainbow angle* – that is, the angle between these rays and the incident radiation.
(b) Determine where an observer should look for the rainbow at sunset when all sun rays are approximately horizontal.
(c) Determine which of the colors, red or blue, will be at the top of the rainbow if the index of refraction for red light is smaller than the index of refraction for blue.
(d) Double reflection is the cause for the secondary rainbow. Determine its angle, its location in the sky relative to the primary rainbow, and the order of the colors from top to bottom.

Problem 2.11

(a) Show that the magnification of a Galilean telescope is determined by the ratio of the focal lengths of its two lenses.
(b) Determine the magnification if one of the lenses has a negative focal length.
(c) Design a Galilean telescope to be used as a beam expander. Assume that the incident beam and the expanded beam are collimated. Determine the ratio of focal lengths for a desired expansion M and the required space between the lenses. Propose approaches to reduce the length of the telescope when a large magnification is required.

Problem 2.12

(a) Derive the ray transfer matrix for a flat mirror.
(b) A ray emerges in a direction θ at point A located at the focus of a lens, passes through the lens, and continues propagating a distance L until it hits a mirror that is parallel to the lens; see Figure 2.27 (overleaf). Using ray transfer matrices, find where the beam will cross the (B, B) plane in its return path and determine the beam's slope.

Research Problems

Research Problem 2.1

Read the paper by Weinstein (1993) or by Kantrowitz and Trimpi (1950) describing focusing schlieren.

(a) Compare the focusing schlieren system to the schlieren system described in Section 2.7.
(b) How can a plane in the test section be selected using the focusing schlieren system?

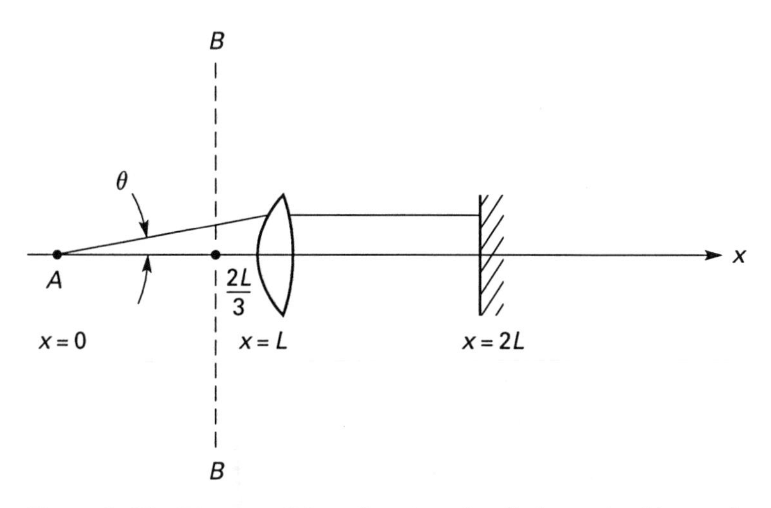

Figure 2.27 Diagram of the refocusing of radiation emitted by a point source.

(c) When imaging a plane at the center of a test section, determine what effect a strong density gradient would have on the image when using focusing schlieren and regular schlieren techniques.

Research Problem 2.2

Read the paper by Cloupeau and Klarsfeld (1973).

(a) In calculating the curve for their Figure 3, the authors assumed that the refractive index of glass is independent of temperature. In reality, the temperature dependence of the refractive index of glass is $\Delta n/\Delta T = 1\times10^{-5}\ \mathrm{K}^{-1}$. Find the error ΔT [°C] that is introduced by not including this dependence when measuring the temperature of chlorobenzene at 31.5°C.

(b) The detection of a particular color is accomplished by installing a bandpass filter in front of the imaging lens. This filter transmits radiation in a range of wavelengths $\Delta\lambda = 100$ Å and blocks all other radiation. What will be the temperature measurement uncertainty introduced by imaging through this filter?

(c) Propose an alternative dispersion curve for pellets to be used with chlorobenzene, one that will enhance the sensitivity of the technique. You may express that curve by an empirical equation or by a table of n versus λ. Using this new dispersion curve, generate a new curve for T versus λ.

3 Maxwell's Equations

And God said, Let there be light:
and there was light.
Genesis 1:3

3.1 Introduction

In the previous chapter we introduced the theory of geometrical optics, a very simplistic analysis of the propagation of radiation describing only the lines that trace the radiation trajectories. In that analysis, the lines, or rays, were not subjected to the effects of diffraction or interference; with the exception of dispersion, color too had no influence on these trajectories. The absolute value of the speed of light had no bearing on the propagation; only its magnitude relative to the speed in free space had to be known, and even that parameter could not be derived directly and had to be retrieved from other sources. Similarly, parameters of the important effect of dispersion could not be derived directly. Attenuation by absorption was outside the scope of geometrical optics, as were other effects related to the nature of radiation such as polarization, coherence, and wavelength. These shortcomings of geometrical optics were to be expected. After all, such fundamental questions as how radiation is created or how it interacts with a particular medium were not asked. Without consideration of these questions, the nature of radiation and the details of its propagation cannot be fully understood.

Historically, the first studies attempting to understand the nature of light, and not merely its patterns of propagation, were made in the seventeenth century. At that time, visible light was the only known mode of radiation. Nevertheless, despite the unavailability of modern experimental methods or equipment, effects of interference and diffraction were observed (for an historical background see Born and Wolf 1975 or Iizuka 1983). Hooke suggested that the effects of diffraction and interference could occur only if light consisted of certain waves that propagate very much like ocean waves or like vibration along an oscillating string. Although the explanation was plausible, Hooke could not identify the physical quantity associated with the propagating waves. Unlike ocean waves, the crests and troughs of light waves in his model were not visible. On the other hand, Newton – who attached more importance to the distinctive rectilinear motion of light – suggested that light must consist of small particles propagating at high speed. Incidentally, this theory was also useful in describing the effect of the polarization of light. However, this corpuscular theory of light was abandoned for lack of evidence, while the wave theory of propagation gained more supporting experimental evidence.

A consistent theory of the physics, formation, and propagation of light and its relation to other modes of radiation emerged only in the late nineteenth century. Following developments in the research fields of electricity and magnetism, Maxwell (1873) was able to consolidate the experience in these two fields into a concise system, the *Maxwell equations*. Because they were designed to consolidate the theories of electricity and magnetism, no relationship between these equations and any of the theories of light propagation was observed. However, Maxwell's equations could be reduced to equations that describe the simultaneous propagation of electric and magnetic fields in a form of waves – *electromagnetic waves*. A clue that these waves might actually be the elusive waves of light predicted by Hooke more than two centuries earlier came when Maxwell found that the propagation speed of the electromagnetic waves calculated from his equations matched almost exactly the measured speed of light. Furthermore, these equations could also quantify observations that were consistent with the wave nature of light. Thus, albeit with some reluctance, Maxwell's equations were accepted as the foundation of the theory of physical optics. The corpuscular theory of light was abandoned until the development of quantum mechanics.

Despite evidence that the wave theory of light could not account for many effects of the interaction of radiation with matter, it was (and still is) regarded as a comprehensive theory of the propagation of radiation. Most physical parameters, such as the speed of light, as well as effects such as diffraction, interference, and even Fermat's principle, can conveniently be deduced from this theory. Only after the discovery of the quantum nature of light could typical phenomena (e.g. absorption and emission) associated with the interaction between radiation and matter be fully explained. We will need the theory of quantum mechanics to explain how lasers operate or to model advanced phenomena of the interaction between radiation and matter. Otherwise, all effects of propagation can be described by the classical electromagnetic theory.

In this chapter we will draw the lines that connect electricity and magnetism with radiation. We will show how the theories of electricity and magnetism merge into a broader picture of electromagnetism. We will start with the theories of electrostatics and magnetostatics and follow with electrodynamics and magnetodynamics. Each of these topics corresponds to one Maxwell equation. When combined, these four equations produce the wave equation of radiation.

The primary objective of the electromagnetic theory is to describe the interaction among electric charges, whether stationary or moving. When a pair of charges is stationary and isolated from any other interaction, only one electric force acts between them. This force, which may be attractive or repulsive, points along their connecting line. In the simplest case, when only one pair is present, only one electric force can exist. When more than two charges are present, each charge is attracted or repelled independently by all the other charges. These electrostatic forces can then be added vectorially to form one resultant for each charge. When the charges begin to move, the interaction becomes time-dependent, as evidenced by changes in the magnitude and orientation of the electrostatic forces. In addition, a new force, the *magnetic force,* is induced by this motion. Thus, the analysis of moving charges must now include two groups of forces, both additive. Although electromagnetic theory fails to explain why these forces exist or how they are communicated between the charges, it is instrumental in describing the magnitude and orientation of the electrostatic and magnetic forces and their time dependence for prescribed, time-dependent distribution of charges.

When a group of charges oscillates harmonically, the forces they induce in space must include oscillating electric and magnetic forces. These magnetic and electric forces, although diminishing with distance, can drive other electric charges into a similar oscillatory motion, thereby increasing their energy. Although the study of the interaction between oscillating charges is only a special case of the electromagnetic theory, it is the core of the physics of optics. However, before these oscillations can be studied, the equations that describe the interaction between stationary charges, moving charges, steady magnetic forces, and time-dependent magnetic forces must be established.

3.2 The Electric Field

The simplest interaction between stationary electric charges is the electrostatic force. Schematically (Figure 3.1), these charges q_1 and q_2 can be represented by two points, located at A and B respectively and separated by a distance r_{12}. When the signs of these charges are opposite, the force induced at B

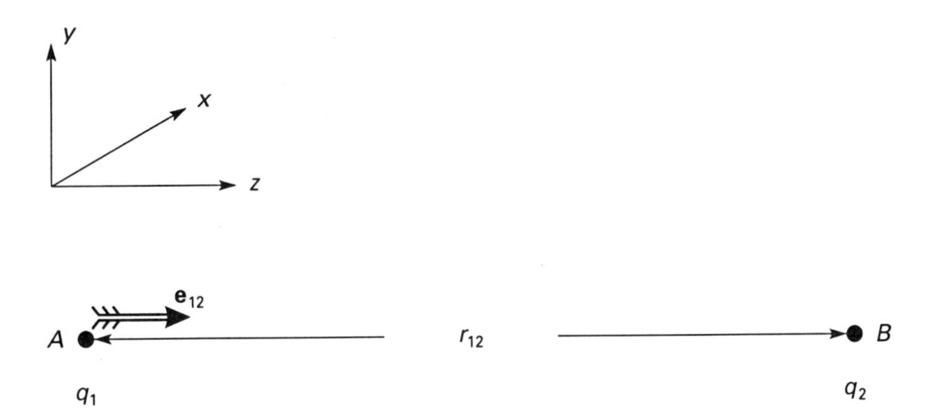

Figure 3.1 Geometrical parameters of the interaction between two electric charges.

by an electric charge at A is attractive and is pointing toward A. Otherwise, the force is repulsive and points away from A. The magnitude of the electrostatic force was found by Coulomb to increase linearly with either q_1 or q_2, and to decrease by the square of their distance r_{12}. By introducing a proportionality factor, Coulomb's equation for the force applied at B by a charge at A is

$$\mathbf{F} = \frac{1}{4\pi\epsilon} \frac{q_1 q_2}{r_{12}^2} \mathbf{e}_{12} \tag{3.1}$$

(Halliday and Resnick 1963), where the *dielectric constant* ϵ is a material constant of the medium between the charges and where $\mathbf{e}_{12}$ is the unit vector pointing from A to B. The coefficient of 4π in the denominator is included for historical reasons. This equation is analogous to the equation for gravity, with the exception that it can represent both attraction and repulsion. Thus, when the charges at A and B are opposite in sign, the force acting at B is negative – that is, pointing opposite to the unit vector $\mathbf{e}_{12}$.

The magnitude of the dielectric constant depends on the medium separating the two charges and on the system of units used to specify the remaining quantities. When the charges are specified in coulombs [C], the forces in newtons [N], and the distance r_{12} in meters [m], the dielectric constant of free space, ϵ_0, may be expressed as

$$\frac{1}{4\pi\epsilon_0} = 8.99 \times 10^9 \text{ N-m}^2/\text{C}^2 \quad \text{or} \quad \epsilon_0 = 8.854 \times 10^{-12} \text{ C}^2/\text{N-m}^2. \tag{3.2}$$

The dielectric constant for most substances is a scalar quantity – its value is independent of the orientation of $\mathbf{e}_{12}$. However, since (3.1) relates two vectors, $\mathbf{F}$ and $\mathbf{e}_{12}$, each having up to three components, it is also possible to represent ϵ in the following tensor form:

$$\bar{\epsilon} = \begin{bmatrix} \epsilon_{xx} & \epsilon_{xy} & \epsilon_{xz} \\ \epsilon_{xy} & \epsilon_{yy} & \epsilon_{yz} \\ \epsilon_{xz} & \epsilon_{yz} & \epsilon_{zz} \end{bmatrix}. \tag{3.3}$$

For isotropic substances, where $\mathbf{F}$ and $\mathbf{e}_{12}$ are collinear, $\epsilon_{xy} = \epsilon_{xz} = \epsilon_{yz} = 0$ and $\epsilon_{xx} = \epsilon_{yy} = \epsilon_{zz}$. In other words, the tensor can be replaced by a single scalar. However, for certain anisotropic substances (such as crystals), the off-diagonal terms do not vanish and the diagonal components of the dielectric tensor may not be equal to each other. Nonetheless, all tensors of the dielectric constant are symmetric and so have at most six different coefficients:

$$\epsilon_{xx} \neq \epsilon_{yy} \neq \epsilon_{zz} \neq \epsilon_{xy} \neq \epsilon_{xz} \neq \epsilon_{yz} \neq 0.$$

Thus, in anisotropic media the force acting between charges along the x axis is not equal to the force acting between identical charges separated by the same distance along the y or z axes. Furthermore, when the off-diagonal terms do not vanish, a component of $\mathbf{e}_{12}$ along one coordinate axis may introduce a force component along another coordinate axis. We will defer further discussion of the tensorlike characteristics of the dielectric constant to Chapter 5. Meanwhile, ϵ will be considered as a scalar.

When the medium between the interacting charges is an electric insulator (i.e., when all the electrons are bound and are not free to move), the interaction between charges q_1 and q_2 in Figure 3.1 can be fully characterized by a noncomplex (i.e. real) dielectric coefficient. However, when the medium between A and B is electrically conducting, the dielectric coefficient must be represented by a complex number:

$$\epsilon = \epsilon' + j\epsilon'', \tag{3.4}$$

where ϵ' is the real component of the dielectric constant; the imaginary part, ϵ'', is related to the mobility of the free charges. The imaginary part of the dielectric constant is also related to the energy loss for propagation of radiation through optical media. For most applications, optical media include glasses, various plastics, and transparent crystals, all of which are electrical insulators. Furthermore, optical losses for propagation through these media are negligible. Therefore, ϵ is a real number for these media. "Lossy" media include metals, semiconductors, and plasmas; for these media, $\epsilon'' \neq 0$.

The electrostatic force in (3.1) is the result of the interaction between two charges at two different locations and the distance between them. Without loss of generality, the three parameters can be sorted by an observer at point B into two groups: the charge q_2 at the location of that observer into one group, and the charge q_1 at point A together with the separation r_{12} into the second group. However, to the observer at B, the two parameters in the second group are inseparable and amount to a single effect that is characteristic of the location of point B: a proportionality factor that determines the magnitude and direction of the electrostatic force applied on q_2. Thus, it may be compelling to represent the electrostatic force experienced by the charge q_2 at any arbitrary location by a map of the spatial variation of the effect of the charge at any point A. This map represents the force that would be applied by point A on a unit charge at any point in space, and includes both the magnitude and direction of that force. This field of forces, acting on a unit charge in the presence of any charge q_1 located at point A, is called the *electric field* $\mathbf{E}(x,y,z)$. Thus, to the observer at B, the electrostatic force is determined by the product of q_2 and $\mathbf{E}(x,y,z)$. Using (3.1), the electric field induced by q_1 located at A is

$$\mathbf{E}(x,y,z) = \frac{1}{4\pi\epsilon}\frac{q_1}{r_{12}^2}\mathbf{e}_r \tag{3.5}$$

(Halliday and Resnick 1963), where $\mathbf{e}_r$ is a unit vector pointing from A toward a point (x,y,z) in space. Thus, $\mathbf{E}(x,y,z)$, in units of newtons per coulomb, is the force experienced by a unit charge at any point (x,y,z). The electric field is more commonly defined in units of volts per meter [V/m], where 1 V/m = 1 N/C. Thus, a charge of 1 C, placed in a field of 1 V/m, is subject to an electric force of 1 N. With that definition, the dielectric coefficient of free space can be modified to include the new units as follows:

$$\epsilon_0 = 8.854\times 10^{-12}\ \mathrm{C^2/N\text{-}m^2} = 8.854\times 10^{-12}\ \mathrm{C/V\text{-}m}.$$

Note that we elected to use the MKSA system of units; other texts (e.g., Born and Wolf 1975) use the Gaussian system. The conversion factors between these systems is given in Appendix A. The electric field is a vector field, so an arrow is assigned to each point in space. The direction of that arrow coincides with the direction of the force applied to a positive unit charge at that point, and the length of that arrow is proportional to the magnitude of that force. Although this is a complete presentation, it can be replaced by an alternative scalar field by evaluating the work done on a unit charge when moving it against the electric field. This is analogous to the raising of a weight in the field of gravity. Thus, just as for gravity, the electrostatic field can be replaced by equipotential lines where the potential is V. Although this potential can be expressed in joules per coulomb, the more commonly used units are volts. The electric field is simply the gradient of the potential field. The work done by a charge q, and hence the energy it gained by moving from a potential V_1 to a potential V_2, is

$$W_{12} = q(V_2 - V_1).$$

When the charge of a single electron, $e^- = 1.602 \times 10^{-19}$ C, is moved through a potential differential of 1 V, the energy of this charge is increased by 1 electron-volt, where 1 eV $= 1.602 \times 10^{-19}$ J.

Until now our analysis of the Coulomb force and the ensuing electric field and potential assumed that the electric charges are concentrated at a point. However, only rarely can charges be expected to be concentrated at a point. Usually, owing to the electrostatic forces, charges spread over surfaces or volumes. Even when distributed, charges generate an electric field and can be subjected to forces induced by existing fields. Generally, when an electric charge is distributed at a density of $\rho(x, y, z)$ C/m^3, Coulomb's equation (eqn. 3.1) can be applied to a differential volume element dV by placing that element – together with its charge – at the center of a spherical envelope with radius r (Figure 3.2). When the size of this element is small relative to r, the charge in the center can be seen as a point charge, and the total electric field flux through the spherical envelope (i.e., the surface integral of the electric field passing through that envelope) is

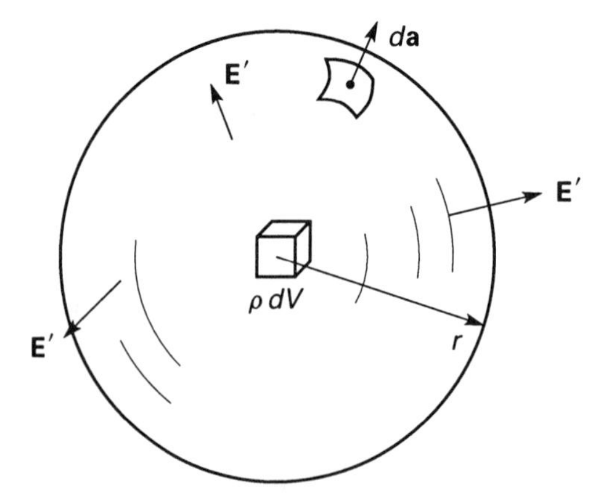

Figure 3.2 Electric field vector emanating from an element of volumetrically distributed charges.

$$\oint \mathbf{E}' d\mathbf{a} = \oint \frac{1}{4\pi\epsilon} \frac{\rho \cdot dV}{r^2} \mathbf{e}_r \, d\mathbf{a}.$$

Here $\mathbf{E}'$ is the field induced at the surface by the infinitesimal charge in dV. This field is uniform anywhere on the spherical envelope. Thus, the area element $d\mathbf{a}$ can be integrated independently. By offsetting the area $4\pi r^2$ of the spherical envelope with the $1/4\pi r^2$ term in the denominator of the second integral, the surface integral of the electric field becomes

$$\oint \mathbf{E}' d\mathbf{a} = \frac{\rho \cdot dV}{\epsilon}.$$

This result is independent of the spherical radius and can therefore be applied to any spherical surface or, according to Gauss's law (Feynman, Leighton, and Sands 1964) to any arbitrarily shaped closed envelope. By adding, through integration, all the charges within a volume, the total field flux passing through an arbitrarily shaped envelope is:

$$\oint \mathbf{E} \, d\mathbf{a} = \frac{\text{net charge inside the envelope}}{\epsilon} = \int \frac{\rho}{\epsilon} \, dV. \tag{3.6}$$

Using the divergence theorem (Hildebrand 1962), the integral equation can be replaced by the mathematically more convenient differential form describing the field in the presence of a distributed charge:

$$\nabla \cdot (\epsilon \mathbf{E}) = \rho. \tag{3.7}$$

Equation (3.7) can be further condensed by introducing the *electric displacement* vector $\mathbf{D} = \epsilon \mathbf{E}$. Thus:

$$\nabla \cdot \mathbf{D} = \rho. \tag{3.8}$$

Example 3.1 One of the most sensitive light detectors is the photomultiplier tube (PMT) (see Engstrom 1980, Kume 1994). The PMT consists of a light-sensitive element, the photocathode, that emits electrons when illuminated. These photoelectrons are accelerated through a potential difference, ΔV, toward an electrode that is biased positively relative to the cathode. When the photoelectrons hit that electrode, much of their kinetic energy is absorbed at the surface and is used to release several secondary electrons. The ratio between the number of secondary electrons and the number of primary electrons is the secondary emission ratio δ. These secondary electrons are accelerated toward yet another electrode, thereby forcing the release of even more electrons. The secondary emission process is repeated several times until there are sufficient electrons for direct measurement. The electrodes in this amplifying chain are the dynodes. After the last dynode, the electrons are collected at an anode where the total current is measured.

Assume that δ is uniform and proportional to $(U_e)^\alpha$ for all dynodes in the chain, where U_e is the kinetic energy of the incident electrons and α is a coefficient determined by the dynode material and its geometrical shape. Find the number of electrons collected at the anode for every electron emitted by the

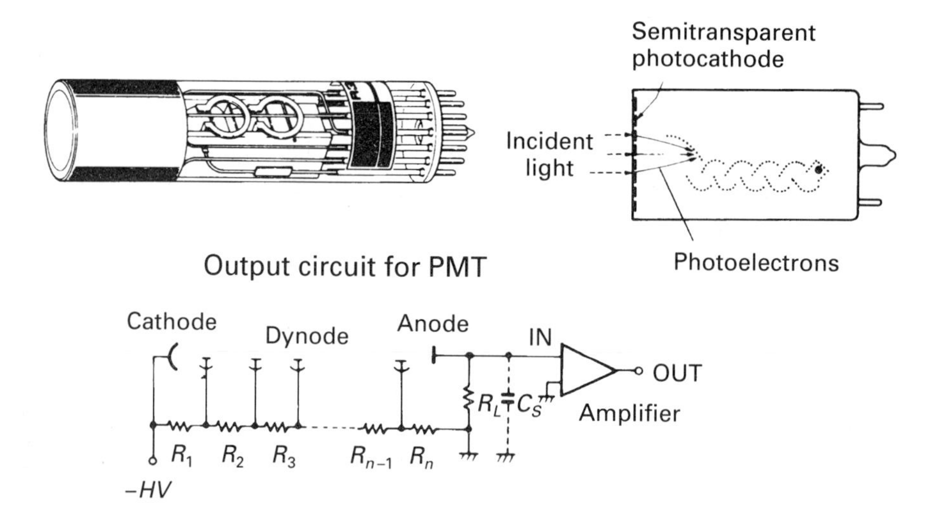

Figure 3.3 Schematic diagram and wiring diagram of a head-on type photomultiplier tube. [Hamamatsu Catalog, January 1988, © Hamamatsu Photonics K.K., Hamamatsu City, Japan]

photocathode in a PMT with $n-1$ dynodes. Also find the sensitivity of the PMT to fluctuations in the supply voltage.

Solution Figure 3.3 presents a typical circuit diagram of the wiring of a PMT. There are total of n resistors that divide the voltage V applied between the photocathode and the anode to the $n-1$ dyodes and to the anode. If the resistance of all resistors is the same, then the potential through which an electron is accelerated while passing between two adjacent dynodes is

$$\Delta V = \frac{V}{n}.$$

The energy of an electron after acceleration across that potential differential is

$$U_e = e\Delta V.$$

The secondary emission ratio can now be calculated as

$$\delta = A(e\Delta V)^\alpha = Ae^\alpha\left(\frac{V}{n}\right)^\alpha,$$

where A is an empirical proportionality coefficient.

The gain G across a chain of $n-1$ dynodes is a measure of the total number of electrons collected by the anode for every electron emitted by the photocathode. Since the secondary emission ratio is uniform for all dynodes, the gain is

$$G = \delta^{n-1} = (Ae^\alpha)^{n-1}\left(\frac{V}{n}\right)^{\alpha(n-1)} = KV^{\alpha(n-1)}.$$

Here all the constants of this equation were lumped into a single coefficient K, a characteristic of the PMT. The gain of the PMT is seen to be a highly nonlinear function of the tube voltage. For a chain of $n-1$ dynodes, G increases as the $\alpha(n-1)$th power of V. For a sensitive PMT employing ten dynodes and for $\alpha = 0.6$, the gain increases as the sixth power of the supply voltage.

Because of this highly nonlinear dependence, measurements by a PMT are very susceptible to fluctuations in the supply voltage. To determine the effect of such fluctuations, assume that the signal s measured at the anode is proportional to the gain. Thus, for a supply voltage perturbation of dV, the perturbation ds in the signal is

$$\frac{ds}{s} = \alpha(n-1)\frac{dV}{V}.$$

For a supply voltage of 1,000 V, a fluctuation of 10 V will result in signal fluctuations of approximately 6%. To avoid this noise, which is additive to other noise sources, PMT power supplies must be extremely well stabilized. ■

Example 3.2 The microchannel plane (MCP) is an alternative device for the amplification of photoelectrons (Wiza 1979). It is used primarily in conjunction with solid-state cameras as an image intensifier. As with the PMT, the gain is obtained by accelerating electrons emitted by a photocathode and using their kinetic energy to force secondary emission of more electrons. However, unlike the PMT, where gain is generated by electrons bouncing off the dynodes, the signal of a photoelectron in the MCP is amplified by passing that electron along a capillary tube or a channel. Electrons emitted by the photocathode are forced by an electric field into the channel. While propagating along the channel these electrons occasionally hit the channel walls, where their kinetic energy is used to force secondary emissions. Similarly, the secondary electrons are also accelerated by the electric field and while racing along the channel they bounce off its walls, thereby inducing further electron emission. By packing numerous channels together, an array of amplifiers is obtained that can be used to amplify electrons emitted by a large area of the photocathode. The electrons emitted at the exit of the MCP are accelerated again toward a phosphoric screen where upon incidence they form a fluorescing image of the original, faint object that was imaged on the photocathode surface. Such image intensifiers are routinely used for night-vision amplifiers, IR imaging, and many scientific applications.

A typical MCP (see Figure 3.4) consists of a flat plate of semiconducting material (e.g., lead glass) with a dense matrix of parallel, penetrating capillary channels. The faces of the MCP at the opposite sides of the channels are coated with thin metallic layers that serve as the electrodes. The typical diameter D of a channel may range from 10 to 100 μm with a length-to-diameter ratio $\alpha = L/D$ ranging from 40 to 100. Since the voltage that accelerates the electrons is applied longitudinally to the channel, the field $\mathbf{E}$ has primarily one component along the channel. When photoelectrons are forced by the field into the channel they are accelerated longitudinally, thereby gaining kinetic energy. However, owing to an initial transverse velocity component, most of the electrons strike the channel walls at least once during their transit, so their kinetic energy is

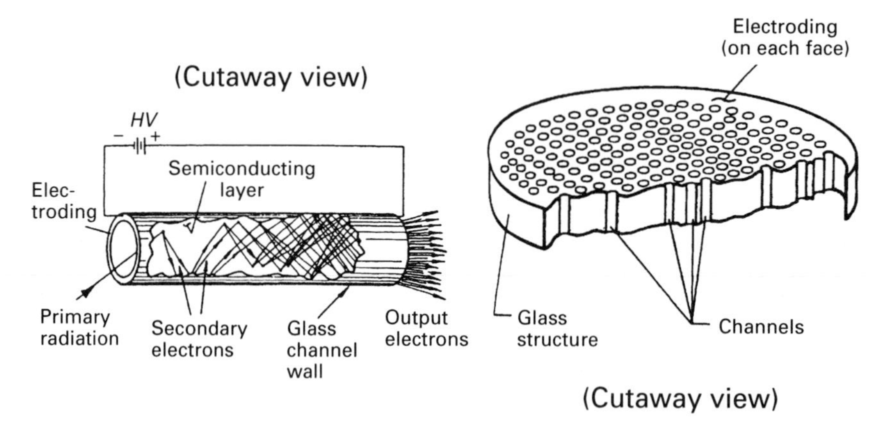

Figure 3.4 Cutaway view of single-channel and multi-channel plate of an intensified electronic camera. [Wiza 1979, © Galileo Electro-Optics Corporation, Sturbridge, MA]

absorbed and used to force the emission of secondary electrons. Because the channel walls are made of semiconducting material, current flow can replenish the charge to compensate for the emission.

In a very simplistic description of the gain process, the secondary electrons are assumed to be emitted perpendicularly to the channel wall; for a cylindrical channel, this is an emission along the radius. Therefore, at emission, the secondary electrons have only a radial velocity component. Assume that the secondary emission ratio of the channel is $\delta = A \cdot V_c^{1/2}$, where V_c is the electron bombardment energy in electron-volts and $A \approx 0.2$ is a proportionality constant. Find the total gain G of a single channel when the voltage between its ends is V_0.

Solution Assume as in Example 3.1 that each electron strike at the wall produces the same gain δ. Therefore, for n wall encounters, the total number of electrons available at the channel exit for every electron entering it is equal to the total gain:

$$G = \delta^n.$$

The number of wall encounters is analogous to the number of dynodes in the PMT. However, unlike the PMT, the number of stages is not necessarily a fixed design parameter. Instead, it depends on the horizontal and tangential velocities of the electrons and the inclination of the plane of their trajectory relative to the major plane of the channel. Therefore, in order to find G, both n and δ must be determined as functions of the channel geometry and the voltage V_0 applied between the channel ends. If all secondary emission electrons are ejected in the radial direction with a kinetic energy V_r (where V_r is expressed in eV) while the radial component of the electric field in the channel is vanishingly small, then from principles of mechanics we can deduce that the time each electron is allowed to travel before striking the opposite side is

$$\tau = \frac{D}{\sqrt{2eV_r/m_e}},$$

where m_e is the mass of the electron and e, the electron charge, is included to convert the energy units to joules. If the electric field is uniform then its lengthwise component can be expressed by dividing V_0 by the channel length L or, alternatively, in terms of the voltage between two points of consecutive wall encounters V_c and the axial distance z between them:

$$E = \frac{V_0}{L} = \frac{V_c}{z}.$$

Since the force acting on the electron, $m_e a = eE$, is constant anywhere in the channel, the axial distance z between two wall encounters is determined by the constant acceleration a. If the initial axial velocity is nearly zero, the distance traveled by an electron between two adjacent wall encounters is

$$z = \frac{1}{2}a\tau^2 = \frac{1}{2}\frac{eE}{m_e}\cdot\frac{D^2}{2eV_r/m_e} = \frac{V_0 D^2}{4LV_r}.$$

The number of wall encounters that can be accommodated along the channel is

$$n = \frac{L}{z} = \frac{4V_r\alpha^2}{V_0}.$$

The secondary emission ratio depends on the bombardment energy V_c, but V_c is the electric potential between two adjacent wall-encounter points. Thus, if the kinetic energy at emission can be neglected relative to V_c, the emission ratio is

$$\delta = AV_c^{1/2} = A(E\cdot z)^{1/2} = A\left(\frac{V_0}{L}\cdot\frac{V_0 D^2}{4LV_r}\right)^{1/2} = \frac{AV_0}{2\alpha V_r^{1/2}}$$

and the channel gain is then

$$G = \left(\frac{AV_0}{2\alpha V_r^{1/2}}\right)^{4V_r\alpha^2/V_0}.$$

This result implies that the gain increases with V_0 up to a maximum, beyond which it declines as the voltage between the ends of the channel increases. This may be expected, since the electron kinetic energy increases with V_0. However, as the velocity of the electrons increases, the axial distance between two wall encounters also increases, thereby reducing the number of wall encounters that can be accommodated along the channel length. A maximum gain is typically reached at approximately 1,000 V, where the gain approaches 10^4 (Wiza 1979). ■

3.3 The Magnetic Field

A phenomenon similar to electrostatic attraction and repulsion forces is the magnetic force. In its simplest form, a magnetic force occurs between two magnets or between a magnet and ferromagnetic substances. Depending on their alignment, magnets may repel or attract each other. The magnetic force depends on the magnetic properties of the interacting magnets, which may be parallel to the electric charge; it varies with the inverse of the square of the distance between the interacting bodies; and it depends on the properties of the medium contained between them.

Although the analogy between the magnetostatic and electrostatic interactions is striking, one important distinction must be made: the magnetic force is not generated by magnetic charges. Therefore, although electric charges consist of positive and negative charges that can be isolated, magnets cannot be separated into north-pole magnets and south-pole magnets. No matter how thinly a magnet is sliced, it always contains both poles. Thus, any closed envelope around a magnet contains no net magnetic "charges." This is in contrast to the picture presented by Figure 3.2, where net electric charges - either positive or negative - could be contained within a closed envelope. Therefore, modeling the effects of magnetic forces can be similar to modeling electric forces applied through electrically neutral media.

To establish the interaction between magnets, we first need to identify the magnetic field. But since we have not yet identified the source of the magnetic force, we will use instead the analogy between the electric force and the magnetic force. Thus, analogous to **E** we define a magnetic field **H** by assigning a vector to each point in space. The vector points away from the north pole of one magnet and represents the force that would be applied on a second magnet with a "unit" magnetic "charge" that would be placed at that point. The units for **H** are amperes per meter [A/m].

Because of the similarity between the characteristics of **E** and **H**, (3.7) (describing the variation of **E** in space) can be modified to describe the spatial variation of **H**. By replacing **E** with **H** and requiring that the medium be magnetically neutral, we obtain for **H**

$$\nabla \cdot (\mu \mathbf{H}) = \text{net magnetic charge} = 0. \tag{3.9}$$

The coefficient μ is the *magnetic permeability*. It is a material constant that accounts for the magnetic properties of the medium in which the magnetic field is established. By requiring that the last equality in (3.9) be zero, we express mathematically our inability to separate the magnetic poles.

To complete the analogy between (3.7) and (3.9), we define a new quantity $\mathbf{B} = \mu \mathbf{H}$ that is analogous to the electric displacement vector **D**. Although some authors extend the analogy to call **B** the "magnetic displacement," it is more commonly known as the *magnetic induction*. In MKSA units, the magnetic induction is measured in webers per square meter. Thus, the units of μ in the MKSA system are webers per ampere-meter. The value of the magnetic permeability of most optical materials is very near its value in free space, $\mu_0 = 4\pi \times 10^{-7}$ Wb/A-m. By replacing the magnetic field vector in (3.9) with the magnetic induction, we obtain

$$\nabla \cdot \mathbf{B} = 0. \tag{3.10}$$

Equation (3.8) for **D** is identical to (3.10) when the charge density inside the closed envelope is zero. This is consistent with our previous observation that net magnetic charges are nonexistent. We will see that the analogy between the electric and magnetic fields extends also to their dynamic behavior.

3.4 Electrodynamics and Magnetodynamics

The distinction between electro- and magnetostatics on the one hand and between electro- and magnetodynamics on the other is sometimes obscure. In some definitions (e.g. Feynmann et al. 1964), electrostatics and magnetostatics can include moving electric charges and moving magnetic fields (e.g., by moving a magnet) as long as the transients of these fields - that is, their time derivatives - are zero. Under such a definition, by simply moving at constant velocity, an electric conductor carrying constant current through a uniform magnetic field would be considered as a static interaction. For our purposes, to be considered static, both electric charges and the magnetic and electric fields must be stationary in space and their magnitude must remain constant. Both (3.8) and (3.10) are consistent with this definition and therefore represent static phenomena. We prefer this more restrictive definition because immobile electric charges and immobile magnetic fields do not interact with each other. In the absence of motion, (3.8) and (3.10) represent two independent and seemingly unrelated phenomena. Only by setting electric charges or magnetic fields into motion does the relation between magnetism and electricity become evident. The motion of charges and electric fields, whether steady or unsteady, is the essence of the electromagnetic theory. Although the distinction between accelerating electric charges and charges moving at constant velocity and the distinction between steady and unsteady magnetic fields is important, it is primarily the motion itself that determines the extent of the interaction between the magnetic and electric phenomena.

Two fundamental experiments are usually considered to demonstrate the formation of magnetic fields, their effect on moving electric charges, and hence the interaction between magnetic and electric effects. These two experiments are related to each other and may be viewed as the equivalent of the mechanical action–reaction rules. In both experiments an electric conductor is suspended in space between two electrically insulated bars (Figure 3.5). In the first experiment, Figure 3.5(a), a galvanometer is connected between the two ends of the conductor and a magnet is placed under it with the north pole pointing up. When the conductor is moved transversely to the field induced by the magnet, an electric current, passing through the moving conductor, is detected by the galvanometer. In the second experiment, Figure 3.5(b), the galvanometer is replaced by a power supply. When electric current is allowed to flow through the conductor by closing the switch, the magnet underneath the conductor is subjected to a force pointing in the direction indicated in the figure. The first experiment represents the induction of electric current by a varying magnetic

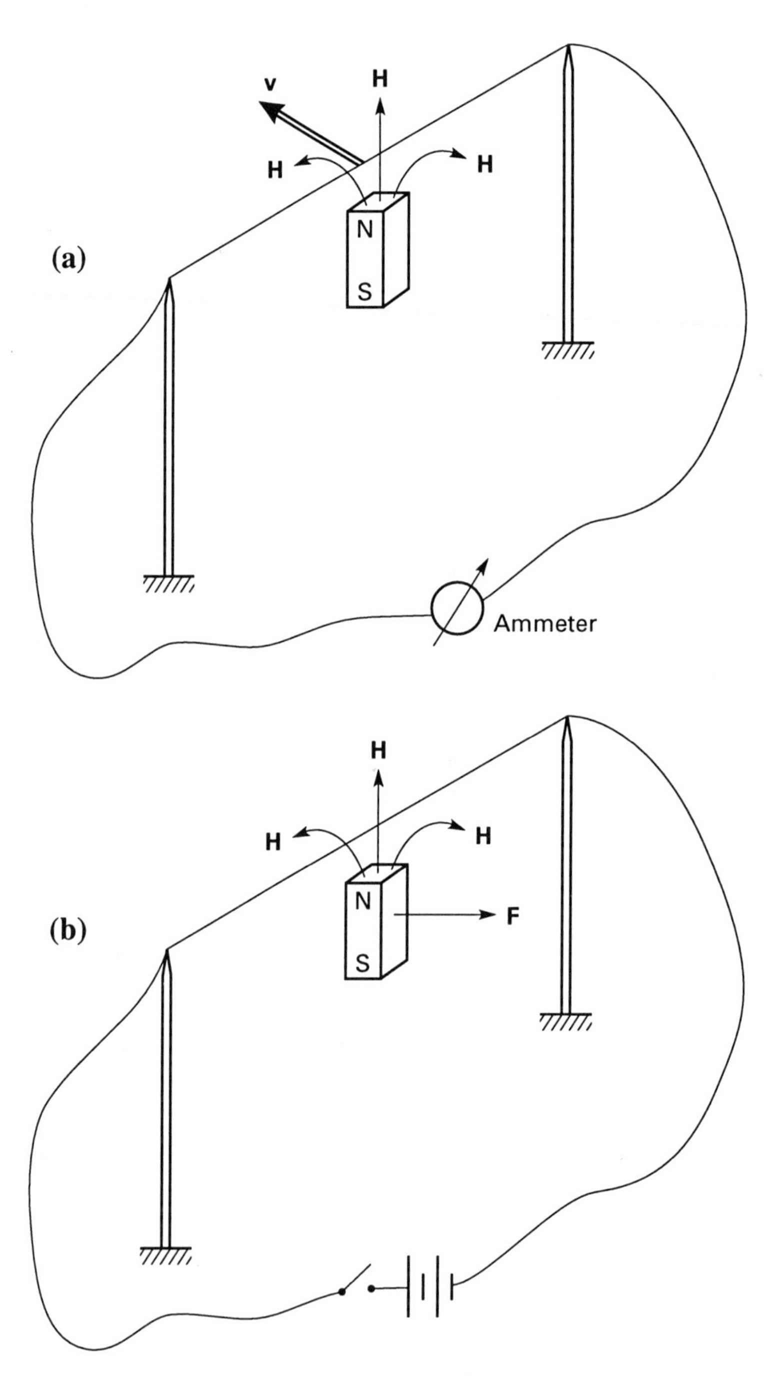

Figure 3.5 Interaction between a magnet and a wire in a closed electrical circuit: (**a**) the conductor is moved relative to the magnet, thereby inducing current in the circuit; (**b**) current is passed through the conductor, thereby subjecting the magnet to a force.

field; the second experiment represents the formation of a magnetic field by the motion of electric charges.

The effect associated with the first experiment can be easily described mathematically by observing that the electric current measured by the galvanometer in Figure 3.5(a) is proportional to the component of the conductor velocity normal to the magnet, and also to the intensity of the magnetic induction. The linear dependence between the induced current and the magnetic field can be demonstrated by adding similar magnets while monitoring the increase in the current. The results of this experiment can be represented by the following empirical vectorial equation:

$$\mathbf{F} = q_0 \mathbf{v} \times \mathbf{B} \tag{3.11}$$

(Halliday and Resnick 1963). Here **F** is the force in newtons applied to the free charges q_0 [C] within the electric conductor, and **v** is their velocity [m/s] acquired by the motion of the conductor relative to a fixed coordinate system. This experiment can be used to find the relation between the units of **B** [$\mathrm{Wb/m^2}$] and other MKSA units. The force developed by the motion of these charges across the magnetic induction is in the direction of the conductor, and establishes the current flow that is detected by the galvanometer. Equation (3.11) successfully predicts the forces applied on electric charges moving through lines of magnetic fields or magnetic induction. It can also be combined with (3.5) to account for the force applied on these charges by an electric field. These two equations are used to evaluate the deflection of electron beams in cathode-ray tubes (CRTs) under the influence of uniform electric and magnetic fields (see Problem 3.2).

Despite its success, (3.11) contains an apparent paradox. When the velocity of the conductor relative to a fixed coordinate system is zero, the force **F** and hence the current in the conductor must vanish even if the magnet is moved relative to the conductor. The problem is not resolved by requiring that the velocity be measured relative to the magnet, because the result of the calculation would then depend on the selection of the coordinate system – hardly a general physical law. However, motion of the magnet relative to the conductor does establish time-dependent variation in the magnetic induction. Thus, by describing the current in the conductor as the result of variations in the magnetic induction surrounding it, we obtain a physical model that is independent of the selection of coordinate system.

In order to derive the alternative to (3.11), consider a rectangular loop drawn through a uniform magnetic induction (Figure 3.6). Although uniform, the extent of the magnetic induction is limited to the domain marked by the dashed lines. When the loop is moved to the left through the magnetic induction (as shown by the arrow), electric current that is proportional to the velocity of the wire is monitored by the galvanometer. This is consistent with (3.11), but now we will consider the rate of change of the total field enclosed by the loop, that is, the time derivative of the magnetic induction flux through the loop. We can

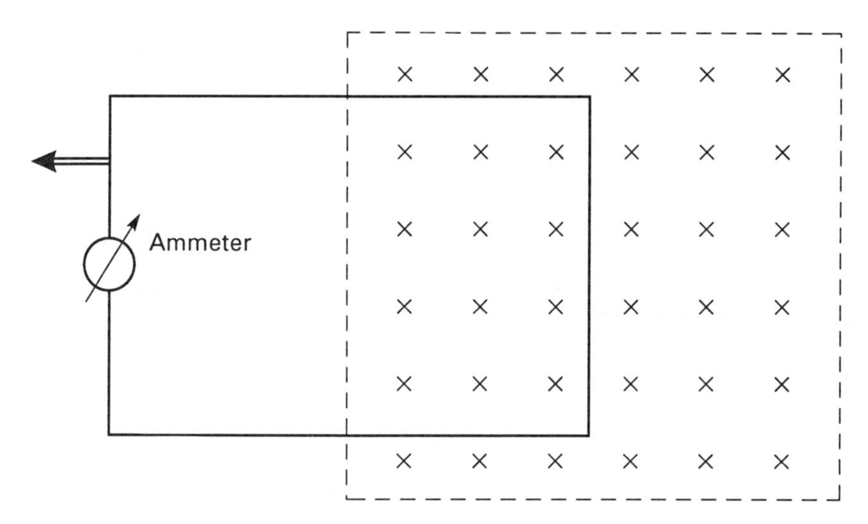

Figure 3.6 Translation of a closed electrical circuit through a uniform magnetic field.

show empirically that, whenever the magnetic flux through the loop varies, there is an electric current flowing around it. However, for current to flow in a closed circuit, the circulation of the electric field (i.e., the line integral of **E** around this circuit) must be nonzero. Otherwise, without net circulation, charges moving "downhill" in one arm of the loop will face an identical "uphill" resisting field in the other arm. These two observations can be summed up by the following empirical expression:

$$\oint \mathbf{E}\, d\mathbf{L} = -\frac{d}{dt}\int \mathbf{B}\, d\mathbf{a} \tag{3.12}$$

(Halliday and Resnick 1963). Equation (3.12) can be used to successfully predict the results of the experiments in Figures 3.5(a) and 3.6, but unlike (3.11), (3.12) is independent of the selection of coordinate systems.

To present (3.12) in a differential form, we invoke Stokes's theorem (Hildebrand 1962). This theorem establishes the equivalence between the contour integral of a physical quantity (i.e. the circulation) and the area integral of the vorticity of that quantity inside the contour. With that theorem, both sides of (3.11) become area integrals and thus the integrands on both sides must satisfy the following differential equation:

$$\nabla \times \mathbf{E} = -\frac{d\mathbf{B}}{dt}. \tag{3.13}$$

This equation shows explicitly that the rate of change of the magnetic induction equals the net vorticity of the electric field, which can – in the presence of free charges – force current flow in electrical circuits. This is the complete description of the phenomenon in which an action by the magnetic induction induces a reaction in the form of an electric current. Intuitively we may expect that the converse is also true: the flow of electric current should induce a change in the magnetic field or the magnetic induction.

An experiment demonstrating the induction of a magnetic field by the motion of electric charge is depicted in Figure 3.5(b). When the electric circuit is closed, the magnet positioned under the conductor experiences a force pointing in the direction of the tangent to an imaginary circle pierced normally by the conductor at the center. The force can be easily measured using an ordinary scale, and it can be shown to increase linearly with the current passing through the conductor and to decay with distance from the wire. The magnet clearly responds to a magnetic field that is established by the moving charges in the conductor. To visualize the pattern of this field, the conducting wire can be passed normal to an insulator (e.g. a plastic sheet) sprayed with iron dust. When the electric circuit is closed, the randomly distributed iron particles are re-arranged in concentric circles, with the wire at the center. The results of these observations have been described by several alternative mathematical models. However, for consistency with our previous results, we will use here the presentation by Maxwell. Using previous work by Ampère, Maxwell showed that the magnetic induction around a contour surrounding the wire can be related to the current density $\mathbf{j}$ [A/m^2] passing through that contour by the following integral equation:

$$\oint \mathbf{H}\, d\mathbf{L} = \int \mathbf{j}\, d\mathbf{a} \quad \text{or} \quad \oint \mathbf{B}\, d\mathbf{L} = \int \mu \mathbf{j}\, d\mathbf{a}. \tag{3.14}$$

The magnetic induction can be calculated around any arbitrary contour, but in many cases the calculation is simplified by the selection of a circular contour. The current density $\mathbf{j}$ must be integrated over the entire area enclosed by the contour. For the experiment in Figure 3.5(b), current is passed only through the conductor, and the area integral in (3.14) equals the current passed through the conductor for any contour surrounding it. However, when the current is not confined to a thin conductor, the result of the area integral will depend on the contour selection.

To reduce (3.14) to a differential form, we again invoke Stokes's theorem to obtain

$$\nabla \times \mathbf{H} = \mathbf{j}. \tag{3.15}$$

Note that (3.15) is similar to (3.13). This similarity is expected, since both equations express a cause-and-effect relation between two related phenomena. Such similarity was also noticed between the equations describing electrostatics and magnetostatics. To explore the full symmetry between the effects of electricity and magnetism, Maxwell assembled the four differential equations describing the static and dynamic phenomena of these two related fields into a single set.

3.5 Maxwell's Equations

Equations (3.8), (3.10), (3.13), and (3.15) form the basis of a comprehensive set of equations that was believed to describe successfully all the effects of electricity and magnetism and the interaction between them. When Maxwell first reduced these equations into their present differential form and compiled them into a single set, all known effects were presented. This compilation was analogous to the derivation of the ideal gas equation from the more specific laws of Boyle and Charles. Thus, all the empirical laws that described only one or two phenomena were now included in the following comprehensive set of equations:

$$\nabla \times \mathbf{E} = -\frac{\partial \mathbf{B}}{\partial t} + [0], \tag{3.16a}$$

$$\nabla \times \mathbf{H} = \left[\frac{\partial \mathbf{D}}{\partial t}\right] + \mathbf{j}, \tag{3.16b}$$

$$\nabla \cdot \mathbf{D} = \rho, \tag{3.16c}$$

$$\nabla \cdot \mathbf{B} = 0. \tag{3.16d}$$

With the exception of the bracketed terms in (3.16a) and (3.16b), the reader can recognize the first two equations as (3.13) and (3.15) and the last two as the equations of electrostatics (eqn. 3.8) and magnetostatics (eqn. 3.10). The symmetry between the effects of electricity and magnetism should now be evident. For each equation describing effects of the electric field there is a counterpart describing effects of the magnetic field. Even the electric charges fit into this symmetric picture: when a term for electric charge (or electric current) is present in one equation, a zero term is present in its magnetic counterpart, representing the absence of magnetic monopoles. Thus, the zero term in brackets in (3.16a) is the magnetic analog to the current density $\mathbf{j}$ in (3.16b). Similarly, the charge density term ρ in (3.16c) is replaced by a zero in (3.16d). The symmetry appears to be complete, with one exception: the time derivative of the magnetic field in (3.16a) did not originally have an analogous field derivative term in its counterpart equation (eqn. 3.16b). This was recognized by Maxwell who proposed, purely for reasons of symmetry, to introduce the $\partial \mathbf{D}/\partial t$ term into (3.16b). With this term in place, a complete system of equations fully describing the behavior of electromagnetic fields was obtained. The presence of this additional term was subsequently justified by numerous experiments (see e.g. Feynmann et al. 1964, vol. II, pp. 1–8).

Maxwell's equations form the basis for the development of the equations that describe the propagation of electromagnetic waves. Historically, the electromagnetic wave equations were derived by Maxwell merely to describe the propagation of oscillating electric or magnetic fields in space. Neither Maxwell nor his peers recognized the relation between the propagation of electromagnetic fields and the propagation of light. Optics and the propagation of electromagnetic waves were at that time considered to be separate and unrelated fields of physics. Only after showing that the propagation velocity of the electromagnetic waves was identical to the already measured speed of light did Maxwell suggest that his results might be more general than expected and hence applicable to the studies of optics.

3.6 A Simple Model of Radiation

Before developing the electromagnetic wave equations and solving them, we may try to visualize how electromagnetic fields are created and how they interact with various media. Using a simple physical model, we will anticipate some of the characteristics of these waves and subsequently use them to interpret the results of our calculations. Invariably, electromagnetic fields are generated by oscillating charges. As their name implies, such fields consist of simultaneously present electric and magnetic waves. These waves carry the frequency of the oscillating charge and propagate indefinitely in space until encountering other charges at a remote location that respond to the field by oscillating at its frequency.

Our simple model consists of a unit charge oscillating at a frequency of ω [rad/s]. In nature this oscillation usually takes place within atoms or molecules. Although quantum mechanics is required to fully understand how these charges oscillate and how their oscillation is related to the emission and absorption of radiation, we may be able to understand many of the effects with our simple classical models. For illustration we simulate in Figure 3.7 the oscillation of electric charges within an electrically neutral molecule by a pair of pendulums, each consisting of one positive and one negative charge and each attached to a separate string and oscillating at opposite phase in the (x, y) plane. Because the charges are opposite and identical in magnitude, the system is electrically neutral. Such a pair of balanced electric charges is called an electric dipole, or simply a *dipole*. We cannot explain with our classical model why the charges do not collapse together or how the energy stored in the oscillation mode can be transferred to another system. However, even without such detail

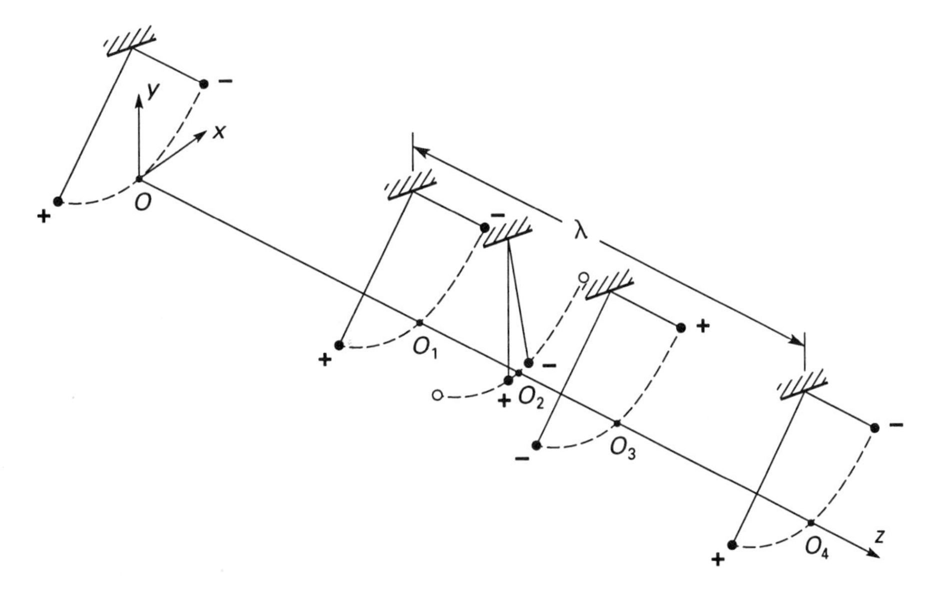

Figure 3.7 Oscillation of electric dipoles.

it is evident that an observer at a distant point O_1 will not detect any net field in the z direction because the attraction by one charge will be identically offset by the repulsion of the other charge. However, a net electric field in the x direction is expected. To evaluate the magnitude of that field, assume that the oscillation amplitude of the pendulum is d. Therefore, when the positive charge of the dipole at the origin is at a position of full amplitude (as shown in Figure 3.7), the electric field points along the line connecting the positive charge with O_1. Although the magnitude of the field is determined by the nearly constant distance R between these points, the x component of that field is determined by the projection of that field on the x axis. When the angle θ between the field vector and the z axis is small, the x projection of the field is

$$E_x \approx \mathbf{E}\cdot\theta = \frac{1}{4\pi\epsilon}\frac{q^+}{R^2}\cdot\frac{d/2}{R},$$

where q^+ is the positive charge. A similar term is obtained for q^-, the negative charge. Since the projections of both fields point in the positive x direction, they can be added and the distinction between the positive and negative charges eliminated. Thus, the net component in the x direction of the electric field induced by a dipole is

$$E_x = \frac{q\cdot d}{4\pi\epsilon R^3} = \frac{p}{4\pi\epsilon R^3}. \tag{3.17}$$

Here a new term, the *dipole moment* $p = q\cdot d$, has been introduced. The dipole moment is often used to measure the extent by which certain charges are naturally polarized or the extent of their polarization when subjected to an external field. This term represents the maximum moment that will be experienced by the dipole when placed in an electrostatic field with an intensity of unity. From (3.17) we see that, in the presence of a dipole, the x component of the electric field increases proportionally with the dipole moment. Therefore, a dipole moment oscillating at a frequency ω induces at O_1 an electric field E_x that oscillates at the same frequency. Thus, (3.17) represents the variation along the z axis of the amplitude of E_x. Away from the z axis, the amplitude of E_x decreases (see Problem 3.3) until it vanishes for points on the x axis.

Because the motion of the two opposite charges of the dipole along the x axis amounts to a net oscillating current, according to (3.15) an oscillating magnetic field must be present simultaneously with the field vector and must be orthogonal to the current vector. Thus an observer at O_1 on the z axis can detect, in addition to the electric field, an oscillating magnetic field pointing parallel to the y axis. Similarly, an observer along the y axis will detect an oscillating magnetic field pointing in the direction of the z axis simultaneously with an oscillating electric field pointing in the x direction. Without calculating its magnitude, it is evident that (with the exception of points along the x axis) the magnetic field is present anywhere simultaneously with E_x, is perpendicular to E_x, and oscillates at the same frequency ω.

By forcing a dipole oscillation in the y direction, a new pair of oscillating **E** and **H** field vectors is obtained. For an observer at O_1 along the z axis, the electric field vector is now oscillating along the y axis while the magnetic field oscillates along the x axis. These two independent oscillating modes that can be observed at O_1 are the *optical polarizations*. Although their frequency or amplitude may be identical, they still remain distinguishable and independent of each other. A third independent oscillating mode of the driver pendulum can exist along the z axis. Usually, the distance between the observer and the driver oscillator is large relative to the oscillation amplitude, so when the pendulum oscillates along the z axis the field induced at O_1 by the positive charge will almost identically cancel the field induced by the negative charge, thereby resulting in no net field along the z axis. Similarly, the magnetic field will be confined to concentric circles in the (x, y) plane. Barring some minor edge effects, no magnetic field will be sensed in the z direction. This can become more evident by calculating the angular dependence of E_x (Problem 3.3). Thus, along any line of sight connecting a source to an observer, only two independent polarizations can exist; they are orthogonal to each other and perpendicular to the line of sight. Any other optical polarization must be their linear combination.

An observer placed at any point along the z axis (e.g. point O_1) can detect these fields by placing an electric dipole, similar to the oscillating pendulum, at the point of detection. If the motion of that dipole in the x direction is unrestricted, the electric field and the magnetic field induced by the driver dipole will force it to oscillate. Normally, the restoring forces of dipoles are linear with their displacement. Therefore, the dipole moment induced at O_1 varies linearly with E_x. If, in addition, the natural frequency of the dipole at O_1 is matched with the natural frequency of the driving dipole, the detection dipole can be forced to oscillate at large amplitudes. Similarly, if the detection dipole is free to oscillate in the y direction then it may be used for the detection of the y polarization. Thus, a detector along the z axis can detect only two independent optical polarizations: one in the x direction and one in the y direction. Any other polarization must then be a linear combination of the two independent polarizations.

Although a single molecule or a single atom can radiate or can be forced to oscillate by incident radiation, realistic light sources and optical materials contain a very large number of dipoles. The density of these dipoles can be represented by their total count within a unit volume, N [#/m^3]. Usually, dipoles are induced by an external electric field and are therefore parallel to each other. If an average dipole moment is $q \cdot \mathbf{d}$ then the vector formed by the addition of all the dipole moments within the unit volume is

$$\mathbf{P} = \frac{\Sigma(\text{dipole moments})}{\text{volume}} = N \cdot q \cdot \mathbf{d}, \tag{3.18}$$

where **P** is the *polarization vector*. The reader is cautioned not to confuse the polarization vector with the polarization of radiation. Although they may be related, the first represents a material parameter while the second represents the direction of the electromagnetic field vectors.

The vector **d**, which represents the separation between the charges of the individual dipoles, is an important parameter, but usually it cannot be evaluated directly. Instead, a material parameter is introduced. Owing to the linear dependence between the polarization vector and the incident field inducing it, we can also write:

$$\mathbf{P} = \epsilon_0 \chi \mathbf{E}. \tag{3.19}$$

Here, ϵ_0 is the dielectric constant in free space and χ is the material constant that represents the ability of dipoles within the medium to respond to a polarizing electric field. This coefficient is the electric *susceptibility* of the dielectric. From the definition of (3.18), the units of the polarization vector are coulombs per square meter, which are also the units of $\epsilon_0 \mathbf{E}$. Therefore, χ is a dimensionless material parameter.

The results of (3.19) can be put in a broader context. Recall that the electric displacement vector was defined as $\mathbf{D} = \epsilon \mathbf{E}$. However, it can also be shown (see Feynmann et al. 1964, vol. II, chap. 10) that

$$\mathbf{D} = \epsilon \mathbf{E} = \epsilon_0 (1 + \chi) \mathbf{E}. \tag{3.20}$$

Therefore, for a given electric field, **P** is the difference between the electric displacement vector in a medium and the electric displacement vector in free space, and χ is the difference between the dielectric constant in a medium and the dielectric constant in free space. We will show in the next chapter that important optical parameters such as the index of refraction or attenuation by metallic media can be determined from χ.

Other parameters of the propagation of radiation – such as the relation between wavelength, frequency, and the speed of light – can also be found from the simple model of the oscillating dipoles in Figure 3.7. Assume that, in addition to the dipole at point O_1, other dipoles are placed at judiciously selected locations along the z axis. If the oscillating field from the driving dipole could propagate at an infinite velocity, then all the dipoles, irrespective of their location, would oscillate in phase. Thus, when one dipole is stretched to its full amplitude, the other dipoles will be fully stretched as well. However, when the propagation velocity c of the field induced by the driver dipole is finite, the phases of the detector dipoles depend on their location. We can thus conceive of a case where the dipole at O_1 is already at the peak amplitude of a new cycle, while the dipole at O_2 is still at the neutral point of the previous cycle and the dipole at O_3 is fully stretched in the opposite phase of the previous cycle. If the wavelength λ is defined as the distance between two adjacent oscillators at the same phase, then the time for travel between these two oscillators is

$$\tau = \lambda / c. \tag{3.21}$$

At that velocity, all the dipoles along the way are encountered with the same phase and, after traveling a distance of λ, the dipole at the end is encountered just after the completion of one full cycle. For this velocity, τ in (3.21) is thus the cycle time of the oscillation, and is related to the oscillation frequency ν by

$$\tau = \frac{1}{\nu} = \frac{2\pi}{\omega}, \tag{3.22}$$

where ω [rad/s] denotes the angular frequency. Combining (3.21) and (3.22), we find that

$$c = \nu \cdot \lambda. \tag{3.23}$$

These simple relations will be derived later from the wave equation.

3.7 Boundary Conditions

In the previous section we saw that oscillating electric charges induce simultaneously oscillating electric and magnetic fields that propagate in space at a finite velocity c. Although we did not discuss how these fields propagate, we did notice that they may drive the electric charges they encounter into a similar oscillation. However, the driven oscillation of these charges must induce a new field that is likely to interact with the incident field. Such interaction may cause the incident field to change its direction of propagation (e.g. reflection or refraction), may force the oscillation into a new phase, or may simply eliminate the entire field by absòrption. Many of these changes occur over a finite distance that may extend from a single to several molecular layers. These finite distances are extremely small relative to most practical scales and in most engineering applications may appear to occur abruptly, as with the effects of reflection or refraction that take place at an interface between two media. Mathematically these abrupt changes can be cast into a set of boundary conditions that must be used when solving Maxwell's equations (eqns. 3.16). These boundary conditions describe the effect on the components of the electromagnetic field of an abrupt change in material properties. Thus, away from these boundaries, solutions of Maxwell's equations describe the propagation of the fields. However, at a boundary, the solution at either side must be matched to the solution at the other side while conforming with a set of conditions: one for each field (**E** and **H**) component parallel to the plane and a second for each vertical component. Thus, for the two fields **E** and **H**, four boundary conditions must be formulated. We will use these four conditions after developing the wave equation for the derivation of the laws of refraction and reflection.

1. *Boundary condition for an electric field perpendicular to an interface*

Figure 3.8 presents the boundary between two different optical media, labeled 1 and 2. An electric field vector is incident perpendicular to the interface from the side of medium 2. We can use Gauss's law (eqn. 3.6) to find the value of the field at the other side of the interface by wrapping the entire surface with an imaginary thin envelope. By selecting a sufficiently thin envelope, only the

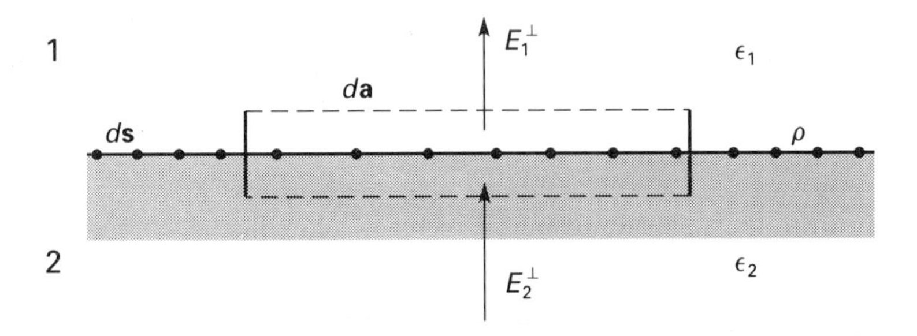

Figure 3.8 Perpendicular components of the electrical field vector at both sides of an interfacial discontinuity.

charges at the interface (or its immediate vicinity) will be included in this evaluation; they may be considered as surface charges with a density of ρ_A C/m^2. Hence, the volume integral in (3.6) can be replaced with a surface integral that describes the total charge at the interface. With that modification, the following form of Gauss's law can be used to evaluate the total field flux through a closed surface containing net surface charges at the interface:

$$\oint \epsilon \mathbf{E}\, d\mathbf{a} = \int \rho_A\, d\mathbf{s},$$

where $d\mathbf{a}$ is an area element of the envelope and $d\mathbf{s}$ is an area element of the interface. Because the selected envelope is thin rclative to its other dimensions, edge effects are negligible and the integration over the envelope surface can take place only over areas parallel to the interface. This leaves in the calculation only the field components that are perpendicular to the interface. Furthermore, without the edges, the extent of the envelope can be made arbitrarily small. Therefore, to satisfy Gauss's law for any area element, the electric displacement vectors must satisfy the following boundary condition:

$$D_1^\perp - D_2^\perp = \rho_A. \tag{3.24}$$

2. *Boundary condition for a magnetic field perpendicular to an interface*

Figure 3.9 presents the boundary between two different optical media, labeled 1 and 2. This is similar to the previous case, except that here a magnetic field vector is incident perpendicular to the flat interface from the side of medium 2. In order to find the value of the field at the other side of the interface, we again wrap the entire surface with an imaginary thin envelope. However, unlike the previous case, the envelope cannot enclose any net magnetic poles. Therefore, the magnetic flux passing through the entire envelope must vanish (eqn. 3.9). Barring edge effects, the magnetic displacement across the bottom and top parts of the imaginary sleeve must be continuous. For most optical materials the magnetic permeability μ equals the permeability of free space. Thus, for optical applications the boundary condition for the normal component of the magnetic field is

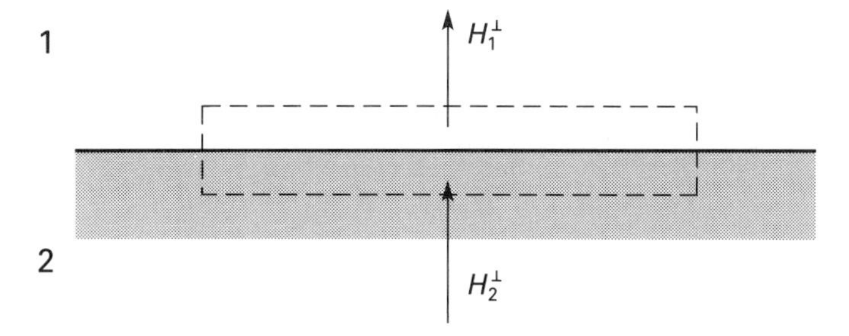

Figure 3.9 Perpendicular components of the magnetic field vector at both sides of an interfacial discontinuity.

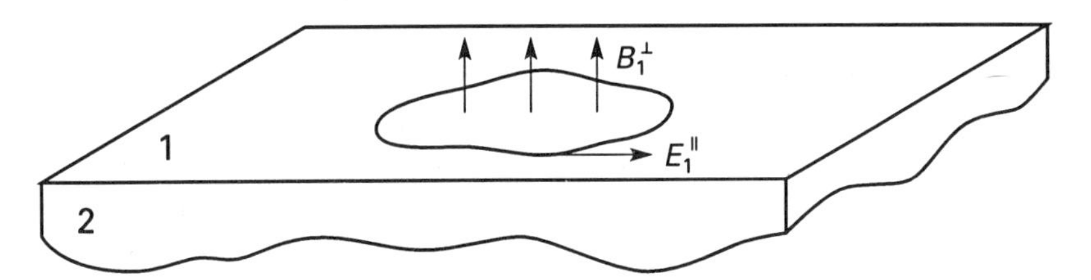

Figure 3.10 Parallel component of the electric field vector at an interfacial discontinuity.

$$H_1^{\perp} = H_2^{\perp}. \tag{3.25}$$

3. *Boundary condition for an electric field parallel to an interface*

Figure 3.10 presents the boundary between two different optical media, labeled 1 and 2. To simplify the illustration, the parallel component of the electric field vector is shown only at one side of the interface (medium 1). Although a vertical field component may also be present, it can be analyzed independently and so its exclusion here does not diminish the generality of our analysis.

From the first of Maxwell's equations (eqn. 3.16a) and also from (3.12), we can see that the perpendicular component of the magnetic induction vector is the only physical quantity at the interface that can affect the value of the parallel electric field component. However, in the previous boundary condition (eqn. 3.25) we found that the normal component of the magnetic induction is continuous across the interface. Therefore, the circulation of the electric field vector around a closed contour (eqn. 3.12) calculated at one side of the interface must be equal to the circulation at the other side. The boundary condition for the tangential component of the electric field vector thus requires that:

$$E_1^{\parallel} - E_2^{\parallel} = 0. \tag{3.26}$$

4. *Boundary condition for a magnetic field parallel to an interface*

The last boundary condition applies to the component of the magnetic field that is tangential to the interface between two media. Figure 3.11 presents a side

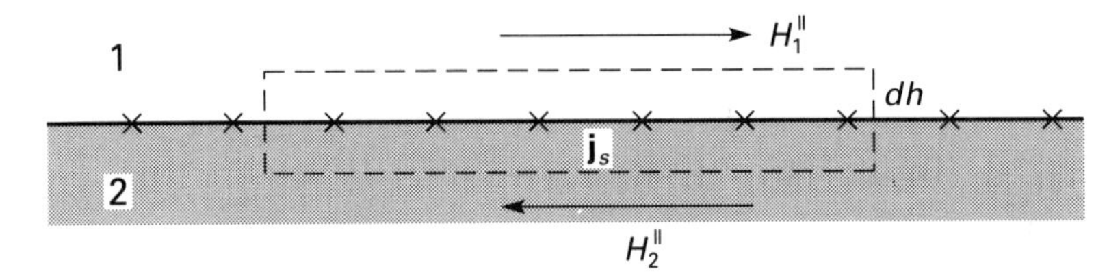

Figure 3.11 Parallel components of the magnetic field vector at both sides of an interfacial discontinuity.

view of the interface separating medium 1 from medium 2. As before, the interface is wrapped by a thin envelope with a height of dh. Assume that the medium enclosed by the envelope carries electric current perpendicularly to the envelope cross section. For a sufficiently thin envelope, the current may be viewed as confined to the interface. Of course, a current flow may also exist outside this envelope, but it is not part of the boundary effects. By selecting from the second of Maxwell's equations (eqn. 3.16b) the components of $\nabla \times \mathbf{H}$ and of $\partial \mathbf{D}/\partial t$ that are parallel to the surface current $\mathbf{j}_s$ (see Problem 3.4), the following relation for the tangential components of the magnetic field is obtained:

$$H_1^{\parallel} - H_2^{\parallel} = \frac{\partial}{\partial t}(D_1^{\parallel} - D_2^{\parallel})\, dh + \mathbf{j}_s,$$

where $\mathbf{j}_s = \mathbf{j}\, dh$. However, since $dh \to 0$, the contribution of the parallel component of the electric displacement vector vanishes, and the boundary condition for the tangential component of the magnetic field vector becomes

$$H_1^{\parallel} - H_2^{\parallel} = \mathbf{j}_s. \tag{3.27}$$

These four boundary conditions, together with Maxwell's equations, can be used to describe the formation of electric and magnetic fields by stationary and moving charges, as well as the propagation of these fields through media with various properties. In the next chapter we will use these equations to derive the wave equations that describe the propagation of electromagnetic waves. The boundary conditions will then be used to derive the laws of reflection, transmission, and refraction.

References

Born, M., and Wolf, E. (1975), *Principles of Optics,* 5th ed., Oxford: Pergamon, pp. 1–7.

Eberhardt, E. H. (1979), Gain model for microchannel plates, *Applied Optics* 18: 1418–23.

Engstrom, R. W. (1980), *Photomultiplier Handbook, Theory, Design, Application,* Lancaster, PA: RCA Co.

Feynmann, R. P., Leighton, R. B., and Sands, M. (1964), *The Feynmann Lectures on Physics,* vol. II, Reading, MA: Addison-Wesley.

Halliday, D., and Resnick, R. (1963), *Physics for Students of Science and Engineering,* part II, New York: Wiley, pp. 555–822.

Hildebrand, F. B. (1962), *Advanced Calculus for Applications,* Englewood Cliffs, NJ: Prentice-Hall, pp. 290–4.
Iizuka, K. (1983), *Engineering Optics* (Springer Series in Optics, vol. 35), Berlin: Springer-Verlag, pp. 1–27.
Kume, H., ed. (1994), *Photomultiplier Tubes, Principle to Application,* Hamamatsu City, Japan: Hamamatsu Photonics K.K.
Maxwell, J. C. (1873), *A Treatise on Electricity and Magnetism,* 2 vols., Oxford University Press.
Wiza, J. L. (1979), Microchannel plate detectors, *Nuclear Instruments and Methods* 162: 587–601.

Homework Problems

Problem 3.1

The current generated at the anode of a photomultiplier tube can be detected by passing it through a load resistor R_L shorting the anode to ground (see Figure 3.12). The voltage developed across that resistor is amplified and detected by either a voltmeter or (if temporal resolution is required) an oscilloscope. In some time-dependent experiments, resolution of better than 10 ns may be required. To meet that requirement, the time constant of the detection circuitry is reduced by using a load resistor with $R_L = 50\ \Omega$. Assume that the PMT has 10 dynodes, each with a secondary emission ratio of $\delta = 5$.

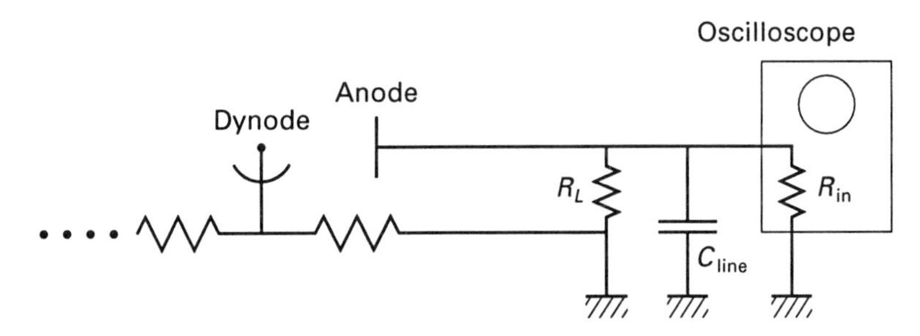

Figure 3.12 Circuit diagram of a photomultiplier tube.

(a) Find the maximum allowable capacitance in picofarads of the detection system for a required time response of 10 ns. (*Answer:* 200 pF.)
(b) Find the average current developed at the anode for each electron emitted by the photocathode if that current is emitted over a period of 10 ns. (*Answer:* 1.56×10^{-4} A.)
(c) Find the voltage developed across the resistor R_L. (*Answer:* 7.8 mV.)
(d) What will be the time resolution and the voltage across the load resistor if $R_L = 1\ \text{M}\Omega$? (*Answer:* $t = 0.2$ ms, $V = 156$ V.)

Problem 3.2

Television tubes and computer monitors use cathode-ray tubes (CRTs) to display an image. In a typical CRT, electrons are accelerated until their kinetic energy is 10 keV. Initially the electrons are moving along the tube axis. To deflect the beam, the electrons are passed between two plates. One plate is grounded while the other is biased either positively or negatively. If the bias is positive, the beam is deflected toward the biased plate; otherwise, the deflection is away from that plate. In addition, the beam may be influenced by the magnetic field of the earth, which at Washington D.C. is 5.7×10^{-5} Wb/m^2. (Assume that the charge of an electron is $e^- = 1.6 \times 10^{-19}$ C and the mass of an electron is $m_e = 9.11 \times 10^{-31}$ kg.)

(a) For a 25-cm tube, find the maximum deflection of the electron beam by the earth's magnetic field. (*Answer:* 5.3 mm.)

(b) Find the maximum deflection angle that can be achieved if the voltage between the deflection plates is 100 V. (*Answer:* 5.7°.)

(c) If the gap between the deflection plates is 1 cm, find the length of the plates for which the maximum deflection calculated in (b) will be achieved. (*Answer:* 2.0 m.)

Problem 3.3

Equation (3.17) shows that E_x decreases with R^3, where R is the distance from the dipole along the z axis. However, E_x also depends on α, the angle between the line of the dipole and the line of sight of an observer located off the z axis. Using the nomenclature of Figure 3.7, show that E_x for an observer located off the z axis is

$$E_x = \frac{p}{4\pi\epsilon}\frac{3\cos^2\alpha - 1}{R^3}.$$

(*Hint:* To avoid vectorial addition of the fields induced by the positive and negative charges, you may use instead the potential ϕ. This potential is a scalar obtained by calculating $\int \mathbf{E}\, d\mathbf{R}$. After arithmetically adding the potential of the positive and negative charges, find $E_x = -\partial\phi/\partial x$.)

Problem 3.4

Derive the difference between the tangential component of the magnetic field at the two sides of an interface carrying surface current of $\mathbf{j}_s$.

Research Problem

Research Problem 3.1

Read the paper by Eberhardt (1979).

(a) Derive Eberhardt's equation (1) for the axial distance traveled by an electron with an initial kinetic energy eV_{0r} in the radial direction and initial kinetic energy eV_{0z} in the axial direction.

(b) Using the term for the gain in Example 3.2, find the voltage V_0 between the two ends of the channel for which the gain is maximum. Also determine that maximum gain. You may assume that $V_{0z} \ll V_{0r}$, $A = 0.2$, and $\alpha = 50$.

(c) In the derivation of Eberhardt's equation (1), the electrons were assumed to be emitted in the main plane (i.e., the plane that includes the channel axis). This allowed the longest trajectory between two wall encounters. However, when the trajectory is in a plane inclined to the main plane by an angle θ, the time of flight decreases. Find the dependence of the secondary emission ratio $\delta(\theta)$ on the angle θ.

4 Properties of Electromagnetic Waves

4.1 Introduction

In the previous chapter we saw that, when electric dipoles are forced to oscillate, they induce an electric field that oscillates at the same frequency. In addition, owing to the motion of the oscillating charges, a magnetic field oscillating at the same frequency is also induced. These simultaneous oscillating fields are the basis for all known modes of electromagnetic radiation. Thus, X-rays, UV radiation, visible light, and infrared and microwave radiation are all part of the same physical phenomenon. Although each radiating mode is significantly different from the others, all modes of electromagnetic radiation can be described by the same equations because they all obey the same basic laws.

Oscillation alone is insufficient to account for electromagnetic radiation. The other important observation is that radiation propagates. It is broadcast by a source and, if uninterrupted, can propagate indefinitely in both time and space. An example of the boundless propagation of electromagnetic waves - whether X-ray, visible, or microwave - is the radiation emitted by remote galaxies. Some of this radiation, generated at primordial times and at remote reaches of the universe, can be detected on earth billions of years later. Evidently, radiation is not limited to the immediate vicinity of the source. Although we know that certain media can block radiation, we find it more astonishing that electromagnetic waves can propagate through free space; unlike electrical currents or sound, conductors are not necessary for the transmission of radiation. Although this property is unique to radiation, some of its other properties are analogous to the propagation of acoustical waves or vibrations in solids. These waves, like the electromagnetic waves, combine propagation with the oscillation of a physical parameter. Thus, by analogy, the description of the propagation of electromagnetic radiation should involve equations similar to those describing the propagation of sound waves or the vibrational modes of solids. Furthermore, since we anticipate that electromagnetic waves are the result of the oscillatory motion of electric charges, we should be able to derive equations describing their propagation from Maxwell's equations.

Although equations for the propagation of electromagnetic waves are likely to be similar to those for acoustic waves, there is an important distinction between the two. Acoustic wave equations describe the propagation of a pressure disturbance, which is a scalar quantity; electromagnetic wave equations describe the propagation of electric and magnetic fields, which are vectorial. Thus, associated with each field (electric and magnetic) we expect to find two equations, one for each of the vector components perpendicular to the direction of propagation.

4.2 The Wave Equations

In order to derive the equations that describe the propagation of electromagnetic waves, we begin with Maxwell's equations (eqns. 3.16). Inspection of these equations reveals that when the magnets or electric charges are static, the fields they induce are independent of each other. That is, the equation for the electric field vector (eqn. 3.16c) does not contain any terms of the magnetic field, and conversely (3.16d) is independent of the electric field. When in motion, however, the magnets or electric charges induce fields that are interdependent. This is apparent in both (3.16a) and (3.16b), where **E** depends on the time derivative of **B**, and where **H** varies with the magnitude or direction of the current flow or with the time derivative of **D**. Nonetheless, an equation that describes the propagation of electric waves is expected to be independent of the terms that include the magnetic field, and similarly the equation of the magnetic waves should not include terms of the electric field. Since only two equations (3.16a and 3.16b) describe the dynamic effect of these two fields, they will be our starting point for the derivation of equations describing the propagation of electric or magnetic waves. Thus, after replacing the magnetic field term with terms that describe the electric field, (3.16a) will yield an equation for the propagation of electric field waves. Similarly, (3.16b) will yield (after reduction) an equation for the propagation of magnetic waves.

We first consider (3.16a). The simplest way of eliminating the magnetic field term from this equation is by obtaining the curl of both sides:

$$\nabla\times\nabla\times\mathbf{E} = -\frac{\partial}{\partial t}(\nabla\times\mathbf{B}) = -\mu\frac{\partial}{\partial t}(\nabla\times\mathbf{H}). \qquad (4.1)$$

Assuming that the magnetic permeability μ is constant, it was placed outside the derivative operators, thereby leaving only the magnetic field to be operated on. However, the term $\nabla\times\mathbf{H}$ in (4.1) can be replaced by the right-hand side of (3.16b), thereby eliminating the magnetic field term. The following equation,

$$\nabla\times\nabla\times\mathbf{E} = -\mu\frac{\partial^2\mathbf{D}}{\partial t^2} - \mu\frac{\partial\mathbf{j}}{\partial t}, \qquad (4.2)$$

is now in the desired form; it contains only terms of the electric field or electric charge. Furthermore, it includes both time and space derivatives of these quantities and so describes both the temporal and spatial variation of the electric field due to the motion of electric charges. Although this equation is complete in itself, it may be simplified by using the following vector identity (see Problem 4.1):

$$\nabla \times \nabla \times \mathbf{A} = \nabla(\nabla \cdot \mathbf{A}) - \nabla^2 \mathbf{A}, \tag{4.3}$$

where $\mathbf{A}$ is an arbitrary vector and $\nabla^2 = \nabla \cdot \nabla$ is the Laplacian operator (Hildebrand 1962, p. 278). With few exceptions (e.g., radiation emitted by a point source), the use of Cartesian coordinates is most convenient. Thus, in the Cartesian coordinate system, operating on any vector $\mathbf{A} = A_x \mathbf{e}_x + A_y \mathbf{e}_y + A_z \mathbf{e}_z$ with the Laplacian yields

$$\nabla^2 \mathbf{A} = \left(\frac{\partial^2 A_x}{\partial x^2} + \frac{\partial^2 A_x}{\partial y^2} + \frac{\partial^2 A_x}{\partial z^2} \right) \mathbf{e}_x + \left(\frac{\partial^2 A_y}{\partial x^2} + \frac{\partial^2 A_y}{\partial y^2} + \frac{\partial^2 A_y}{\partial z^2} \right) \mathbf{e}_y + \left(\frac{\partial^2 A_z}{\partial x^2} + \frac{\partial^2 A_z}{\partial y^2} + \frac{\partial^2 A_z}{\partial z^2} \right) \mathbf{e}_z.$$

Thus, the left-hand side of (4.2) can be replaced by

$$\nabla \times \nabla \times \mathbf{E} = \nabla(\nabla \cdot \mathbf{E}) - \nabla^2 \mathbf{E}.$$

However, when the medium in which $\mathbf{E}$ propagates is homogeneous (i.e., when all the spatial derivatives of ϵ vanish), and when the medium does not contain any free charges (i.e., $\rho = 0$), the first of these terms is $\nabla \cdot \mathbf{E} = 0$ (eqn. 3.16c). With these vector identities, (4.2) can thus be reduced to

$$\nabla^2 \mathbf{E} = \mu \frac{\partial^2 (\epsilon \mathbf{E})}{\partial t^2} + \mu \frac{\partial \mathbf{j}}{\partial t}. \tag{4.4}$$

This is the wave equation that describes the propagation of an electric wave. It does not specify what caused the field or how the field can be annihilated, but it accurately predicts the magnitude and direction of $\mathbf{E}$ at any point in space or time. To use this equation for the solution of problems with cylindrical or spherical symmetry, the Laplacian ∇^2 must be expressed in those coordinates.

Because most optical elements consist of uniform media, the assumption that ϵ is constant is usually justified. The second assumption, that is, that the density of unbalanced electric charges is $\rho = 0$, merits additional consideration. This condition is met in free space and of course in all electrically neutral media, whether dielectric or conducting. However, the condition may not be satisfied when a medium is electrically charged, unless we can show that whenever such a charge appears it dissipates faster than any characteristic time of the wave – for example, the time for one oscillation cycle. To determine which material properties control the dissipation rate of free charges, we examine (3.16b), the only one of Maxwell's equations that includes a term for the electric current. By calculating the divergence of (3.16b) and using the vectorial identity

$\nabla\cdot\nabla\times\mathbf{H}=0$, we can eliminate the magnetic field term and leave this equation with terms for only the electric field and current:

$$\nabla\cdot\nabla\times\mathbf{H}=\nabla\cdot\dot{\mathbf{D}}+\nabla\cdot\mathbf{j}=0. \tag{4.5}$$

The last term, $\nabla\cdot\mathbf{j}$, represents the dependence of the current density on the pointwise material properties, which can be obtained from Ohm's law evaluated at the microscopic level. On the macroscopic level, Ohm's law represents the relation between V, I, and the resistance R. On the microscopic level it relates vectorially the terms $\mathbf{E}$, $\mathbf{j}$, and the specific conductivity σ [mho-m] (where 1 mho = 1/ohm). Thus, on the microscopic level Ohm's law is

$$\mathbf{j}=\sigma\mathbf{E} \tag{4.6}$$

(Halliday and Resnick 1963, pp. 682–8). This equation, together with the equations for the displacement vectors $\mathbf{D}=\epsilon\mathbf{E}$ and $\mathbf{B}=\mu\mathbf{H}$, specify the material properties of the optical medium. By using Ohm's law on the microscopic level, we not only introduce the material properties that control the conduction of current but also replace $\mathbf{j}$ with the electric field, thereby reducing (4.5) to the following single-variable differential equation:

$$\nabla\cdot\dot{\mathbf{D}}+\frac{\sigma}{\epsilon}\nabla\cdot\mathbf{D}=0,$$

where again we used the condition $\epsilon=$ constant and introduced the condition $\sigma=$ constant. Finally, using (3.16c), the divergence of the electric displacement vector $\mathbf{D}$ can be replaced with the charge density ρ to form the following scalar equation:

$$\dot{\rho}+\frac{\sigma}{\epsilon}\rho=0. \tag{4.7}$$

Equation (4.7) can be readily solved to show the dissipation of any initial charge distribution ρ_0:

$$\rho=\rho_0 e^{-t/\tau},$$

where the time constant for the dissipation of the charge is $\tau=\epsilon/\sigma$ [s]. For highly conducting media (e.g. metals or plasmas), $\tau\approx 10^{-18}$ s. This is significantly shorter than the characteristic time of most electric field waves, which for optical radiation oscillate at a frequency ν of approximately 10^{15} Hz. Thus, even if the medium is initially charged, most of the charge is dissipated well before the completion of one oscillation cycle. The dissipation time for dielectrics is longer, but in most applications optical dielectric media are free of charges.

Although conductors may appear to be free of net electric charge when the frequency of the incident electric field is $\nu<1/\tau$, the current density $\mathbf{j}$ may not vanish. Electrical currents may exist within dissipating pockets of electric charges or may be induced by external fields. In dielectrics, on the other hand, electric currents are usually small and can be neglected. Thus (4.4) is applicable to analyzing the propagation of electric fields both through conductors and

through dielectrics: for propagation through dielectrics, the current term may be neglected; for propagation through conductors, the term with $\mathbf{j}$ must be retained.

To present (4.4) in a more general form with the current term included, the microscopic Ohm's law (eqn. 4.6) can be used again. Replacing $\mathbf{j}$ with $\sigma\mathbf{E}$ introduces conductivity, and the equation is reduced to the following form with a single dependent variable:

$$\nabla^2\mathbf{E} = \mu_0\epsilon\frac{\partial^2\mathbf{E}}{\partial t^2} + \mu_0\sigma\frac{\partial\mathbf{E}}{\partial t}. \tag{4.8}$$

Equation (4.8) is the final form of the *wave equation for the electric field.* The magnetic permeability of free space, μ_0, was introduced to characterize optical media (for which the magnetic permeability is approximately that of free space). When σ is sufficiently small, the last term can be neglected and the solution represents propagation through dielectric media. Otherwise, a complete solution must be obtained.

The assumptions required for the derivation of this equation include:

1. the dielectric constant and the specific conductivity are uniform ($\nabla\epsilon = 0$ and $\nabla\sigma = 0$ for a homogeneous medium) and are independent of time ($\partial\epsilon/\partial t = 0$ and $\partial\sigma/\partial t = 0$);
2. the medium does not contain any free charges ($\rho = 0$), or the time for the dissipation ρ is short relative to the time of one oscillation cycle;
3. the magnetic permeability of optical media is equal to the magnetic permeability of free space ($\mu = \mu_0$); and
4. implicit in Assumption 3 is that the magnetic permeability is uniform in space ($\nabla\mu = 0$ for a homogeneous medium) and in time ($\partial\mu/\partial t = 0$).

Although (4.8) appears to describe exclusively the behavior of electric fields, we should remember that this field was induced by oscillating charges. By virtue of their oscillatory motion, these charges must also induce a varying magnetic field. Therefore, whenever an oscillating electric field is present, one may expect to observe a magnetic field with similar characteristics (e.g., frequency or propagation velocity). As we shall soon see, even the vectorial orientation of the magnetic field is expected to be related to the orientation of the electric field. Hence, any equation for the propagation of the magnetic field that is derived from (3.16b) will yield results similar to those obtained from (4.8). We will therefore solve only for the propagation of the electric field and subsequently show that a magnetic wave propagates simultaneously with the electric wave.

4.3 Solution of the Electric Field Wave Equation

Assume that a solution of the electric wave can be obtained by separation of variables (Hildebrand 1962, p. 430), where a possible solution takes the

following form:

$$\mathbf{E}(\mathbf{r}, t) = \mathbf{E}(\mathbf{r})f(t), \quad \text{where } \mathbf{r} = x\mathbf{e}_x + y\mathbf{e}_y + z\mathbf{e}_z.$$

Substituting this solution into (4.8) and solving for $f(t)$, we have

$$\mathbf{E}(\mathbf{r}, t) = \text{Re}[\mathbf{E}(\mathbf{r})e^{-i\omega t} + \mathbf{E}^*(\mathbf{r})e^{i\omega t}].$$

This solution describes an electric field oscillating at an angular frequency ω [rad/s], where $\omega = 2\pi\nu$ and ν are defined by (3.22). Any oscillation frequency can serve as a valid solution, so other conditions must be specified. Although the electrical field in this solution can be represented by a complex variable, only the real part of it can be used to describe a physical electric wave. The solution must therefore include $\mathbf{E}^*(\mathbf{r})$, the complex conjugate of $\mathbf{E}(\mathbf{r})$.

When this solution of the time dependence of the electric field vector is substituted into (4.8), the spatial dependence of $\mathbf{E}$ can be obtained from the following differential equation:

$$\nabla^2\mathbf{E}(\mathbf{r}) = -\omega^2\mu_0\epsilon\mathbf{E}(\mathbf{r}) - i\omega\mu_0\sigma\mathbf{E}(\mathbf{r}) = -\mathbf{k}^2\mathbf{E}(\mathbf{r}),$$

where

$$\mathbf{k}^2 = \omega^2\mu_0\epsilon + i\omega\mu_0\sigma \tag{4.9}$$

is the *propagation vector*. In Cartesian coordinates, $\mathbf{k} = k_x\mathbf{e}_x + k_y\mathbf{e}_y + k_z\mathbf{e}_z$. This vector depends on the material properties of the medium in which the electric field propagates, as well as on the field frequency. In the most general form, each of the material properties may be represented by a tensor. However, with few exceptions (e.g., the interaction with optical crystals), all optical media will be considered as isotropic and will be represented by scalar parameters. In conducting media where $\sigma \neq 0$, $\mathbf{k}$ is a complex variable, whereas in dielectrics $\mathbf{k}$ is real.

The solution for $\mathbf{E}(\mathbf{r})$ includes spatial harmonics, similar to the time-dependent solution. Combining both the spatial and temporal solution for $\mathbf{E}$ we obtain

$$\mathbf{E}(\mathbf{r}, t) = \text{Re}[\mathbf{E}_0 \exp\{-i\omega(t - t_0) \pm i\mathbf{k}\cdot(\mathbf{r} - \mathbf{r}_0)\}]. \tag{4.10}$$

Equation (4.10) describes the temporal and spatial propagation of an electric wave, where $\mathbf{E}_0$ is the amplitude of the electric field (to be determined from the initial and boundary conditions of the problem) and t_0 and $\mathbf{r}_0$ are the time and location (respectively) where the peak amplitude is reached. It can be readily seen from this solution that if the coordinate $\mathbf{r}$ is held constant - that is, if the field is monitored over a period of time by a stationary observer located at $\mathbf{r}$ - then the field appears to oscillate at a frequency of ω. Alternatively, if it were possible to record instantaneously the distribution of the vectors of the oscillating field over a certain distance, a pattern similar to that in Figure 4.1 would be obtained. There the field at the origin, $\mathbf{E}_i$, is pointing in the positive direction, but its momentary value is less than the peak value of the field or the amplitude, $\mathbf{E}_0$, which is seen to be reached a certain distance away from the origin. It is also seen that the peaks, or the troughs, are reached at regular intervals

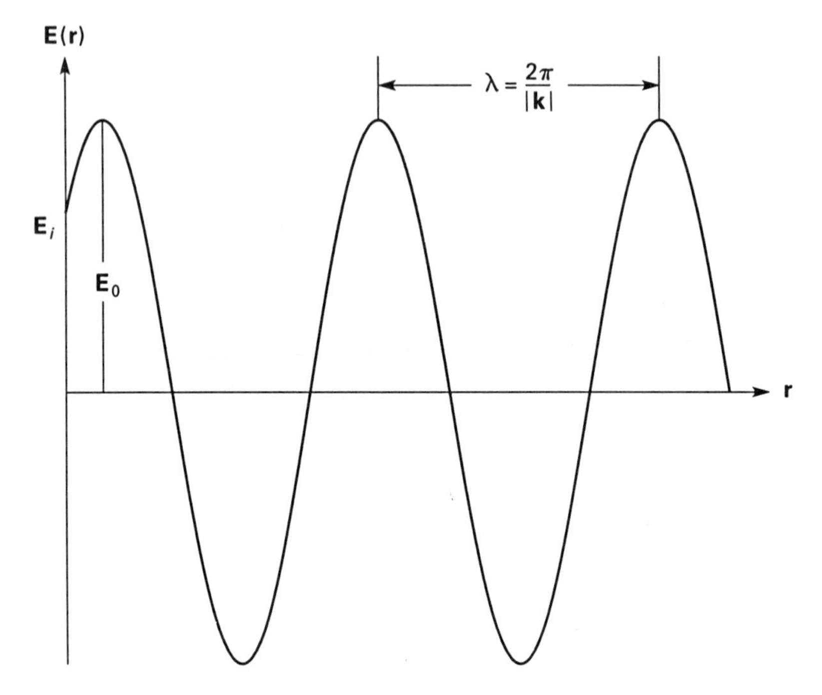

Figure 4.1 Spatial distribution of the electric field associated with radiation.

separated by $\lambda = 2\pi/|\mathbf{k}|$, the wavelength of this oscillation. For the regular harmonic oscillation described by (4.10), the value of $\mathbf{E}_i$ can be related to the amplitude $\mathbf{E}_0$ by the *phase angle* ϕ. Therefore, with the inclusion of the phase angle, the solution is:

$$\mathbf{E}(\mathbf{r}, t) = \text{Re}[\mathbf{E}_0 \exp\{-i(\omega t \mp \mathbf{k}\cdot\mathbf{r} + \phi)\}]. \tag{4.11}$$

The choice of a negative sign for the exponential term is arbitrary. Selection of the sign of this term affects only ϕ.

In most engineering applications, the value of ϕ has very little significance. The oscillation frequency of optical radiation is much higher than the frequency response of any electronic system, and the wavelength, which is usually in the micron or submicron range, is shorter than most dimensions of engineering relevance. Therefore, analysis of radiation fields very rarely includes the value of ϕ. However, there do exist applications for which the effect of the phase angle is critical. Although its actual value may not be important, it may be required that ϕ remain uniform transversely and unchanged over a certain period of time. Radiation that meets these requirements is said to be transversely and longitudinally *coherent*. Thus, for one-dimensional propagation in the z direction (Figure 4.2), transverse coherence implies that for any plane $z = z_0$, the phase is identical at all points in that plane. When the propagation is not one-dimensional, the transverse coherence requirement implies that a geometrically simple surface with $\phi = \text{constant}$ can be assigned. Such surfaces are called *wavefronts*. Clearly, the wavefront for one-dimensional propagation is a plane perpendicular to the propagation vector. For propagation out of a point source,

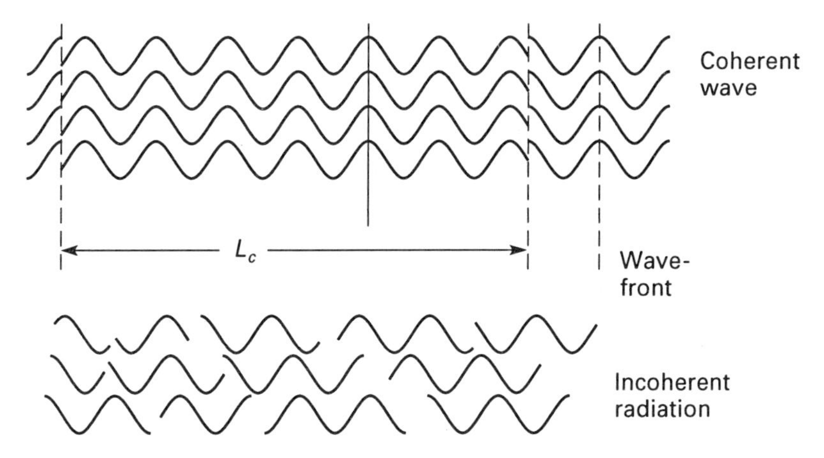

Figure 4.2 Comparison between coherent and incoherent monochromatic wavetrains.

transversely coherent fronts are spherical. When a radiation source is noncoherent, the value of the phase angle at any point in the beam is random.

The second coherence requirement, that is, that the phase angle remain constant over a period of time, is equivalent to the condition that the oscillation continues regularly without any abrupt interruptions or sudden phase changes. Even if a beam is transversely coherent, the phase angle may change abruptly after a certain number of oscillations. However, until ϕ is changed abruptly, all the wavefronts, whether plane or curved, are separated by the same wavelength λ. This regular sequence of wavefronts is called a coherent *wavetrain*. Usually, after an abrupt change in ϕ, a new group of wavefronts (a new wavetrain) is established. However, the phase angle of the new train is but randomly related to the phase angle of the previous group. Therefore, in applications where a coherent superposition of two beams is required (i.e., a superposition in which the phase angles of both components influence the outcome), that superposition must take place within the same wavetrain. Thus, when one beam is split into two legs and the legs are recombined after propagation through different paths, the difference in distances traveled by the two legs may not exceed the length of the wavetrain. To establish this length, we define a new parameter, the *coherence length* L_c:

> *Coherence length is the distance traveled by radiation between two consecutive changes in the phase angle. It is proportional to the coherence time* t_c*, which is the time between two consecutive changes in the phase angle with* $L_c = c_0 t_c$.

The radiation emitted by most light sources is considered to be incoherent. However, even the least coherent source has a finite coherence length. Typically, radiation from incandescent light sources has transverse and longitudinal coherence lengths of less than 1 μm. For most applications requiring coherence, this length is too short and the radiation is considered to be incoherent. Lasers

are used to generate more coherent radiation. With a well-designed laser, coherence lengths of a few meters can be obtained with transverse coherence extending to most parts of the beam cross section.

4.4 The Magnetic Field Wave

The magnetic field wave is induced by the motion of the oscillating dipole charges and is simultaneously propagating with the electric field wave. Since both fields are induced by the same dipoles (see Section 3.6), one may expect that the solution of the propagation of the wave of one field will have a simple relation to the solution of the waves of the other field. For example, the frequency of the magnetic field, just like the frequency of the oscillating electric field, must be ω. Thus, if the time-dependent solution of $\mathbf{H}$ takes the same form as the solution of $\mathbf{E}$, it may be possible to obtain a solution for the magnetic field wave by simply replacing the electric field term $\mathbf{E}(\mathbf{r}, t)$ in (4.11) with $\mathbf{H}(\mathbf{r}, t)$ and the amplitude term $\mathbf{E}_0$ with a magnetic amplitude term $\mathbf{H}_0$. Of course, this proposed solution must be compatible with the solution of the electric field wave (eqn. 4.11). To determine the conditions for such compatibility, the solutions for the magnetic field wave and for the electric field wave are introduced into Maxwell's equations (eqns. 3.16a and 3.16d), which yield the following vectorial equations:

$$\mathbf{k} \times \mathbf{E}(\mathbf{r}, t) = \omega\mu_0 \mathbf{H}(\mathbf{r}, t), \tag{4.12a}$$

$$\mathbf{k} \cdot \mathbf{E}(\mathbf{r}, t) = 0, \tag{4.12b}$$

where the following vectorial identities were used:

$$\nabla \times \mathbf{E}(\mathbf{r}, t) = i\mathbf{k} \times \mathbf{E}(\mathbf{r}, t),$$

$$\nabla \cdot \mathbf{E}(\mathbf{r}, t) = i\mathbf{k} \cdot \mathbf{E}(\mathbf{r}, t),$$

$$\frac{\partial \mathbf{H}(\mathbf{r}, t)}{\partial t} = -i\omega \mathbf{H}(\mathbf{r}, t).$$

From equations (4.12) is can be readily seen that the magnetic field vector is perpendicular to both the electric field and the propagation vectors. In addition, from the vectorial relation of (4.12b) it can be seen that the propagation vector is perpendicular to the electric field vector. However, since $\mathbf{H}(\mathbf{r}, t) \perp \mathbf{E}(\mathbf{r}, t)$, $\mathbf{H}(\mathbf{r}, t) \perp \mathbf{k}$, and $\mathbf{E}(\mathbf{r}, t) \perp \mathbf{k}$, one may conclude that the three vectors are mutually perpendicular. Figure 4.3 illustrates the orientation of the three vectors relative to each other, where the right-hand convention was used to determine the orientation of $\mathbf{H}$ relative to $\mathbf{E}$. From the scalar multiplication of $\mathbf{k}$ with $\mathbf{r}$ in (4.11) we learn that the propagation vector is pointing along the direction of the propagation of the electromagnetic field. However, since both $\mathbf{H}(\mathbf{r}, t)$ and $\mathbf{E}(\mathbf{r}, t)$ are perpendicular to $\mathbf{k}$, the two field vectors must be restricted to a plane perpendicular to the direction of the propagation. Thus, the electric field and magnetic field vectors of a plane wave are parallel to the wavefronts and are tangential to the wavefront surface when the front is curved. This mode of propagation,

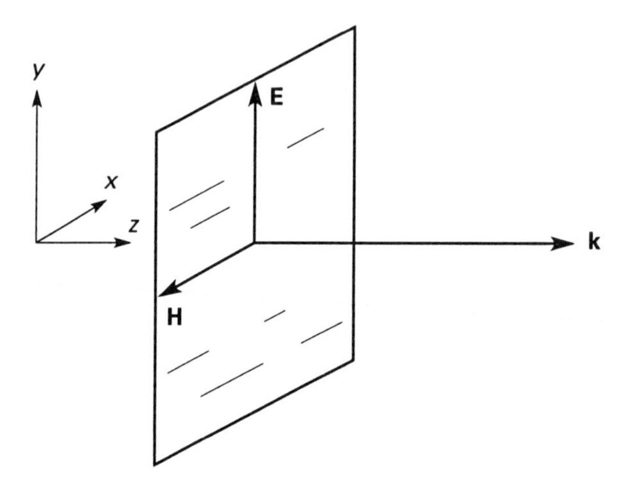

Figure 4.3 Three vectors that define the propagation of TEM electromagnetic waves.

the *transverse electromagnetic* (TEM) *mode,* exists almost exclusively in all media. Although the TEM mode restricts both field vectors to a plane, their pointing direction within that plane can be arbitrary. The two independent orthogonal directions within that plane are the polarizations of the waves. Thus, for propagation in the z direction, there may be x and y polarizations of the electric field and their linear combinations, together with orthogonal counterparts of the magnetic field.

Exceptions to these most prevalent propagation modes can exist when $\rho \neq 0$ or when the medium through which the electromagnetic field is propagating is inhomogeneous (i.e., $\epsilon \neq$ constant). Both cases represent a departure from the assumptions that led to the formulation of (4.4). Using (3.16c) it can be shown that when $\rho \neq 0$, the propagation vector is no longer normal to the electric field vector. However, from (3.16d) it is seen that the magnetic field may still remain perpendicular to the propagation vector and the wave may therefore be considered as *transverse magnetic*. Similarly, a non-TEM mode is often seen in propagation through stratified media, that is, when $\epsilon \neq$ constant along one dimension only (e.g. the z coordinate) while remaining constant in all other directions. Solution of this mode may include either TM (transverse magnetic) or TE (*transverse electric*) modes (Born and Wolf 1975, pp. 51–5).

Equation (4.12a) may also be used to determine the magnitude of the magnetic field vector relative to the electric field vector. In that equation the two magnitudes were seen to be proportional. Thus, in free space where $\epsilon = \epsilon_0$,

$$|\mathbf{H}(\mathbf{r}, t)| = \frac{1}{\eta_0}|\mathbf{E}(\mathbf{r}, t)|, \quad \text{where } \eta_0 = \sqrt{\frac{\mu_0}{\epsilon_0}} = 376.73. \tag{4.12'}$$

Here η_0 denotes the *impedance of free space.* Without explicitly solving for $\mathbf{H}$, one may conclude that in order to maintain its magnitude (through time and space) relative to $\mathbf{E}$, the magnetic field must also behave in a wavelike manner with a wavelength of $\lambda = 2\pi/|\mathbf{k}|$. To maintain the same frequency and wavelength, the solution of the magnetic field wave must be identical, as proposed, to the solution of the electric field wave, with the exceptions that the magnitude of the field vector is determined by the proportionality constant η and its orientation is orthogonal to **E**. Thus, the solution of the electric wave equation can be viewed now as a general solution that can be used to describe the propagation of both the magnetic and the electric field waves.

4.5 One-Dimensional Propagation in Free Space

Equation (4.11) is the general solution of the wave equation for propagation through homogeneous, charge-free media. It contains all the relevant material properties and can be solved for three-dimensional propagation through arbitrarily shaped homogeneous media and for all possible polarizations, that is, with the electric or magnetic field vectors pointing in any direction that is normal to the propagation vector. However, to visualize the important propagation parameters of electromagnetic radiation, a simpler solution is preferred in which the propagation is restricted to only one dimension and the material properties represent free space. For simplicity, the propagation will be modeled as occurring along the z axis, thereby requiring only one component to describe the propagation vector. In addition, we introduce the conductivity of free space, $\sigma = 0$, and the dielectric constant $\epsilon = \epsilon_0$. With these simplifications, the propagation vector of free space is

$$\mathbf{k}_0 = \omega\sqrt{\mu_0 \epsilon_0}\,\mathbf{e}_z. \tag{4.13}$$

After executing the scalar multiplication between $\mathbf{k}_0$ and the position vector $z\mathbf{e}_z$ in the z direction and assuming that $\phi = 0$, (4.11) reduces to

$$\mathbf{E}(z,t) = \mathrm{Re}[\mathbf{E}_0 \exp\{-i(\omega t \mp k_0 z)\}]. \tag{4.14}$$

Although the solution for the propagation is one-dimensional, $\mathbf{E}(z,t)$ may include two components – both perpendicular to the z axis. This can be shown explicitly by

$$\mathbf{E}_0 = E_{0x}\mathbf{e}_x + E_{0y}\mathbf{e}_y,$$

where E_{0x} is the x component (x polarization) of the electric field and E_{0y} is the y polarization. The vectorial representation of $\mathbf{E}_0$ distinguishes this solution of the wave equation from scalar solutions such as the propagation of acoustic waves, where the amplitude of the wave is a scalar quantity but the propagation is still represented by a vector. To further simplify our analysis, it may be assumed that the electromagnetic wave is linearly polarized – that is, of the two possible polarizations, only one component (e.g. E_{0x}) exists. Although not explicitly presented, a magnetic polarization H_{0y} exists simultaneously with E_{0x}.

When discussing the solution of the wave equation, it was shown that the time-dependent part of the solution describes an oscillation that can be observed by a stationary observer, while the spatial part represents a sinusoidal

Figure 4.4 Surfer riding on the crest of an ocean wave. The velocity of this surfer while riding the crest is the phase velocity of that wave.

distribution of the field vector that would be recorded by "freezing" time – for example, by an instantaneous snapshot (Figure 4.1). However, both of these parts conceal an important element of the waves: their propagation. To witness that propagation, an observer must be moving with the wave in both time and space. An excellent illustration of such motion is the surfer who rides on the crest of an ocean wave (Figure 4.4). To move with the wave, the surfer struggles to maintain his position on the crest. As long as he is successful, he can claim that his velocity is identical to the propagation velocity of that wave. Maintaining a fixed position relative to the crest amounts to keeping a constant phase in (4.14), that is, holding the exponential term constant by the following condition:

$$\omega t \mp k_0 z = \text{constant}. \tag{4.15}$$

Thus, the velocity of an observer following one point, or a phase, along the propagation of the wave is the *phase velocity*. The phase velocity can be readily found from the derivative of (4.15):

$$v_p = \pm\frac{dz}{dt} = \pm\frac{\omega}{k_0} = \pm\frac{1}{\sqrt{\mu_0 \epsilon_0}}. \tag{4.16}$$

A positive v_p corresponds to the negative sign in (4.15) and in the exponent of (4.14), and represents propagation in the positive direction of the z axis. A negative v_p therefore represents propagation in the reverse direction. These propagation modes are called, accordingly, the forward and backward propagation waves. Since coordinate axes are usually selected to point in the direction of the propagation, most solutions must include only the forward propagation. The solution for v_p includes the two universal coefficients μ_0 and ϵ_0, so the numerical value of v_p in free space is also a universal coefficient. This is c_0, *the speed of light in free space*:

$$v_p = c_0 = \frac{1}{\sqrt{\mu_0 \epsilon_0}} = \frac{1}{\sqrt{(4\pi \times 10^{-7})(8.854 \times 10^{-12})}} = 2.998 \times 10^8 \text{ m/s}. \quad (4.17)$$

Although the theory of relativity will not be discussed here in any detail, one of its results was to show that neither energy nor matter could propagate, relative to any coordinate system, faster than c_0. To determine if the phase velocity obeys this requirement for propagation through other media, the solution of the wave equation is repeated using new material coefficients that represent the medium in question. Such analysis may also provide an insight into the properties of optical media.

4.6 One-Dimensional Propagation in a Nonlossy Isotropic Medium

The principle that the speed of light is a variable depending on the material properties of the medium through which it travels has long been recognized and has become an integral part of the theories of geometrical optics. However, it is only with the theories of physical optics that this principle can be evaluated directly from measurable material parameters. Once again, such derivation requires solving the wave equation (eqn. 4.8), with this modification: the parameters that describe free space are replaced with parameters that specify the medium. Three parameters need to be considered: the conductivity σ, the dielectric constant ϵ, and the magnetic permeability μ.

In the present solution, the medium is again assumed to be nonconducting (i.e., $\sigma = 0$). Although the primary result of this choice is to specify the medium as a dielectric, it will be seen in Section 4.9 that the attenuation, or energy loss, for the propagation of radiation through a medium increases with σ. Therefore, by the selection of $\sigma = 0$, the medium is generally characterized as *nonlossy*: radiation can propagate through it without any attenuation.

The second material parameter to be specified is the dielectric constant. In the derivation of the wave equation, the medium was assumed to be homogeneous, that is, ϵ was uniform for the entire medium. However, here the choice of ϵ is restricted even further by requiring that the medium be isotropic. With that restriction, a wave propagating in the z direction with the electric field linearly polarized in the x direction ($E_{0x} \neq 0$, $E_{0y} = 0$) will be subject to the same value of ϵ as a wave linearly polarized in the y direction ($E_{0x} = 0$, $E_{0y} \neq 0$). More generally, the condition of isotropic material implies that ϵ is identical for all polarizations and directions of propagation. When this requirement is met, the solution of the propagation of a plane wave requires consideration of only one propagation component and one polarization vector, even though the calculated phase velocity can represent propagation in any direction. Finally, for the magnetic permeability it is again assumed that $\mu = \mu_0$.

When these parameters are introduced into the wave equation, a solution in the form of (4.11) is obtained. The procedure for solution is identical to that for propagation in free space. But with a new value for ϵ, the vector of radiation propagating in the z direction becomes

$$\mathbf{k} = \omega\sqrt{\mu_0 \epsilon}\,\mathbf{e}_z.$$

Thus, the magnitude of the propagation vector in a nonlossy medium is different from its magnitude in free space (eqn. 4.13). On the other hand, the oscillation frequency ω appearing in the solution for the propagation in free space (eqn. 4.11) is independent of the material properties and therefore remains unchanged.

Using the same analysis that led to the derivation of phase velocity in free space, it is found that the phase velocity c in an isotropic nonlossy medium is

$$v_p = c = \frac{1}{\sqrt{\mu_0 \epsilon}} = \frac{1}{\sqrt{\mu_0 \epsilon_0}}\sqrt{\frac{\epsilon_0}{\epsilon}} = \frac{c_0}{n}, \tag{4.18}$$

where n is a parameter defined by

$$n = \sqrt{\frac{\epsilon}{\epsilon_0}}. \tag{4.19}$$

With this definition of n together with (3.22) and (3.23), the equation for the propagation vector can be rewritten in the following form:

$$\mathbf{k} = \frac{\omega n}{c_0}\mathbf{e}_z = \frac{\omega}{c}\mathbf{e}_z = \frac{2\pi n}{\lambda_0}\mathbf{e}_z = \frac{2\pi}{\lambda}\mathbf{e}_z, \tag{4.20}$$

where $\lambda = \lambda_0/n$ is the wavelength in a medium with an index of refraction n. This compact form of the propagation vector is convenient because it includes only one dimensionless material property (n) instead of the dimensional parameters ϵ and μ.

Equation (4.18) defines the phase velocity in any nonlossy isotropic medium relative to its velocity in free space. This equation is identical to (2.5), where the index of refraction was defined as a factor by which the speed of light in free space was to be divided in order to obtain the speed of light in a specified medium. Therefore, the parameter n defined by (4.19) must be the index of refraction of the medium as defined by geometrical optics.

The result presented in (4.19) shows that the index of refraction, which by itself is a parameter of the medium, is derived from another fundamental parameter ϵ. Recall that for isotropic, nonlossy media, the dielectric coefficient ϵ was defined, using the electric susceptibility (eqn. 3.20), by

$$\epsilon = \epsilon_0(1+\chi).$$

Therefore, the index of refraction depends also on the electric susceptibility χ of the medium, through which (see eqn. 3.19) there is also a dependence between the index of refraction and the polarization vector. Although for most optical applications χ is positive, it may generally have any value, positive or negative (see Problem 4.2). Thus, when $\chi > 0$, $\epsilon/\epsilon_0 > 1$ and $n > 1$. However, when $\chi < 0$, the ratio between the dielectric constants is less than unity and hence (eqn. 4.19) $n < 1$. This case is of particular interest since it implies, through (4.18), that the phase velocity in a medium where $\chi < 1$ exceeds the phase velocity

in free space! This, of course, would appear to violate the theory of relativity. However, before we determine whether a phase velocity higher than c_0 violates the theory of relativity, we would like to find under what conditions χ will be forced below unity.

For a simple approach that relates χ with an identifiable material behavior, recall that the polarization vector of a medium is proportional to the incident electric field (eqn. 3.19) and that χ is part of the proportionality coefficient. Thus, as χ inceases, the medium becomes more polarizable even if the magnitude of the polarizing field remains unchanged. To illustrate this, each dipole within the medium may be represented by a harmonic oscillator in which the restoring spring constant q may depend on coulombic forces or on the molecular or atomic properties of the dipole. The driving force of the oscillator, $e\mathbf{E}$, is a coulombic force obtained by the multiplication of the polarizing field with the charge of an electron. For an oscillating mass m and for a damping coefficient γ, the separation r between the dipole charges can be calculated from the following equation:

$$m\ddot{r}+\gamma\dot{r}+qr=e\mathbf{E}. \tag{4.21}$$

Thus, when electromagnetic radiation passes through a medium composed of such oscillators, the forced oscillation is at the field frequency; that is, the separation r between the charges must be an oscillating term. After finding the charge separation within individual oscillators, the polarization vector of the forced oscillation can be calculated from (3.18). Clearly, when the driving frequency coincides with a resonance frequency of the oscillating dipole, the polarization vector rapidly increases (or decreases) without accompanying changes in the amplitude of the incident field. Any variation in the polarization vector, regardless of variations in the incident field amplitude, can take place only by variations in χ (eqn. 3.19). Thus, if the driving frequency coincides with an internal resonance of the medium then the electric susceptibility, and accordingly the index of refraction, undergo extensive changes. Terms describing the *dispersion* – that is, the variation of the index of refraction with the incident frequency – have been derived (Born and Wolf 1975) for a damped harmonic oscillator in a gaseous medium driven by a coulombic force induced by an incident field $\mathbf{E}'=\text{Re}[\mathbf{E}_0'e^{i\omega t}]$. Normally, the density of the dipoles is sufficiently large to perturb the incident field amplitude. Therefore, the field amplitude at any dipole within the medium is somewhat different from the incident amplitude, and the driving field that must be used is $\mathbf{E}'\neq\mathbf{E}$. The charge separation r for a dipole with only one resonant frequency $\omega_0=\sqrt{q/m}$ driven by the field $\mathbf{E}'$ is (Born and Wolf 1975, p. 93)

$$r=\frac{e\mathbf{E}'}{m(\omega_0^2-\omega^2)-i\gamma\omega}$$

(see Problem 4.3). Using (3.18) and (3.20) together with the equation for the index of refraction (eqn. 4.19), a formula for the dispersion can be obtained:

$$n^2-1 \approx \frac{2A}{m(\omega_0^2-\omega^2)-i\gamma\omega},$$

where A is a lump coefficient that depends on the density of the medium and the dipole properties. At low densities and when $n \to 1$, the term n^2-1 can be replaced by $2(n-1)$, thereby reducing the dispersion equation to the following simplified form:

$$n = 1+\frac{A}{m(\omega_0^2-\omega^2)-i\gamma\omega}. \tag{4.22}$$

Note that the index of refraction is a complex variable. Typically the resonance frequency $\omega_0 > 10^{14}$ Hz. Therefore, when ω is far from ω_0, the imaginary term is negligible relative to the real term and the index of refraction can be approximated by a real number. However, as ω approaches ω_0, the real term decreases relative to the imaginary term in the denominator, which can no longer be neglected. Although not yet discussed (see Section 4.9), the imaginary part represents attenuation of the radiation passing through the medium; hence this term is necessary to evaluate such losses. However, even near resonance, the real part of n still represents the ratio between the speed of light in the medium and the speed of light in free space. Figure 4.5 presents a typical dispersion curve of a medium with only one resonance frequency. Although all media have numerous resonance frequencies, they are sufficiently separated from each other that presentation of the dispersion by one resonance frequency does not diminish the generality of the analysis. The index of refraction is seen to increase slowly with frequency at both sides of the resonance frequency. This trend is interrupted only within a narrow frequency range around ω_0, where n decreases steeply with frequency. Since this behavior deviates from the normal trend it is known as the *anomalous dispersion*. Since the anomalous dispersion occurs when $\omega \approx \omega_0$, it coincides with the frequency range where attenuation

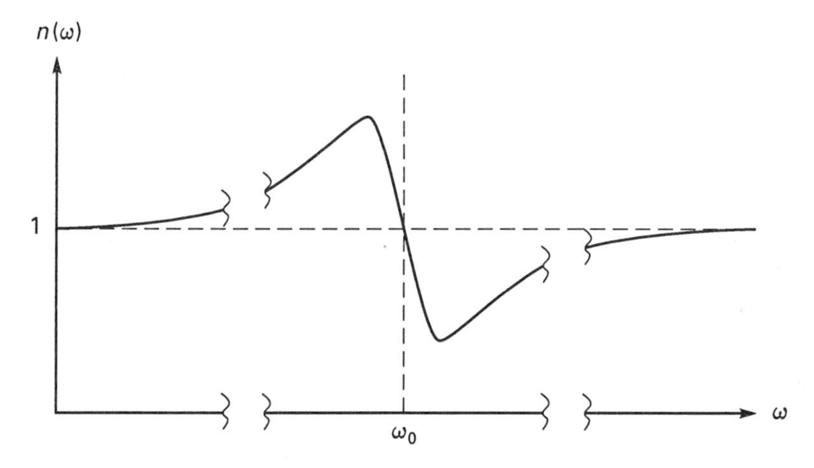

Figure 4.5 Typical dispersion curve, showing a region of anomalous dispersion. Most substances have several such regions coinciding with their absorption lines.

by absorption is high (Born and Wolf 1975, p. 93). The frequency range where absorption is high is an important spectroscopic parameter of the medium, and is often called the *absorption line*. The location and range of the anomalous dispersion is likewise an important characteristic of the medium.

For most optical materials (i.e. glass and quartz) and at optical wavelengths, the index of refraction $n > 1$. Hence $v_p < c_0$, as predicted. However, as is evident from (4.22) and Figure 4.5, n may be less than unity in the vicinity of a resonance frequency, thereby forcing the phase velocity beyond the phase velocity in free space. This may appear to be a violation of the relativity principle. The following example demonstrates that, although the phase velocity of any wave (not necessarily electromagnetic) may be faster than the speed of light in free space, no energy or matter propagates with it and so the theory of relativity is not violated.

Example 4.1 In a hypothetical experiment, 19 students are positioned in equally spaced stations along the highway connecting Sun City with Moon City. Each student is given a frictionless pendulum with a period of $T = 0.5$ s and an accurate watch. Before the experiment begins, the students are instructed to start the oscillation of their pendulums at the same amplitude but at different times. The student at Sun City is instructed to start first at $t = t_0$, the student next to him at $t_0 + T/12$, the next student at $t_0 + 2T/12$, and so on until the last student at Moon City starts her pendulum. After the students are assigned to their stations and the instructions are handed out, no communication is allowed between the students. When the time is $t = t_0 + T$, the pendulum at Sun City is just about to begin a new cycle, the pendulum of the thirteenth student is just about to start, and the pendulum in Moon City is still at rest. An aerial photograph of all the stations (see Figure 4.6) obtained at time $t = t_0 + 3T/2$ shows the momentary positions and the velocity directions of all the pendulums. Both the positions and velocities of the pendulums are harmonically distributed. Since the pendulums are frictionless, this wave pattern can oscillate indefinitely at a frequency of 2 Hz. If the distance between the two cities is 4.5×10^8 m, find the phase velocity of this pendulum wave.

Solution Although this oscillation has some of the characteristics of a wave, it does not have an identifiable source. Even though the instructions to the students were designed to give the aerial photographer the impression that a wave is propagating from Sun City to Moon City, the real trigger that started the chain of events is not a disturbance at Sun City but rather the clocks in each of the independent stations. Since the students could not communicate with each other, the apparent coordination between all the stations is the result of successful planning and not of a real event at Sun City.

To calculate the phase velocity of this oscillation, we need only find the velocity that an observer at A must have in order to cover the distance L_{AB} between stations A and B within the time of one period. For the parameters of this problem,

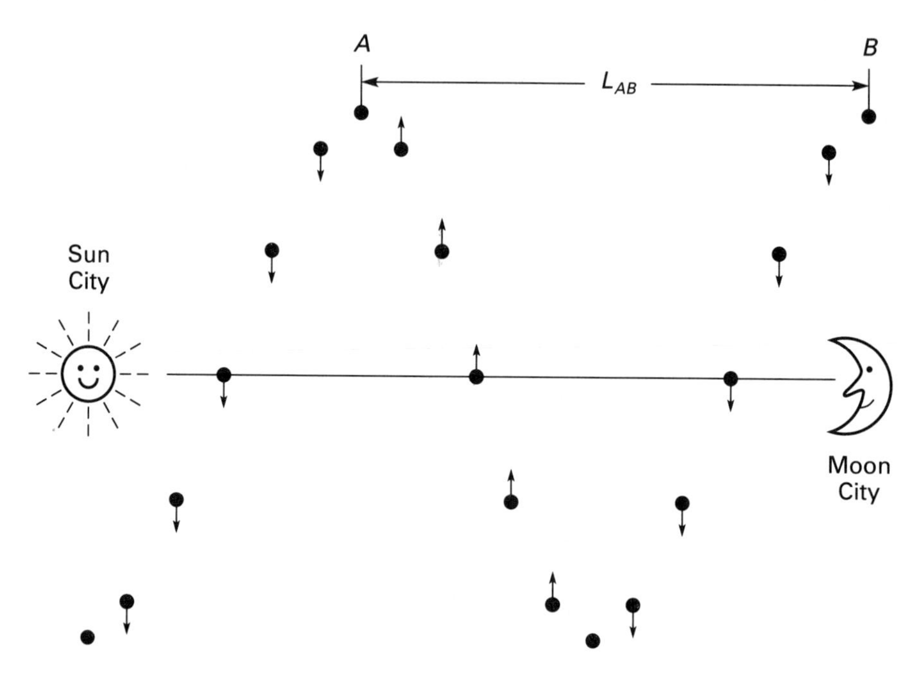

Figure 4.6 Hypothetical experiment, consisting of an array of synchronized oscillators, that was designed to illustrate the meaning of phase velocity.

$$L_{AB} = \frac{2}{3}(4.5 \times 10^8)\ \text{m},$$

$$v_p = \frac{L_{AB}}{0.5} = 6 \times 10^8\ \text{m/s} = 2c_0.$$

Thus, to arrive on time to witness a new peak amplitude at B, the observer needs to move at twice the speed of light in free space. While we know that it is impossible to send a message or a messenger from A to B at such a speed, no fundamental physical law was violated by coordinating the oscillation of the pendulums. This result simply demonstrates that, by judicial selection of the oscillation frequency and the apparent wavelength for the oscillation, any phase velocity could be obtained. The apparent motion of the phase is merely an illusion. It does not convey energy, matter, or information. A more valuable parameter that describes the speed of light must be the velocity at which an electric or magnetic perturbation created at one point propagates toward another point. Because all modes of information broadcasting require some modulation of a fundamental frequency, we will need to analyze first how frequencies of waves are superposed. ■

4.7 Superposition of Frequencies

The difficulty we encountered in calculating the speed of light resulted from a flawed assumption: radiation was viewed as a *monochromatic* wave, that is,

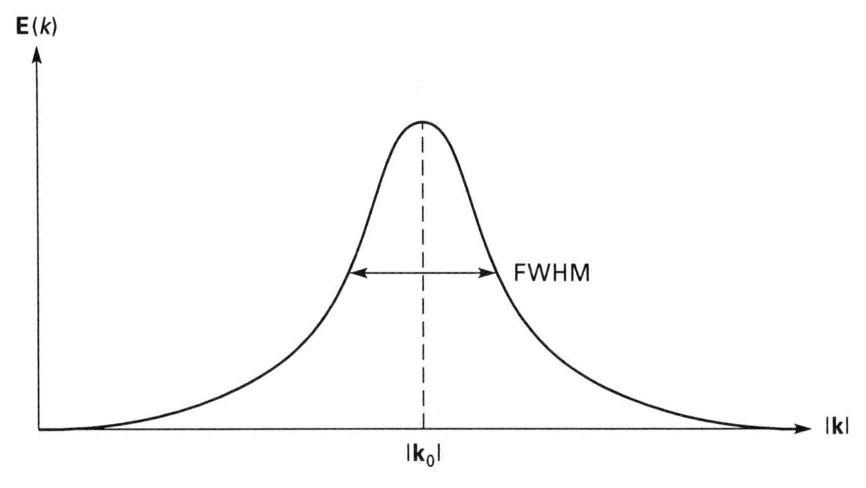

Figure 4.7 Spectral distribution of a narrowband wave with finite spatial extent. The frequency is represented by the magnitude of the propagation vector $|\mathbf{k}| = \omega/c$.

as a wave containing only a single frequency. Although in many applications – particularly communication, spectroscopy, and certain techniques of velocity measurement – it is desirable to have truly monochromatic radiation, we find that fundamental physical laws prevent us from obtaining it. Invariably, even the purest monochromatic radiation is made up of a range of frequencies that are usually clustered around a primary frequency, which is usually the center frequency. Figure 4.7 presents such a distribution of frequencies, where $|\mathbf{k}|$ is used instead of ω to describe the spatial distribution of frequencies. This distribution, which often appears in spectroscopy as an absorption or emission line, can be characterized by two parameters: the propagation vector $|\mathbf{k}_0|$ at which the electric field amplitude $\mathbf{E}(k)$ reaches a maximum, and the bandwidth (or linewidth), which is usually measured by the full width at half maximum (FWHM). For symmetric distribution, $|\mathbf{k}_0|$ is located at the center of the line. From (4.20) we have $|\mathbf{k}| = \omega n(\omega)/c_0$, so the frequency dependence of $\mathbf{k}$ also includes the effects of dispersion; therefore, $\mathbf{k}$ varies nonlinearly with frequency.

Several mechanisms can combine to broaden the spectral bandwidth of radiation. The theory of quantum mechanics has developed satisfactory explanations for most line-broadening phenomena (see Sections 10.3 and 10.7). They include the effects of atomic and molecular collisions, the effects of their motion, interaction of electromagnetic waves with acoustic waves or with other electromagnetic waves, and more. However, the most fundamental reason for broadening is the limited duration of all emission processes (see Section 10.3). Invariably, all sources have a limited time to complete emission: they begin emission after being excited energetically and cease emitting when their energy is dispersed. Even when emission appears to last indefinitely, it is marred by abrupt interruptions such as brief pauses or sudden changes in ϕ. These are the interruptions that limit the coherence length of radiation. Thus, even for an extremely coherent source with a coherence length of 10 m, the time between two interrupting events is $10/c_0 = 33.3$ ns. Mathematical description of the effect of such perturbations of the emission process on the frequency content of the radiation requires Fourier analysis of the spectral distribution of radiation.

Radiation, like many other oscillating physical quantities, can be described in two independent domains – for example, the time domain and the frequency domain. Thus, in the time domain, radiation is described by the variation (with time) of the electric field amplitude, whereas in the frequency domain it is described by a spectrum similar to Figure 4.7. The description of these functions is complete in either domain. However, depending on the application, one presentation may be preferred over the other. For measuring the duration of a laser pulse we would prefer a presentation in the time domain, whereas for spectroscopy we would prefer a presentation in the frequency domain. Fourier integrals, or Fourier transforms, are used to transform the distribution from the t (time) domain to the ω (frequency) domain. Alternatively, for a one-dimensional (1-D) propagation, since the distance traveled by radiation is proportional to time, any temporal modulation also corresponds to a longitudinal modulation (i.e., modulation along the beam path). Thus, for 1-D propagation, transformation from the z to the $|\mathbf{k}|$ domain is analogous to the transformation from t to ω domains, with the exception that it also includes the effect of dispersion. Therefore, for propagation through dispersive media, this is expected to be a more complete presentation.

For 1-D propagation, the equations for the z–$|\mathbf{k}|$ Fourier transform include two transformations: one from the z domain to the $|\mathbf{k}|$ domain, and the other back to the z domain (Jackson 1975). For simplicity, we abandon here the vectorial notation for the propagation vector. However, we still preserve the vectorial notation $\mathbf{E}$ to indicate that the electric field may represent one of two polarizations. For a known spectral distribution of the incident radiation, the transformation from the k domain to the z domain is given by

$$\mathbf{E}(z) = \frac{1}{\sqrt{2\pi}} \int_{-\infty}^{\infty} \mathbf{E}(k) e^{ikz} dk, \tag{4.23}$$

where $\mathbf{E}(z)$ is the electric field at any point z along the propagation axis and $\mathbf{E}(k)$ is the amplitude of a component of the electric field with a propagation vector k. Conversely, when the spatial distribution of the incident field is known, the transformation from the z to the k domain becomes

$$\mathbf{E}(k) = \frac{1}{\sqrt{2\pi}} \int_{-\infty}^{\infty} \mathbf{E}(z) e^{-ikz} dz. \tag{4.24}$$

The time dependence of the electromagnetic wave $e^{i\omega t}$ was implied in both transformations.

The spectral content of a sinusoidal wave with an infinite extent along the z axis and without any perturbation (i.e., with an infinite coherence length) is only one frequency. Thus, the z-to-k Fourier transform must result with a single spatial frequency component k_0 in the k domain, represented by the $\delta(k-k_0)$ function. This result is intuitive, but it can also be obtained directly by integration of (4.24) using $\mathbf{E}(z)/\mathbf{E}_0 = e^{ik_0 z}$:

$$\frac{1}{2\pi}\int_{-\infty}^{\infty} e^{-i(k-k_0)z}\,dz = \delta(k-k_0). \tag{4.25}$$

Similarly, when the spectral content of the radiation is infinite, the longitudinal modulation according to (4.23) must be an infinitesimally narrow pulse propagating along the z axis at the speed of light. Of course, the duration of the spatially narrow pulse approaches zero. Mathematically, this pulse may be described as

$$\frac{1}{2\pi}\int_{-\infty}^{\infty} e^{ik(z-z_0)}\,dk = \delta(z-z_0). \tag{4.26}$$

Equations (4.25) and (4.26) describe two extreme cases of the longitudinal modulation of radiation. At one extreme, radiation is emitted continuously and (although physically impossible) the coherence length is infinite. This requires only one frequency component in the k domain. At the other extreme, radiation is shaped into an infinitesimally brief pulse that requires an infinitely wide spectral content for its formation. This extreme is being approached by the use of subnanosecond laser pulses. As an illustration, a laser pulse with a duration of 1 ps appears as a burst of light that occupies 0.3 mm along its propagation path. This spatially (as well as temporally) narrow pulse travels through free space at the speed of light. To form such a pulse, the source must emit at a range of frequencies $\Delta\nu > 1.6\times 10^{11}$ Hz. Further shortening this pulse would require further spectral broadening. From these observations at both ends of the modulation range, it may appear that the product of the pulse duration and its spectral width (or its longitudinal stretch and the range of the k vectors) is a constant parameter. This may be shown directly by Fourier transformations, where the product of Δk and Δz - which are the rms (root-mean-square) deviations from the average values of z and k (see Figure 4.8) - must meet the following condition (Jackson 1975):

$$\Delta z \Delta k \geq 1/2. \tag{4.27}$$

Although both Δz and Δk are defined for the distribution function of the irradiance, we will use $\mathbf{E}(k)$ and $\mathbf{E}(z)$ to illustrate them until we discuss the irradiance (Section 4.10). In practical applications, the value of this product, which depends on the shape of the pulse (see Problem 4.4), exceeds the projection of (4.27). This result is similar to the Heisenberg uncertainty principle (see Section 8.6).

4.8 The Group Velocity

In the previous section we saw that radiation propagates in packets, or groups of frequencies. The spectral shape (distribution of frequencies) within these packets depends on the emission features of the source, and can be characterized by the bandwidth and by the centerline frequency ω_0 or the propagation vector $k_0 = \omega_0 n(\omega)/c_0$. From the Fourier analysis we also saw that the longitudinal (temporal) modulation of the radiation is determined by the spectral distribution, and vice versa. Therefore, when radiation is modulated by

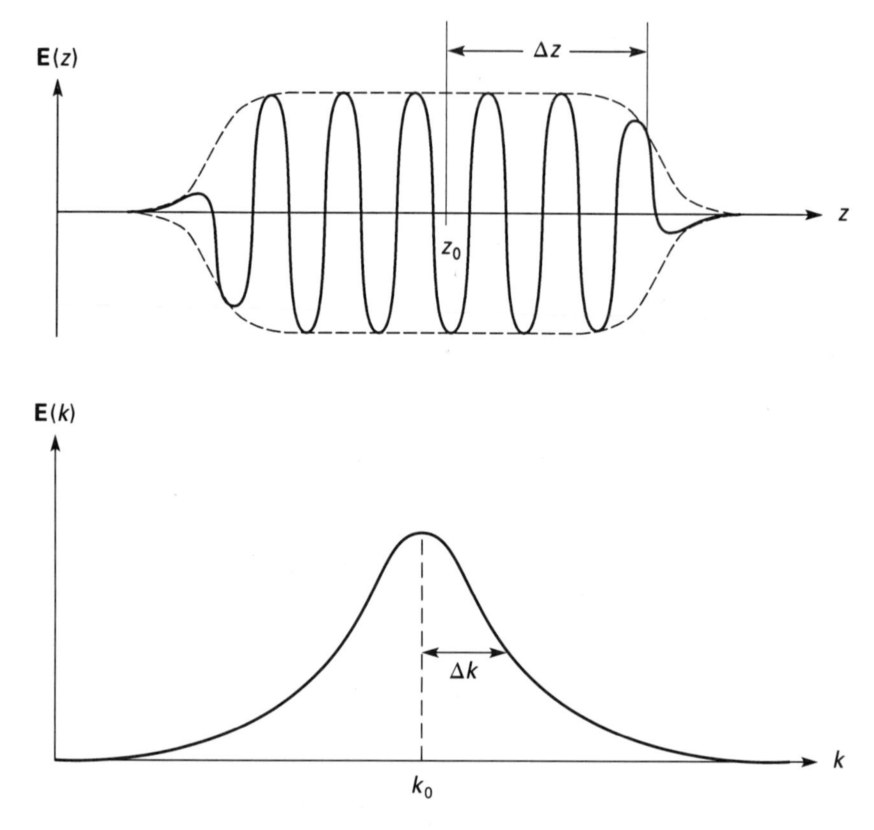

Figure 4.8 Correspondence between the spatial and spectral distribution of a pulsed wave.

varying its temporal or spatial distribution, the spectral content changes to account for the new spatial distribution. Of course, if we expect the modulation to propagate undistorted then the spectral packet (the group of frequencies) must also propagate undistorted. But because certain frequencies travel faster through dispersive media than others, the shape of the packets in the temporal or spatial domain can be distorted, particularly following propagation through long optical paths or through highly dispersive media. When the shape of the group is distorted, many of its essential characteristics may be lost. An important example is the propagation of brief laser pulses. Such pulses are used for spectroscopic studies, material processing, and nonlinear interaction with gases and crystals; more importantly, they present the potential for high-rate communication. By avoiding distortion, a 1-ps pulse can be used for a communication rate of up to 10^{12} Hz, which is a thousand times faster than the current rate of $\sim 10^9$ Hz. Thus, as the following example demonstrates, if a wave packet is initially shaped as a short pulse then the dispersion of the carrier medium must be kept small to avoid spreading the pulse.

Example 4.2 A high-quality fiber is used to transmit a train of laser pulses, each 100 ps long, at a rate of 1 GHz. As they exit the fiber, the pulses are

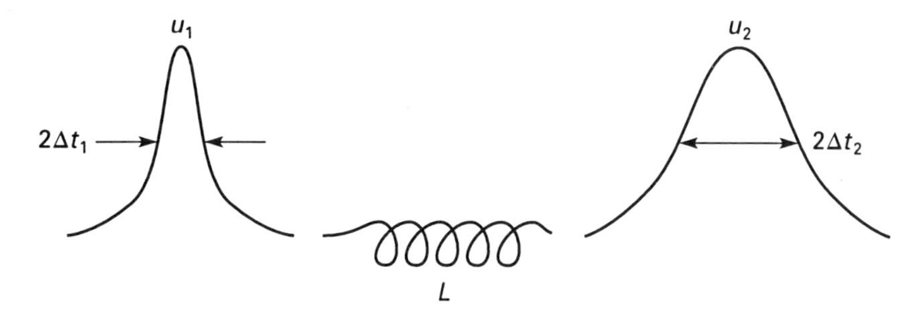

Figure 4.9 Broadening of a short pulse induced by propagation through a long dispersive optical fiber.

passed through an optical device (a second harmonic generator or SHG) that converts their wavelength from 1,064 nm to 532 nm. The efficiency of this conversion process – that is, the percentage of the incident energy that is converted to 532 nm – increases quadratically with the incident power. The specified dispersion factor of the fiber is $D = 1.5$ ps/nm-km. Determine the maximum length of the fiber that can be used before the conversion efficiency by the SHG decreases by 50%.

Solution Figure 4.9 illustrates the temporal distribution of a laser pulse with an energy of u_1 entering the fiber. Note that, according to the definition used to derive (4.27), all deviations from the mean are expressed by the rms. Therefore the pulse duration is $2\Delta t_1 = 100$ ps. When the pulse emerges at the other end of the fiber, its duration has extended to $2\Delta t_2$ and its energy has diminished by transmission losses to u_2. To determine the power, we need a detailed description of its temporal distribution. This distribution is usually a complex function that cannot be described in simple mathematical terms. Instead, the power can be approximated by dividing the total pulse energy by its full width at half maximum as follows:

$$P_1 = \frac{u_1}{2\Delta t_1}.$$

Similarly, the power at the exit of the fiber is

$$P_2 = \frac{u_2}{2\Delta t_2}.$$

In order to determine the effect of the fiber dispersion on the pulse duration, we must find its spectral content. From (4.27) we see that the length of the pulse, the longitudinal separation between the leading and the trailing edges, is limited by the spectral content of the pulse. For propagation in free space ahead of the fiber, the limiting spectral content of the pulse can then be determined from its duration using the following modification of (4.27):

$$\Delta z \Delta k = (c_0 \Delta t)\left(\frac{\Delta\omega}{c_0}\right) = \Delta t \Delta\omega \geq \frac{1}{2}.$$

Thus, for a pulse duration of 100 ps, the frequency content must be

$$\Delta\omega \geq \frac{1}{2\Delta t} = 10^{10} \text{ rad/s.}$$

For a nominal wavelength of $\lambda = 1{,}064$ nm, the incident frequency (eqns. 3.22 and 3.23) is 1.772×10^{15} rad/s. The expected range of frequencies, $\pm 10^{10}$ rad/s, is split evenly between the two sides of the nominal frequency. Thus, the range of wavelengths covered by the pulse (eqns. 3.22 and 3.23) is

$$\lambda_1 = 1{,}064 \pm 0.006 \text{ nm.}$$

To assure that the conversion efficiency of the SHG remains at least half that of the undispersed pulse, the power at the exit of the fiber must be maintained at a level above $1/\sqrt{2}$ of the incident power. Assuming that energy losses for transmission through the fiber are negligible, this condition determines the longest pulse duration at the exit of the fiber: $\Delta t_2 < \sqrt{2}\Delta t_1$. Thus, the maximum allowed length of the fiber is

$$L = \frac{2\Delta t_2 - 2\Delta t_1}{D \times \Delta\lambda} = \frac{100(\sqrt{2}-1)}{1.5 \times 2 \times 0.006} = 2{,}301 \text{ km.}$$

That such an exceptionally long fiber may be used without adverse dispersion-related effects indicates that dispersion may often have little effect on the spectral distribution, particularly when the duration of the incident pulse is longer than the 100-ps pulse in this example. However, as pulses become shorter and shorter and the distance for their transmission through dispersive fibers increases, their distortion may reduce the allowable pulse rate or the peak power. ■

This example demonstrated that the duration of a short pulse may be extended by propagation through dispersive media. Although energy is preserved, its distribution is spread over a longer duration, thereby reducing its peak power. On the other hand, with low dispersion, propagation through long distances is possible without significant distortion. To preserve its original spatial and temporal shape, all frequency components of a wavepacket must emerge from a dispersive medium in the same order they entered it, and with the same temporal and spatial distribution. The speed of propagation of the entire packet must then be the propagation velocity of the pulse. This is also the propagation velocity of information transmission or the velocity of propagation of a radio broadcast. To determine the propagation velocity of a packet, the *group velocity,* we must first describe the entire packet using its spectral distribution and follow the motion of each frequency component through time and space.

When a group of waves is traveling along the z axis, the total electric field at any point can be calculated from the Fourier transformation of the spectral distribution, similar to the integration in (4.23). To determine the velocity of that group, its position both in space and time must be tracked. Hence (4.23) is modified to include both the temporal and spatial distributions in the following integration:

$$\mathbf{E}(z,t)=\frac{1}{\sqrt{2\pi}}\int_{-\infty}^{\infty}\mathbf{E}(k)e^{-i[\omega(k)t-kz]}\,dk. \tag{4.28}$$

As in (4.23), $\mathbf{E}(k)$ describes the spectral distribution of the propagating radiation. For simplicity we assume that it is time-independent. That is, the shape of the spectral line is unchanged with time and the only time dependence of the field is due to the harmonic oscillation, which in turn may be perturbed by dispersion. To express the effect of dispersion on propagation in the time domain, ω was shown in the integral to be a function of k. Although dispersion is defined as the variation of either n of k with ω, introduction of the dispersion through ω will simplify our subsequent analysis. When the dispersion is small (i.e., when the changes in n and k with ω are small), the frequency distribution around a center frequency ω_c can be presented by the following expansion:

$$\omega(k)=\omega_c+\left.\frac{d\omega}{dk}\right|_c(k-k_c)+\cdots, \tag{4.29}$$

where k_c is the propagation vector associated with ω_c. Introducing this expansion into (4.28), for the propagating wavepacket we have

$$\mathbf{E}(z,t)=\frac{\exp\left\{-i\left[\omega_c-k_c\left.\frac{d\omega}{dk}\right|_c\right]t\right\}}{\sqrt{2\pi}}\int_{-\infty}^{\infty}\mathbf{E}(k)\exp\left\{-i\left[\left.\frac{d\omega}{dk}\right|_c t-z\right]k\right\}dk. \tag{4.30}$$

The equation now contains an amplitude term and a phase term. Since the amplitude term $\mathbf{E}(k)$ is time-independent and the bracketed phase term outside the integral is independent of k, the bracketed phase term inside the integral must remain constant while propagating through a dispersive medium if the spectral distribution is to remain unchanged. Thus,

$$\left.\frac{d\omega}{dk}\right|_c t-z=\text{constant}.$$

This is similar to the formula required for calculating the phase velocity (eqn. 4.15). However, here the requirement amounts to a condition that the entire wavepacket remain undistorted. From the differential of this condition we find the following group velocity term v_g for propagation through media with small dispersion:

$$v_g=\left.\frac{d\omega}{dk}\right|_c, \tag{4.31}$$

where the derivative is evaluated around ω_c.

For propagation through a nonlossy medium, the group velocity can be obtained from the direct derivative of the propagation vector (eqn. 4.20) as follows:

$$dk=\frac{n}{c_0}\,d\omega;$$

or, for $n = 1$,

$$v_g = \frac{d\omega}{dk_0} = c_0.$$

Thus, in free space the group velocity and the phase velocity are identical and both equal the speed of light. This is to be expected because there is no dispersion in free space; all frequency components in the wavepacket propagate at c_0 and hence the entire packet propagates at c_0 as well.

Although the group velocity coincides with the phase velocity when dispersion vanishes, all media are dispersive. Therefore, these two velocities are not identical for any propagation that is not through free space. The relation between v_g and v_p can be computed from the derivative of (4.16), where k is substituted for k_0:

$$v_g = \frac{d(kv_p)}{dk} = v_p + k\frac{dv_p}{dk}. \tag{4.32}$$

To illustrate the variation of v_g with dispersion, consider the dispersion curve (Figure 4.5). The index of refraction is seen to increase slowly with frequency, except where anomalous dispersion occurs around ω_0. There n is decreasing very rapidly with frequency. Away from the point of anomalous dispersion, the linear expansion of (4.29) and the result for v_g (eqn. 4.31) are valid. Therefore, outside the range of anomalous dispersion, the derivative term in (4.32) is negative (see Problem 4.5) and $v_g < v_p$ as expected for all values of n. This result is illustrated in Figure 4.10, where the variation of k with ω has been derived from the dispersion curve (Figure 4.5). At low and high frequencies, where $n \approx 1$, the slope of the curve and hence the group velocity approach c_0. As n increases, the slope of the curve and the group velocity both decrease. This is true as long as $\omega \ll \omega_0$ or $\omega \gg \omega_0$. Thus, away from ω_0, the group velocity

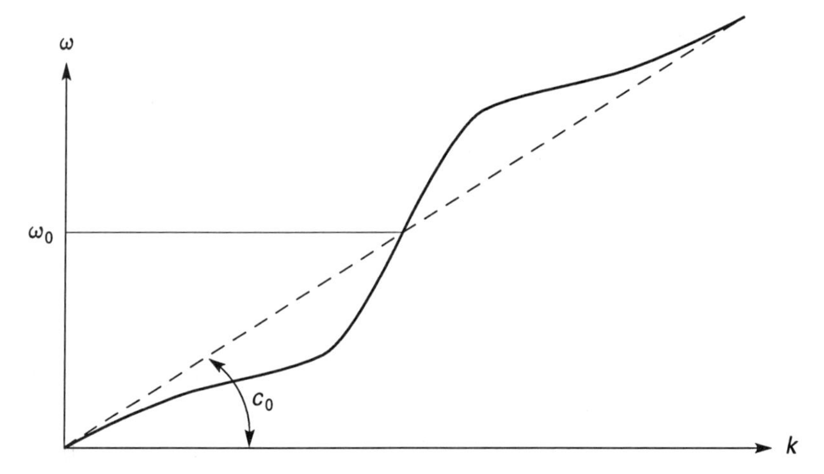

Figure 4.10 Variation of k with ω for propagation through a medium with anomalous dispersion.

does not exceed c_0. However, as ω approaches ω_0, the slope of the curve in Figure 4.10 increases sharply and the group velocity appears to exceed c_0. Although this may again seem to violate the theory of relativity, recall that the linear expansion of (4.29) was valid only when dispersion was small. For larger dispersion, higher-order terms had to be included in that expansion. Inclusion of these terms will once again result in a group velocity that is slower than c_0. Analysis of the group velocity for frequencies in the vicinity of the anomalous dispersion is beyond the scope of this book.

4.9 Propagation through Isotropic Lossy Media

Until now, the propagation of radiation was considered to be loss-free. With the exception of minimal reflection losses at interfaces, solutions did not include a loss term and therefore could be used to describe unlimited propagation. Although optical devices are expected to introduce negligible losses, in many engineering applications the losses cannot be fully eliminated and so gradually diminish the energy throughput, particularly when the optical path is long. In Section 4.6, where the index of refraction was shown to include an imaginary part (eqn. 4.22), the concept of attenuation in propagation was touched upon briefly. To illustrate the effect of such an imaginary term (which at times may be nonnegligible) on the propagation and how it contributes to the loss, we rewrite the solution of the wave equation (eqn. 4.11) for one-dimensional propagation along the z axis. The following result,

$$\mathbf{E}(z,t) = \text{Re}[\mathbf{E}_0 \exp\{-i(\omega t \mp kz)\}], \tag{4.33}$$

is similar to (4.14) (propagation in free space), with the exception of the propagation vector k which now represents propagation through the lossy medium. In (4.9) we saw that, for any medium with a dielectric constant ϵ and with a specific conductivity σ, the propagation vector is a complex variable. Note that the imaginary term introduced by σ is independent of the radiation frequency and exists even when ω approaches zero, whereas the imaginary term associated with the index of refraction (eqn. 4.22) is nonnegligible only near ω_0. Nevertheless, the consequences of these imaginary terms are identical and can be incorporated into the wave equation in a similar manner. To illustrate the effect of the imaginary term, we combine (4.18) and (4.19) with (4.9) to yield, for the propagation vector,

$$k = \frac{\omega n}{c_0}\sqrt{1 + \frac{i\sigma}{\omega\epsilon}}, \tag{4.34}$$

where n may also be a complex variable (eqn. 4.22). However, when the radiation frequency ω is far from any resonance frequency ω_0 or when $\omega \to 0$, only the conductivity contributes to the complex term of k. This situation is typical of propagation through conducting media (e.g., semiconductors or metals). For weakly conducting media we may expand the square-root term in (4.34) and retain the first-order term, and so obtain the following simplified equation:

$$k \approx \frac{\omega n}{c_0}\left[1+\frac{i\sigma}{2\omega\epsilon}\right]. \tag{4.35}$$

Substituting this result into (4.33) results in the following equation for propagation through a conducting medium:

$$\mathbf{E}(z,t) = \mathrm{Re}\left[\mathbf{E}_0 \exp\left\{-i\left(\omega t - \frac{\omega n}{c_0}z\right)\right\}\right] \exp\left\{-\frac{\sigma n}{2\epsilon c_0}z\right\}. \tag{4.36}$$

To simplify the interpretation of this result, only the propagation in the positive z direction was included by keeping the negative sign in front of the coordinate term. We see that, in addition to the oscillatory term, (4.36) carries an exponentially decaying term. A similar term (with a positive sign) will exist for a backward-propagating wave. However, for such propagation $z<0$ and the exponential term still represents a decay. Therefore, independently of the direction of propagation, the exponential term represents a field that diminishes along the propagation path. The rate of decay (in units of 1/length) is determined by the coefficient $\alpha = \sigma n/\epsilon c_0$. This is the *extinction* coefficient. When $\sigma \to 0$, the effect of this term remains significant only when z is large.

For strongly conducting media such as copper (for which $\sigma = 5.9\times10^7$ mho-m, versus $\sigma = 10^{-16}$ mho-m for quartz), the approximation of (4.35) may not be valid. Instead, it is assumed that the real and imaginary parts of k are approximately equal. Thus,

$$k \approx (1+i)\sqrt{\frac{\omega n^2\sigma}{2c_0^2\epsilon}}. \tag{4.37}$$

For these highly conducting media, the field decays to $1/e$ of its incident amplitude within a distance of $\delta = \sqrt{2c_0^2\epsilon/\omega n^2\sigma}$, where δ is the *skin depth*. For metallic media where $\mu \neq \mu_0$, the skin depth may also be written as $\delta = \sqrt{2/\mu\sigma\omega}$. For a copper surface illuminated by an incident frequency of 10^{15} Hz, $\delta = 4$ nm. Thus, the field of an incident visible light decays within a distance that is significantly shorter than one wavelength.

Although (4.36) was derived under the assumption that the complex part of the index of refraction is negligible, this may not be valid for some applications. Inspection of (4.22) shows that when $\omega \to \omega_0$ (i.e., when the radiation frequency approaches the resonance frequency of the oscillator), the index of refraction becomes imaginary. Thus, even if $\sigma = 0$, the propagation vector is a complex number when the radiation frequency coincides with the oscillator's resonance frequency. The frequency range (or absorption line) within which the complex term of the index of refraction contributes significantly to the attenuation of the propagating radiation is limited. However, owing to their dependence on molecular or atomic parameters, many such absorption lines may exist, some of them overlapping. In gases at atmospheric conditions, the typical width of an isolated line is 1 cm^{-1} (30 GHz). For visible light this corresponds to approximately 0.025 nm. The absorption lines of common molecules and all the atoms are well known and can be found in the literature (for atoms see e.g. Moore 1971). Thus, to avoid attenuation, the wavelength of the incident radiation must be selected to fall into a spectral range that is sufficiently removed from all absorption lines.

4.10 The Energy of Electromagnetic Waves

Although we recognized from the outset of this book that electromagnetic waves carry energy, we have not yet determined which of the field quantities contribute to that energy. Although electromagnetic energy terms can be derived directly from the rate of work per unit volume, $\mathbf{j}\cdot\mathbf{E}$, done by an electric field on charge elements, here we will use results derived elsewhere (Jackson 1975, pp. 236–8). However, to illustrate the role of the various terms in that solution, we will use the first law of thermodynamics formulated for a control volume (Fox and McDonald 1992, p. 152). In its simplest form,

$$\dot{Q}-\dot{W}=\frac{\partial}{\partial t}\int_{cv} u\,dV+\int_{cs} u\mathbf{V}\,d\mathbf{A},$$

the law describes the conservation of energy in its various modes, where $\dot{Q}$ and $\dot{W}$ are the rates of heat and work transferred across all the boundaries of the control volume and u is the specific internal energy (energy per unit volume). When the net transfer across the boundary is nonzero, the internal energy of the medium contained in the volume must change. This is expressed by the first and second terms on the right-hand side of the equation. The first term represents the rate of change of the internal energy within the control volume. The second term represents the energy flux across the boundaries of that volume carried by fluid flow at a velocity $\mathbf{V}$. With the use of the divergence theorem (Hildebrand 1962), the first law of thermodynamics can be reduced to the following differential form:

$$\dot{q}-\dot{w}=\frac{\partial u}{\partial t}+\nabla\cdot(u\mathbf{V}). \tag{4.38}$$

By analogy, each of the terms of (4.38) can be shown to correspond to a term representing either electrostatic or electromagnetic energy in a control volume (Jackson 1975, pp. 236–8). By this analogy, the rate of total work per unit volume done by the electric field on the charges is $\mathbf{j}\cdot\mathbf{E}$. Similarly, the first term on the right-hand side must represent the total rate of change of the electromagnetic energy stored within a unit volume. To complete this analogy, u must represent the stored electromagnetic energy density, which is

$$u=\frac{\epsilon}{2}\mathbf{E}^2+\frac{\mu_0}{2}\mathbf{H}^2 \tag{4.39}$$

(Jackson 1975). Finally, the last term in (4.38) represents energy flux across the boundary of the volume. This is the energy carried by an electromagnetic wave as it propagates. This energy flux is represented by *Poynting's vector,* $\mathbf{S}$, which is analogous to the $u\mathbf{V}$ term in (4.38). Poynting's vector may be expressed explicitly in terms of $\mathbf{E}$ and $\mathbf{H}$ as

$$\mathbf{S} = \mathbf{E} \times \mathbf{H} \tag{4.40}$$

(Jackson 1975). The units of $\mathbf{S}$ in the MKSA system are W/m^2 (note that in Gaussian units $\mathbf{S} = c/4\pi(\mathbf{E} \times \mathbf{H})$). The units of $\mathbf{S}$ are identical to the units of either the emittance (eqn. 1.1) or the irradiance (eqn. 1.2). Therefore, depending on the direction of this vector relative to the area element, we may interpret it as either the emittance or the irradiance. Combining these energy terms into an equation similar to (4.38), we obtain the equation for the conservation of electromagnetic energy:

$$-j \cdot \mathbf{E} = \frac{\partial u}{\partial t} + \nabla \cdot \mathbf{S}. \tag{4.41}$$

Note that the heat dissipation term is absent from (4.41). This implies that the otpical medium is adiabatic, as expected in most optical applications. In the absence of free electrostatic charges within a control volume (i.e., $\mathbf{j} = 0$) and if the internal energy is unchanged (i.e., when the absorption of radiation is negligible), the term $\nabla \cdot \mathbf{S}$ vanishes, meaning that the irradiance remains unchanged as it passes through this element.

Both (4.40) and (4.41) represent time-dependent energy functions. This dependence includes both the oscillation of the electromagnetic field as well as the relatively slower time-dependent variations of the amplitude. Included in this slower component are the temporal distribution of energy within a fast (e.g. 1-ps) laser pulse or the temporal fluctuations in the irradiance of a light bulb. All these variations are extremely slow relative to the time of one oscillation cycle of visible light, which is approximately 10^{-15} s. Thus, to eliminate the time dependence associated with the oscillation of the field, we may integrate $\mathbf{S}$ over a period T that is long relative to the time of one cycle while short relative to the characteristic time of the illumination. If we then use (4.12′) to replace $\mathbf{H}$ with $\mathbf{E}$ we have:

$$\mathbf{S} = \text{Re}\left[\frac{1}{T}\int_0^T \frac{\mathbf{E}^2}{\eta}\, dt\right]\mathbf{e}_s = \frac{1}{2\eta}\mathbf{E}_0 \cdot \mathbf{E}_0^* \mathbf{e}_s = \frac{1}{2\eta}|\mathbf{E}_0|^2 \mathbf{e}_s, \tag{4.42}$$

where the asterisk represents the complex conjugate and $\eta = \sqrt{\mu_0/\epsilon}$. Performing the integration in (4.42) introduces a factor of 1/2 on the right-hand side, so the irradiance can be evaluated directly from the field amplitude $\mathbf{E}_0$. Similar integrals can be obtained for the stored energy.

The unit vector $\mathbf{e}_s$ in (4.42) is pointing in the direction of Poynting's vector. For a TEM wave this coincides with the direction of $\mathbf{k}$. Therefore, for propagation through free space or in isotropic materials, energy and radiation propagate collinearly. This is expected, and is consistent with daily experience and intuition. However, some radiative modes are not necessarily TEM. Non-TEM modes may include surface waves, propagation through anisotropic materials, or propagation through media containing free charges. In all these cases the assumptions that led to the development of (4.4) break down. Thus, for non-TEM propagation, the energy flow and the propagation direction do not coincide. This causes the "walk-off" effect, which can be observed in the diminishing of the electromagnetic field and the irradiance.

Both (4.41) and (4.42) are vectorial and therefore describe not only the propagation direction but also the polarizations of the electric and magnetic fields. However, from the scalar product of (4.42), it is evident that the energy of one polarization of radiation is uncoupled from the energy of the other polarization. A similar observation can also be made for the stored energy (eqn. 4.39). In free propagation, all the energy can be carried by one polarization or the other without any exchange. This aspect of the propagation of light led Newton to conclude that light consists of corpuscules - one type of particle for each polarization.

A notable characteristic of the energy of electromagnetic waves is its quadratic dependence on the electric or magnetic field vectors. Aside from its mathematical importance, this quadratic dependence assures that energy is conserved at an interface where reflection and transmission occur (see Section 5.2 and Problem 5.3) or that energy is conserved when two beams interact and form interference fringes (see Section 6.2 for a discussion of interferometry). An important group of modern applications in which this quadratic dependence gives rise to unusual physical effects is the interaction of coherent radiation with matter and the interaction between coherent beams. To illustrate the effect of the former we consider the following example. (The interaction of two coherent sources will be discussed in Chapter 6.)

Example 4.3 Scattering of radiation accounts for most of the visible effects in the atmosphere, such as the blue sky or white clouds. But aside from their beauty, these light-scattering effects are extremely useful as tools for various gas and liquid flow measurements. Two notable processes are *Rayleigh scattering* and *Mie scattering*. In both, the wavelength of the scattered radiation is either identical to or only minimally shifted (usually by no more than 2 cm^{-1}) relative to the incident radiation. Rayleigh scattering is the scattering by individual molecules, whereas Mie scattering is by particles that are smaller than a few wavelengths of the radiation. Both Rayleigh and Mie scattering are coherent with the incident radiation. Therefore, special consideration must be given to the addition of the electric fields emitted by multitudes of scatterers.

Assuming that the scattered radiation is coherent with the incident radiation, find the variation of the irradiance of Rayleigh scattering by gas molecules with their density N. The spacing between the gas molecules is approximately λ. For comparison, find also the variation of the irradiance of Mie scattering using the diameter d of a single scattering particle when $d \ll \lambda$.

Solution Note that the problem does not state the coherence properties of the incident radiation. This suggests that although the scattered radiation is coherent with the incident radiation, the result of the analysis must be independent of the coherence of the source. Thus, the dependence on N for Rayleigh scattering or on d for Mie scattering should be independent of the nature of the illuminating source. This means that the scattering of solar radiation, where the coherence length is less than 1 micron, should then be similar to the scattering of laser radiation, where the coherence length may exceed several meters. Thus, when solar radiation is scattered, the coherence length of the scattered radiation is also of the order of 1 μm. When more coherent radiation is scattered, the coherence length may accordingly be longer. Since the only distinction between the molecules in the gas phase (where Rayleigh scattering takes place) and those forming the particles (responsible for Mie scattering) is the intermolecular spacing, all differences in their scattering characteristics must be explained by these geometrical differences and the coherence properties of the scattering.

Consider first the Mie scattering by submicron particles, where $d \ll \lambda$ and $\lambda \approx 1\ \mu$m. Since the molecules in the submicron particle are densely packed, the spacing between any two molecules is much less than λ. Thus, even when illuminated by a wavetrain with a length of approximately λ, all molecules of the particle are forced to scatter by the same wavetrain. Furthermore, since they are so closely packed, all molecules "see" approximately the same phase angle of the incident radiation, thereby scattering waves all with approximately the same phase angle. Thus, the resultant field of the scattering by all molecules of the particle is simply the algebraic sum of the fields of the individual molecules. The scattering efficiency by an individual molecule is a complex function of its properties, the polarization of the radiation, and the direction of observation relative to the direction of incidence (Van De Hulst 1957). Nonetheless, the scattered field is proportional to the incident field and can be represented by the proportionality coefficient α. Thus, for an incident field $\mathbf{E}_0$, the scattering E_s by a cluster of N_p molecules within a small particle is

$$E_s = \sum_{N_p} \alpha \mathbf{E}_0 = N_p \alpha \mathbf{E}_0.$$

Because the irradiance of the incident and scattered radiation (I_0 and I_s, respectively) increases quadratically with the incident field, the preceding summation can be used to obtain the following relation between I_0 and I_s:

$$I_s \propto N_p^2 I_0. \tag{4.43}$$

The most notable consequence is that, when coherent fields emitted by several identical sources are additive, the irradiance of the resultant field increases quadratically with the number of these sources. This quadratic dependence is even more noticeable when compared to the linear dependence between I_0 and I_s. Thus, the coherent addition of the fields serves as a reinforcement of the emission.

Equation (4.43) can be modified to represent the variation of the scattered irradiance with the size of the scattering particle. Since N_p is proportional to the volume of the particle, it may be replaced with d^3 for nominally spherical particles with a diameter d:

$$I_s \propto d^6 I_0.$$

This rapid increase in the scattering irradiance with particle size is typical with small particles. However, as the diameter increases, molecules in one part of the particle no longer scatter coherently with molecules in other parts, and the exponent decreases until it reaches a quadratic dependence for large particles. This quadratic dependence can be derived from considerations of geometrical optics, and represents the effect of the cross-sectional area of the particle that intercepts the incident radiation.

In contrast with the enhancement of the scattering process by coherent addition, the irradiance by a multitude of incoherent scatterers increases only linearly with their density. Rayleigh scattering by gas molecules, which are randomly located relative to an arbitrary origin, can serve as an illustration for such incoherent addition. Although the scattering by each molecule is coherent with the source, the large and random separation between the scatterers that is typical for gases puts each scatterer at a random location along the incident wavetrain. Therefore, irrespective of the coherence properties of the source, the phases of the scattered fields are determined by the location of the scatterers along the incident wavetrain. Even when the coherence length of the incident radiation is long relative to the intermolecular spacing, the random location of the molecules randomizes the phases of the scattered fields. Thus, superposition of the various fields is not simply additive; the random phases must be considered as well. Instead, the irradiance by each scatterer can be evaluated separately by

$$I_s' = \alpha^2 I_0,$$

where I_s' is the irradiance emitted by a single scatter. Because energy is always additive, the total irradiance is simply

$$I_s \propto N I_0.$$

Thus, the irradiance resulting from addition of the emission of incoherent sources increases linearly with their number.

This striking difference between the scattering efficiency of closely and loosely spaced molecules is visible in many atmospheric phenomena and engineering applications. The Rayleigh scattering by atmospheric molecules is the cause for the blue sky. However, when the skies are covered with a thin layer of clouds, the white light scattered by the mist in the clouds often appears much brighter than the sky in the background. This is despite the large disparity between the thickness of the atmosphere that contributes to Rayleigh scattering and the thickness of the cloud layer. ■

In this chapter we discussed the propagation of radiation using the electromagnetic wave theory. Although the solution of the wave equation was one-dimensional, the results were sufficient to determine the concepts of energy associated with the waves, the index of refraction, and propagation losses. However, some of the most important consequences of this theory have not yet been discussed. Polarization, interference, and diffraction – which are direct consequences of the wave theory – will be discussed in subsequent chapters, where they can be presented along with some relevant engineering applications.

References

Born, M., and Wolf, E. (1975), *Principles of Optics,* 5th ed., Oxford: Pergamon, pp. 51–5, 90–4.

Feynman, R. P., Leighton, R. B., and Sands, M. (1964), *The Feynman Lectures on Physics,* vols. I and II, Reading, MA: Addison-Wesley.

Fox, R. W., and McDonald, A. T. (1992), *Introduction to Fluid Mechanics,* 4th ed., New York: Wiley, pp. 152–6.

Halliday, D., and Resnick, R. (1963), *Physics for Students of Science and Engineering,* part II, New York: Wiley, pp. 684–8.

Hildebrand, F. B. (1962), *Advanced Calculus for Applications,* Englewood Cliffs, NJ: Prentice-Hall, pp. 278–80.

Jackson, J. D. (1975), *Classical Electrodynamics,* 2nd ed., New York: Wiley, pp. 67–8, 299–303.

Moore, C. E. (1971), *Atomic Energy Levels as Derived from the Analysis of Optical Spectra* (National Standard Reference Data Series NSRDS-NBS 35, vols. I–III), Washington, DC: National Bureau of Standards.

Nakatsuka, H., and Grischkowsky, D. (1981), Recompression of optical pulses broadened by passage through optical fibers, *Optics Letters* 6: 13–15.

Van De Hulst, H. C. (1957), *Light Scattering by Small Particles,* New York: Wiley, Chapters 4 and 6.

Homework Problems

Problem 4.1

Prove the vectorial identity of (4.3).

Problem 4.2

Review Feynman, Leighton, and Sands (1964, vol. II, chap. 10) and describe a physically plausible situation for which $\chi < 0$.

Problem 4.3

Solve (4.21) for a dipole moment with an oscillating mass m, restoring force constant q, and damping γ subjected to a field $\mathrm{Re}[\mathbf{E}'e^{-i\omega t}]$.

Problem 4.4

Find the value of $\Delta z \Delta k$ for a pulse in which the longitudinal distribution of the field is Gaussian, that is, where $\mathbf{E}(z) = \mathbf{E}_0 \exp\{-a^2 z^2\}$. *Hint:* Find the Fourier transform of the field and obtain the irradiance distribution (eqn. 4.42) in both the z and k domains. (*Answer:* 1/2.)

Problem 4.5

Show that, when $dn/d\omega > 0$, the derivative term $dv_p/dk < 0$ in (4.32).

Problem 4.6

Prove that modulating the amplitude at frequency α of a carrier wave at frequency β can be effected by the superposition of two simultaneously propagating waves at frequencies α and β, by showing that the following complex equation holds for any real α and β:

$$2\,\mathrm{Re}[e^{i\alpha}]\,\mathrm{Re}[e^{i\beta}] = \mathrm{Re}[e^{i(\alpha+\beta)} + e^{i(\alpha-\beta)}].$$

Research Problem

Research Problem 4.1

Read the paper by Nakatsuka and Grischkowsky (1981), which describes a method to compensate for dispersion-induced broadening of a short pulse propagating through a long optical fiber. The pulse is generated by a mode-locked laser, and its duration is measured by an autocorrelation technique. (Mode-locked lasers will be discussed in Section 12.8, but details of the laser and the autocorrelation method are not relevant to the present problem.) The wavelength of the laser beam coincides with one of the absorption lines of sodium (Na). By passing the beam through a cell containing Na, the pulse can be recompressed. By adjusting parameters of the cell, such as its length or the density of the Na vapor, the recompression is matched to the broadening by the fiber.

(a) Explain the principles of the recompression process.

(b) What should be the full bandwidth of the 3.3-ps pulse when its spectrum is limited by Fourier transformation? (Note that the frequency unit in the paper is $1\ \text{cm}^{-1} = 3 \times 10^{10}$ Hz.) Compare your result to the measured bandwidth.

(c) Find the dispersion factor D of the fiber in units of ps/nm-km and compare it to the dispersion factor of the fiber described in Example 4.2 of Section 4.8.

(d) Find the dispersion factor (and sign) of the sodium vapor cell.

(e) How would the dispersion factor of the cell vary if the density of the vapor were doubled?

(f) If passed through only the fiber, which of the frequency components of the pulse will emerge first? Which component would emerge first if passed through only the Na vapor cell?

(g) Which side of Figure 3 in the paper represents the absorption line center of Na?

5 Propagation and Applications of Polarized Light

5.1 Polarization of Radiation

In the previous chapter it was shown that electromagnetic radiation is formed by the simultaneous oscillation of electric and magnetic field vectors; both vectors are mutually orthogonal and both are confined to a plane that is perpendicular to the propagation vector. Therefore, both the magnetic and electric field vectors can have no more than two independent components. For the propagation illustrated in Figure 4.3, the electric field vector **E** had only one component, pointing in the y direction. An additional, independent component that could exist in the x direction was not shown there. Similarly, **H** could have two components, H_x (shown in the figure pointing in the negative x direction) and H_y (not shown). When both components are present, the total electric field is

$$\mathbf{E}(z,t) = [E_{0x}(z,t)e^{-i\phi_1}\mathbf{e}_x + E_{0y}(z,t)e^{-i\phi_2}\mathbf{e}_y]e^{-i(\omega t \mp kz)}. \tag{5.1}$$

Thus, the total electric field at any point and at any time is the outcome of the superposition of two independent fields, or polarizations, each having the same frequency and propagation vector but not necessarily the same amplitude or phase angle ϕ. Furthermore, the phase or amplitude of each component may vary along the propagation path independently of the other. Nevertheless, they are indistinguishable spectrally. And, as demonstrated by (4.42), the energy carried by one component is independent of the energy carried by the other. The combination of the amplitudes and phase angles of the two independent components is sufficient to characterize all possible polarizations of radiation. Although the two independent polarization components are spectrally indistinguishable, they may still be separated by optical devices that are sensitive to the polarization characteristics of radiation.

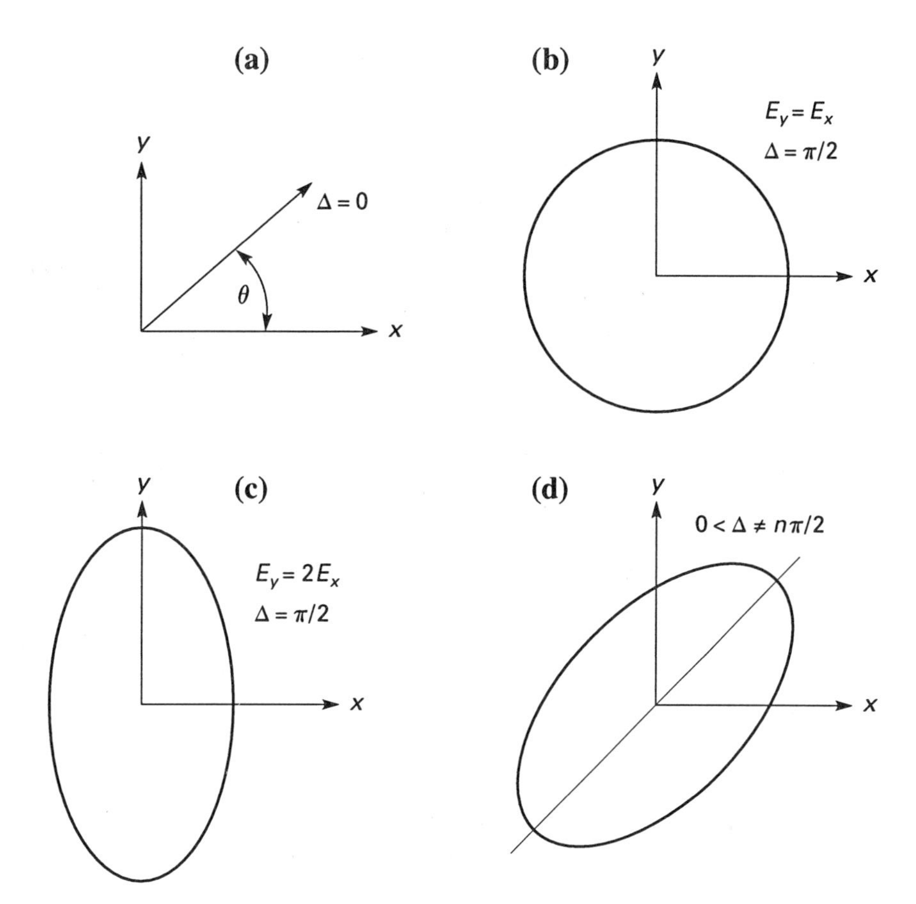

Figure 5.1 Polarizations induced by phase delay between the x and y components of the electric field of radiation: (**a**) linear; (**b**) circular; (**c**) and (**d**) elliptical.

Mathematically, the simplest polarization of radiation is obtained by combining the two orthogonal components of the polarization, E_x and E_y, each having the same phase angle; the ratio of their amplitudes is time- and space-invariant. This configuration is illustrated in Figure 5.1(a). The total field is the vectorial sum of its components, and forms an angle of $\theta = \tan^{-1}(E_y/E_x)$ with the positive x axis. Both components oscillate at a frequency of ω and their vectorial sum oscillates at the same frequency, so the angle θ is time-invariant even if the amplitudes of both components of the polarization vary, provided that their ratio remains unchanged. This is called *linear polarization*. The term *plane polarization* may also be used to describe time- and space-invariant polarization. With the latter terminology, the plane is defined by the polarization vector and the propagation vector $\mathbf{k}$.

Clearly, there can exist only two independent linear polarizations for any propagating beam. Often radiation contains a random assembly of linear polarizations, each having its own amplitude, orientation θ, and phase angle ϕ. Such radiation is considered to be unpolarized. The term *random polarization* is used to describe linearly polarized radiation in which the orientation of the polarization vector changes randomly from time to time, even though it is linear at any given moment.

A more general description of polarization includes the superposition of two field components, each having a different phase angle and possibly a different amplitude. Within this general group of polarizations, there exists a special category where the difference $\Delta = \phi_1 - \phi_2$ remains time-invariant and is independent of the coherence properties of the source – that is, where all wavetrains along the propagation path and at all times have the same phase difference Δ between the polarization components. Without such restriction (i.e., if Δ is allowed to fluctuate randomly), the polarization reverts to the random mode. The simplest combination, where $\Delta \neq 0$ is invariant, includes two polarizations each having the same amplitude but with $\phi_1 - \phi_2 = \pi/2$. Figure 5.1(b) illustrates the sum of these components. It can be shown (see Problem 5.1) that the tip of the resultant field vector, when viewed in the direction of the propagation, traces out a circle while rotating clockwise at a frequency ω. Combined with the propagation, the field vector can be seen as forming a circular helix. The direction of rotation of the field vector when $\Delta < 0$ is in the counterclockwise direction. It can also be shown (Problem 5.2) that if two circularly polarized fields with opposite spins and identical amplitudes are superimposed, the resultant field is linearly polarized.

The most general group of nonrandom polarizations is presented by the combination of $\Delta \neq 0$ and $E_x \neq E_y$. Of course, both Δ and E_x/E_y are assumed to be time and space invariants. This is *elliptical polarization*. This group includes as subcategories both linear polarization ($\Delta = 0$ and π) and circular polarization ($\Delta = \pi/2$ and $3\pi/2$ and $E_x = E_y$). By varying Δ and the relative amplitudes of the x and y components of the field vectors, we can control the orientation of the ellipses' major axes, their relative magnitude, and the direction of spin. Figures 5.1(c) and 5.1(d) each illustrate a case of elliptically polarized radiation. In the first, the ellipse is formed when $\Delta = \pi/2$ while the amplitudes are uneven. The second ellipse is formed by combining two uneven amplitudes and $\Delta \neq n\pi/2$.

In practical applications, imperfections in optical elements distort the polarization characteristics of radiation. This may be particularly troubling when the incident polarization is expected to be circular or linear. Owing to such imperfections, even the purest polarization modes must be considered as elliptical polarizations, and the extent of their purity is specified by the ratio of their axes. Highly linearly polarized radiation may have a ratio of $E_x/E_y = 10^{-4}$. Although the field of the minor axis in such highly linear polarization may appear negligible relative to the major axis, in high-power applications and when discrimination of one polarization relative to another may be required, the effect of this minor component may become significant.

Manipulating the polarization of radiation is an important option available for controlling its propagation. Polarization selection techniques can be used to attenuate or to block radiation, to direct beams into selected propagation channels, or to pick out one beam while rejecting another. The devices that can be used for these applications can generally be grouped into two categories: (a) those that act as polarization filters – that is, devices that can discriminate between polarizations and transmit one while rejecting the other; and (b) devices that can change the state of polarization of the transmitted radiation (e.g., convert linear polarization into circular polarization and vice versa). Alternative groupings distinguish between passive devices with permanent polarization characteristics and active devices whose polarization characteristics can be switched on or off either electronically, acoustically, or mechanically.

Most devices for manipulating polarization consist of optical materials that transmit radiation but have a certain anisotropy that is either permanent or induced. Because of this anisotropy, the propagation characteristics (e.g. velocity or transmission) of one field component differ from those of the other component. Polarizers that are used to separate one polarization component of radiation from the other are usually designed to preferentially transmit (or reflect) one field component relative to the other. Such polarizers may depend on either the reflection or transmission characteristics of their medium. On the other hand, devices designed to alter the polarization characteristics of the transmitted radiation consist of materials that have a different index of refraction for each field vector, thereby delaying the propagation of one component relative to the other; such materials are called *birefringent*. The property of *birefringence* can be permanent or induced. Section 5.2 discusses the reflection and transmission characteristics of homogeneous and isotropic media; Section 5.3 covers homogeneous anisotropic media.

5.2 Transmission and Reflection at Interfaces of Homogeneous and Isotropic Media

When radiation is incident upon an interface separating two homogeneous and isotropic media, a portion of the radiation is transmitted and the balance is reflected. Assuming that there are no losses at the interface (e.g., by scattering or by absorption by surface impurities), the combined energy of the reflected and transmitted components must be equal to the incident energy (see Problem 5.7). From (4.42) is is evident that the energy of one polarization component is independent of the energy of the other component; that is, there is no energy exchange between the polarizations at the interface. Therefore, the reflection and transmission characteristics of each polarization can be determined separately using the boundary conditions for Maxwell's equations (eqns. 3.24–3.27). Recall that these conditions determine the magnitude of the electric and magnetic field component parallel and normal to the interface. For simplicity we assume that the surface is uncharged, that is, $\rho_a = 0$. To allow for equivalent treatment of the magnetic field, the surface current must also vanish; $\mathbf{j}_s = 0$. These two requirements are compatible with the properties of most optical media. Of course, more general treatments that are beyond the scope of this book may allow for $\rho_a \neq 0$ and $\mathbf{j}_s \neq 0$.

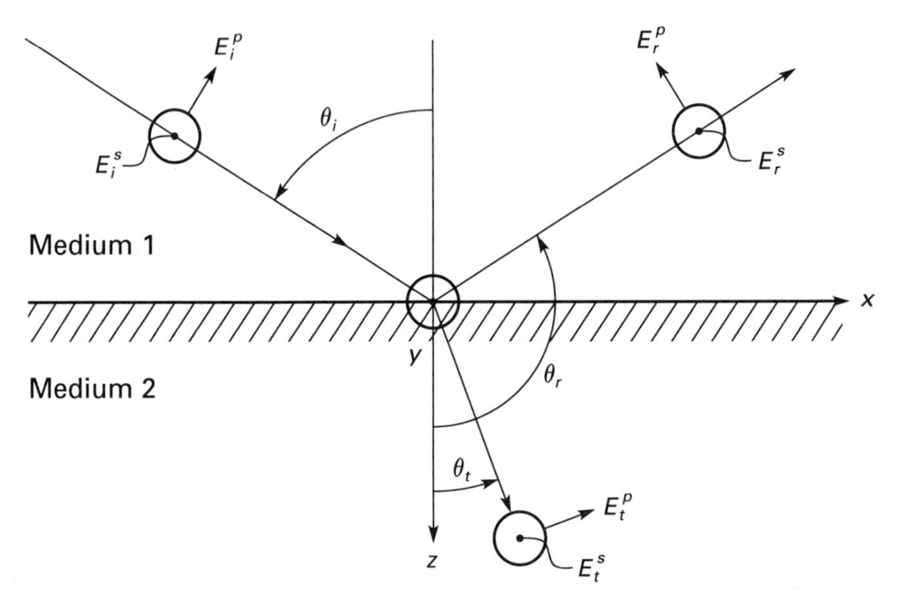

Figure 5.2 Reflection and refraction of the s and p polarizations at an interface between two media.

Figure 5.2 presents schematically the incidence of a single ray on the interface separating two homogeneous, isotropic, nonconducting, charge-free media. The resulting refraction and reflection rays are also illustrated. These three rays are confined to a single plane, called the *incidence plane*. The positive x and z axes are marked by arrows, while the positive y direction (pointing out of the page) is indicated by a circle. The incidence angle θ_i is measured counterclockwise from the negative branch of the z axis. The two independent polarizations of the incident ray are marked by E_i^p and E_i^s, where the i subscript symbolizes the incident field, the p superscript marks the polarization component that is in the incidence plane, and the s superscript marks the polarization component pointing in the positive y direction. Note that the p and s components together with the propagation direction form an orthogonal system that follows the right-hand convention. Similarly, the transmitted components are E_t^p and E_t^s and the reflected components are E_r^p and E_r^s. To maintain the right-hand convention upon reflection also, the direction of the positive p component of the reflected ray is defined by an arrow pointing toward the incident ray. The refraction and reflection angles, θ_t and θ_r respectively, are measured in the counterclockwise direction from the positive branch of the z axis.

The relation between the incident, reflected, and transmitted polarizations requires the solution of four unknowns ($E_t^p, E_t^s, E_r^p, E_r^s$) in terms of the incident field components and the incidence and refraction angles. Since the boundary conditions for the electric field can produce only three independent equations – one equation each for the x, y, and z components of the field – the solution must also include the magnetic field. However, from (4.12′) it can be found that $H^p = -\sqrt{\epsilon_0/\mu_0}\cdot E^s$ and similarly $H^s = -\sqrt{\epsilon_0/\mu_0}\cdot E^p$. Hence the boundary conditions for the magnetic field expressed in terms of the electric field can also be used.

Of the six equations that can be written using the three components of each of the fields, only four are independent. Thus, the following equations for the transmission and reflection of each of the incident polarizations can be obtained from the boundary conditions for the x and y components of the two fields:

$$E_t^p = \frac{2n_1 \cos\theta_i}{n_2 \cos\theta_i + n_1 \cos\theta_t} E_i^p, \tag{5.2a}$$

$$E_t^s = \frac{2n_1 \cos\theta_i}{n_1 \cos\theta_i + n_2 \cos\theta_t} E_i^s, \tag{5.2b}$$

$$E_r^p = \frac{n_2 \cos\theta_i - n_1 \cos\theta_t}{n_2 \cos\theta_i + n_1 \cos\theta_t} E_i^p, \tag{5.2c}$$

$$E_r^s = \frac{n_1 \cos\theta_i - n_2 \cos\theta_t}{n_1 \cos\theta_i + n_2 \cos\theta_t} E_i^s. \tag{5.2d}$$

(See Born and Wolf 1975, pp. 38–47.) These are Fresnel formulas, where n_1 and n_2 are the indices of refraction of medium 1 and 2, respectively. Using the law of refraction (eqn. 2.6), the indices of refraction can be replaced with $\sin\theta_i$ and $\sin\theta_t$, thereby reducing (5.2) to the following form, which is written in terms of the incident fields and the geometrical parameters only:

$$E_t^p = \frac{2 \sin\theta_t \cos\theta_i}{\sin(\theta_i + \theta_t)\cos(\theta_i - \theta_t)} E_i^p, \tag{5.3a}$$

$$E_t^s = \frac{2 \sin\theta_t \cos\theta_i}{\sin(\theta_i + \theta_t)} E_i^s, \tag{5.3b}$$

$$E_r^p = \frac{\tan(\theta_i - \theta_t)}{\tan(\theta_i + \theta_t)} E_i^p, \tag{5.3c}$$

$$E_r^s = -\frac{\sin(\theta_i - \theta_t)}{\sin(\theta_i + \theta_t)} E_i^s. \tag{5.3d}$$

Several important observations can be readily made from these equations. The simplest corresponds to the case where $n_1 = n_2$. This is when the interface is either an imaginary surface dividing a single medium or a real surface separating two different media with the same index of refraction. For such an interface θ_i is clearly equal to θ_t, and the amplitude of the reflected field (see eqns. 5.3c and 5.3d) vanishes while the transmitted field equals the incident field. By contrast, it is evident from these equations that if $n_1 \neq n_2$ then a portion of the incident field must be reflected, where the extent of the reflection increases with θ_i. This reflection is of concern in the design of complex optical devices. Such spurious reflection can interfere with the performance of a device by creating unwanted beams, and is also the cause for attenuation of transmitted radiation. For normal incidence on an air–glass interface where $n_1 = 1$ and $n_2 = 1.5$, the reflected field is 20% of the incident field. This corresponds to a loss in transmission of 4% of the incident irradiance. For transmission through a glass slab, the loss through both interfaces amounts to approximately 8%. In optical systems with multiple elements, attenuation by reflection losses may become significant.

The effect of the interface on the phase angles of the reflected and transmitted fields relative to the incident field is of considerable importance. Since the trigonometric functions in these equations are real (with the exception of total internal reflection), the phases of the transmitted and reflected polarizations are shifted by only 0 and π relative to the incident polarizations. Thus, at the interface, the phases of the transmitted polarizations remain identical to the phases of the incident polarizations, while the phases of the reflected polarizations depend on the relative magnitudes of θ_i and θ_t. When $\theta_i > \theta_t$ (i.e. $n_2 > n_1$), the signs of the p components of the incident and the reflected fields are identical (eqn. 5.3c). Notice, however, that by definition the positive p components of the incident and reflected rays point toward each other. Thus, for normal incidence and when the incident and reflected amplitudes are the same (i.e., pure reflection), these two fields cancel each other and the field component tangential to the interface is zero at both sides (cf. eqn. 3.26). Similarly, from (5.3d) it can be seen that if $\theta_i > \theta_t$ then the phase of the reflected s component is reversed relative to the phase of the incident s component. This implies that the reflected s component points in the opposite direction of the incident s component. The opposite is true, for both the s and p components, when $\theta_t > \theta_i$ (i.e., when $n_1 > n_2$).

The reflection at an interface depends not only on the angle of incidence but also on the indices of refraction at both sides. Thus, by judicial selection of these parameters, reflection may be minimized or enhanced. To determine the conditions for total reflection at an interface, one simply looks for conditions of zero transmission. From (5.2a) and (5.2b), transmission of both the s and p polarizations is seen to vanish when $\theta_i = \pi/2$. However, this configuration of grazing incidence is unique and presents very few technical opportunities. Alternatively, to eliminate transmission independently of the incidence angle one may require that $n_2 \to \infty$. From (4.18) it is clear that when $n_2 \to \infty$, $c \to 0$; that is, propagation is not possible in that medium. Thus, at normal incidence, the amplitudes of the reflected and transmitted fields are equal. However, since the field behind such a mirror approaches zero, the boundary condition for tangential electric field components (eqn. 3.26) requires that the field in front of the mirror be zero as well. This is possible only if the incident and reflected field vectors point opposite to each other at all times – that is, if normal reflection of the s polarization introduces a phase shift of $\Delta\phi = \pi$ while the normal reflection of the p polarization is with $\Delta\phi = 0$. Any point where the field is consistently zero, such as the mirror interface, is called a *node*. Other nodes can be found for normal reflection at planes parallel to the interface and separated by λ (see Problem 5.7). Since the location of these nodes is independent of time, the wave formed by the superposition of the incident and reflected field is called a *standing wave*. Standing waves often occur inside laser cavities.

Although both transmission and reflection depend on the incident polarization, only reflection presents a genuine opportunity for polarization selection by an isotropic medium. From (5.3c) it can be seen that when $\theta_i + \theta_t = \pi/2$, the reflected field of the p component vanishes while the reflected s component remains finite. Thus, when the incidence and transmission angles meet this condition, the radiation reflected by an interface is linearly polarized. Using (2.6), we can show that this condition is met when

$$\tan\theta_i = \tan\theta_B = \frac{n_2}{n_1}. \tag{5.4}$$

This unique incidence angle θ_B is the *Brewster angle.* For incidence at Brewster's angle the reflected radiation is linearly polarized, but only a small percentage of the incident energy is reflected by the interface. Therefore, using a reflection at the Brewster angle is a relatively inefficient approach for polarization selection. Instead, anisotropic media are normally used.

Reflections at the Brewster angle do, however, have other applications. For example, sunglasses can be equipped with polarizers. Although any orientation of the polarizers can reduce transmission of the unpolarized light by 50%, one orientation can reduce it even further. During the bright hours of the day, a large portion of sunlight is incident at approximately θ_B, so the s polarization component is dominant in reflection and scattering from surfaces such as roads and cars. Thus, by orienting the polarizers of such sunglasses so that their axes are orthogonal to the reflected polarization, much of the glare from surfaces can be eliminated.

The design of laser systems also takes advantage of the polarization selectivity of the reflection at θ_B. In most laser systems, the beam oscillates between two parallel mirrors while passing repeatedly through the laser medium, which acts as an amplifier of light. The gain medium may consist of a crystal or an electric discharge tube capped at both ends by transparent windows. To minimize reflection losses (which increase with the number of round trips between the two mirrors), the ends of many laser gain media are cut such that the incidence is at θ_B (see Problem 5.5). When oscillation begins, both the p and the s polarizations may be present. However, for incidence at θ_B, the p polarization ideally does not suffer any reflection losses while the losses of the s polarization amount to a measurable percentage. Despite this seemingly slight advantage, the gain process favors the p polarization, which accumulates energy at the expense of the s polarization. After only a few passes, the p polarization dominates, the laser beam becomes linearly polarized, and reflection losses in subsequent passes through the gain medium are minimized. Without Brewster windows, the polarization of laser beams is random.

Finally, Fresnel formulas may also be used to describe total internal reflection (see Section 2.3). When $n_1 > n_2$, there exists a critical incidence angle $\theta_i = \theta_{\text{crit}}$ where $\theta_t = \pi/2$; this is when the maximum value of $\sin\theta_t = 1$ is reached. Because the transmission angle cannot increase any further as the incidence angle increases, the incident beam is reflected at the interface. Mathematically this is represented by introducing a complex function for $\sin\theta_t$. When such a function is introduced into (5.3a) and (5.3b), it can be shown (see Born and Wolf 1975, pp. 47–51 or Jackson 1975, pp. 282–4) that the reflected field undergoes a phase change but its amplitude equals the incident amplitude. More surprising is the observation that, although a beam could not be transmitted, there does exist an electric field in medium 2. This is a direct consequence of the boundary conditions for the electric field (eqns. 3.24 and 3.26), where continuity of the field, or the displacement vector, across the boundary is stipulated. Depending on the incident polarization, two field components, E_x and E_y, may exist with an amplitude that decays exponentially in the z direction within the distance of a few wavelengths. This rapidly decaying field, called the *evanescent wave,* can propagate in the x direction only and is therefore not a TEM wave. Evanescent waves exist at the air side of any prism face (e.g., the penta prism in Problem 2.6) where total internal reflection takes place. Although it carries no energy, the evanescent wave can be coupled across other interfaces into energetic modes. By bringing two such prism interfaces into "intimate" contact – that is, by reducing the distance between the two interfaces to less than one wavelength – the evanescent wave penetrates the second interface and hence energy from one prism can be coupled into the second. With this technique, surface waves from a prism where total internal reflection occurs can also be launched into metallic surfaces (Otto 1968, Chabal and Sievers 1978), or portions of the incident beam can be transmitted into other dielectric optical elements. For a review of the applications of total internal reflection and "frustrated" total internal reflection, see Zhu et al. (1986).

5.3 Propagation of Polarized Radiation through Anisotropic Media

Equations (5.2) or (5.3) show that reflection and transmission at the interface between two homogeneous and isotropic media discriminate between the two incident polarizations. Thus, depending on the angle of incidence, the reflection or transmission of one polarization can be enhanced relative to the other. While this may serve as an avenue for the separation of polarization components, the efficiency is poor: at Brewster's angle the amplitude of the reflected s polarization is approximately 38% of the incident amplitude, corresponding to only 14% of the incident energy. Thus, although the reflected radiation may be linearly polarized, the energy of the reflected beam is reduced considerably. These numbers demonstrate that interfaces have a significant effect on the state of polarization of reflected and transmitted radiation, but this effect is insufficient for polarization discrimination or manipulation. Instead, optically anisotropic media are used. Depending on the application, devices made of such materials can be used to separate the polarization components of radiation, rotate the direction of the polarization of linearly polarized radiation, and transform the state of polarization from linear to circular or elliptical and vice versa. Furthermore, devices can be designed to perform these functions on command and so become, in effect, optical switches.

The underlying property of anisotropic media is their ability to alter the polarization of transmitted radiation without significant loss of energy. Using the analogy of a harmonic oscillator, the oscillation of the medium's oscillators is induced along a line that is not collinear with the line of the optical polarization. Thus, the electric displacement vector **D** is not parallel to **E**. From the condition that $\mathbf{D} = \epsilon \mathbf{E}$, it follows that the dielectric constant ϵ must be represented by a tensor (eqn. 3.3) to account for the condition that $\mathbf{D} \nparallel \mathbf{E}$. Owing to the symmetrical properties of radiation, it is possible to show that the tensor of the dielectric constants is symmetric; that is, $\epsilon_{ij} = \epsilon_{ji}$ where $i, j = x, y$, or z (Yariv 1975, pp. 82–90). Thus, the total number of different dielectric constants cannot exceed six. For further simplification, the coordinate system can be aligned with the principal axes of the tensor, reducing the equation for the electric displacement vector to

$$\mathbf{D} = \begin{bmatrix} \epsilon_{xx} & 0 & 0 \\ 0 & \epsilon_{yy} & 0 \\ 0 & 0 & \epsilon_{zz} \end{bmatrix} \mathbf{E}. \tag{5.5}$$

It is evident from (5.5) that the dielectric properties of a medium can be fully characterized by three independent dielectric constants (which may be real or imaginary) when the directions of the principal axes are known. When these coefficients are real, the medium is nonlossy. However, since they are not equal to each other, the medium is birefringent – it may have more than one index of refraction.

To illustrate propagation through a birefringent medium, consider the following simple configuration (Figure 5.3) of propagating along the z axis of a birefringent plate. The principal axes of the plate are aligned with the coordinate axes, and the indices of refraction along these axes are n_1, n_2, and n_3. The linear polarization at the $z = 0$ plane is defined by the two components, E_x and E_y, that oscillate at the same phase. The inclination of that polarization vector is determined by the relative magnitudes of E_x and E_y. When $n_1 > n_2$, the propagation velocity of the x component of the polarization is slower than the propagation velocity of the y component. Therefore, as both polarizations

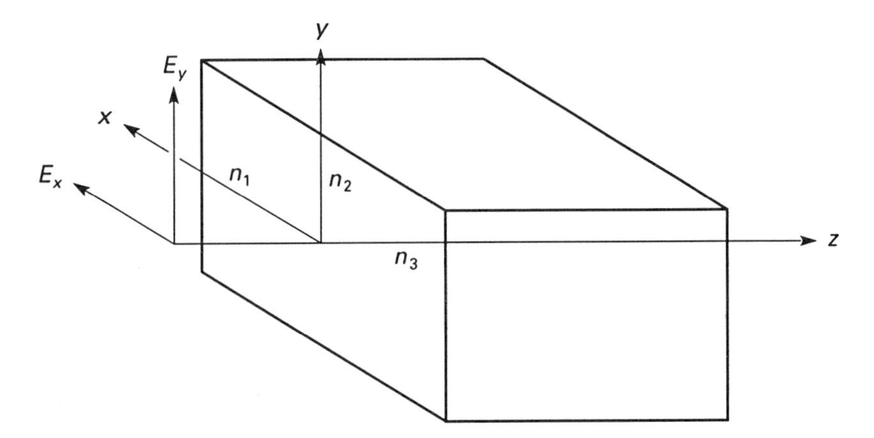

Figure 5.3 Indices of refraction along the three axes of a birefringent optical element.

propagate through the medium, the x component is delayed relative to the y component by an amount that depends on the length L of the medium and the difference $n_1 - n_2$ between the indices of refraction. If the delay is exactly $m\lambda$ (where m is an integer) then the two components emerge at the opposite side of the medium with the same phase, and the inclination of the combined polarization is identical to the inclination of the incident linear polarization. However, when m is not an integer, the superposition of the two components at $z = L$ no longer matches the incident polarization. Of particular interest are plates that are cut to lengths that induce delays of

$$(n_1 - n_2)L = \lambda/4 \quad \text{and} \quad (n_1 - n_2)L = \lambda/2. \tag{5.6}$$

These plates are called, accordingly, *quarter-* and *half-wave retardation plates*. Retardation plates can be fabricated using permanently isotropic materials such as mica, quartz, or other crystalline materials, or by inducing anisotropy by either applying an intense electric field along a selected direction or stressing

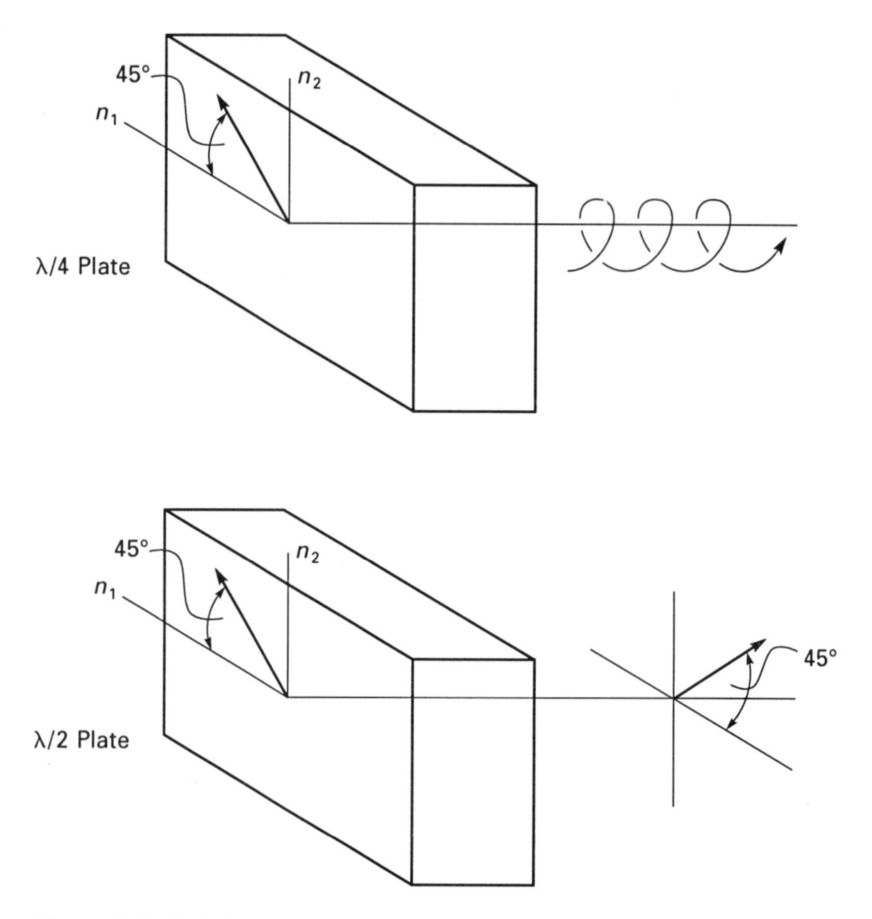

Figure 5.4 Polarization behind quarter-wave (λ/4) and half-wave (λ/2) plates for linearly polarized radiation, with the polarization vector incident at 45° to the x axis.

the material along a selected axis. When a permanently birefringent medium is used for retardation, the length is adjusted by the manufacturer to obtain the desired delay between the two polarization components. However, owing to dispersion, when the length is fixed the required retardation will be realized over only a limited range of wavelengths. Other permanently birefringent retardation plates, consisting of sheets of polymer material, are available but are useful only when precision is not required. When the anisotropy is induced (see Section 5.4), the extent of retardation and its timing can be controlled, thereby presenting an advantage for light-control applications.

Although the use of retardation plates is not limited to any particular configuration, they are normally positioned such that the incident linear polarization is oriented at 45° to the x axis so that $E_x = E_y$. For a $\lambda/4$ plate, the delay between E_x and E_y is $\pi/2$ and the emerging polarization is circular; see Figure 5.4(a). Similarly, for a $\lambda/2$ plate the exit polarization is linear but is rotated by 90° (Figure 5.4(b)).

In applications where retardation between two polarizations is induced, anisotropy is required only in the (x, y) plane (Figure 5.3), that is, where $n_1 \neq n_2$. The index of refraction along the z axis may match either n_1 or n_2. Crystals with such partial isotropy ($n_1 \neq n_2 = n_3$ or $n_2 \neq n_1 = n_3$) are called *uniaxial*. The two axes with the identical indices of refraction are the *ordinary* axes, while the third axis is the *extraordinary* axis. The indices of refraction of such crystals are accordingly labeled n_o for the ordinary axes and n_e for the extraordinary axis. For more information and data on optical crystals, see Dmitriev (1991).

An important use of permanently birefringent media is for polarization selection. Figure 5.5 presents a typical polarizer. This the *Glan prism,* which consists of a uniaxial crystal (e.g. calcite) cut into two prisms that are held together along their diagonal surfaces with a small gap in between. The extraordinary axis is parallel to the x axis while the oridinary axes are aligned along the y and the z axes. The latter is also the optical axis of the prisms. When $n_e > n_o$, the angle between the inclined planes of the prisms and the optical axis is cut to

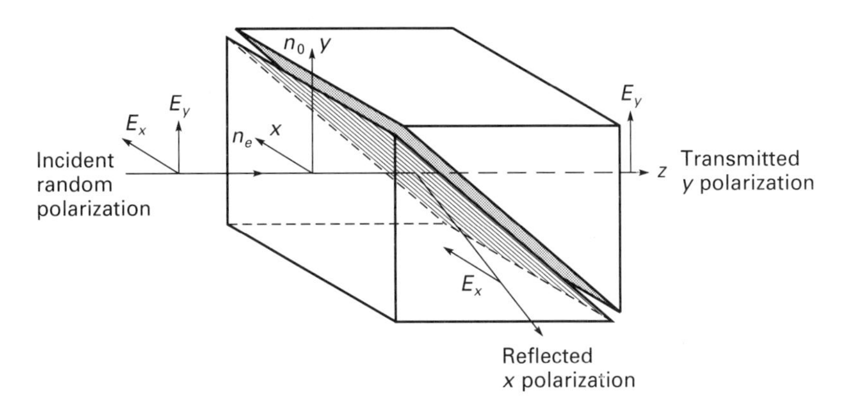

Figure 5.5 Separation of radiation into its polarization components by a Glan prism.

exceed θ_{crit} for the x polarization of the incident beam but to remain below θ_{crit} for the y polarization. Thus, when radiation consisting of two polarization components is passed through the first prism, the E_x component is totally reflected at the inclined interface with no transmission in the z direction. The second polarization, E_y, is refracted through the diagonal surface according to Snell's law (eqn. 2.6). Thus, the transmitted radiation is purely polarized along the y axis. A fraction of E_y is also reflected in the direction of the reflected E_x polarization, so along that direction the polarization selectivity is poor. Although the first prism is sufficient to separate the polarizations, a second prism is needed to correct for the undesired refraction effects suffered by the transmitted beam. By keeping the gap between the two prisms small, the lateral shift of the polarized beam can be minimized while its original propagation angle is recovered by the second prism. With the exception of reflection losses at the interfaces, the E_y polarization component emerges with no losses. Typical discrimination for these prisms is approximately 10^{-5}; that is, the energy in the minor component is only 0.001% of the energy in the major component. When the space between the prisms contains only air, the polarizer can be used for high-power applications at a wide range of wavelengths. Glan prisms are most often used to separate one polarization from randomly polarized radiation. However, they can also be used as variable attenuators for linearly polarized radiation. By rotating the prism around the optical axis, the irradiance of the transmitted radiation can be varied continuously from approximately 100%, when the incident polarization is parallel to the ordinary axis, to nearly 0%, when it is pointing in the direction of the extraordinary axis.

Although the applications presented here require that the extraordinary axis of a uniaxial crystal be aligned with one of the incident polarizations, other applications require that the extraordinary axis point in the propagation direction or be inclined at an angle θ relative to it (Figure 5.6). While one polarization, E_x, remains aligned with one of the ordinary axes, the other polarization,

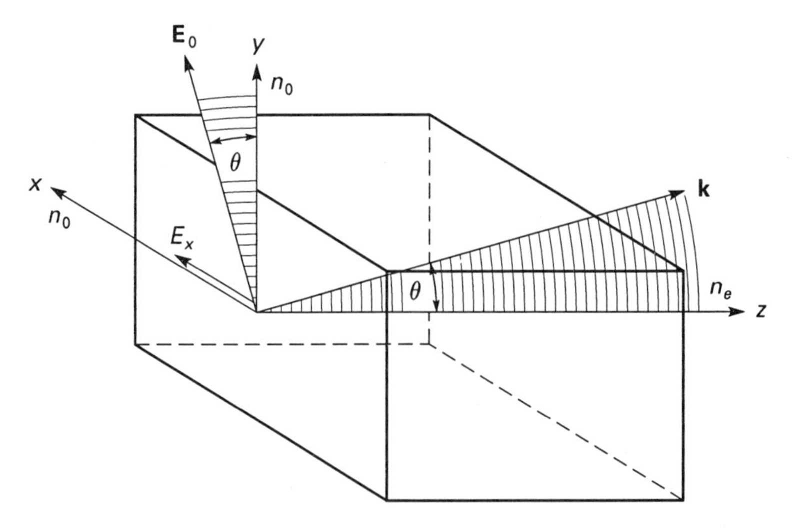

Figure 5.6 Propagation through a uniaxial crystal.

E_θ, is oriented parallel to an axis with an index of refraction that is a combination of n_o and n_e and a function of θ. Equations that describe the variation of the index of refraction with the propagation direction are available for general anisotropic media; see Yariv (1975). Here we present, without derivation, only the equation for propagation through a uniaxial crystal:

$$\frac{1}{n_e^2(\theta)} = \frac{\cos^2\theta}{n_o^2} + \frac{\sin^2\theta}{n_e^2}. \tag{5.7}$$

At $\theta = 90°$, when propagation is along one of the oridinary axes, one of the indices of refraction is n_e, whereas both indices are n_o for $\theta = 0°$. Equation (5.7) is known as the *index ellipsoid* for uniaxial crystals.

5.4 The Electro-Optic Effect

When a d.c. electric field is applied to certain media that otherwise may be isotropic, their birefringence changes as the voltage increases, thereby influencing the polarization characteristics of radiation passing through them. This is the *electro-optic* effect. When associated with crystals, the effect is named after Pockels. However, electro-optic responses can also be obtained in liquids and noncrystalline solids. This is the most common method of modulating radiation electronically or of rapidly switching high energies of radiation.

The Pockels cell, a simple but useful application of the Pockels effect, is illustrated in Figure 5.7. The cell consists of a uniaxial crystal enclosed at its two opposite faces by a pair of optically transparent electrodes. Initially, the ordinary axes are aligned with the x and y directions. However, as a d.c. field is applied by the electrodes along the extraordinary axis, the principal x and y axes of the crystal rotate to x' and y' and the magnitude of the indices of refraction along these axes changes so that $n_{x'} \neq n_{y'}$. On the other hand, both n_e

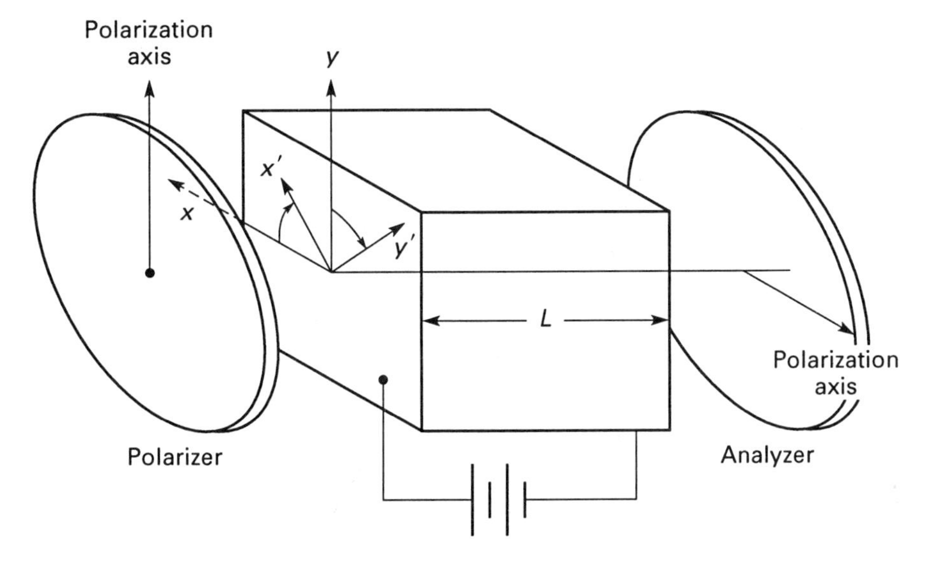

Figure 5.7 Typical configuration of a Pockels cell.

and the orientation of the extraordinary axis remain unchanged. The field can be increased until the new axes are rotated by 45°. The difference between the new indices of refraction, $n_{x'}$ and $n_{y'}$, is proportional to the external field E_z. Therefore, the delay between the $E_{x'}$ polarization component in the x' direction and between $E_{y'}$ in the y' direction increases as $E_z L$, where L is the crystal length (Das 1991). Depending on the type of uniaxial crystal, its length, and the voltage applied across its optical faces, Pockels cells can be designed for quarter-wave or half-wave retardation (Goldstein 1968). In the application presented in Figure 5.7, the cell is placed between two polarizers, the first of which is oriented for transmission of the y polarization. The second polarizer - the *analyzer* - transmits polarization in the x axis only. In the absence of an external field, the polarized radiation that was transmitted by the polarizer is not affected by the Pockels effect and is blocked by the analyzer at the opposite side of the cell. However, when a voltage that corresponds to $\lambda/2$ retardation is applied, the polarization of the incident field is rotated by 90° and can be transmitted by the analyzer. Commercially available Pockels cells can switch radiation within 1 ns and at rates of approximately 100 KHz while holding off high energy pulses.

A more familiar system that depends on the electronic switching of light via retardation is the liquid crystal display (LCD), which is commonly used for pocket calculators or watches (Shields 1991). Unlike Pockels cells, LCDs are very compact (0.5–5 μm thick), require low voltage (0–5 V) for switching, and operate well with white light, that is, the entire range of the visible spectrum. However, unlike Pockels cells, they cannot be used to block or transmit high-power radiation.

The term "liquid crystal" is somewhat misleading in that crystals are normally conceived of as being solids. What this term implies is that - although the medium is in liquid phase - under certain conditions the amorphous structure of the liquid can be modified to become more structured. These liquids contain long chains of molecules that tend to align themselves with the structure of surfaces that they contact or with an electric field vector. Therefore, when a liquid crystal is contained between two optical windows, the molecules in the liquid layer at one side are aligned with the surface structure of the adjacent window while the molecules on the opposite side follow the structure of the other window. By buffing the windows before assembly, any desired orientation can be obtained. Because of this molecular alignment, the liquid crystal can act as a retardation device and can be designed to generate half-wave retardation. Unlike Pockels cells, this retardation is present in the absence of any external field. However, when a d.c. voltage is applied across two transparent electrodes that are located at the two opposite windows, all molecules (including those adjacent to the windows) are aligned with the field and the retardation characteristics are eliminated. A typical LCD includes such a liquid crystal cell that is placed between a polarizer and an analyzer in a manner similar to the arrangement of Figure 5.7. In the absence of an external field, the liquid crystal retarder rotates the polarization of the polarized light which in turn can pass through the crossed polarizers. Viewed through the analyzer, the unbiased device appears transparent. However, when the external voltage is switched on, the retardation by the liquid crystal is eliminated and radiation that was passed by the polarizer is blocked by the analyzer. Viewed through the analyzer, the cell now appears dark.

5.5 The Photoelastic Effect

Anisotropy that causes birefringence can also be introduced by mechanical stresses in solids. These stresses may originate in the manufacturing process of optical or transparent materials, may result from inadvertent loading, or may be induced intentionally to create a desirable optical effect. Depending on the nature of the medium and the stress, the mechanical anisotropy can be temporary or permanent. This mechanically induced birefringence is the *photoelastic* effect. Photoelastic effects are often observed in optical windows where permanent residual stresses are introduced by uneven cooling during manufacture. Because the effect can deteriorate the polarization purity of transmitted radiation, high-quality optical windows are annealed. Although of no practical consequence, extruded plastic sheets of polymeric materials also present some photoelastic effect. The most important use of this effect is for optical imaging of the stress distribution of mechanically loaded models (Zandman, Redner, and Dally 1977). In this application, the photoelastic effect is induced intentionally.

Because the mechanically induced birefringence results from a change in the dielectric coefficient along one or more axes, description of a birefringent object - whether the birefringence is permanent or temporary - requires that the dielectric constant be represented by a tensor ϵ_{ij}. Similarly, a three-dimensional state of stress requires representation by a tensor σ_{ij}. This means that the dependence between the dielectric constant and the stress tensors must be described by a fourth-rank tensor q_{ijkl}. Fortunately, only two independent coefficients are needed to describe the dielectric constant or a mechanically stressed isotropic medium. Similarly, the indices of refraction of such stressed media may be described by only two stress-optic coefficients (Das 1991, pp. 182–3). If an isotropic object with an unperturbed index of refraction n_o is simultaneously compressed along the x, y, and z axes by σ_{xx}, σ_{yy}, and σ_{zz}, then the change in its index of refraction for radiation polarized along the x axis is

$$\Delta n_x = -\frac{n_o^3}{2}(q_{11}\sigma_{xx}+q_{12}\sigma_{yy}+q_{12}\sigma_{zz}). \tag{5.8}$$

Similar equations can be written for Δn_y and Δn_z. For fused silica, the measured *stress-optical coefficients* are $q_{11} = 5.65\times10^{-14}\ \mathrm{Pa}^{-1}$ and $q_{12} = 2.75\times10^{-12}\ \mathrm{Pa}^{-1}$ (Primak and Post 1959). Thus, when an optical medium of length L is subjected to planar loading σ_{xx} and σ_{yy}, the optical path difference ΔL_p for the x and y polarizations propagating in the z direction is:

$$\Delta L_p = (n_x - n_y)L = CL(\sigma_{xx}-\sigma_{yy}), \tag{5.9}$$

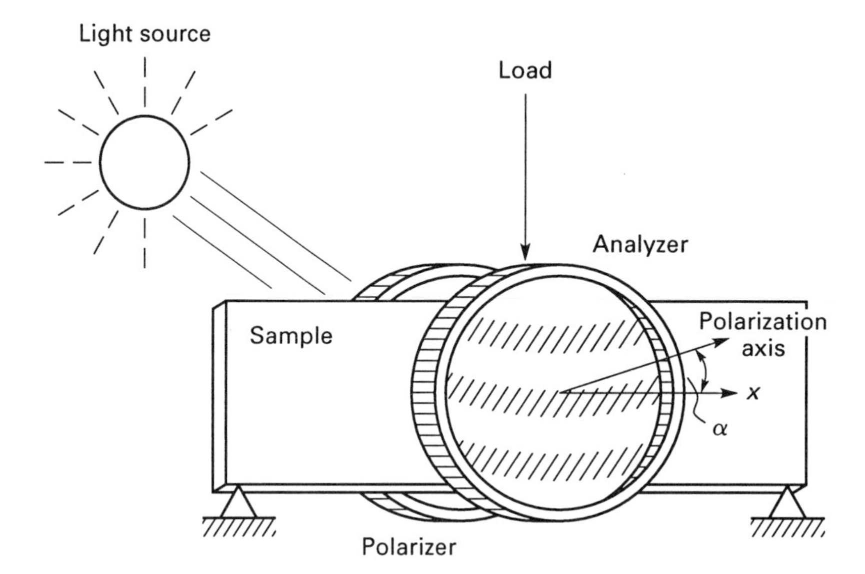

Figure 5.8 Experimental set-up for imaging the isoclines of a stress field using the photoelastic effect.

where $C = -n_o^3(q_{11} - q_{12})/2$ is the relative stress-optic coefficient. It is clear that, depending on L and the stress difference, any retardation between the phases of the x and y polarizations can be induced by the photoelastic effect. This can be visualized by placing a stressed transparent medium between the polarizers of a *polariscope* (Figure 5.8). The polariscope assembly is similar to the pair of polarizers used in conjunction with the Pockels cell. However, unlike the Pockels cell, the orientation of the principal axes in the mechanically loaded sample is not known a priori. In complex loading situations, where the orientation of the principal stresses is not uniform throughout the entire field of view, the orientation of the principal birefringence axes likewise varies. Therefore, the polarizers of the polariscope can be rotated relative to each other and relative to the sample. If the axis of the polarizer in Figure 5.8 forms an angle α relative to the x axis then monochromatic radiation transmitted through it is linearly polarized, with the polarization vector also forming an angle α with the x axis. When the stressed sample in the figure is illuminated by this linearly polarized light, the transmitted polarization is transformed everywhere into an elliptical polarization – except at locations where $\Delta L_p = m\lambda/2$ (m an integer). When m is even the polarization direction remains unchanged, whereas for odd m the linear polarization is rotated by $\pi/2$. Thus, when viewed through the analyzer, the image is interspersed by dark and bright fringes. When the analyzer axis is parallel to the polarizer axis, the dark regions represent loci where simultaneously (a) the inclinations of the principal stress axes are parallel and perpendicular to the polarizer axis and (b) the difference between the principal stresses is a constant defined by an odd value of m (i.e., resulting in the rotation of the polarization by $\pi/2$). Similarly, the regions where m is even can be viewed by rotating the analyzer by 90°. From (5.9), the stress differential is related to m (the *fringe order*) by

$$m\lambda = CL(\sigma_{xx} - \sigma_{yy}). \tag{5.10}$$

Although (5.10) can be used to determine the difference between the principal stresses, the exact value of m must be known. Unfortunately, when stress concentration occurs, the density of fringes at a point often exceeds the resolution limit of the imaging system and the exact value of m cannot be determined. Thus, the primary value of this visualization technique is to define regions where the principal stresses are uniformly oriented. These regions are called the *isoclines*. Other families of isoclinic fringes can be obtained by rotating the entire polariscope assembly. Polariscopes are often used to detect residual stresses in optical elements or to study the stress distribution in mechanical models subjected to complex loading.

An alternative photoelasticity method employs a combination of a polarizer and a $\lambda/4$ plate to illuminate the sample with circularly polarized light. The transmitted light is then analyzed by a second assembly of a polarizer and $\lambda/4$ plate. This system is the *circular polariscope* (see Research Problem 5.1). The circular polariscope is assembled so that, when the photoelastic effect does not perturb the circular polarization, the radiation passing through the second $\lambda/4$ plate is restored to linear polarization and the linearly polarized light is blocked by the analyzer. Otherwise, the radiation transmitted by the test object is also transmitted through the analyzer. Thus, dark regions represent areas where the stress differential between the principal axes introduced a retardation of $\Delta L_p = m\lambda$. However, since the radiation that illuminates the sample is circularly polarized, this method is insensitive to the orientation of the principal axes. Fringes obtained in the circular polariscope are called *isochromates*.

Advanced techniques for photoelastic visualization of the strains of realistic (i.e., not optically transparent) models use photoelastic coatings (Zandman et al. 1977) to visualize the strain. In such experiments, the polarization status of the light reflected from the loaded object is analyzed by the analyzer.

5.6 Ellipsometry

Measurements of the state of the polarization of light upon reflection at an interface can be used to calculate some of the interface parameters. If the interface is uncoated, the indices of refraction of the media separated by the interface can be measured. If the medium is coated by a thin film or by multiple layers, their thickness and indices of refraction can be measured as well. From these results, the nature (or even the identity) of the layers that form the coating can be determined. Although there exist numerous methods for these measurements, all are comprised by the rubric *ellipsometry* (Azzam and Bashara 1977).

To illustrate the potential of ellipsometry, consider one of the simplest configurations (Figure 5.9). In this arrangement, a monochromatic unpolarized

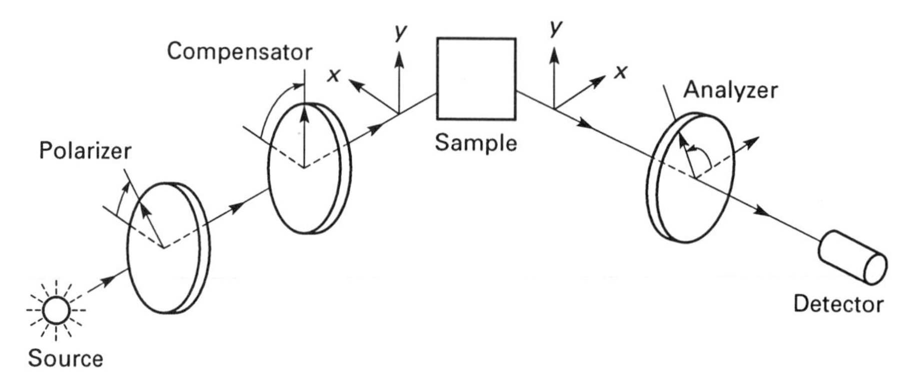

Figure 5.9 Typical experimental set-up for ellipsometry.

light is used to illuminate the sample. The light is passed through a polarizer and a retarder (or *compensator*) that may be either a $\lambda/4$ or $\lambda/2$ plate. The elliptically polarized light is reflected off the sample surface and analyzed by a second polarizer - the analyzer - which can be rotated around its axis until the reflected radiation is blocked. The angle of the analyzer or its azimuth, the azimuths of the polarizer and of the compensator, and the incidence angle can all be independently set and measured.

After passing through the first leg of optical elements, the incident radiation has two polarization components, E_x^i and E_y^i. Using the notation of (5.1), these polarization components may be described by

$$E_x^i = E_{0x}^i e^{-i\phi_1} \quad \text{and} \quad E_y^i = E_{0y}^i e^{-i\phi_2}, \tag{5.11}$$

where both the temporal and spatial dependence have been omitted. Thus, the state of the incidence polarization is defined by four parameters: the amplitudes E_{0x} and E_{0y} and the phases ϕ_1 and ϕ_2. In most applications, only the relative amplitudes and the phase difference need to be measured. Thus, the state of the incident polarization can be expressed by the polarization variable χ_0, where

$$\chi_0 = \frac{|E_{0x}^i|}{|E_{0y}^i|} e^{-i(\phi_1 - \phi_2)} \tag{5.12}$$

(Azzam and Bashara 1977). If the absolute amplitude and phase need not be determined, then clearly the state of the polarization can be uniquely specified by the following independent parameters:

$$|\chi_0| = \frac{|E_{0x}^i|}{|E_{0y}^i|}, \quad \arg(\chi_0) = \phi_1 - \phi_2.$$

After reflection at the interface, the new state of polarization includes two components, E_x^r and E_y^r, which can be expressed using notation similar to that of (5.11) and (5.12). Thus, the parameters $|\chi_r|$ and $\arg(\chi_r)$ can be used to specify the reflected polarization when the amplitude and absolute phase need not be specified.

To find the number of independent parameters needed to determine the state of polarization of the reflected beam, consider the two independent com-

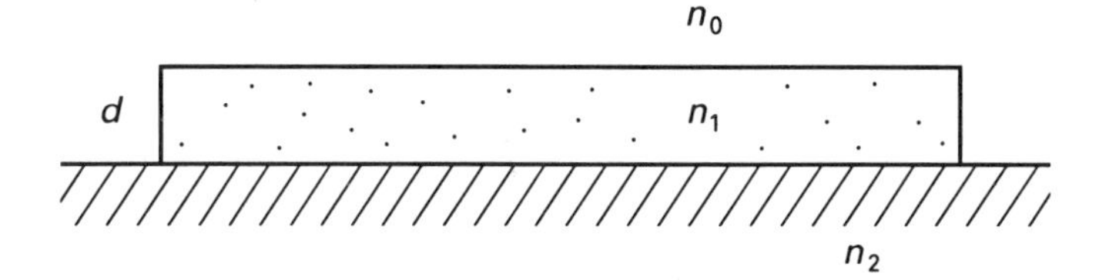

Figure 5.10 An interface between isotropic media coated with a single layer of refractive medium.

ponents of the incident polarization as being the elements of a two-dimensional vector and (similarly) the two reflected polarization components as elements of another two-dimensional vector. These two vectors can be related by the following matrix equation:

$$\begin{vmatrix} E_x^r \\ E_y^r \end{vmatrix} = \begin{vmatrix} T_{11} & T_{12} \\ T_{21} & T_{22} \end{vmatrix} \begin{vmatrix} E_x^i \\ E_y^i \end{vmatrix} \tag{5.13}$$

(Azzam and Bashara 1977). This is the *Jones matrix* representation of the transformation of the incident polarization. In its most general form (i.e., when the polarization vectors are presented as in eq. 5.11), the elements T_{ij} of the Jones matrix are complex, thereby requiring eight independent parameters to specify the transformation. However, if the absolute amplitude and absolute phases of the incident and reflected light do not need to be determined (as in eq. 5.12), then the Jones matrix can be reduced to only three independent terms by dividing each element by T_{22}. Thus, the matrix elements reduce to T_{11}/T_{22}, T_{12}/T_{22}, T_{21}/T_{22}, and 1.

For reflection at an isotropic interface coated by a single layer of depth d (Figure 5.10), there may exist at most seven independent parameters; these include the real and complex components of the indices of refraction n_0, n_1, and n_2 of the three media, as well as the film thickness. In most applications $n_0 = 1$, so there may actually be no more than five independent parameters that need to be measured by determining the polarization transformation at three different azimuth angles. Some surfaces may be specified by even fewer parameters.

In a typical measurement, the azimuth angles of the polarizer, compensator, and analyzer are set to completely extinguish the radiation incident at the detector. This is *null ellipsometry* (Azzam and Bashara 1977). The measurement may also require adjustment of the retardation of the compensator, thereby raising the number of measured parameters to four. Other configurations, involving different angles of incidence or other wavelengths, may also be used. With these alternative configurations, anisotropy or spectral properties of the reflector and its coating may be determined.

References

Azzam, R. M. A., and Bashara, N. M. (1977), *Ellipsometry and Polarized Light,* Amsterdam: North-Holland.

Born, M., and Wolf, E. (1975), *Principles of Optics,* 5th ed., Oxford: Pergamon, pp. 51–5, 90–4.

Chabal, Y. J., and Sievers, A. J. (1978), Surface electromagnetic wave launching at the edge of a metal film, *Applied Physics Letters* 32: 90–2.
Das, P. (1991), *Lasers and Optical Engineering,* New York: Springer-Verlag, pp. 172–5.
Dmitriev, V. G. (1991), *Handbook of Nonlinear Optical Crystals,* New York: Springer-Verlag.
Goldstein, R. (1968), Pockels cell primer, *Laser Focus* 4: 21–7.
Jackson, J. D. (1975), *Classical Electrodynamics,* 2nd ed., New York: Wiley, pp. 67–8, 299–303.
Otto, A. (1968), Excitation of nonradiative surface plasma waves in silver by the method of frustrated total reflection, *Zeitschrift für Physik* 216: 398–410.
Primak, W., and Post, D. (1959), Photoelastic constants of vitreous silica and its elastic coefficient of refractive index, *Journal of Applied Physics* 30: 779–88.
Shields, S. E. (1991), Liquid crystals: unusual materials with a picturesque future! *Optics & Photonics News* 2(April): 58–9.
Yariv, A. (1975), *Quantum Electronics,* New York: Wiley, pp. 82–90.
Zandman, F., Redner, S., and Dally, J. W. (1977), *Photoelastic Coatings,* Ames: Iowa State University Press and Society for Experimental Stress Analysis.
Zhu, S., Yu, A. W., Hawley, D., and Roy, R. (1986), Frustrated total internal reflection: a demonstration and review, *American Journal of Physics* 54: 601–7.

Homework Problems

Problem 5.1

(a) Show that, when $E_x = E_y$ and $\phi_1 - \phi_2 = \pi/2$, the tip of the polarization vector forms a circle while rotating at a frequency of ω.

(b) Show that, when $E_x \neq E_y$ and $\phi_1 - \phi_2 = \pi/2$, the tip of the polarization vector forms an ellipse while rotating at a frequency of ω. Determine the orientation of the principal axis of the ellipse and state how it can be changed.

(c) What polarization will be obtained when $\phi_1 - \phi_2 \neq \pi/2$? (Consider both $E_x = E_y$ and $E_x \neq E_y$.)

Problem 5.2

Show that the superposition of two circularly polarized fields with opposite spin directions can result with a linearly polarized field. Determine the conditions required to obtain purely linear polarization and the direction of the linear field vector relative to the positive x axis shown in Figure 4.3.

Problem 5.3

Prove that energy is conserved at an interface where both reflection and transmission take place. Show also that the energy of each polarization is conserved independently.

Problem 5.4

The equation for Brewster's angle can be derived using intuitive considerations. Assume that reflection occurs by radiation emitted by dipoles that are excited by the incident field at the interface. The orientation of the dipole oscillation coincides with the incident field orientation immediately after refraction; however, the direction of the reflected beam and its field are determined by the rules of specular reflection. Thus, reflection may not be possible when the refracted field is parallel to the propagation vector of the reflected beam. With these considerations, prove (5.4).

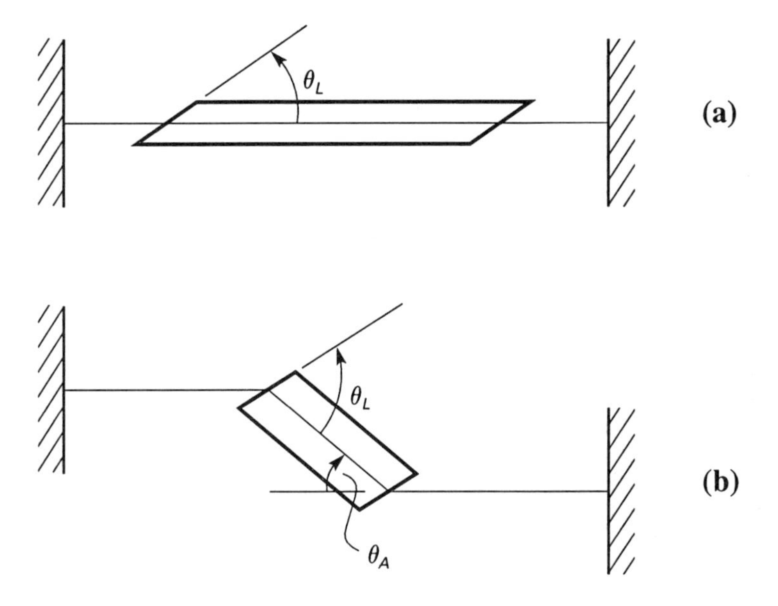

Figure 5.11 Brewster angles of laser media for **(a)** a gas-filled discharge tube and **(b)** a solid crystalline rod.

Problem 5.5

In order to minimize reflection losses, the interfaces of laser media are tilted so that the beam oscillating between the laser mirrors is incident at the interface at Brewster's angle. When the laser medium consists of a gas ($n \approx 1$) contained in a tube, only the end windows must be tilted; see Figure 5.11(a). However, when the laser medium is a solid (e.g. a crystal), both the ends of the crystal and the crystal itself must be tilted to allow for refraction (Figure 5.11(b)).

(a) Find the angle θ_L between the window face ($n = 1.5$ for the window) and the tube axis for a gas laser.

(b) Find the angle θ_L between the faces of a laser crystal and its axis, as well as the angle θ_A between the crystal axis and the laser axis, when $n = 1.5$ for the crystal.

Problem 5.6

Show that successful transmission of the y polarization through the Glan prism in Figure 5.5 requires the extraordinary index of refraction n_e to be larger than the ordinary index of refraction n_o. What is the angle θ between the inclined surface and the horizontal plane when $n_e = 1.55$? (*Answer:* $\theta > 49.8°$.)

Problem 5.7

An electromagnetic wave is reflected normally by a fully reflecting mirror.

(a) What is the field of the s component of the polarization and of the p component at the interface?

(b) Write the 1-D wave equation for the incident beam, the reflected beam, and the total electromagnetic field in the vicinity of the mirror.

(c) What type of wave is described by the total electromagnetic field? Explain.

Research Problem

Research Problem 5.1

Read the first chapter of Zandman et al. (1977), or any other text describing photoelasticity.

A slab of photoelastic material is calibrated by stretching it in a tensile machine. The calibration consists of monitoring the light transmission through a linear polariscope illuminated by a collimated beam at $\lambda = 632.8$ nm. The slab is stretched by applying a load that can be varied continuously from 0 to 25 kg. The cross section of the slab is 0.3×3 cm^2. The wide side of the slab is facing the light source.

(a) Plot (freehand) the variation of the transmitted intensity with the applied load when $\alpha = 0°$.

(b) It is observed that when $\alpha = 45°$, the points of maximum transmission appear at intervals of 3 kg. Plot the variation of the transmitted intensity with the applied load.

(c) Find the value of C, the relative stress-optic coefficient, as defined by the following equation:

$$n_1 - n_2 = C(\sigma_1 - \sigma_2).$$

(d) Assume that C is independent of the incident wavelength. Draw the calibration curve for an incident radiation at $\lambda = 514.5$ nm and when $\alpha = 45°$.

(e) A circular polariscope consists of two $\lambda/4$ plates and two polaroids. They are mounted with one $\lambda/4$ plate and one polaroid on the same frame, with the polaroid axis forming an angle of 45° with the $\lambda/4$ plate axes. To identify the polaroid side, the pair is placed over a reflecting metallic surface. How will the reflection look (i) when the $\lambda/4$ plate is facing the metallic surface and (ii) when the polaroid is facing the metallic surface. Explain.

6 Interference Effects and Their Applications

6.1 Introduction

The discussion in Section 4.10 on the scattering by gas molecules and by submicron particles illustrated the rules of superposition of radiation from several sources. Without much detail, the analysis there showed that the irradiance resulting from such superposition depends on the coherence properties of the sources: when the radiation emanating from several sources is coherent, the fields are additive; if incoherent, only the energies are additive. The distinction between these two modes of addition is important in view of the quadratic dependence (eqn. 4.42) between the irradiance and the electric field. Thus, the analysis of the superposition of radiation emitted by incoherent sources requires only the summation of the irradiance from all sources at a point. No consideration of the frequencies or the phases of the interacting fields is needed. On the other hand, the irradiance that results from the superposition of radiation from a multitude of sources that are coherent with each other depends on the spatial and temporal distribution of the interacting fields, on their phases, and on their frequencies. Thus, before such irradiance can be determined, the distribution of the combined fields must be found. The spatial and temporal distribution of the irradiance is then obtained from the field distribution using (4.42). Here we discuss the details of the superposition of coherent electromagnetic fields. Such detailed analysis can be simplified when considering the superposition of only two beams obtained by splitting one beam emitted by a single source. The results of this analysis, although limited in scope, can be readily generalized. Furthermore, many engineering applications depend on the superposition of only two coherent sources. Therefore, the results of our analysis can be directly implemented for such applications.

6.2 Interference by Two Beams

To obtain two separate beams that are mutually coherent, a single beam with sufficiently long coherence length is split into two parts. After splitting, each part may proceed through a separate optical path. When the two parts are recombined, the energy distribution at the point of superposition may be significantly different from the distribution in the original beam. The new distribution depends on the propagation characteristics along each path. Of primary interest here are the effects of each path on the phase and the frequency of each beam. The former may affect the energy distribution longitudinally and transversally in the region of overlap between the beams. The resultant pattern, which includes visually apparent dark and bright fringes, is time-independent and is called an *interference* pattern. The second effect, induced by small frequency changes along one or both of the paths of the beams, may introduce time-dependent fluctuations of the energy at the overlap region. This effect, which may be viewed as a time-dependent interference, is called *heterodyning*. These two related phenomena offer important engineering applications. Here we present the governing equations for both phenomena.

A simple device that can be used to illustrate both the effects of interferometry and heterodyning is the Michelson–Morley interferometer (Figure 6.1). It was originally designed to measure the aether speed. This rather simple interferometer consists of a beam splitter illuminated by a collimated beam and two mirrors. The incident beam is split into two parts. One part propagates toward mirror M_1, where it is reflected back toward the beam splitter. A portion of this returning beam is transmitted by the beam splitter and is incident on the screen with a field intensity $\mathbf{E}_1$. The field of the second part, which reaches the screen after reflection by mirror M_2 and after a reflection by the beam splitter, is $\mathbf{E}_2$. The length of the optical paths traveled by the first and second beams are L_1

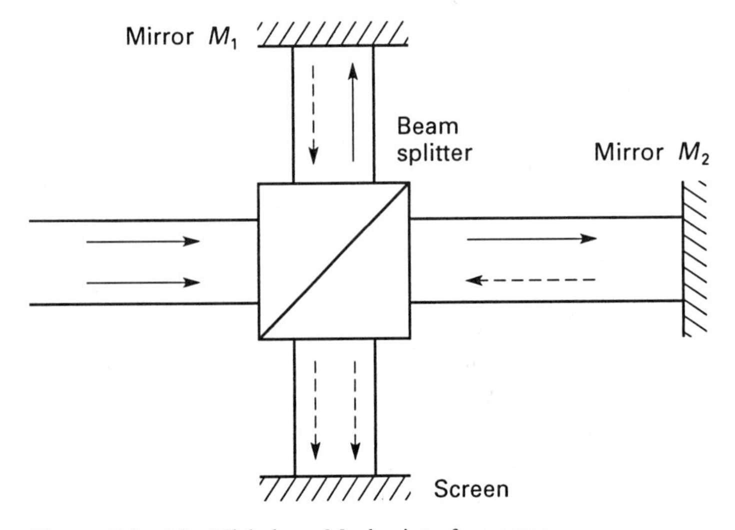

Figure 6.1 The Michelson–Morley interferometer.

and L_2, respectively. These optical paths include the distances traveled between the beam splitter and the mirrors, the optical paths within the beam splitter, and finally the distance along the common path between the beam splitter and the screen. To determine the irradiance I at the screen, we must first find the functions that describe the combined distribution of the two incident fields. Modifying (4.11) to describe a one-dimensional propagation along L_1 and L_2, we obtain the following equations for forward propagation along each of the paths:

$$\begin{aligned} \mathbf{E}_1 &= \mathbf{E}_{01} e^{-i\omega t} e^{-i\phi_1} e^{ikL_1}, \\ \mathbf{E}_2 &= \mathbf{E}_{02} e^{-i\omega t} e^{-i\phi_2} e^{ikL_2}, \end{aligned} \tag{6.1}$$

where the frequencies of the two beams were assumed to be identical. Using (4.42), the irradiance at the screen obtained by superposition of the two beams is

$$I = \frac{1}{2\eta}(\mathbf{E}_1 + \mathbf{E}_2)(\mathbf{E}_1 + \mathbf{E}_2)^*, \tag{6.2}$$

where the asterisk represents the complex conjugate. After substituting (6.1) for $\mathbf{E}_1$ and for $\mathbf{E}_2$, we have

$$\begin{aligned} I = \frac{1}{2\eta}[&\mathbf{E}_1\mathbf{E}_1^* + \mathbf{E}_2\mathbf{E}_2^* \\ &+ \mathbf{E}_{01}\mathbf{E}_{02}^* e^{-i(\phi_1-\phi_2)} e^{ik(L_1-L_2)} + \mathbf{E}_{02}\mathbf{E}_{01}^* e^{-i(\phi_2-\phi_1)} e^{ik(L_2-L_1)}]. \end{aligned} \tag{6.3}$$

Although the time-dependent terms $\exp\{i\omega(t_2-t_1)\}$ and $\exp\{i\omega(t_1-t_2)\}$ were part of the integration used to derive this irradiance, the large value of ω assures that – after integration over the time T of a few periods – the result becomes time-independent for all values of t_1 and $t_2 \gg T$. When we discuss the effect of heterodyning, this time-independence will have to be reconsidered.

The result of (6.3) shows that the irradiance of the superimposed beams depends on the difference between their respective initial phases, $\phi_1 - \phi_2$. Thus, when the coherence length of the incident beam exceeds $|L_1 - L_2|$, their initial phases remain equal, despite passing through two different paths, and hence $\phi_1 - \phi_2 = 0$. This is a necessary requirement for successful interference between the radiation from two sources. To meet this requirement when the coherence length is short, the lengths of the two arms of the interferometer must be carefully adjusted to maintain a difference in the optical paths that is shorter than the coherence length. The restrictions on the difference between the path lengths of the two arms is usually met when using lasers, where typical coherence lengths range from a few millimeters to several meters. (For example, the coherence length of a typical argon ion laser is 4 cm but can be extended to a few meters by the addition of an intracavity etalon.) Thus, when the interferometer is illuminated by a beam with a coherence length that exceeds $|L_1 - L_2|$, the irradiance is

$$I = I_1 + I_2 + 2\sqrt{I_1 I_2}\cos k(L_1 - L_2), \tag{6.4}$$

where I_1 and I_2 are the irradiances of the two parts of the split beam. Equation (6.4) represents a spatially varying distribution of the irradiance at the region where the two beams overlap. For perfectly collimated beams (i.e., for plane waves) and for collinear propagation, this spatial distribution is longitudinal. However, more realistically, when the beams are slightly divergent and not perfectly collinear, the distribution of the irradiance has a transverse structure also. As a result of this distribution there exist, within the overlap region, points with minimum irradiance I_{min} and points with maximum irradiance I_{max}. If $I_1 = I_2$ then the irradiance at the minima, or the dark fringes, is $I_{min} = 0$ and at the maxima is $I_{max} = 4I_1$. Otherwise, $I_{min} \neq 0$. Thus, when $I_1 \neq I_2$, the contrast between the dark and the bright fringes is diminished. This contrast can be quantified by the following *visibility function*:

$$V = \frac{I_{max} - I_{min}}{I_{max} + I_{min}}. \tag{6.5}$$

The visibility function approaches its maximum value of unity when the irradiance of both beams is equal. Irrespective of V, the fringe pattern obtained by the interaction between two beams with the same frequency is time-independent and is representative of the initial shape of the wavefronts of the interacting beams, as well as of the disturbances encountered along their individual paths. This presents many opportunities for engineering and scientific measurements that will be discussed in the following sections.

The result of the interference between two coherent beams was presented by (6.4). However, it often happens that two incoherent beams, or beams with a coherence length comparable to $|L_1 - L_2|$, are superimposed. To determine the effect of such interaction, (6.4) is rewritten while also including the term $\phi_1 - \phi_2$:

$$I = I_1 + I_2 + 2\sqrt{I_1 I_2}\cos[k(L_1 - L_2) - (\phi_1 - \phi_2)]. \tag{6.6}$$

The phase of each of the beams varies randomly at a rate that exceeds 10 GHz. Therefore, the phase difference between the two beams results in a rapid oscillation of the cosine term. Practically, with such rapid and random variation, only the average irradiance can be detected. Since the average of the cosine term is zero, the irradiance at the point of interaction is simply the sum of the irradiance of the two incident beams without any visible interference pattern. This is similar to the result obtained in Section 4.10 for the superposition of incoherently scattered radiation, where the total irradiance was seen as the sum of the irradiance from all scatterers. Of course, when the path difference between the two beams is comparable to the coherence length, interference may take place between some of the longer wavetrains and so result in a partially visible interference pattern. Thus, as the difference between the path lengths increases, the visibility (eqn. 6.5) of the fringe pattern diminishes until the fringes gradually disappear.

6.3 Heterodyning

In the analysis of the previous section, the frequencies of the two interfering beams were assumed to be identical. Because the frequency of radiation is only rarely changed by reflection or refraction, this assumption is generally valid. However, when one or both reflecting surfaces is moving, the frequency of one or both beams may be slightly shifted. When this occurs, the superposition of the two fields also includes a time-dependent term that results in an oscillation of the irradiance. This oscillation is detectable anywhere within the region of overlap, and its frequency is equal to the difference between the frequencies of the interacting beams. Measurement of this frequency difference is used for flow velocimetry or for the detection of speeding cars by police radars. This process of heterodyning can be combined with interference to form both spatially and temporally varying fringe patterns.

To illustrate the heterodyning process, and particularly the process that results in a frequency shift, consider again the Michelson–Morley interferometer (Figure 6.1). However, mirror M_1 is now allowed to move toward the beam splitter at a velocity $\mathbf{V}$. Thus, the initial path length L_1 along this arm decreases during a time t by $2\mathbf{V}t$. Including this term in (6.1) and lumping together all the time-dependent terms yields

$$\mathbf{E}_1 = \mathbf{E}_{01} e^{-i(\omega - 2\mathbf{k}\cdot\mathbf{V})t} e^{-i\phi_1} e^{ikL_1},$$

$$\mathbf{E}_2 = \mathbf{E}_{02} e^{-i\omega t} e^{-i\phi_2} e^{ikL_2}.$$

The additional time-dependent term appears as an increase of $\Delta\omega$ in the incident beam frequency:

$$\Delta\omega = -2\mathbf{k}\cdot\mathbf{V}. \tag{6.7}$$

Because the velocity of M_1 and $\mathbf{k}$ are both vectors, the frequency shift (which is a scalar) must be presented by a scalar multiplication. This velocity-dependent frequency change is the *Doppler shift*. The negative sign in (6.7) indicates that the shift is positive when M_1 is moving toward the incoming radiation and negative when moving away from it. Thus, for the example presented here, the Doppler shift observed at M_1 is positive with a magnitude of:

$$\Delta\omega = 2\frac{|\mathbf{V}|\omega}{c}. \tag{6.8}$$

We can show that (6.8) combines two shifts: one associated with the motion of M_1 toward the incoming beam, and the other by the motion of M_1 toward the stationary screen (see Problem 6.6). Although relativistic effects were neglected, this result accurately represents almost all phenomena with engineering importance.

Of course, the frequency of the radiation reflected by the stationary mirror M_2 is unshifted. Nonetheless, calculation of the superposition of the two fields at the screen must include the effect associated with the Doppler frequency shift of the first field. When the propagation of the overlapping beams is collinear, the exponential term of $k(L_1 - L_2)$ that would appear in their superposition may be omitted if only the effect of the frequency is analyzed. In addition, $\phi_1 = \phi_2$ when the coherence requirement is met. From (6.2) we therefore have, for the irradiance of the two superimposed beams with frequencies of ω_1 and ω_2,

$$I = I_1 + I_2 + 2\sqrt{I_1 I_2}\cos(\omega_1 - \omega_2)t. \quad (6.9)$$

Equation (6.9) was obtained by time integration of the superimposed fields over a period of T, which is long relative to one optical cycle but short relative to the characteristic time of the process (eqn. 4.42). Thus, the terms I_1 and I_2 are independent of the field oscillation. On the other hand, when the frequency difference $(\omega_1 - \omega_2) \ll \omega_1$ or ω_2, the associated characteristic time is long and so the term containing this difference does not average out in the integration. Thus, the resultant irradiance has an oscillating time variation even if I_1 and I_2 are constant. This is the effect of *heterodyning*. It can be observed when the frequency of one beam is shifted by the motion of one source such as M_1 (eqn. 6.7). The time-dependent irradiance at the region where the two beams overlap depends on the mirror velocity as follows:

$$I = I_1 + I_2 + 2\sqrt{I_1 I_2}\cos(-2\mathbf{k}\cdot\mathbf{V})t \quad (6.10)$$

(cf. eqns. 6.7 and 6.9). From this result it is evident that a detector placed at any point within the overlap region of these two coherent beams will detect a signal that consists of two components: a constant component with an irradiance of $I_1 + I_2$, modulated by an oscillating component at a frequency of $2\omega|\mathbf{V}|/c$. This frequency is proportional to the absolute value of the velocity component of the moving mirror that is directed along the incoming beam. The more general case of noncollinear velocity and propagation vectors will be discussed in Section 6.7.

The amplitude of the oscillating term in (6.9) depends on the irradiance of both beams. Thus, even if the irradiance of the frequency-shifted beam I_1 is low, the amplitude of the oscillating component may be amplified by selecting a large I_2. Such amplification of the oscillating term by combining radiation from a faint remote source with radiation from a more intense local source is an important characteristic of the heterodyning effect. It is used in astronomy – to amplify the signal gathered from remote stars – and in devices used by motorists to detect police radar. In both applications, radiation that simulates the unshifted reflection by M_2 in the Michelson–Morley interferometer (Figure 6.1) is generated by a source adjacent to the detector: the *local oscillator*. In the absence of other sources, the output of the detector simply depends on the irradiance of the local oscillator. When radiation from a remote source is detected (e.g., when radiation emitted by a police radar is collected by the antenna of the detector), it is combined by the detector with the radiation emitted by the local oscillator. The signal generated by the radiation from both sources contains, in addition to the two irradiance terms, an oscillating term (eqn. 6.9) that can be isolated electronically. Thus, even if the signal from the remote source is relatively weak, its superposition with the local oscillator generates a detectable oscillating output that indicates the presence of radiation from the remote source. Note that the use of a local oscillator for the detection of faint radiation requires that both sources, the local and the remote, be coherent. This is true for both stellar radiation (see Section 7.2) and microwave radiation. The coherence length of the local oscillator is accordingly a design parameter of the detection system.

A wider application of the heterodyning effect is the measurement of the velocity of moving objects. Several techniques have been developed for the measurement of the velocity of gas flows or moving objects. Velocities of aircraft and cars are measured using radar frequencies that have longer coherence lengths. The velocity of gas flows is usually measured using visible laser beams. A description of one technique, the laser Doppler velocimeter, will be presented in Section 6.7.

In this section we showed that the effect of heterodyning is identical to interferometry, with the exception that the former is a time-dependent phenomenon while the latter is spatial. This becomes evident by noticing that (6.10) could also be obtained directly from (6.4) by introducing the time dependence of $(L_1 - L_2)$ associated with the motion of mirror M_1 (see Problem 6.4). A more complete description of interference effects can be found in Hecht (1990). In the following sections we will discuss some engineering applications that depend on interference and heterodyning.

6.4 Interferometry

Interferometry is one of the most sensitive methods for measuring small displacements or small variations in the length of an optical path induced by motion of objects, changes or irregularities in surface structure, or changes in the index of refraction along an optical path. Interferometers are used to analyze the curvature of lenses and the displacement of fine machining devices, or to visualize flow fields with a sensitivity that can reach a fraction of the wavelength used for the measurement.

The effect of splitting a coherent monochromatic beam and recombining its parts on the distribution of the combined irradiance was discussed in the previous section. It was shown (eqn. 6.4) that when the two parts of the split beam travel through different optical paths without any frequency change, the irradiance of the combined beams depends on the difference between the lengths of these paths. At points where the difference $|L_1 - L_2| = m\lambda$ (where m is an integer), the irradiance reaches a maximum; at points where $|L_1 - L_2| = (2m+1)\lambda/2$, the irradiance is at a minimum. Thus, the distribution of the irradiance of the recombined beams appears to be modulated with bright and dark interference fringes that represent a map of the optical path differences. An exception (without any technical consequences) is the collinear recombination of planar wavefronts, where only one fringe – bright, dark, or gray – can exist longitudinally and transversely. An optical path difference can be induced by transmitting one beam through a medium with either a different index of refraction or a longer geometrical length than the other, by bouncing one beam from a curved surface while passing the other through an unperturbing path, or by any other method that either delays one of the beams relative to the other or distorts the shape of its phase front. Because a difference of less than λ in the optical paths is sufficient to induce a noticeable variation in the irradiance distribution, minor variations in the index of refraction of gases, or a slight motion of an object, can be detected as a change in the pattern of the interference fringes. Holography is one of the most notable applications of interference effects. However, holography depends also on diffraction effects so our discussion of holography will be deferred to the next chapter (Section 7.5). Some of the simpler applications of interference effects are discussed next.

Interferometers are the most direct application of the effects of interference. By modifying the Michelson–Morley interferometer (Figure 6.1), we can create a device that is useful for measurements of minute displacements or for mapping the contours of curved objects. When mirror M_1 is replaced with a test object, the interference pattern recorded on the screen (or directly by a camera placed at the location of the screen) will depend on the location of the object, its angle relative to the incident beam, and its shape. For illustration, consider a variation of the Michelson interferometer that is used to determine the slope and position of a flat object O relative to a reference plane (see Figure 6.2). Ideally, a plane wave is used for the illumination. The beam is divided into two arms: the object beam used to illuminate the object and the reference beam in the second arm. For this measurement, the object represented by M_1 is expected to be specularly reflective. The effect of interferometry by diffuse reflection will be discussed in Section 6.5. Following reflection at the end of each arm, the beams are recombined by the beam splitter and their combined irradiance is recorded.

The equations describing the fringe pattern at the screen, following reflection by the tilted object, can be obtained directly by determining the wave

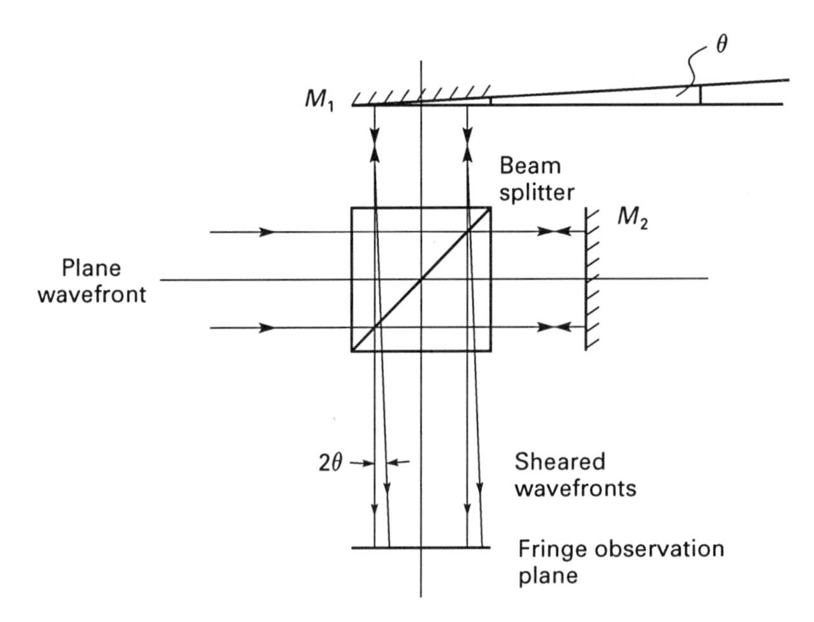

Figure 6.2 Interference induced by tilting one mirror in the Michelson–Morley interferometer. [Jones and Wykes 1989, © Cambridge University Press]

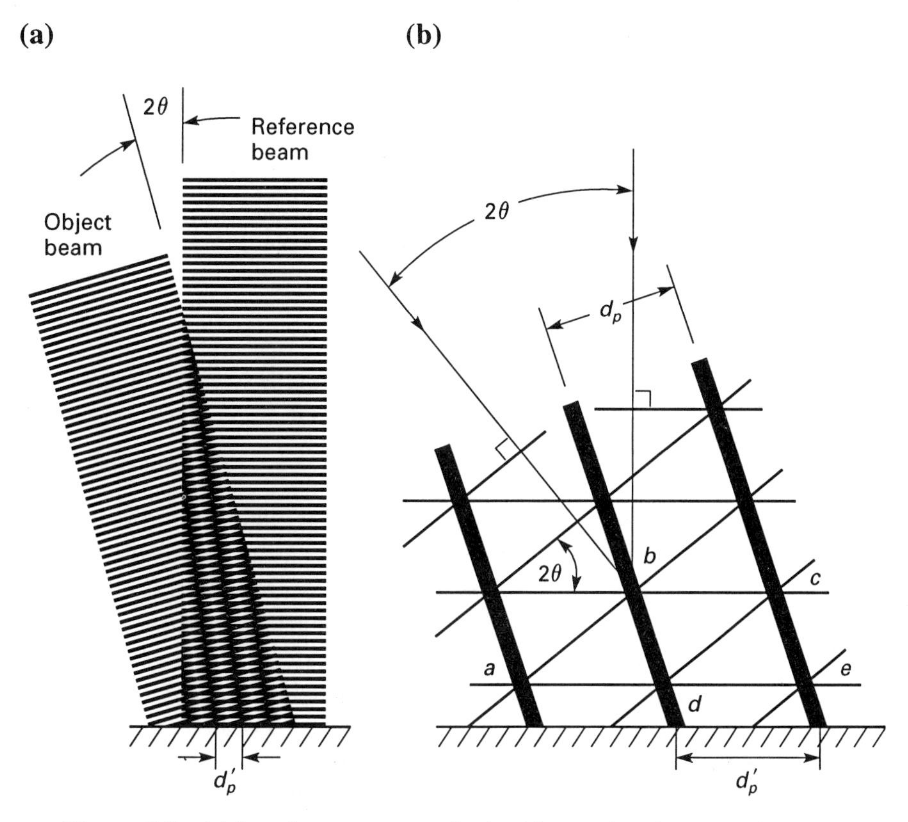

Figure 6.3 (**a**) Interference pattern obtained by overlapping two coherent plane waves. (**b**) Close-up view showing the corresponding geometrical parameters.

functions for each of the interacting beams at the screen, superimposing them, and calculating the irradiance to obtain an expression similar to (6.4). Alternatively, a geometrical presentation of the wavefronts of the two beams may serve better to elucidate the interaction pattern and its essential characteristics. Figure 6.3(a) illustrates the overlap between the plane wavefronts of the reference beam and the object beam at the screen. Each wavefront is represented by a straight line extending transversely across the beam. The regular spacing between these parallel lines represents the wavelength λ. The loci of the intersection points between the wavefronts of the two beams form a geometrical pattern of bright and dark strips that are parallel to the bisector of the angle between the intersecting beams. Although the strips in the figure result from the geometrical overlap between the lines of the two grids, they accurately simulate the geometry of interference fringes caused by the intersection of the wavefronts of the two beams. Thus, the same geometrical considerations needed to determine the spacing between the strips in Figure 6.3(a), their orientation, or their shape can be used to describe the geometry of the interference fringes formed by the intersection of two coherent beams.

To predict from this geometrical analogy the structure of interference patterns formed by the intersection of two collimated beams, we use a close-up of several wavefronts (Figure 6.3(b)). Since interference patterns are formed by the intersection of wavefronts, one fringe can be represented in Figure 6.3(b) by line *bd* and another by line *ce*. The distance d_p between these fringes is one half the length of the diagonal *ac* of rhombus *abcd*. When the angle between the surface of the object in Figure 6.2 and the object beam is θ, the angle between the object beam and the intersecting reference beam is 2θ; this is also the apex angle of the rhombus. Thus, the distance between two adjacent fringes is

$$d_p = \frac{\lambda}{2 \sin \theta}. \tag{6.11}$$

The distance between these fringes after projection on the screen in Figure 6.2 is

$$d_p' = \frac{d_p}{\cos \theta} = \frac{\lambda}{\sin 2\theta}. \tag{6.12}$$

This distance between adjacent fringes is an excellent indicator of the tilt angle of the planar object relative to the normal plane. It is a particularly useful measurement when $\theta \ll 1$. Even more useful is the measurement of the shape of curved surfaces such as lenses or mirrors. Using the fringe pattern, which may include concentric circles or other geometries, the shape of the object can be calculated by assuming that immediately after reflection the wavefronts mimic the shape of the reflecting surface. However, more practically, the fringe pattern can be compared to a reference pattern that represents a desired shape. Thus, the fringe pattern created by a newly manufactured lens can be compared to the fringe pattern obtained with a standard of that lens as an object. With available digital imaging technology, such measurements are now quite practical.

Alternatively, a Michelson–Morley interferometer can be used to measure the relative position of objects. When the object is translated along the object beam by a distance d_1, the optical path along that beam increases by $2d_1$. Accordingly, the wavefronts of the object beam (e.g., front *ab*) translate by a distance $2d_1$ away from the screen. This results in a translation to the left of each fringe by a distance of

$$d_2 = \frac{2d_1}{\sin 2\theta}. \tag{6.13}$$

Usually, $\theta \ll 1$ and so a small translation of the object results in a large translation of the fringes. In many applications, the distance traveled by an individual fringe exceeds many times d_p. In this case – instead of measuring d_2 directly – the number of fringes passing through a point on the screen can be counted, either visually or electronically using a photodetector. Measurements within a fraction of fringe are possible. For each passing fringe, $d_1 = \lambda/2$.

Although interferometry is a sensitive method for measurement of minute translations, it is also extremely sensitive to vibrations. Thus, most interferometric measurements require that the optical system be rigidly mounted and vibrationally isolated from its surroundings, often by pneumatic flotation of

the optical tables. Air currents may also introduce uncontrollable jitters of the interference fringes, so sensitive interferometers are enclosed in airtight boxes.

Another application of interferometry is the measurement of the density of gases and its distribution, particularly of airflows or flames. This method is used for airflow visualizations, where density variations can be induced by small variations in pressure or temperature. Equation (4.22), which was derived to show the dispersion by absorbing species, includes a coefficient A that was assumed to depend on the density of the medium and on the molecular or atomic properties that characterize this medium. More specifically, in the absence of nonlinear effects, this coefficient increases linearly with the density of the oscillators. Thus, to describe the density dependence of n, we introduce a new wavelength-dependent coefficient, α. With this coefficient, (4.22) can be reduced to show the following density dependence of the index of refraction:

$$n = 1 + \alpha N \tag{6.14}$$

(see Born and Wolf 1975). Here N is the gas density expressed by the number of molecules in a unit volume. Values for α are available in numerous handbooks. For air illuminated by radiation at $\lambda = 488$ nm, $\alpha = 1.092 \times 10^{-23}$ cm^3 (Weast and Astle 1980). Using the Loschmidt number N_0, which is the standard density for air at 1 atm and 0°C, we have $N = N_0 = 2.69 \times 10^{19}$ cm^{-3}. The index of refraction is then $n = 1 + 2.94 \times 10^{-4}$. Assuming that (at moderate temperatures and pressures) air behaves as an ideal gas, the index of refraction at 1 atm and 15°C is found to be $n = 1 + 2.78 \times 10^{-4}$. The difference between the indices of refraction at these temperatures is only 1.6×10^{-5}. Although this difference is extremely small, it can increase the optical path of a $\lambda = 488$-nm wave traveling through a distance of approximately 3 cm by approximately 1λ. Such a minute change in the optical path will result in the translation of interference fringes by d_p. This (or even smaller) change(s) in the index of refraction can be easily measured by interferometric techniques, which can be implemented for flow or combustion diagnostics.

Figure 6.4 presents a typical interferometer used for visualization of gas flows. This is the Mach–Zender interferometer. As with the Michelson–Morley interferometer, one beam is split and then recombined before projecting it on a screen; however, the beam's two parts – the object arm and the side arm (reference beam) – do not retrace their paths. The object beam of this interferometer passes through the test section only once. Although this reduces the sensitivity of the system, it simplifies considerably the interpretation of its results. Normally, the reference beam is passed uninterrupted to the recombining beam splitter. However, to compensate for distortions that may be introduced by test-section windows, compensating elements (that are identical to and at the same temperature as the test-section windows) may be added along the side arm. The fringe pattern obtained at the screen, where the two beams are recombined, depends only on the variations in the index of refraction along the path of the object beam through the test section. This technique is used to visualize flow structures transverse to the beam, in a manner similar to flow

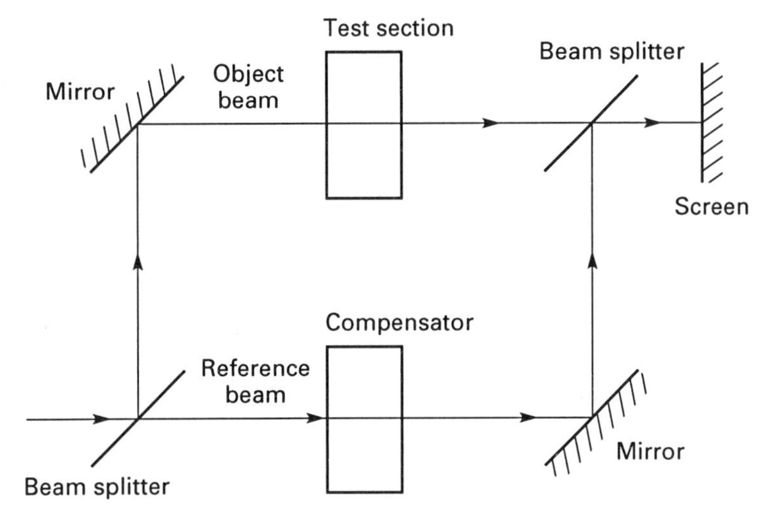

Figure 6.4 The Mach–Zender interferometer.

visualizations by the schlieren and shadowgraph techniques (Section 2.7). Images obtained by interferometry represent an integration along the beam path and therefore cannot be used to resolve structures longitudinal to the beam. However, unlike the schlieren and shadowgraph techniques (where only the first and second orders, respectively, of the density gradients can be imaged), interferometry offers a method for visualization of the density distribution itself.

6.5 The Fabry–Perot Etalon

The analysis of the interference by two beams (Section 6.2) revealed that these effects are very sensitive to variations in the optical paths of each beam and to the wavelength of the interacting beams. Thus, applications of interferometry may include either devices for sensing changes in the optical paths or devices for sensing or selecting wavelengths. In the previous section we showed how interferometry may be used for the detection of motion or fluid flow. Here we discuss a type of interferometer that is designed to measure or select the wavelength of the radiation passing through it: the Fabry–Perot etalon. This interferometer consists of a flat plate (Figure 6.5) illuminated by a plane wave. The multiple reflections at the interface of the plate create a unique interference pattern that results in a wavelength-dependent transmission and reflection of light.

The Fabry–Perot etalon in Figure 6.5 consists of an isotropic homogeneous plate of thickness d with an index of refraction n_2. It is illuminated uniformly by a monochromatic plane wave at a wavelength λ incident at an angle θ_i. Although the indices of refraction above and below the plate may be different from each other, we consider here the simpler case of a plate embedded in a uniform medium with an index of refraction n_1. The incident beam is illustrated in Figure 6.5 by two parallel rays, I_1 and I_2, and the planar wavefront

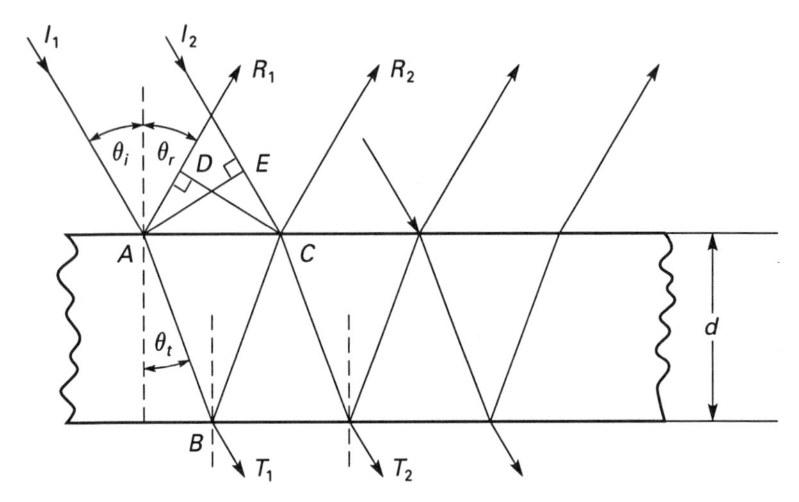

Figure 6.5 Reflected and transmitted rays at the faces of a Fabry–Perot etalon.

AE that is perpendicular to these rays. Upon incidence, a portion of each ray is reflected and the remainder is transmitted by refraction. The reflection constant $r = \mathbf{E}_r/\mathbf{E}_i$ (i.e., the fraction of the incident field amplitude that is reflected) and the transmission constant $t = \mathbf{E}_t/\mathbf{E}_i$ can be calculated from (5.2) for each of the incident polarizations using the ratio of the indices $n = n_2/n_1$ and θ_i. The reflected rays R_1 and R_2 are marked in the figure together with their wavefront CD. The refracted rays propagate through the slab to the next interface, where they encounter another reflection and refraction with the corresponding coefficients r' and t'. The secondary reflection of ray I_1 is illustrated in the figure by the line BC. This and the other secondary reflections undergo once again a reflection and refraction at the upper interface. Although the amplitude of subsequent reflections is diminishing, theoretically there may exist an infinite number of reflected and refracted rays. Furthermore, although the figure includes only two rays, in reality an incident beam can be divided into an infinite number of rays. Thus, if the the width of the incident beam exceeds the dimension AC that is traveled horizontally by the first reflected ray, then each incident ray must encounter at the interface other rays associated with the secondary reflections in the slab. This was illustrated in the figure by positioning ray I_2 at point C, where it encounters the first reflection of ray I_1. Thus, the amplitude of the field propagating in the direction of R_2 must include the sum of the amplitudes of the reflected ray R_2, the transmitted portion of ray BC, and the sum of the amplitudes of the infinite remaining rays associated with the secondary reflections of the rays that are incident to the left of I_1. To determine the total amplitude, the reflection and transmission coefficients at both interfaces must be determined, together with the phase difference δ between rays R_2 and BC. The phase difference between higher-order reflections and R_2 is an integral multiple of δ.

To determine the phase difference δ between R_2 and BC, note that CD is the wavefront of the reflected beam enclosed by rays R_1 and R_2. Therefore the phase of R_2 at point C is identical to the phase of R_1 at point D. The phase

differential δ can hence be determined from the difference between two optical paths, AD and ABC, as follows:

$$AD = AC\sin\theta_i = 2d\tan\theta_t\sin\theta_i;$$

$$ABC = \frac{2nd}{\cos\theta_t}.$$

Using the difference between these paths, $\Delta = ABC - AD = 2nd\cos\theta_t$, for δ we obtain

$$\delta = k_0\Delta = \left(\frac{2\pi}{\lambda_0}\right)2nd\cos\theta_t, \tag{6.15}$$

where λ_0 is the incident wavelength measured in the external medium. Each subsequent passage through the medium increases the phase difference by an additional increment δ. Therefore, the phase difference between R_2 and the Nth reflection is $N\delta$.

The magnitude of the electromagnetic field at any point on the plate is determined by superposition of the fields of all the rays that interact at that point. In turn, the amplitude and phase of each contributing ray is determined by the number of reflection and transmission events it underwent. Thus, the amplitude of the first reflection is $r|\mathbf{E}_i|$, the amplitude of the first secondary reflection that involved refraction into and out of the upper surface and reflection at the bottom surface is $r'tt'|\mathbf{E}_i|$, and the amplitude of the second secondary reflection is $r'^3tt'|\mathbf{E}_i|$. Thus the amplitude of the Nth-order secondary reflection is $r'^{2N-1}tt'|\mathbf{E}_i|$. However, since $r' = -r$ (eqn. 5.3) for any incident polarization and for all incidence angles, the total reflected field can be described by the following converging series:

$$\mathbf{E}_r = \mathbf{E}_i(r - rtt'e^{-i\delta} - r^3tt'e^{-2i\delta}\cdots - r^{2N-1}tt'e^{-Ni\delta}).$$

The resultant wave is determined from the sum of this series (Hecht 1990):

$$\mathbf{E}_r = \mathbf{E}_i\left[\frac{r(1-e^{-i\delta})}{1-r^2e^{-i\delta}}\right], \tag{6.16}$$

where the temporal and spatial dependence of the incident and reflected fields are implied. Using (4.42), the reflected irradiance is

$$I_r = I_i\frac{[2r/(1-r^2)]^2\sin^2(\delta/2)}{1+[2r/(1-r^2)]^2\sin^2(\delta/2)}. \tag{6.17}$$

When the slab consists of a nonabsorbing medium, the total energy in the reflected and transmitted beams equals the incident energy: $I_i = I_r + I_t$. The following term for the transmitted irradiance may thus be obtained by subtraction:

$$I_t = I_i\frac{1}{1+[2r/(1-r^2)]^2\sin^2(\delta/2)}. \tag{6.18}$$

Both the transmitted and reflected irradiance are seen to depend on δ – the phase difference introduced by one transition through the slab – and on the

reflection coefficient r. To illustrate the effect of these parameters, consider the simple case of $\delta = 2m\pi$. For normal incidence this condition corresponds to $2nd = m\lambda_0$; that is, the optical pathlength of a round trip through the plate is an integer multiple of the incident wavelength. For this phase difference, $I_r = 0$. That is, independently of r, the reflected irradiance is zero. Thus, the combination of all the secondary reflections is sufficient to reinforce the primary transmitted ray, so that $I_t = I_i$. Hence, by carefully matching d, n, and θ_t (eqn. 6.15), efficient transmission of the incident radiation can be forced while reflection losses are minimized. This technique is used to select the wavelength of lasers (Hercher 1969). By inserting a thin etalon inside the laser system, transmission of selected wavelengths can be reinforced while other wavelengths are suppressed (see Research Problem 6.2).

This property of thin etalons is also used for the development of antireflection (AR) or highly reflective (HR) coatings of optical elements. Typical AR or HR coatings consist of several thin layers of transparent dielectric media with indices of refraction that are different from the indices of the substrate or the external medium. The design of such coating – deciding the number of layers, selecting the material for each layer, its thickness, the wavelengths and incidence angles at which optimal results are expected – involves calculations that are similar to those leading to (6.17) and (6.18).

The design of a dielectric coating with specific properties requires consideration of the phase shift δ and the reflection coefficient r. To determine the role of r, notice that in both (6.17) and (6.18) r is introduced by the same term:

$$F = \left(\frac{2r}{1-r^2}\right)^2 = \frac{4R}{(1-R)^2}, \tag{6.19}$$

where $R = I_r/I_i = r^2$ is the *reflectivity* of the interface. Figure 6.6 illustrates the variation of transmitted irradiance with δ for $F = 360$, 1.77, and 0.17, corresponding (respectively) to $R = 0.9$, 0.25, and 0.04. The first two reflectivities require surface preparation, but the last reflectivity occurs naturally for most visible wavelengths at the interface between glass ($n = 1.5$) and air. Thus, the upper curve in Figure 6.6 represents the variation with the incident wavelength of the transmission through a glass slab with $n = 1.5$ and a thickness d. As expected, the transmission for normal incidence approaches unity when $2nd = m\lambda_0$. For example, full transmission of radiation at $\lambda = 500$ nm through a medium with $n = 1.5$ can be obtained when $d = 1/6$ μm and $m = 1$. The next wavelength that will be transmitted effectively through such a thin film is $\lambda = 250$ nm, for which $m = 2$. Since $\lambda = 500$ nm, which appears green, is the only wavelength in the visible range that is transmitted effectively by this thin sheet, it will have a green hue. Similarly, thin oil films spread over water appear colorful owing to such interference effects.

This phenomenon of color-selective transmission can be used for spectrally filtering radiation. The simplest device for this application is an etalon consisting of two reflectors of reflectivity R separated by an air or vacuum gap. Alternatively, it may consist of a solid transparent plate coated on both sides with

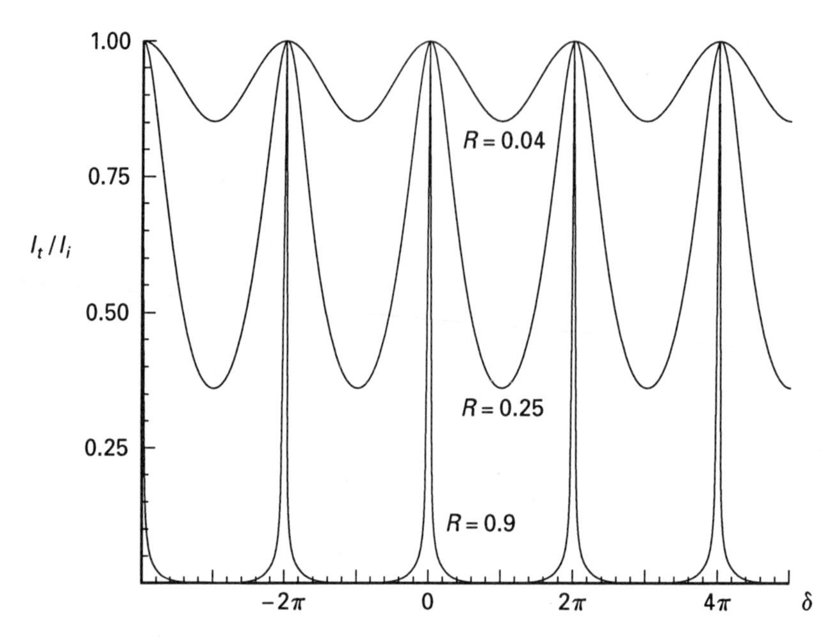

Figure 6.6 Transmission spectra of etalons with surface reflectivities R of 0.04, 0.25, and 0.9.

a reflecting layer of reflectivity R. In either configuration, the maximum transmission at normal incidence is reached when the optical path of a round trip between the two reflectors equals an integral multiple of the incident wavelength (Figure 6.6). Although transmission at that wavelength is independent of reflectivity, the spectral selectivity of the device can be improved by increasing R. Thus, at $R = 0.9$ ($F = 360$) the spectral range in which transmission is possible is much narrower than the range when $R = 0.25$ or 0.04. The value of δ at which the transmission is 50% can be used as a measure of the transmission bandwidth of the etalon mode. From (6.17) we have that $\delta \approx \pm 2/\sqrt{F}$ for $T = I_r/I_i = 1/2$. The range (or *bandwidth*) of this transmission is then

$$\delta_0 \approx 4/\sqrt{F}.$$

Although F can serve as a parameter to estimate the spectral resolution of an etalon, the *finesse* Φ is more commonly used. It represents the number of transmission lines that can be accommodated between two adjacent transmission peaks, and is defined as

$$\Phi = \frac{2\pi}{\delta_0} = \frac{\pi\sqrt{F}}{2} \tag{6.20}$$

(Born and Wolf 1975, p. 328). The spectral separation between two adjacent transmission peaks (i.e., the frequency difference) is called the *free spectral range* of the etalon. To determine its value, consider two adjacent transmission modes expressed by their wavelengths λ_{01} and λ_{02} in free space:

$$\frac{1}{\lambda_{01}} = \frac{m}{2nd\cos\theta_t},$$

$$\frac{1}{\lambda_{02}} = \frac{m+1}{2nd\cos\theta_t}.$$

Subtracting the two equations yields

$$\Delta\bar{\nu} = \frac{1}{\lambda_{01}} - \frac{1}{\lambda_{02}} = \frac{1}{2nd\cos\theta_t}; \tag{6.21}$$

when d is measured in centimeters, the free spectral range is given in units of cm^{-1}. Note that the free spectral range decreases with the slab thickness. Thus, free spectral range for transmission through moderately thick slabs may not be sufficient to resolve these lines; both the transmission and the reflection appear to be independent of the incident wavelength.

When the incidence angle θ_i is changed without varying any other parameter, maximum reflectivity can be obtained only at selected angles. Thus, when a conically diverging circular monochromatic beam illuminates a thin etalon, the projection of the transmitted radiation on a screen consists of well-defined concentric rings (see Figure 6.7). The width of the rings is defined by the finesse of the etalon and the bandwidth of the incident radiation. The separation between adjacent rings is analogous to the free spectral range. Thus, when the free spectral range and the finesse of the etalon are known, the linewidth of the incident radiation can be estimated from the thickness of a ring relative to the ring spacing. Such etalons are used to select the wavelength of tunable lasers and also to monitor their monochromacity.

More complex etalons can be obtained by stacking several layers of dielectric substances of varying thickness and index of refraction. Depending on the configuration, reflection or transmission can be obtained with extremely high

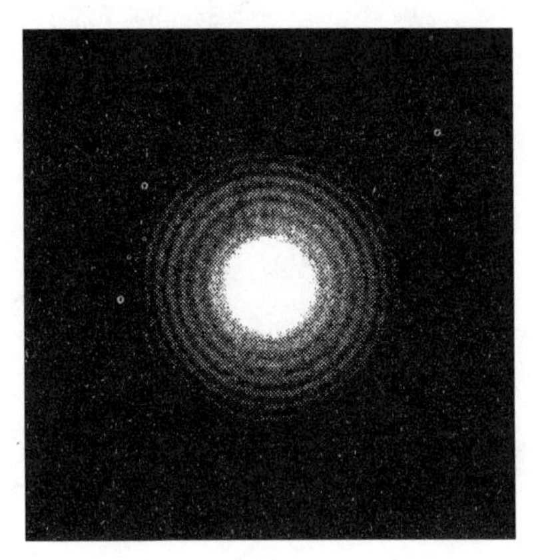

Figure 6.7 Interference rings obtained by transmission of a He–Ne laser beam through an etalon. Note the speckle pattern imposed on the rings; this too is an interference effect (see Section 6.6).

spectral selectivity or reflectivity. Thus, reflectivity in excess of 99% is possible over a range of wavelengths. Such reflectors, or *dielectric mirrors,* are often used to reflect high-power laser beams. Not only are the reflection losses minimized, but the unreflected portion of the radiation is transmitted, thereby minimizing the energy deposited in the reflecting layer. Thus, by reducing the absorption, the potential for heating of the absorbed layer is reduced and so the threshold for optical damage is extended. For comparison, the reflectivity of metallic mirrors is in the vicinity of 97%, where 3% of the radiation is deposited in the thin reflective layer. Metallic mirrors are thus susceptible to optical damage even by moderate laser energies.

Dielectric coatings can also be used to obtain optical filters with unique characteristics. Narrowband transmission ranging from 1 nm bandwidth and up can be obtained for wavelengths ranging from 190 nm to 20 μm. Alternatively, cutoff filters can be used to transmit a range of wavelengths above or below a given threshold. The cutoff point and its slope can be specified separately.

6.6 Speckle Pattern Interferometry

The images of objects illuminated by a laser beam appear significantly different from their images obtained when illuminated by incoherent radiation. The difference is particularly evident when the surface of the object is not smooth, when most of the radiation is scattered rather than specularly reflected. Figure 6.8 presents for comparison an image of a coin illuminated by an argon ion laser (Figure 6.8(a)) and by an incoherent source (Figure 6.8(b)). The first image appears very "grainy"; the surface is marred by a random pattern of dark and bright spots. These spots are called *speckles*. By contrast, the image

Figure 6.8 Comparison of the images of the same object when illuminated with **(a)** coherent and **(b)** incoherent light. Note the apparent speckle pattern induced by the coherent illumination.

obtained with illumination by incoherent light is uniform, with a resolution that is limited only by the imaging optics and the properties of the recording medium. Although the speckle pattern associated with the laser illumination is random, it is reproducible. Thus, when the laser source and the camera are kept at the same location and the camera aperture is unchanged, each recording of the image will contain the same random pattern of speckles. This is in contrast to other random imaging noise, such as film grain, that varies from recording to recording. It is easy to show that the speckle pattern results from the coherence - and not from the monochromaticity - of the illuminating laser source. This suggests that interference effects are the cause for this phenomenon. Since speckles appear only when the illuminated surface is nonspecular (i.e., when most of the radiation is scattered rather than reflected), it can be argued that the interference effects must originate at the surface and are in fact a complex interferogram of the surface structure. Thus, the random speckle pattern is formed by the random surface structure and remains reproducible, provided the surface structure and its location relative to the illumination and recording systems are unchanged. Although the speckle pattern presents difficulties when using laser light for imaging, it presents many opportunities for studies of surface structures, strain or other surface distortion measurements, and some flow diagnostics applications. Here we discuss only one application of speckle patterns: measurements of mechanical strains.

To illustrate the formation of speckles, consider the microscopic structure of the illuminated surface. The topography of surfaces that are considered to be diffusive reflectors consists of an irregular array of bumps and troughs with a characteristic size that is comparable to the wavelength of the incident radiation (Figure 6.9). This is in contrast to specular reflectors, where the characteristic size of surface structure elements is small relative to the wavelength of the

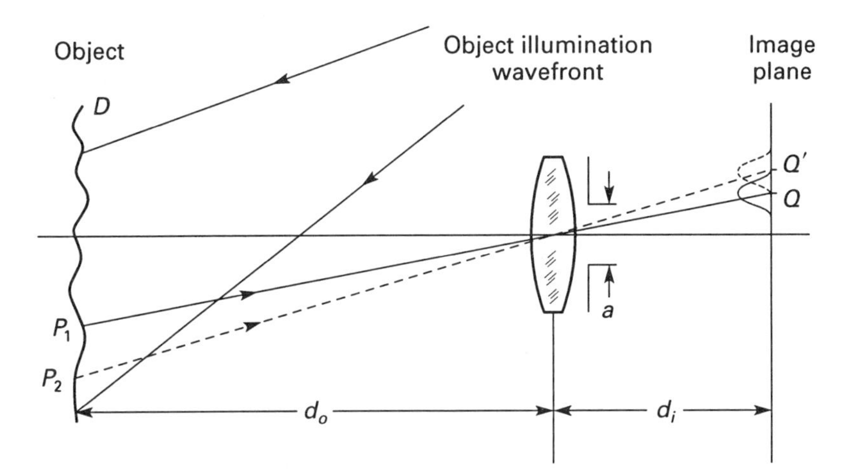

Figure 6.9 Speckle pattern formation at the image plane of a lens. Two rays scattered at points P_1 and P_2 of the surface are projected to points Q and Q', where they form a speckle. [Jones and Wykes 1989, © Cambridge University Press]

incident radiation. Thus, when the wavefront of the incident radiation is planar, the wavefront of a reflection by a specular reflector remains planar (see Problem 6.2). By contrast, the wavefront after a reflection by a diffuse reflector is modified by the reflecting surface and appears, like the surface, to have a random structure. Thus, the diffuse reflection may be considered as consisting of infinite rays all pointing in various directions. This is illustrated in Figure 6.9 by two rays originating at the surface, collected by a lens with an aperture a, and projected on the surface of a recording element that may consist of a photographic film or detector array of an electronic camera. The phases of these rays at the recording element are determined by the illumination and imaging parameters and by the structure of the scattering site. Thus, together with other rays that are scattered from that site, they form by interference (and diffraction) a unique pattern that depends on the shape of that site. Images of other sites will of course have other interference patterns.

When the surface scattering site is moved by a distance that is short relative to a characteristic distance (to be defined shortly), the interference pattern will undergo little change. Thus, if the entire surface is moved by such a short distance, the overall speckle pattern remains essentially unchanged. However, with a larger translation, the new speckle image changes significantly and becomes uncorrelated with the previous image. The correlation between two sequential images is a necessary condition that must be met for most speckle imaging applications. When achieved, it presents an opportunity for the measurement of minute distortions or mechanical strains.

The characteristic size of a speckle is an important parameter that determines the range of application of speckle techniques for measurements of strain or surface distortion. Although evaluation of this parameter requires the use of concepts from the theory of diffraction (to be discussed in the next chapter), it is assumed that the reader already has a preliminary understanding of the effect of diffraction. Two steps, both diffraction-limited, are associated with the formation of a speckle at the image plane. The first, collection of rays by the lens (Figure 6.9), is characterized by the lens aperture a and the distance d_o between the object and the lens. The second, the formation of the image by the lens, is characterized by the same lens aperture and by the distance d_i to the image plane. For simplicity, consider first the image formation. The sharpest image is obtained when the wavefront, immediately past the lens, is planar. In the absence of diffraction this front can be focused into a well-defined dot. However, diffraction cannot be neglected. Therefore, the radius of the smallest spot that can be obtained by a circular lens is (eqn. 7.11):

$$r_i = \frac{1.22 d_i \lambda}{a}. \tag{6.22}$$

When an electronic camera is used for recording, the smallest element of the image is the pixel. Thus, (6.22) determines the minimum pixel size needed to record a single speckle. Or, for a given pixel size, (6.22) defines the recording parameters (e.g. the lens aperture) required to resolve the speckle. Using the principle of reciprocity, the same analysis may be used to show that the radius r_o of the area on the surface site that contributes to the image of one speckle is

$$r_o \approx \frac{1.22 d_o \lambda}{a}. \tag{6.23}$$

Thus, when the site is moved by less than r_o, the shape of the recorded speckle remains correlated with its original shape.

Several correlation techniques are used to measure surface strains using speckle imaging techniques. In all techniques, a reference speckle image of a surface or part of it is recorded. A second image is recorded after the surface is strained, and the two images are compared using a selected correlation technique that depends on the illumination configuration. When the maximum translation of any point on the surface is less than r_o, this correlation analysis can provide the strain distribution of the surface (Jones and Wyke 1989).

For illustration, consider a speckle interferometry technique using simultaneous illumination of the surface by two mutually coherent beams (Figure 6.10). When the beams are obtained by splitting a laser beam, the path-length difference between them may not exceed the coherence length of the laser. At the region of overlap between the two beams they interfere with each other, thereby marking the surface with planes of alternately bright and dark fringes with spacing determined by (6.11). For illumination by an argon ion laser at 514.5 nm and for $\theta = 3°$, the distance between two adjacent fringes is approximately 5 μm. For comparison, the size of a single speckle site recorded by a 5-mm lens aperture and with $d_o = 400$ nm is 25 μm. Thus, an individual speckle site is illuminated by approximately five interference fringes. With this illumination, the brightness of a recorded speckle depends not only on the surface structure but also on its location relative to the interference fringes. Thus, if a

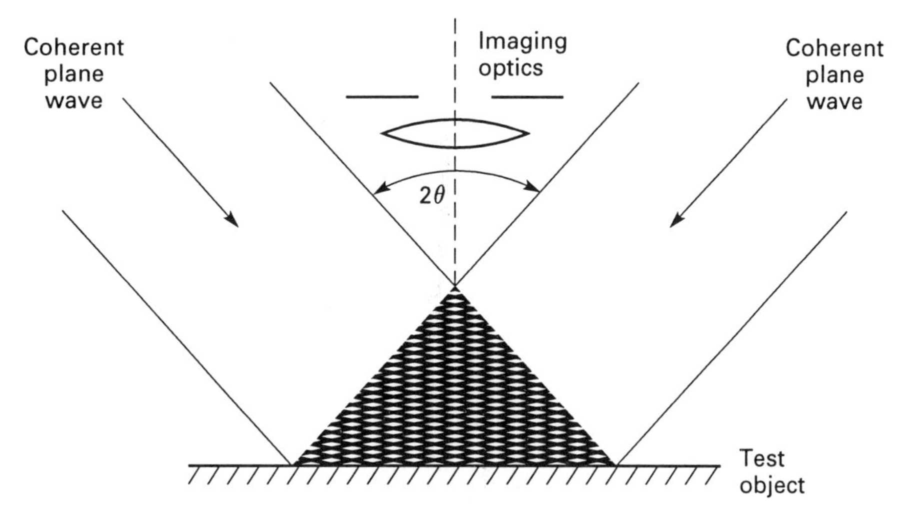

Figure 6.10 Surface speckle interferometry using simultaneous illumination by two coherent beams.

site is covered by five bright and six dark fringes, its speckle will appear darker than if the same site is illuminated by six bright and five dark fringes. An image recorded with two beams illuminating the surface includes a superposition of the speckle pattern as well as the interference fringe pattern.

In a typical experiment, the surface is subjected to a strain after a reference image is recorded digitally. Thus, after distortion by strain, any site that was translated by one entire interference fringe will cast a speckle with the same brightness as the speckle in the reference image. On the other hand, the brightness of a speckle that was translated by only half a fringe increases (or decreases) by the maximum possible amount. Thus, when the reference image is subtracted – pixel by pixel – from the image of the strained object, and the absolute value of the difference image is displayed, the object will appear to be marked with bright and dark strips. The dark strips correspond to sites that had the same brightness in both the reference and the strained-surface images and therefore correspond to a site translation by one interference fringe. The bright strips, on the other hand, correspond to the points where the speckle site was translated by half an interference fringe. Measuring the distance d_s between two speckle strips, the strain can be determined by $\epsilon = d_p/d_s$. Note that out-of-plane translation or translation along the fringe plane results in no change in the images. Therefore, these measurements yield only the component of the strain perpendicular to the interference fringes.

This technique is relatively simple to implement, yet it offers very high sensitivity. For a separation of 5 mm between two adjacent speckle fringes obtained when the interference fringes of the surface are separated by $d_p = 5$ μm, the measured strain is $\epsilon = 10^{-3}$. A 10-μ strain sensitivity is possible by interpolation between speckle fringes. Furthermore, this sensitivity is obtained without the need to process the surface. Therefore, unlike strain gage measurements, this technique is considered to be nonintrusive. However, several difficulties prevent its wide-range application. Since the maximum distance that any point can travel may not exceed the characteristic size of the site, the dynamic range of the method is limited. Hence, for an $L = 5$-cm object subject to uniform strain, the maximum strain may not exceed d_o/L. For $d_o = 25$ μm this corresponds to 500-μ strain. Furthermore, vibrations that introduce a uniform translation of the entire object may cause decorrelation of the speckle image, which limits application of the technique to vibration-free environments.

6.7 Laser Doppler Velocimetry

Laser Doppler velocimetry (LDV) is probably the most notable example of the impact lasers have made on the field of flow diagnostics. Since its original demonstration (Yeh and Cummins 1964), the technique has evolved into a sophisticated method that allows the simultaneous measurement of three velocity components of gas or liquid flows or the velocity of solid surfaces. The data can be collected and digitized at a rate sufficient to reveal turbulence parameters, instantaneous velocity vectors (magnitude and direction), average flow velocity, various histograms, and (in two-phase flows) even the size of particles or bubbles – all with a spatial resolution of less than 100 μm. Despite the wealth of information generated by this technique, its principle of operation is very simple and can be explained by using concepts of interferometry or heterodyning.

In the first demonstration (Yeh and Cummins 1964), polystyrene spheres with a uniform diameter (mono-dispersed) of 0.56 μm were suspended in water and illuminated by a He–Ne laser radiating at $\lambda_0 = 632.8$ nm. Although the laser light was scattered by both the fluid and the particles, the Mie scattering by the particles was dominant. In addition, these particles were sufficiently small to accurately follow the flow velocity. Therefore, the radiation scattered by these mono-dispersed particles was Doppler-shifted (eqn. 6.7). By measuring the Doppler shift, it was possible to determine the scattering particles' velocity as well as the flow velocity. The difficulty that had to be overcome before implementing this technique was measurement of the Doppler shift. Direct measurement of this shift requires two independent measurements of the laser frequency (before and after the shift), both in excess of 4×10^{14} Hz and both beyond the time response of any existing electronic device. Furthermore, the Doppler shift itself is extremely small relative to the laser frequency. For illustration, the maximum reported velocity in the experiment was 0.05 cm/s. At this velocity the maximum Doppler shift is approximately 790 Hz. Thus, even if direct frequency measurement were possible, it had to be accurate to within at least twelve significant digits to make the Doppler shift noticeable. Even at higher speeds (e.g. supersonic airflows) the Doppler shift is approximately 1 GHz, which requires measurements of the laser frequency at an accuracy of five significant digits in order to determine velocity at an accuracy of only one digit! To overcome these difficulties, the Doppler shift must be measured relative to a reference frequency, for example, by determining the frequency shift of the scattered radiation relative to the frequency of the absorption line of an atomic or molecular species (Hiller and Hanson 1985; Shimizu, Lee, and She 1983). In the demonstration by Yeh and Cummins (1964), the Doppler shift was measured relative to the frequency of the incident laser itself using a heterodyning technique (eqn. 6.9). By combining the Doppler-shifted radiation with unshifted radiation on the surface of the photocathode of a PMT, a signal was obtained with a component oscillating at a frequency that expressed the frequency difference between the incident and Doppler-shifted radiation. This much lower frequency could be readily measured. This approach is more universal than the use of a spectral absorption line for a reference frequency because it does not require matching the incident wavelength with the spectral features of any absorbing species.

Before describing methods where heterodyning is used for the measurement of the Doppler shift, we must show how this shift relates to the velocity of the scattering particle. In most experiments, the scattering particle moves relative both to a stationary observer and to a stationary light source. Therefore, the frequency shift measured by the observer must be a linear combination

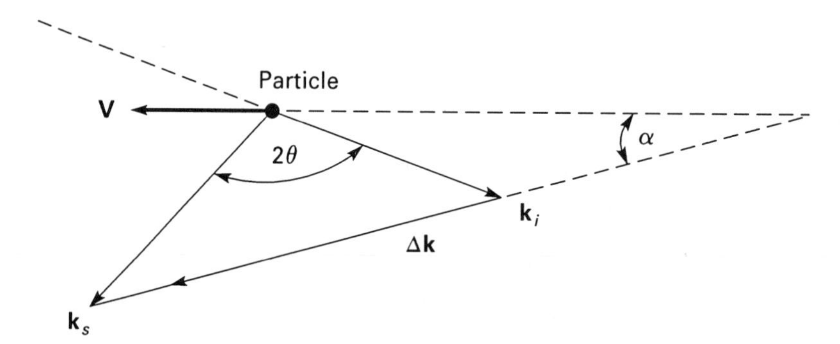

Figure 6.11 Geometry of the scattering of a laser beam by a moving particle.

of the shift induced by the motion of the scatterer relative to the light source and a second shift induced by its motion relative to the observer. This is similar to the shift induced by mirror M_1 in Figure 6.1, where the effect of its motion relative to the source and relative to the screen were combined into one term (eqn. 6.7). The first component, the Doppler shift seen by a particle moving at a velocity $\mathbf{V}$ (Figure 6.11) and illuminated by a beam with a propagation vector $\mathbf{k}_i$, is (see Problem 6.6):

$$\Delta\omega_i = -\mathbf{k}_i \cdot \mathbf{V}.$$

The negative sign indicates that the Doppler shift is positive when the particle is propagating opposite to the laser beam. This equation represents only one relative motion, and accordingly its magnitude is only one part of the shift predicted by (6.7). Since the frequency scattered by the particle must be identical to the frequency seen by it, the frequency of the scattered radiation is $\omega_i + \Delta\omega_i$. However, this is not the frequency seen by a stationary observer analyzing the scattered radiation. Because the scattering particle is a moving source, the Doppler shift associated with its motion must also be included. Thus, when seen by a stationary observer, the scattered radiation will undergo a second Doppler shift of magnitude

$$\Delta\omega_s = \mathbf{k}_s \cdot \mathbf{V},$$

where $\mathbf{k}_s$ is the propagation vector of the scattered radiation. Here the sign of the shift is positive. That is, when the particle is moving toward the observer the shift is positive; otherwise, it is negative. From these equations we see that when the particle moves perpendicularly to the laser beam or to the line of sight of the observer, one or both Doppler shifts vanish. The total frequency shift – the frequency seen by the stationary observer – is of course the sum of these two shifts:

$$\Delta\omega = \Delta\omega_s + \Delta\omega_i = (\mathbf{k}_s - \mathbf{k}_i) \cdot \mathbf{V}. \qquad (6.24)$$

This is a general representation of the total Doppler shift induced by a moving scatterer. Equation (6.7) where $\mathbf{V}$, $\mathbf{k}_i$, and $\mathbf{k}_s$ are counterpropagating collinearly ($2\theta = 180°$ and $\alpha = 0°$) is a special case of this general result. For the geometrical configuration in Figure 6.11, the frequency shift is:

$$\Delta\omega = 2|\mathbf{k}_i||\mathbf{V}|\sin\theta\cos\alpha = 2\omega_i\left|\frac{\mathbf{V}}{c}\right|\sin\theta\cos\alpha \quad \text{or}$$
$$\Delta\nu = \frac{2|\mathbf{V}|}{\lambda}\sin\theta\cos\alpha, \tag{6.25}$$

where $|\mathbf{k}_i| \approx |\mathbf{k}_s|$ was assumed. The observed Doppler shift is proportional to the projection of the velocity vector on the vector $\mathbf{k}_i - \mathbf{k}_s$, which in turn is perpendicular to the bisector dividing the angle between the two propagation vectors.

To use heterodyning for the measurement of the Doppler shift, the scattered radiation must be combined with a fraction of the incident radiation. When measured by a photodetector, the combined irradiance includes two components (eqn. 6.9): a time-invariant d.c. component made of the sum of the two combined amplitudes, and an a.c. harmonic component oscillating at a frequency of $\omega_i - \omega_s$. Using simple electronic circuitry, the a.c. component can be separated and the frequency can be measured directly. Implicit in (6.9) is an assumption that the initial phases of both fields are equal, $\phi_i - \phi_s = 0$. Thus, for heterodyning, the incident and scattered frequencies must be coherent; the optical path between them may not exceed the coherent length of the laser beam.

Several configurations have been developed (see Drain 1980, pp. 65–9, 88–97) for measurement of the Doppler shift. In all configurations, a frequency-shifted electromagnetic field is combined with an unshifted field to generate a heterodyned signal at the detector. Although heterodyning can be used to explain the operation of all LDV systems, the concept of interferometry may also be used, particularly when the system design requires the intersection of two beams. Employing interferometric concepts for the analysis of LDV systems permits the use of such geometrical concepts as shown in Figure 6.3 for visualizing many of the system parameters.

To illustrate the duality between the two concepts, consider Figure 6.12, where a typical crossed-beam LDV system is illustrated. This system (and similar derivatives of it) is widely used, and its principles of operation are often explained using both approaches (Durst and Whitelaw 1971). In this configuration,

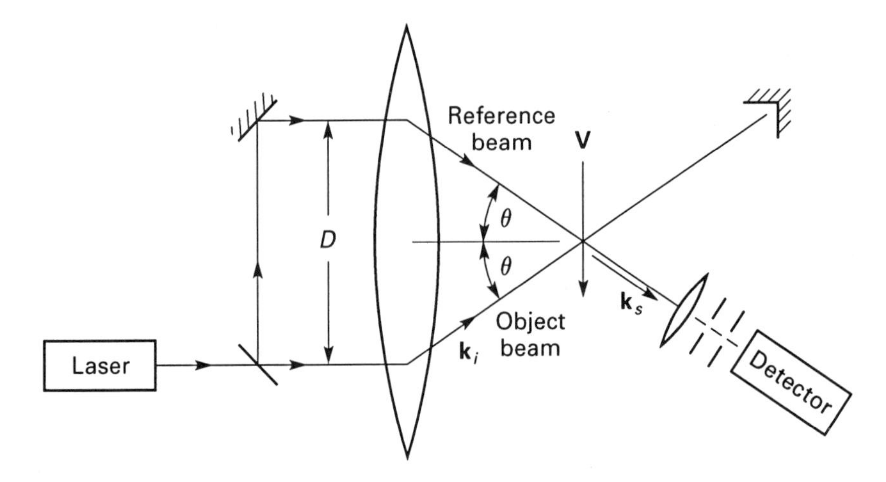

Figure 6.12 A reference-beam laser Doppler velocimeter.

a laser beam is split into two components: a bright beam, the *object* beam, that is used to illuminate the scatterers, and a weaker beam, the *reference* beam, that is used as the heterodyning reference signal. To avoid secondary reflections or interference effects, the beam splitter may consist of a *pellicle* beam splitter (a thin film stretched over a support bracket) or of a Glan prism (Figure 5.5). The two beams are then focused by a large-aperture lens. The intersection angle 2θ between the two beams is determined by the focal length of the lens and the distance D between the beams when intercepted by the lens. When particles are passed through the volume defined by the intersection of both beams, they scatter radiation simultaneously from two sources. Thus, although the incident frequencies of both beams are identical, the scattering angles and hence the Doppler shifts are different. Therefore, radiation collected from this volume contains two frequency components, each associated with the scattering from one of the beams. To simplify interpretation of the measurements, assume that the detector is placed along the reference beam. Furthermore, masks are placed along the entrance aperture of the detector to limit the field of view to the intersection (or probe) volume. Thus, radiation collected by the lens and passed through the masks can include only (a) radiation scattered from the object beam at the probe volume, (b) radiation scattered from the reference beam anywhere along its path, and (c) unscattered radiation of the reference beam itself. Using (6.24) with $\mathbf{k}_i$ as the propagation vector of the object beam, it is seen that, for this detection, only the radiation scattered from the object beam is Doppler-shifted. For scattering off the reference beam, $\mathbf{k}_R - \mathbf{k}_s = 0$ and hence $\Delta\omega = 0$. Similarly, since the radiation transmitted directly along the reference beam is unscattered, $\Delta\omega = 0$. Thus, at the detector, the electromagnetic fields include only two frequency components, with a difference of

$$\nu_D = \frac{2V\sin\theta}{\lambda}, \tag{6.26}$$

where V is the component of the scatterer velocity orthogonal to the bisector of the incident beams and $\nu_D = \omega_D/2\pi$ is the Doppler frequency in hertz. When the irradiance of the reference beam is attenuated sufficiently to avoid saturation, the output of the detector contains both a d.c. and an a.c. component, with the a.c. component oscillating at a frequency of ν_D (eqn. 6.9). Of course, for heterodyning to occur, the radiation scattered off the reference beam must be coherent with radiation scattered off the object beam. Thus, even if the two beams are mutually coherent (i.e., their initial phases ϕ are equal), the scattering off the object beam must occur at some point along the reference beam, within the intersection volume. Otherwise, the phase of radiation scattered off the object beam will not be correlated with the phase of the reference beam. This requirement is met by use of the masks.

The set-up described here is similar to that used by Yeh and Cummins (1964), and is known as the *reference-beam configuration*. Clearly, detection is limited to a direction along the reference beam. Otherwise, the reference irradiance may be insufficient for mixing with the Doppler-shifted signal. This is not

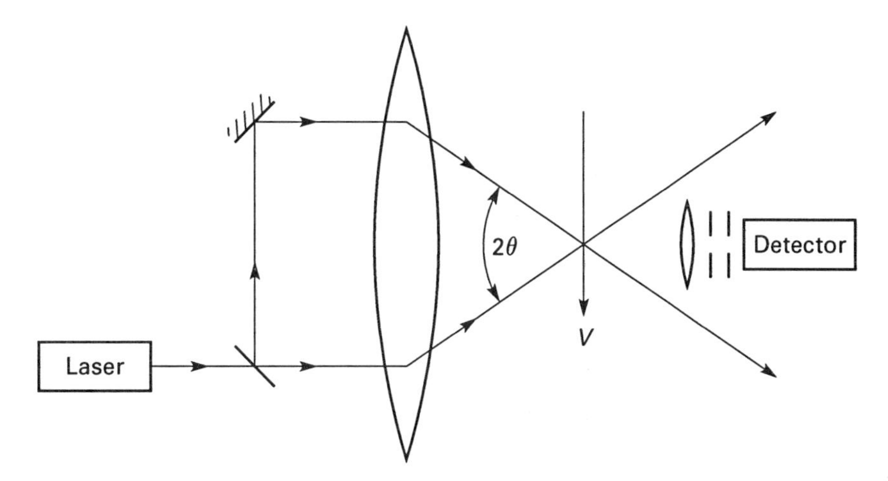

Figure 6.13 A dual-beam laser Doppler velocimeter.

a serious drawback in open flows, but flows confined to wind tunnels or chambers would require two access windows. An alternative approach that allows flexibility in the placement of the detector requires only two slight modifications. This configuration (Figure 6.13) is known as the *dual-beam* or *fringe anemometry* (Rudd 1969). Here the incident beam is split into two beams, each having approximately the same irradiance. The detector is then moved such that its line of sight does not coincide with any of the laser beams. However, the imaging lens and the masks are kept in order to limit viewing to the probe volume, which again is defined by the intersection of the beams. In this configuration, scattering from both beams is detected simultaneously and both are Doppler-shifted. This is in contrast to the scattering off the reference beam in the previous set-up, which occurred without any Doppler shift. If the interacting beams are mutually coherent, radiation that is scattered by a particle within the intersection volume off one beam is coherent with the radiation scattered off the other beam. The scattering may appear to originate from a single source but it actually contains two frequencies, which (after heterodyning by the detector) will include an a.c. component with a frequency that is the difference between the two. This result can be shown (see Problem 6.8) to be identical to the result of (6.26). However, measurements with this configuration can also be explained using the concept of interference, which is more amenable to geometrical interpretation.

Figure 6.14 presents a section of the volume of intersection between the two laser beams (both with a diameter of d_b) through the plane that includes their intersecting axes. The wavefronts of both beams, which in an ideal design are nearly planar, are represented by parallel lines transverse to the beams. This is similar to the interaction between the two coherent beams in Figure 6.3(a), and accordingly it results in a similar interference pattern. The fringes are seen to form planes that are normal to the plane of the figure and parallel to the bisector. Thus, although the wavefronts themselves are not visible, the planes

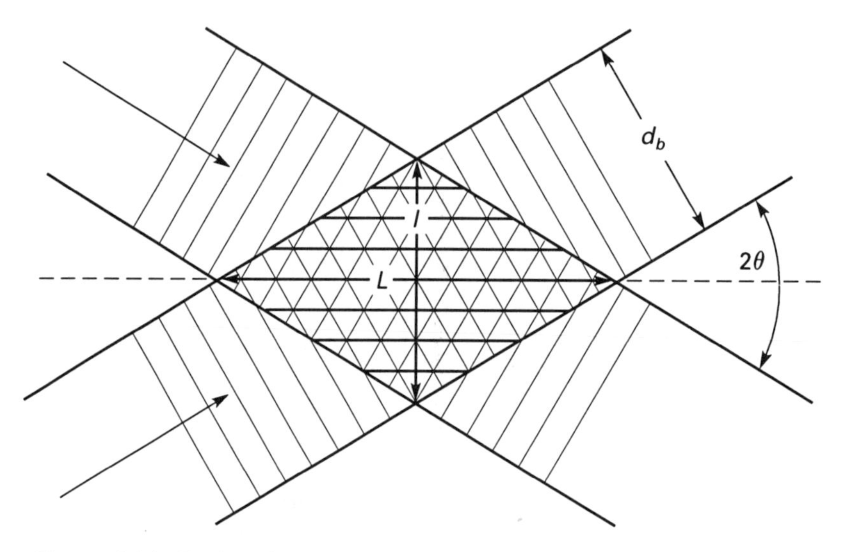

Figure 6.14 Probe volume of a dual-beam LDV. The interference fringes formed by the intersection of the two beams are shown by the lines connecting the intersecting wavefronts.

formed by their mutual interaction are visible and appear as alternately dark and bright sheets. The spacing between these sheets is given by (6.11). A particle with a velocity component V perpendicular to these planes crosses these fringes at a rate of

$$\nu_D = \frac{V}{d_p} = \frac{2V\sin\theta}{\lambda}.$$

Thus, as the particle passes through this intersection volume, the radiation it scatters is modulated by ν_D. This frequency is identical to the frequency obtained from considerations of heterodyning (eqn. 6.26). Of course, radiation can be scattered by particles anywhere along the laser beams, but only scattering by particles crossing the intersection volume and having a velocity component perpendicular to the interference sheets is modulated. The modulation can, however, be detected from any location. This description of the interaction between two laser beams is convenient for the evaluation of design parameters such as the optimal size of the scattering particles, the location of the detector, or the spatial resolution of the system.

The spatial resolution is usually a major concern in point measurements. Here it is determined by the size of the probe volume. Whereas the actual resolution is defined by the three-dimensional shape of the intersection between two cylindrical beams, the limiting resolution is determined from the diamond shape in Figure 6.14. From simple geometrical considerations, the length L and the width l of the shape are

$$L = \frac{d_b}{\sin\theta}, \qquad l = \frac{d_b}{\cos\theta}. \tag{6.27}$$

Often the diameter d_b of the intersecting beams is a loosely defined parameter. As will be shown in Chapter 13, it depends on the distribution of the irradiance transverse to the laser beam, which is a complex function of the laser system emitting it. Until we find a mathematical expression for this distribution, we will consider only the empirical value of d_b by measuring the diameter to the point where the irradiance falls to $1/e^2$ of its peak value. Although the irradiance outside this arbitrarily set diameter may not be negligible when a bright laser is used for measurements, most detection systems can be set to reject signals that are below a threshold level, thereby limiting electronically the effective beam diameter.

The resolution is finer transverse to the beams and coarser along the longer dimension of the diamond. Although the resolution can be improved by focusing the beams, certain limitations must be met. Aside from limitations associated with diffraction effects or the focusing capabilities of existing lenses, the dimensions of the intersection volume must be sufficiently large to accommodate several interference fringes. Otherwise, the frequency cannot be confidently determined, particularly in turbulent flows where the velocity fluctuates. To illustrate the consequences of this limitation, consider the following example.

Example 6.1 Find the smallest diameter of an argon ion laser beam to be used for a dual-beam LDV system when $2\theta = 10°$ and when at least ten fringes are required for the measurement. What is the spatial resolution of the system? Repeat the calculation for $2\theta = 20°$.

Solution For this laser, $\lambda = 514.5$ nm. From (6.11),

$$d_p = \frac{514.5 \times 10^{-9}}{2 \sin 5°} \times 10^6 = 2.95\ \mu\text{m}.$$

To accommodate ten fringes, the width of the probe volume must be $l = 10d_p =$ 29.52 μm. Thus, from (6.27), $d_b = l \cos\theta = 29.40\ \mu$m and $L = 338\ \mu$m.

When $2\theta = 20°$, (6.11) and (6.27) yield the following results: $d_p = 1.481\ \mu$m, $l = 14.81\ \mu$m, $d_b = 14.58\ \mu$m, $L = 84\ \mu$m. Note that as the intersection angle increases, both the fringe spacing and the dimensions of the intersection volume decrease. However, as the spacing between the fringes decreases, the modulation frequency of the detected signal increases. Thus, at high velocities, when the induced modulation frequency is inherently high, small intersection angles may be required to keep the modulation frequency within the frequency response of the detection system. This selection, in turn, can limit the spatial resolution. ■

In the preceding example, the smallest beam diameter that meets certain system requirements was calculated. However, other considerations (such as diffraction) may require the use of larger beam diameters, thereby limiting the resolution of the system. For ideally shaped laser beams, the wavefronts may be considered as planar only at the focal point of the beam; this point is called

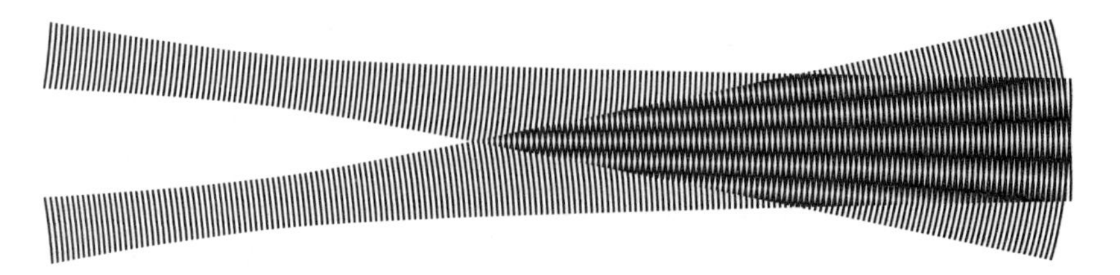

Figure 6.15 Interference pattern obtained by intersection outside the Rayleigh range of two laser beams. [Durst and Stevenson 1976, © Optical Society of America, Washington, DC]

the *waist*. However, there is a narrow range ahead and behind this point where the shape of these fronts is still sufficiently planar. This is the *Rayleigh range* or the *confocal parameter* of the beam, which is related to the wavelength and the diameter of the beam by

$$b = \frac{\pi d_b^2}{2\lambda} \tag{6.28}$$

(Kogelnik and Li 1966; see also eqn. 13.19). Hence, when the beam diameter $d_b = 29.32$ μm, as in the first part of the example, the confocal parameter is 2,632 μm. Although this is larger than the projected value of L, the beams must be carefully aligned to assure their intersection within the confocal parameter. The confocal parameter for the second part of the example is $b = 648$ μm, which is also larger than L. However, here the short range of the confocal parameter may cause the beams to intersect outside their confocal parameters. Figure 6.15 (Durst and Stevenson 1976) illustrates the fringe pattern formed when two beams intersect outside their confocal parameter range. It is evident that the interference fringes are no longer parallel, so now the modulation frequency of the scattered signal depends not only on the particle velocity but also on its trajectory inside the volume. To avoid such distortion, the Rayleigh range must exceed alignment limitations. Thus, the diameter of the beams and hence the spatial resolution of the system may exceed the values predicted by Example 6.1.

The beam diameter, the location of the intersection volume along the beams, and the intersection angle are the first parameters to be determined. However, the design of an entire system must also include specification of the particles seeded in the flow and the location of the detector. In the first demonstration of the technique (Yeh and Cummins 1964), the flow was artificially seeded by mono-dispersed spherical particles. Although seeding the flow is still the more common practice, in many applications an attempt is made to utilize naturally occurring particles. For example, in measurements using natural water there may be a sufficient density of particles for consistent measurements, but in other flows seeding may be required. Thus, the size of the seed particles, their shape, their material, and their density must all be determined. Of course, the primary requirement is that the particles faithfully follow the flow velocity, particularly when large velocity gradients occur. Thus, the inertia of the particles must be kept as low as possible, so their characteristic size must be small and their weight must be low. On the other hand, particles that are too small may be affected by the Brownian random motion of the fluid molecules. Although a detailed analysis is required for each combination of fluids and particles, in most applications the size of the particles is kept between 0.5 and 5 μm. This size is also smaller than or comparable to the spacing between fringes of most systems, thereby providing good modulation depth of the Doppler signal. The selection of the particle material is motivated not only by the velocity distribution of the flow but also by the fluid properties. Thus, the particles must be chemically inert and capable of sustaining the temperature variations of the flow. In addition, to prevent buoyancy, the density of the particles must be matched to the fluid density. Although not all these requirements can be met, most commonly used materials include polystyrene or glass for water or cold air flows, while SiO_2 or smoke are used to seed hot air or combusting flows.

The size of particles used for seeding the flows for LDV measurements is comparable to the wavelength of the incident radiation. This is the regime in which the irradiance of the scattered radiation is controlled by the Mie scattering process (Born and Wolf 1975). Although detailed discussion of Mie scattering is beyond the scope of this book, its main results show strong dependence on the size of the particle (a d^6 dependence was predicted in Section 4.10) and the direction of the scattering. Thus, the maximum scattering irradiance is observed in the forward direction. From the forward direction the irradiance falls rapidly as the angle between the detection line of sight and the incident beam increases. For example, when $2\pi d/\lambda = 6$ the irradiance of the back scattering is approximately 100 times smaller than the forward scattered irradiance (Van De Hulst 1957, p. 152). The precise behavior depends on the particle shape, its size, and its optical properties.

Since the frequency of modulation of the scattered radiation is independent of the location of the detector, its position must be determined primarily by optical access of the test section and by the scattering irradiance. Thus, when the power of the incident laser is high (e.g. 0.1 W), the back-scattered irradiance may be sufficient for detection and the detector may view the scattering through the entrance port. This is particularly useful when window access is limited. On the other hand, when the power of the incident beams is relatively low (e.g. less than 5 mW when a He–Ne laser is used), the scattering irradiance may exceed the detection limit only if collected in the forward direction. When the test section is enclosed, two windows are required for forward scattering detection: one for entrance of the beams and the second at the opposite side for detection.

One of the primary considerations in LDV system design is the selection of the data acquisition and analysis system. For most flows, the data capture rate must meet the rate of particles crossing the intersection volume. For uniform seeding this rate is controlled by the flow velocity. Thus, for slow flows, electronic systems with slow data capture rate and slow frequency response may be used (e.g., storage oscilloscopes). However, in fast flows, both the Doppler frequency and the data rate are high. Furthermore, owing to turbulence, the frequency induced by individual particles may change abruptly while being recorded. To accommodate for this experimental complexity, modern LDV systems include components for high-rate digital analysis of the scattered signal. With such analysis, some parameters of the optical system can be further optimized. For example, the signal scattered by an exceptionally large particle can be recognized and rejected. This serves to eliminate velocity measurements of large or heavy particles that, owing to their inertia, may not accurately follow the flow. Another option included in many systems is the choice of the number of cycles in the signal modulation that must be recognized by the system before a measurement is accepted. Thus, signals induced by particles passing at the edges of the intersection volume, where the number of fringes is small, are rejected. This, of course, serves to reduce the size of the measurement volume and hence improve the resolution. Similarly, comparison of the frequency associated with the first few cycles of the signal modulation with the frequency of subsequent cycles may serve to distinguish particles that cross the probe volume uninterrupted from particles that were interrupted by other scatterers during the measurement period. On the other hand, such distinctions may be obscured by the frequency changes induced by turbulence. More advanced systems now fully digitize the signal and subsequently process it to recognize discrepancies or to generate histograms of velocity. These and other features of available systems will not be discussed here; for more information see Hepner (1994).

Another alternative, currently used for velocity measurements in advanced test facilities, is designed for the simultaneous measurement of two or three velocity components. Such systems are needed to study two- or three-dimensional flows with rapidly varying velocity. Examples include injection of fuel into combustion chambers, mixing of supersonic jets, or simply highly turbulent flows. In all these examples, the consecutive measurement of individual components may not produce the magnitude and direction of the desired momentary velocity component, which can be obtained only by simultaneous measurement of all three components at a point. Several approaches are used for these measurements. Usually, two or three pairs of beams are required to form three sets of interference fringes, each set with its plane oriented along one of the coordinate axes and with their volumes of intersection overlapped at one point. Thus, the scattering of a particle passing through these beams corresponds to two or three velocity components. Although the concept is simple, the design requires that each of the intersecting beam pairs be distinguishable from the others. Otherwise, the scattering associated with one velocity component will not be distinguishable from the scattering associated with the other components. If only two velocity components are desired, both beams may have the same color while their polarizations are mutually orthogonal. By employing two detectors, each equipped with a polarizer, the two velocity components can thus be easily distinguished. When three velocity components are needed, two or three different colors are used. Consequently, the detectors are equipped with bandpass filters to distinguish between colors associated with the scattering from each beam pair. Normally, multiline argon ion lasers are used for these applications. This laser can readily emit at least two bright beams at well-separated colors.

Because of its wide range of application, the LDV technique is more developed than most other optical flow diagnostics system. Numerous manufacturers offer full systems that include the laser source together with units for optical delivery, detection, and data acquisition and analysis. Systems are available for measurements of a single component in a slow flow, for three-component measurement in supersonic or turbulent flow, or for simultaneous measurement of the size and velocity of the scattering particles (Bachalo and Houser 1984). Furthermore, many manufacturers also supply systems for seeding the particles as well as fully established criteria for their selection. However, despite its success, LDV could not be developed into a system that provides a two- or three-dimensional image of velocity distribution. For such distributions to be obtained with LDV, the intersection volume must be translated and measurements must be obtained individually at each point. Thus, other techniques have been developed for imaging of the planar distribution of flow velocity. These include particle imaging velocimetry (PIV) (Grousson and Mallick 1977), filtered Rayleigh scattering (Miles and Lempert 1990), and planar laser-induced fluorescence (Klavuhn, Gauba, and McDaniel 1992). Some of these techniques will be discussed in subsequent chapters.

References

Bachalo, W. D., and Houser, M. J. (1984), Phase/Doppler spray analyzer for simultaneous measurements of drop size and velocity distributions, *Optical Engineering* 23: 583–90.

Born, M., and Wolf, E. (1975), *Principles of Optics,* 5th ed., Oxford: Pergamon, pp. 256–370.

Drain, L. E. (1980), *The Laser-Doppler Techique,* Chichester: Wiley.

Durst, F., and Stevenson, W. H. (1976), Visual modeling of laser Doppler anemometer signals by moiré fringes, *Applied Optics* 15: 137–44.

Durst, F., and Whitelaw, J. H. (1971), Optimization of optical anemometers, *Proceedings of the Royal Society of London A* 324: 157–81.

Grousson, R., and Mallick, S. (1977), Study of flow pattern in a fluid by scattered laser light, *Applied Optics* 16: 2334–6.

Hecht, E. (1990), *Optics,* 2nd ed., Reading, MA: Addison-Wesley, pp. 333–88.

Hepner, T. E. (1994), State-of-the-art laser Doppler velocimeter signal processors: calibration and evaluation, AIAA Paper no. 94-0042, American Institute of Aeronautics and Astronautics, Washington, DC.

Hercher, M. (1969), Tunable single mode operation of gas lasers using intracavity tilted etalons, *Applied Optics* 8: 1103–6.

Hiller, B., and Hanson, R. K. (1985), Two-frequency laser-induced fluorescence technique for rapid velocity-field measurements in gas flows, *Optics Letters* 10: 206–8.

Jones, R., and Wykes, C. (1989), *Holographic and Speckle Interferometry,* Cambridge University Press.

Klavuhn, K. G., Gauba, G., and McDaniel, J. C. (1994), OH laser-induced fluorescence velocimetry technique for steady, high-speed, reacting flows, *Journal of Propulsion and Power* 10: 787–97.

Kogelnik, H., and Li, T. (1966), Laser beams and resonators, *Applied Optics* 5: 1550–67.
Miles, R. B., and Lempert, W. (1990), Two-dimensional measurement of density, velocity, and temperature in turbulent high-speed air flows by UV Rayleigh scattering, *Applied Physics B* 51: 1–7.
Rudd, M. J. (1969), A new theoretical model for the laser Dopplermeter, *Journal of Scientific Instruments* (2) 2: 55–8.
Shimizu, H., Lee, S. A., and She, C. Y. (1983), High spectral resolution lidar system with atomic blocking filters for measuring atmospheric parameters, *Applied Optics* 22: 1373–81.
Van De Hulst, H. C. (1957), *Light Scattering by Small Particles,* New York: Wiley, Chapters 4 and 6.
Weast, R. C., and Astle, M. J., eds. (1980), *CRC Handbook of Chemistry and Physics,* 60th ed., Boca Raton, FL: CRC Press.
Yeh, Y., and Cummins, H. Z. (1964), Localized fluid flow measurements with an He-Ne laser spectrometer, *Applied Physics Letters* 4: 176–8.

Homework Problems

Problem 6.1

(a) Two monochromatic plane waves intersect at an angle of 2α. Use the equation for the propagation of plane waves (eqn. 4.14) to determine the spacing between the interference fringes formed by these beams.

(b) A collimated beam is incident perpendicularly on a flat mirror. Describe the field obtained by the superposition of the incident and reflected fields and show that the locations of the nodes formed by interference are independent of time.

Problem 6.2

Show that the shape of an incident wavefront is unchanged upon reflection off a specularly reflecting surface. You may assume that any wavefront is made of a superposition of plane waves. Show that plane waves remain planar after such a reflection.

Problem 6.3

When two beams interfere, the distribution of their irradiance is presented by (6.4). From this equation it can be shown that the energy distribution is nonuniform. Thus, at a dark fringe the energy flux may be zero. Although this may appear to be a violation of the first law of thermodynamics, in reality energy is conserved. Use integration to show that the average spatial distribution of the energy is conserved even though locally the flux may be higher or lower than the average.

Problem 6.4

Derive (6.10), which describes heterodyning, from (6.4) by introducing the time dependence of $L_1 - L_2 = 2\mathbf{V}t$ induced by the motion of the mirror M_1 of the Michelson–Morley interferometer (Figure 6.1).

Problem 6.5

A beam of width d is incident on a flat mirror at an angle θ_i with the reflector introducing a phase shift of $\Delta\phi$. Derive the field obtained by the superposition of the reflected and incident fields. Compare with Problem 6.1(b).

Problem 6.6

Derive (6.7), which describes the Doppler shift associated with the motion of mirror M_1, by assuming that M_1 intercepts more wavefronts as it heads toward the incoming beam than it does when it remains stationary. Over a period of Δt, the number of additional wavefronts is ΔN. Find the change in frequency due to the motion of M_1 toward the incoming beam and due to the motion of M_1 toward the stationary screen. Show that the two shifts are additive.

Problem 6.7

An etalon is designed to monitor the spectral output of a tunable KrF laser. The laser emits at a nominal wavelength of $\lambda = 248$ nm and a bandwidth of 0.3 cm^{-1}. For the measurement, a small portion of the beam is split and focused by a lens. The etalon is placed past the focus at a point where the beam is diverging. The transmitted beam is projected on a phosphor-coated screen. When the laser operates at a narrowband mode, the projection should appear as concentric, well-defined rings. As the laser is tuned, the rings move inward or outward; otherwise, they remain steady. Find the thickness of the etalon and the reflectivity R necessary to observe well-defined fringes. What reflectivity will assure that the width of the rings is dominated by the laser bandwidth and not by the etalon characteristics?

Problem 6.8

Using the heterodyning concept, show that the frequency of the signal emitted by a detector placed between the two laser beams in Figure 6.13 is independent of the direction of the line of sight.

Research Problems

Research Problem 6.1

Read the discussion of Fabry–Perot etalons in Born and Wolf (1975, pp. 323–41).

(a) The colors of peacock feathers are generated by thin dielectric films deposited on a surface. Assume that the peacock desires that his (not hers – she is quite bland) feathers be green ($\lambda = 500$ nm). What should be the thickness of the film when the index of refraction is 1.4 and the feathers are observed at normal incidence?

(b) What will be the color when viewed at 45°?

(c) Estimate the finesse Φ of the film.

(d) How well does the peacock control his production line? (Remember that if the film thickness varies slightly, the color will change significantly.)

Research Problem 6.2

Read the paper by Hercher (1969). A typical laser consists of two mirrors facing each other; one is highly reflecting while the other is partially transmitting. The laser gain medium is enclosed between these mirrors. As radiation oscillates between these mirrors, it is amplified by the gain medium while a portion is emitted through the partially transmitting mirror. This paper discusses the effect of the insertion of an etalon between these mirrors on the output of the laser.

(a) The laser mirrors themselves may also be seen as an etalon. Therefore, the cavity enclosed between these mirrors can support only certain wavelengths, which Hercher calls *axial modes.* If the distance between the mirrors is 0.5 m and their reflectivity is $R_1 = 0.99$ and $R_2 = 0.8$, find the free spectral range and the finesse of the laser cavity when $\lambda_0 = 500$ nm.

(b) Assume that the gain medium can amplify radiation only within a range of 0.2 cm^{-1}. Find how many axial modes can oscillate simultaneously in this laser and the bandwidth of each.

(c) Derive equation (2) of the paper by starting with our (6.15). You may assume that $\theta_t \ll 1$ and that λ_0 as defined by Hercher is the resonant wavelength of the etalon for $\theta_t = 0$.

(d) Derive equation (7) of the paper.

(e) What condition must the etalon meet in order to assure that only one axial mode is supported by the combined laser–etalon system?

(f) Usually, the wavelength that is selected by the etalon does not exactly match the wavelength of the axial mode of the laser or the etalon mode, so assume that the axial cavity mode that is nearest to an etalon resonance is selected for oscillation. For an intracavity etalon with $n = 1.5$ and $d =$ 3 mm, what must the tilt angle be in order to force an argon ion laser to radiate at exactly $\lambda = 514.5$ nm?

(g) One of the problems of lasers with intracavity etalons is "mode hopping." In response to slight vibrations of the etalon or slight temperature changes in the cavity, the selected axial mode may change abruptly. Assume that a laser system is 0.5 m long and contains air at 25°C. The index of refraction of air is described by (6.14). The laser includes an intracavity etalon as described in part (e). Determine the laser-cavity temperature change that would be necessary to force the oscillation into the next cavity axial mode.

7 Diffraction Effects and Their Applications

7.1 Introduction

Of the three phenomena that result from the wavelike nature of light - polarization, interference, and diffraction - the third is the most puzzling. It does not render itself to intuitive explanation, since intuition suggests that light propagates in straight lines. Diffraction, however, allows for light under certain conditions to travel "around corners." Because of this effect, light may be detected at points that could not be reached by straight rays. This effect also prevents indefinite propagation of collimated beams; invariably, after a certain distance, collimated beams appear to diverge. Similarly, when a focusing lens designed using considerations of geometrical optics is employed to focus radiation, the spot size at the focus cannot be reduced below a defined limit. In these examples, diffraction is seen to pose limitations on the application range of many optical devices. Thus, imaging resolution is reduced by the diffraction limits of lenses, power delivery by collimated laser beams is limited by their divergence, and the application of masks for processing semiconductor chips with photolithographic techniques is limited by diffraction induced by the minute pattern of the masks.

However, there exist numerous applications where diffraction presents an advantage. One example is the diffraction grating used for spectral separation of radiation (see Section 7.3). Another example is the advent of Fourier optics. This relatively new technology is based on the diffraction-limited imaging properties of lenses. It is now used, mostly in simple cases, to obtain the spatial Fourier transform of objects (Goodman 1968, p. 77). With the advent of fast and efficient computational techniques, these transforms may be used in the future for automated recognition of patterns, such as an object on a manufacturing line. Discussion of Fourier optics is beyond the scope of this text. However, the reader is referred to Goodman (1968) and Iizuka (1983) for details.

Here we describe briefly the diffraction effect and its consequence in three applications: the diffraction grating, moiré interferometry, and holography.

7.2 The Diffraction Effect

The reasons for diffraction are well understood. It can be modeled using integral equations of the electric field that were originally developed by Kirchhoff and later refined by Rayleigh and Sommerfeld (see Goodman 1968 or Jackson 1975, chap. 9). These equations are very general and (except for simple cases) cannot be solved without approximations. One such approximation is the Fresnel approximation, which describes the distribution of the electromagnetic field in the vicinity of a disturbance such as a narrow slit. This approximation is called, appropriately, the *near-field* approximation. Alternatively, the Fraunhoffer approximation is used to describe the field distribution in the far field. Here we will use instead a very limited approach, originally proposed by Huygens (1690). Although solutions of diffraction problems using Huygens's principle have limited scope and require a case-by-case analysis, they convey more clearly the physical fundamentals of the diffraction effects and are therefore more suitable for our objectives.

Huygens's principle states that

> *every unobstructed point on a wavefront can serve as a point source that emits spherical waves at the same wavelength as the original front.*

The secondary spherical waves are emitted into the domain that the original front was entering (i.e., the secondary waves do not propagate backwards). Thus, at any point within the diffraction domain, the electric field is determined by the superposition of the amplitudes and phases of the fields induced by all the point sources. This is similar to the analysis of the interference of two beams (Section 6.2), with the exception that here an infinite number of sources are included instead of only two. The phase and amplitude of the wave of each source at the point of interaction is determined from the solution of the wave equation (eqn. 4.8) for emission by a point source.

Although solution of (4.8) for a point source can be obtained directly using spherical coordinates, it is possible to determine its main results from simple physical observations. Any solution of the propagation of waves must include a term for the time dependence of the phase and the amplitude, as well as a term for their spatial variation. Since the time dependence of the solution is independent of the choice of the coordinate system, we need only consider the spatial variation of the phase and amplitude. For a spherical wave, the phase is constant on the surface of any sphere centered at the point source, but varies along the radii centered at that point as e^{jkr}. This is similar to the spatial oscillation of plane waves along the propagation axis. The radial variation of the amplitude can be obtained from considerations of energy conservation by observing that the irradiance is constant on any concentric spherical surface around the

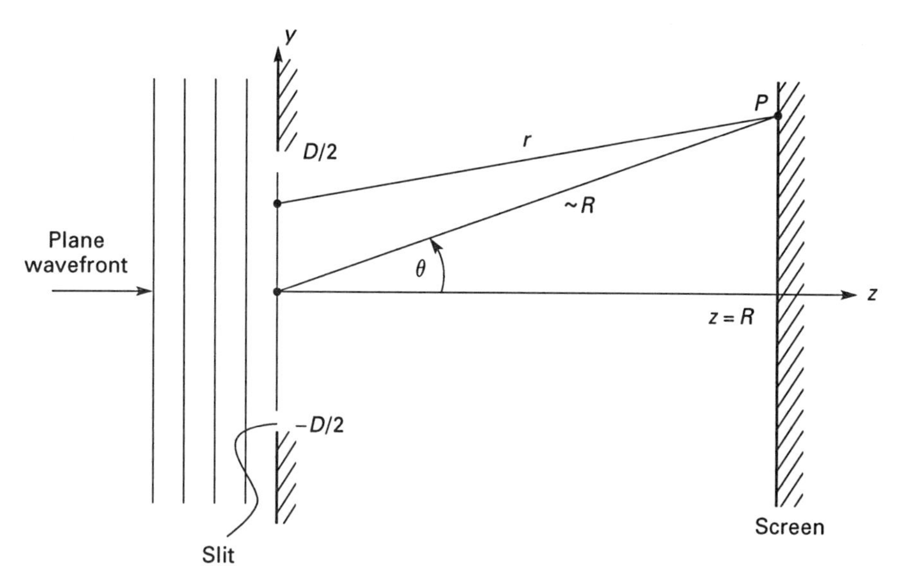

Figure 7.1 Top view of a slit and a screen illuminated by a coherent plane wave, showing two wavelets (marked by •) heading toward point P on the screen.

source. Therefore, to conserve energy, the irradiance must diminish inversely with the surface area or as $1/r^2$ along any radius. From (4.42) it is seen that if the irradiance diminishes as $1/r^2$, the field must diminish inversely with r. Combining these observations for the phase and amplitude terms, we obtain the following r dependence for the field of a spherical wave originating at a point source:

$$\mathbf{E}(r) = \frac{\mathbf{e}_0}{r} e^{-i(\omega t - kr + \phi)}, \tag{7.1}$$

where $\mathbf{e}_0$ is a field amplitude term. As before, the vectorial representation of $\mathbf{E}(r)$ and $\mathbf{e}_0$ allows the independent analysis of each polarization.

To illustrate the application of Huygens's principle, consider a uniform plane wave propagating in the positive z direction (Figure 7.1). The wavefront is incident on a rectangular slit, centered at the origin of a Cartesian coordinate system. The slit is infinitely long in the x direction (which points into the page) and has a width D in the y direction. To determine the distribution of the irradiance on a screen parallel to the slit and located at a distance R from it, assume that D is small relative to the transverse coherence length of the incident wavefront. This assumption implies that, along any line parallel to the y axis, the initial phase ϕ of the field immediately ahead of and past the slit is uniform. The transverse coherence in the y direction is illustrated in Figure 7.1 by depicting the wavefronts with straight lines. Such an assumption of coherence cannot be made for the field distribution along the length of the slit. Owing to its infinite length, it must be assumed that in the x direction the slit length exceeds the transverse coherence length of the radiation and that ϕ varies randomly along any line parallel to the x axis. Hence, fields are additive when propagating in the y direction, whereas in the x direction only the irradiance is additive. With these assumptions, the analysis may be started by selecting a section of the slit with a unit length in the x direction and determining the distribution of the irradiance at the screen by radiation passing this section alone. The total irradiance is obtained by integrating the irradiance through the entire length of the slit.

To simplify the analysis, only two secondary waves, or *wavelets,* are shown in the figure to originate from the slit and propagate toward point P: one at the origin and the second at an arbitrary point with coordinates $(0, y, 0)$. The distribution in the y direction of the electric field induced at P can be found by adding the fields of all the wavelets there. From the results in Section 6.2, one may expect the wavelets to interfere with each other, thereby forming regions where the irradiance vanishes and other regions where the interference results in bright illumination. Of course, the interference at a point on the screen depends on the amplitude of each field component and its phase, which in turn vary with the distance between the source and point P. The field induced at P by the field of the wavelets that emerge from an element of the wavefront in the slit plane with a width dy and a length of unity in the x direction is

$$d\mathbf{E} = \frac{\epsilon_0}{r} e^{-i(\omega t - kr + \phi)}\, dy, \tag{7.2}$$

where ϵ_0 is the field amplitude of a wavefront element (and thus different from $\mathbf{e}_0$ – the field amplitude of a point source) and r is the distance from the slit element to P. The location of P on the screen relative to the emitting element is specified by the distance r. It can also be specified by R (the z coordinate of P) and the angle θ, which is measured from the positive z axis to the ray connecting the origin with P. For representation of r in terms of R and the coordinate y, we use the cosine theorem:

$$r = R\left[1 + \left(\frac{y}{R}\right)^2 - 2\frac{y}{R}\cos\left(\frac{\pi}{2} - \theta\right)\right]^{1/2}.$$

After expansion into a series and neglecting terms of the order $(y/R)^3$ and higher, we have

$$r = R\left[1 - \frac{y}{R}\sin\theta + \frac{1}{2}\left(\frac{y}{R}\right)^2 \cos^2\theta + \cdots\right].$$

For points near the z axis (i.e., for $y/R \ll 1$), the field amplitude can be well approximated by using only the first term of this series. Thus the value of r in the denominator of (7.2) can be replaced with R. On the other hand, the phase of the field, which appears in the exponent of (7.2), is more sensitive to small changes. Thus, to describe the phase more accurately, the approximation for r in the exponent of (7.2) must also include the second term of the series. The third term may be neglected only if its largest contribution to the exponential term in (7.2) is negligible. The maximum contribution is made, of course, by points at $y = \pm D/2$. Therefore, the third term may be neglected when

$$\frac{k}{2}\frac{(D/2)^2}{R} \ll 1 \tag{7.3}$$

(Goodman 1968, p. 61). This is the *Fraunhofer limit* for the far-field approximation. The total field $\mathbf{E}(R,\theta)$ in the Fraunhofer limit, induced at P by the entire width D of the slit (but only by a length of unity in the x direction), is obtained by integrating (7.2) along the y axis:

$$\mathbf{E}(R,\theta) = \mathrm{Re}\left[\frac{\epsilon_0 e^{-i(\omega t+\phi)}}{R}\int_{-D/2}^{D/2} e^{ik(R-y\sin\theta)}\,dy\right]. \tag{7.4}$$

Integrating (7.4) and using (4.42), the irradiance $I'(R,\theta)$ at point P induced by a section of unit length of the slit is found to be

$$I'(R,\theta) = I'(R,0)\left[\frac{\sin\left(\frac{kD}{2}\sin\theta\right)}{\frac{kD}{2}\sin\theta}\right]^2 = I'(R,0)\,\mathrm{sinc}^2\left(\frac{kD}{2}\sin\theta\right), \tag{7.5}$$

where $I'(R,0) = (1/2\eta)(\epsilon_0 D/R)^2$ is the amplitude per unit length of the irradiance at $\theta = 0$, and where $\mathrm{sinc}\,\beta = \sin\beta/\beta$.

To obtain the irradiance at the screen when the entire slit is illuminated, the irradiance $I'(r,\theta)\,dx$ that is passed through a section dx is integrated along the entire slit length, where r is now the distance from the element dx to the screen (Figure 7.2). For an arbitrarily located segment dx, the distance to any

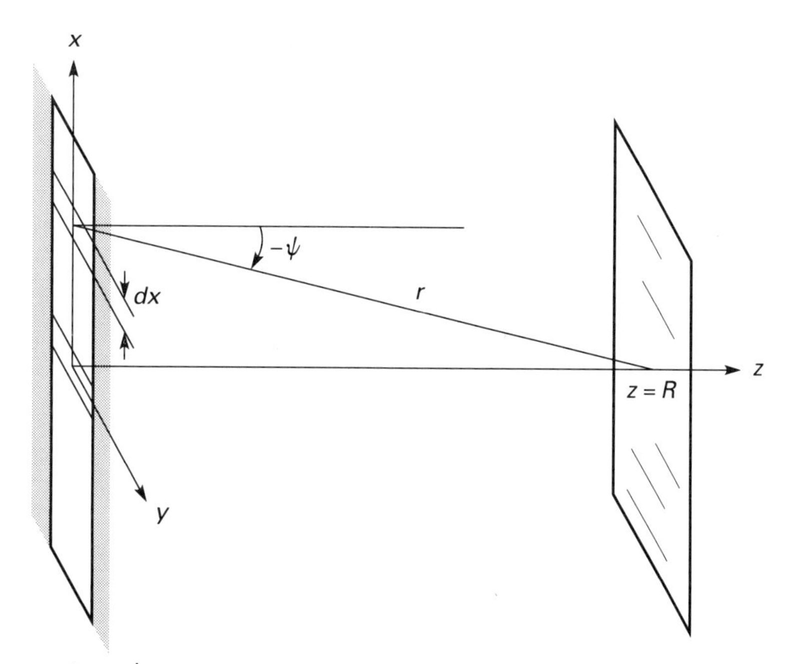

Figure 7.2 A long rectangular slit and the screen where diffraction through that slit is observed.

point on the screen that is also in the (x, y) plane can be expressed by R and the angle ψ as follows:

$$r = \frac{R}{\cos\psi}.$$

Integration of the irradiance, rather than the field, represents an assumption that the transverse coherence of the field is short relative to the slit dimension in the x direction. This is acceptable in view of the infinite length of the slit. By introducing $dx = -r\,d\psi$, the integral for the irradiance by the entire slit becomes

$$I(R,\theta) = -\int_{-\pi/2}^{\pi/2} \frac{1}{2\eta}\left(\frac{\epsilon_0 D}{R/\cos\psi}\right)^2 \operatorname{sinc}^2\left(\frac{kD}{2}\sin\theta\right)\frac{R}{\cos\psi}\,d\psi. \tag{7.6}$$

The integration limits are consistent with the infinitely long–slit assumption. After integration, we obtain the following for the irradiance due to the illumination of an infinitely long slit:

$$I(R,\theta) = I(R,0)\operatorname{sinc}^2\left(\frac{kD}{2}\sin\theta\right), \quad \text{where } I(R,0) = \frac{(\epsilon_0 D)^2}{\eta R}. \tag{7.7}$$

The distribution of the illumination by diffraction through an infinitely long slit is presented in Figure 7.3 by the variation of the normalized irradiance I/I_0 with the argument of the sinc function $\beta = (kD/2)\sin\theta$. The most notable effect is that the shape of this pattern is independent of the distance R

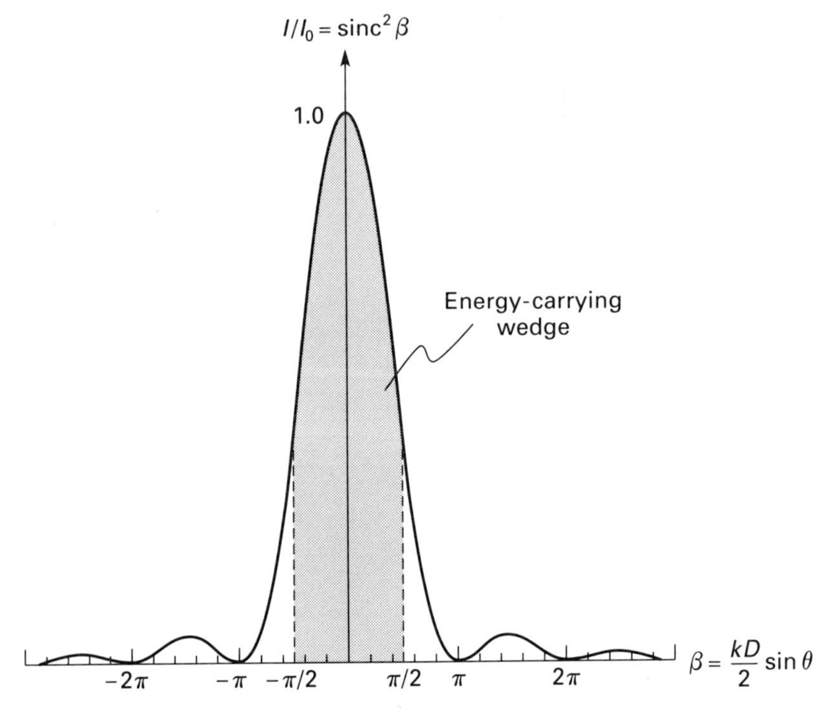

Figure 7.3 Diffraction pattern obtained by the illumination of a narrow, infinitely long slit.

and varies only with the angle θ. Thus, although the linear dimensions lateral to the slit increase proportionally with distance, the shape of the diffraction structure is preserved. It is apparent that most of the radiation is incident within the central lobe, where $-\pi < \beta < \pi$. At both sides of this lobe there are alternately dark and bright fringes with diminishing intensity. The central lobe can be viewed as an infinitely long wedge of light delineated at both sides by the first dark fringe and diverging at an angle of

$$2\theta = 2\sin^{-1}\frac{2\pi}{kD} = 2\sin^{-1}\frac{\lambda}{D}. \tag{7.8}$$

For small angles, the divergence is $\theta \approx \lambda/D$; that is, the divergence angle increases linearly with λ and inversely with the width of the slit. The divergence angle may also be defined as the apex angle of the wedge that carries most of the energy – for example, $-\pi/2 < \beta < \pi/2$ (Figure 7.3). With that definition, the divergence angle is

$$\Delta\theta = 2\theta_{\beta=\pi/2} \approx \lambda/D. \tag{7.8'}$$

Although similar to the definition of (7.8), this interpretation is different and covers the entire apex angle of the wedge. As D increases, the divergence angle decreases and the beam emitted through the slit approaches collimation. On the other hand, when $D \to 0$ the function sinc $\beta \to 1$. Thus, the radiation emitted from an infinitely long but infinitesimally thin slit is independent of θ. Such emission may be viewed as a cylindrical wave emitted by a line source. This result is consistent with Huygens's principle, where each point on the wavefronts at the entrance to the slit was considered to be a source of spherical wavelets. Thus, a linear array of such spherical wavelets must add up to form a cylindrical wavefront. Finally, the result of (7.7) also shows the diminishing of the irradiance inversely with the distance from the source when $\theta = 0$ or when $D \to 0$.

Similar equations can be obtained for diffraction through a rectangular slit of finite length. However, since the transverse coherence may now also exceed the slit length, we must now pursue an integration in the x direction similar to (7.4) instead of the phase-independent integration of (7.6). The result (Hecht 1990) is a diffraction pattern with independent variations in both the x and y directions.

The same analysis can also be extended to describe diffraction through a circularly symmetric aperture of radius a illuminated by a uniform plane wave with an amplitude $\mathbf{E}_0$. The results can be presented in terms of the first Bessel function,

$$I = I(0)\left[\frac{2J_1(ka\sin\theta)}{ka\sin\theta}\right]^2 \tag{7.9}$$

(Hecht 1990), where $I(R,0) = (1/2\eta)(\mathbf{E}_0 a/R)^2$. The radius ϱ of the first minimum is at

$$\varrho = 1.22\frac{R\lambda}{2a}. \tag{7.10}$$

The results of (7.8) and (7.10) suggest that to minimize the divergence of a beam (i.e., to generate a well-collimated beam), the aperture size of all the optical elements as well as the beam diameter must be increased. Although this is generally true, recall that eventually the aperture size of these elements exceeds the transverse coherence of the incident radiation. Therefore, the assumption made before integrating (7.4) – that the phase term $e^{-i\phi}$ is constant through the entire width of the slit – may not be valid, and the phase term must be kept inside the integral sign. Alternatively, the initial phase ϕ may be assumed to be constant over a limited range, and the limits of the integration must match that range. With integration limited to the transverse coherence length, the effective slit size appears much smaller than the actual slit size, and the divergence angle of the packet is determined by its coherence length rather than the slit size. Thus, when passed through the same aperture, incoherent radiation is expected to diverge faster than radiation with transverse coherence length that matches or exceeds the dimensions of that aperture. This is evident from observations on earth of stellar radiation. To overcome astronomical distances, stellar radiation must propagate with minimal divergence. Thus, although most of the radiation emitted by remote stars or galaxies is incoherent, only the minute fraction that happens to be coherent can survive the long travel and be detected on earth. The remainder is lost in space due to its rapid divergence. This is apparent from the twinkling of stars, which is caused by interference effects induced by density gradients in the earth's atmosphere. Laser beams can be collimated (similarly to starlight) into moderately diverging beams. However, owing to the limited size of the available optical elements, even extremely coherent laser beams exhibit a finite divergence (see Problem 7.1).

The results for diffraction by a rectangular slit or a circular aperture describe the divergence of radiation away from these openings. However, the principle of reciprocity (discussed in Section 2.3) enables the same equations to be used for describing radiation that converges toward these openings. Thus, when radiation with an irradiance distribution described by (7.9) is focused, the pattern obtained at the focal spot will be a well-defined, uniformly illuminated circle. Such a lens illumination pattern that includes concentric circular fringes surrounding a bright spot can be obtained by passing uniformly distributed radiation through a mask. However, in most applications we may expect to focus a circular or rectangular beam that is bright at the center and dimmer at the edges. If this illumination pattern is similar to the distribution of the central lobe of the illumination pattern of (7.5) or (7.9), then the spot at the focus will still resemble the rectangular or circular shape of the aperture, albeit with dimensions somewhat larger than the size of the corresponding aperture and with the edges not defined sharply. If, in addition, the transverse distribution at the focusing lens deviates significantly from the distribution of the central lobe in (7.5) or (7.9), further distortion of the focal spot is expected. Thus, the smallest size of a pattern that can be obtained by focusing an ideally distributed illumination pattern is determined by (7.7) or (7.10), where ϱ in (7.10) represents the radius of the focusing lens and a is the radius of the ideal spot size at the focus. This size is also the resolution limit of a lens. Thus, when used for focusing this is the smallest spot that can be obtained, whereas for imaging it is the smallest object that can be resolved. For a circular lens of diameter d_L and focal length f, the resolution limit, based on (7.10), is

$$\delta = 1.22\frac{f\lambda}{d_L} = 0.61\frac{\lambda}{\mathrm{NA}}, \tag{7.11}$$

where δ is the radius of the smallest spot that could be projected (or resolved) by the lens under the most ideal conditions and where NA denotes the numerical aperture (see eqn. 2.14). The result of (7.11) was used previously to determine the resolution of a camera lens (eqns. 6.22 and 6.23). Thus, the smallest object that can be resolved by a microscope of NA $= 0.5$ is approximately the size of the wavelength used for illumination. For visible light, the resolution limit can approach 0.5 μm.

This resolution limit presents difficulties in the manufacturing of highly integrated electronic chips. Some of the largest chips (megachips) contain 16 and 64 megabits (Basting 1991), which are etched on the surface of the semiconductor at a spatial resolution of 0.3 μm. Some of these circuits are manufactured using a microlithography process. In this process, a mask is imaged on the chip substrate, which was previously coated with a photographic material - the *photoresist*. After development of the photographic material, the areas on the chip that were unexposed are etched away chemically to form the required structure. Clearly, the resolution of the image on the chip depends on the quality of the focusing lens, its NA, and the wavelength of the incident radiation. With the recent development of KrF excimer lasers, illumination at an ultraviolet wavelength of 248 nm became possible with a resolution of approximately 1 μm (eqn. 7.11) over a field of 4 mm (Sarbach and Kahlert 1993). Unfortunately, at this wavelength, the dispersion of the quartz lens used for the imaging is large relative to its dispersion at visible wavelengths. To avoid undesired distortion of the image due to chromatic aberration, the laser is operated at a narrowband mode, that is, the bandwidth of the laser beam is maintained below 0.003 nm (Basting 1991).

These and the previous results of this section detail an effect that limits the resolution of imaging devices, reduces the focusing potential of lenses, or deteriorates the quality of optically projected patterns. Almost all high-resolution optical applications are affected by diffraction. However, while in some applications diffraction effects impose difficult limitations, the effect may also be harnessed to produce useful results. One device incorporating desirable diffraction effects, the diffraction grating, will be discussed next.

7.3 The Diffraction Grating

Several applications rely on the diffraction effect for successful results. Holography (Section 7.5) is possibly the most notable of such applications. A typical hologram consists of an extremely fine and complex array of dark and

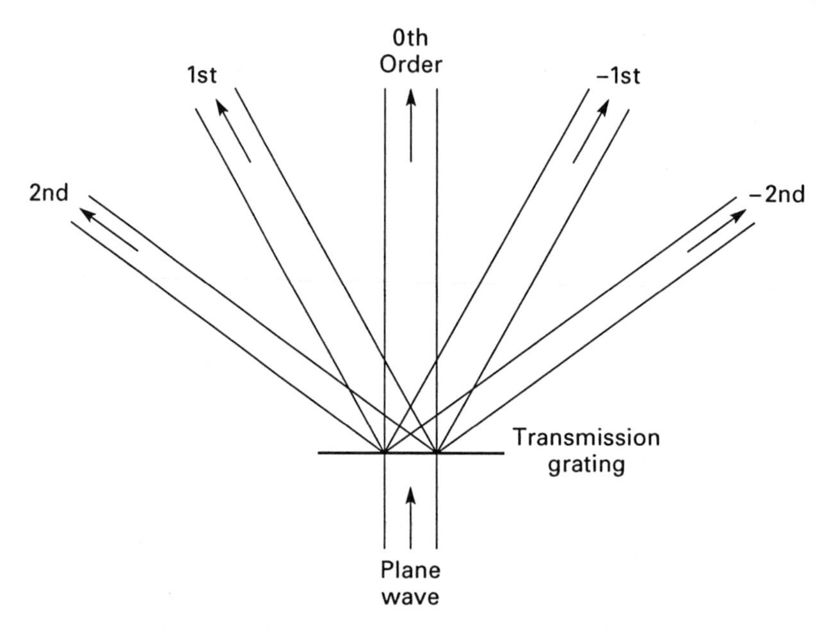

Figure 7.4 Diffraction orders behind a transmission diffraction grating illuminated by a collimated beam.

transparent lines recorded on a photographic plate. When the plate is illuminated, diffraction through this pattern reconstructs a true, three-dimensional image of the object that was recorded on the hologram. Although rarely seen as such, diffraction gratings may be viewed as simple holograms consisting of parallel lines. When the grating is illuminated by a collimated beam, the diffracted pattern consists of multiple collimated beams pointing at angles that depend on the incident wavelength. Figure 7.4 illustrates a typical transmission grating illuminated from below by a collimated monochromatic beam. Immediately behind the grating, the beam appears to remain undisturbed. However, at the far field it is split into several beams, each still collimated and monochromatic. To distinguish between these beams, they are assigned an order; the undisturbed part of the incident beam is labeled as the 0th-order beam, the beam immediately to its left is the 1st-order, the next is the 2nd-order, and so on. The beams to the right of the incident beam are labeled as negative orders. The angle between these secondary beams depends on the wavelength of the incident beam and on the grating properties. Owing to this wavelength dependence, gratings may be used as dispersion elements, in spectroscopic devices or lasers, for spectral separation of radiation. Unfortunately, we are not yet able to demonstrate the relation between holography and diffraction gratings. Instead, the opposite approach is taken; using Huygens's principle, we will describe the principles of gratings and then use these results to subsequently describe some basic principles of holography (Section 7.5).

The underlying principle that defines the operation of gratings is diffraction through an infinite slit. By calculating the illumination by radiation through a single slit (or groove) (eqn. 7.4) and combining the results with the illumination emanating from the other slits (or grooves) of the grating, the transmission (or reflection) characteristics of the grating can be obtained. This general analysis (Born and Wolf 1975, pp. 401–7) is desirable for accurate modeling of gratings. However, the graphical approach adopted here reveals most of their properties using simple mathematical terms.

Although the following description can be equally applied for transmission and reflection gratings, the latter will be illustrated here. These gratings are often used for wavelength selection in tunable lasers or for spectral analysis of radiation in monochromators and spectrometers. Figure 7.5 presents a typical reflection grating illuminated by a plane wave incident at an angle θ_i that is measured between the normal to the grating and the incident propagation vector

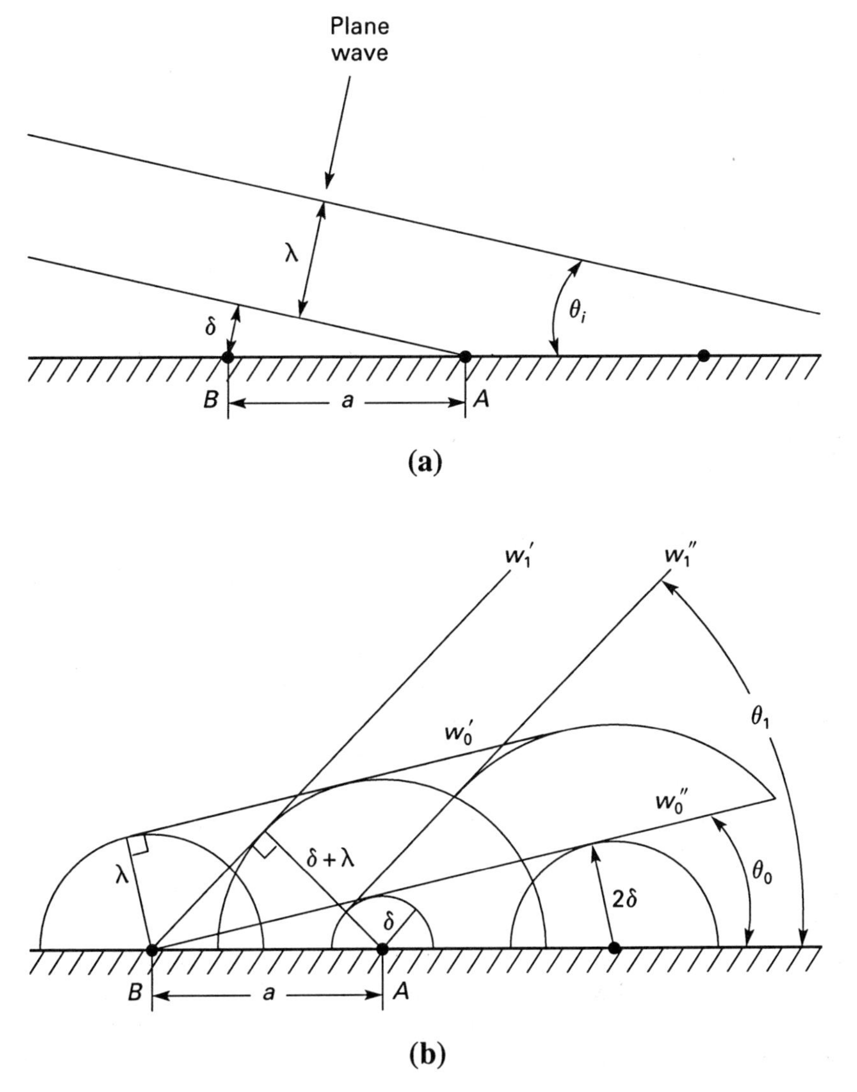

Figure 7.5 Microscopic view of the diffraction by a reflection grating, showing (**a**) illumination by a plane wave and (**b**) cylindrical wavefronts scattered by several grooves.

(or, as in Figure 7.5(a), between the grating plane and the wavefronts). The grating is depicted by a horizontal line that represents an infinitely wide and long flat reflecting surface marked with equally spaced parallel grooves. Owing to its infinite width, this ideal grating contains an infinite number of grooves, each marked by a • that represents an infinitely long line normal to the page. Although the shape, width, and depth of the grooves influence certain properties of the grating, the dispersion of the grating depends only on their density. Thus, for analysis of dispersion, only the grating pitch a, the spacing between adjacent grooves, must be defined; the grooves can be assumed to be infinitesimally narrow. The incident plane wave is marked by a series of parallel lines, each depicting a wavefront, separated by a distance λ that represents the incident wavelength. As before, it is assumed implicitly that the incident radiation is monochromatic and coherent. Although radiation is never perfectly coherent, often the coherence length is much longer than the pitch. In addition, even if the incident radiation is multichromatic, the dispersion resolves it into monochromatic components that can be analyzed separately.

At the moment shown in Figure 7.5(a), the first wavefront coincides with groove A. Groove B will not be illuminated by this wavefront until it has traveled the distance marked by δ. Meanwhile, radiation intercepted by groove A is diffracted. Because of the infinitesimally narrow dimensions of the groove, that diffracted radiation expands away from the grating surface as a cylindrical wave. This is consistent with Huygens's principle, as well as with observations in the previous section regarding diffraction by an infinitely long yet infinitesimally narrow slit. Thus, by the time groove B is illuminated by the first wavefront, the cylindrical wave originating at A has already expanded to a radius δ. Of course, more cylindrical waves are formed by other wavefronts that are intercepted by the grating. Figure 7.5(b) illustrates the array of cylindrical waves immediately after the incidence of the second wavefront on groove B. This wavefront has already developed into a cylindrical wave, with radius δ around A and 2δ around the groove immediately to the right of A. Meanwhile, the first wavefront has developed into a cylindrical front with a radius of $\delta+\lambda$ around A and radius of λ around B. Although Figure 7.5(b) presents only one fully developed cylindrical front around B with a radius of λ, a second front with an infinitesimally small radius has already been produced.

To simplify Figures 7.5(a) and 7.5(b), they include only two incident wavefronts and their resultant diffracted fronts. It is easy to imagine, however, that when a train of wavefronts is diffracted by the infinitely wide grating, the space above each groove will contain a train a cylindrical waves with radii that increase by an increment of λ and with a delay of δ between the train of one groove and the train of the groove to its right. This infinite array of cylindrical waves can be viewed as wavelets that, according to Huygens's principle, make up a new wavefront. This is illustrated in Figure 7.5(b) by the tangents to the depicted cylindrical fronts. Two sets of tangent planes are presented. One set, w_1' and w_1'', are the tangents to the cylindrical fronts of adjacent grooves with phase shift of $\delta+\lambda$; the other set, w_0' and w_0'', are the tangents to adjacent cylindrical fronts with a shift of only δ. These planes are also tangent to other wavefronts to the right of point A, and when extended they include previously formed cylindrical wavefronts, all with the same phase increment. As these cylindrical wavelets propagate away from the grating, their curvature decreases and they merge with the tangential planes, which can be seen as propagating along their own normal. Note that the planes within each set (i.e., w_1' and w_1'') are parallel to each other and are separated by λ; thus they have the characteristics of a collimated beam with wavelength λ. Other beams that belong to sets of wavelets with the general phase shift of $\delta + m\lambda$ between the cylindrical fronts of adjacent grooves emerge from the grating at an angle θ_m, which from simple geometrical arguments is

$$a \sin\theta_m = \delta + m\lambda, \tag{7.12}$$

where m is the order of the diffracted planar wave and $m+1$ is the number of wavefronts that emerged from A before the cylindrical front of this diffraction order at B begins to form. Hence m may be any integer, positive or negative. When m is positive it corresponds to $\theta_m > 0$, thereby defining the counterclockwise direction as positive for the diffraction angle. When the relation $\delta = a \sin\theta_i$ is used (see Figure 7.5(a)), (7.12) can be reduced to the following standard form:

$$\sin\theta_m - \sin\theta_i = m\lambda/a. \tag{7.13}$$

This is the grating equation. Clearly, for the set of wavefronts in Figure 7.5(b) labeled w_0' and w_0'', $m = 0$ and therefore $\theta_m = \theta_i$. Thus, these wavefronts propagate in the direction of the specularly reflected beam, independently of the grating parameters or the incident beam properties. On the other hand, for any value of $m \neq 0$, the propagation direction θ_m of the diffracted planar wavefront increases directly with λ and inversely with a.

Although there exist numerous applications where gratings are used primarily to split a single monochromatic beam into an array of collimated beams (see e.g. Laufer 1984), their primary use is to spectrally disperse radiation. Thus, when the incident beam is not monochromatic, each mode appears as a fan that consists of the superposition of the planar wavefronts of each color component. When illuminated with sunlight, each mode appears as colorful as a rainbow. By placing a slit along the path of one mode, a monochromatic component can be selected from the otherwise nonmonochromatic beam. The irradiance of this spectral component can be measured by placing a PMT behind the slit. This is the *spectrometer.* By turning the grating and continuously monitoring the PMT output, the spectrum of radiation can be measured. Alternatively, an entire spectrum can be measured instantaneously by removing the slit and using an array of detectors or a camera for recording. This is the *monochromator.*

Several parameters determine the spectral purity of the monochromatic component that is selected by a spectrometer. One necessary parameter for the evaluation of the spectral resolution of the slit is the angular dispersion of the mth mode of the grating, $\boldsymbol{D} = d\theta_m/d\lambda$. This can be calculated directly from the following derivative of the grating equation (cf. eqn. 7.13):

$$\boldsymbol{D} = \frac{d\theta_m}{d\lambda} = \frac{m}{a\cos\theta_m}. \tag{7.14}$$

The dispersion increases with the order of the mode and with the groove density, $1/a$. However, from (7.13) is it evident that the grating pitch must exceed λ, thereby limiting the angular dispersion. For a grating with $a < \lambda$, the effective pitch will be the first integral multiple of the existing pitch that meets the condition $\sin\theta_m < 1$ (see Problem 7.2). Thus, for gratings with $a < \lambda$, the angular dispersion may be inferior to the angular dispersion of gratings with $a > \lambda$. Owing to this restriction, a typical grating to be used in the IR ($\lambda \approx 5\ \mu$m) has a groove density of 100 lines/mm, whereas a UV grating ($\lambda \approx 300$ nm) may have up to 2,400 lines/mm. Alternatively, the angular dispersion may be improved by the selection of a high diffraction order. Although the energy content of each mode cannot be calculated by the present model, it is evident that for energy to be conserved, the irradiance of each of the diffraction orders must be lower than the incident irradiance. Furthermore, detailed calculation (Born and Wolf 1975, pp. 401–7) shows that, for most gratings, the irradiance decreases rapidly with m. Thus, attempts to enhance $\boldsymbol{D}$ by increasing m must be weighed against the resulting decreasing in irradiance. These considerations usually limit the useful orders to $m \le 2$.

Although most gratings are effectively limited (by considerations of irradiance) for use at $m \le 2$, certain applications where high spectral dispersion is needed may require the use of gratings at higher orders. In these specialized applications, which may include wavelength selection in tunable lasers or fine spectroscopy, operation at only one mode may be necessary together with high irradiance. To achieve these objectives, gratings can be designed to force much of the incident irradiance into a selected order. Figure 7.6 presents a cross section of such a grating. Since the grooves in this grating appear as stairs, it is known as an *echelon* (Born and Wolf 1975, pp. 407–12). The angle of each of the inclined faces of the grooves is selected so that the specular reflection of the incident beam from these faces coincides with θ_m, the angle of the mode selected for enhancement. Thus, radiation is channeled into this mode by two mutually reinforcing mechanisms: diffraction and specular reflection. This comes at the

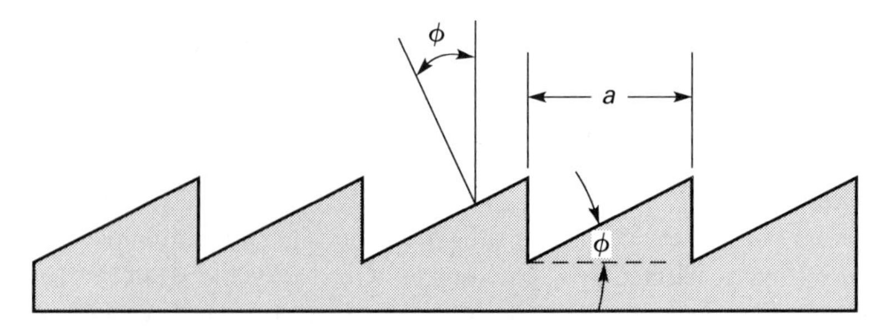

Figure 7.6 Cross section of a blazed diffraction grating (an echelon).

expense of other modes where the irradiance falls below the level obtained by diffraction alone. The enhanced order is usually much brighter than any other mode (including the 0th-order mode that occurs by specular reflection) and is therefore called the *blazed* mode. Thus, when a grating is designed to blaze the mth-order mode when $\theta_i = 0°$, the *blaze angle* ϕ – that is, the angle between the normal to the inclined faces and the normal to the grating plane – is

$$\phi = \tfrac{1}{2}\sin^{-1}(m\lambda/a). \tag{7.15}$$

As the following example shows, a grating designed to blaze the mth-order mode when illuminated normally ($\theta_i = 0°$) by a given wavelength may be used to enhance the diffraction of another mode when illuminated by a different wavelength. Similarly, grating designed for blazing a particular mode for normal incidence may have a different mode blazed when $\theta_i \neq 0°$.

Example 7.1 The blaze angle of a grating with 1,000 lines/mm is $\phi = 30°$. At what incidence angle will the negative second order be blazed when the incident wavelength is 532 mm? What is the blazing order when the incidence is at the same angle but $\lambda = 266$ nm?

Solution Figure 7.7 shows a section of the blazed grating along with a ray of the incident beam and a ray of the diffracted beam, which may also be viewed as specularly reflected at the groove surface. For such specular reflection, it is clear that

$$\theta_i - \phi = \phi - \theta_m.$$

Combining this with the grating equation (eqn. 7.13), for θ_i we have

$$\sin(2\phi - \theta_i) - \sin\theta_i = m\lambda/a.$$

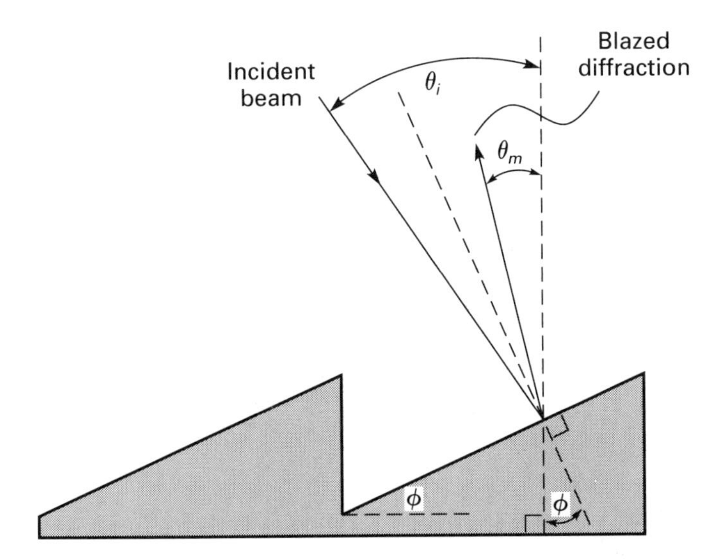

Figure 7.7 Geometrical parameters of diffraction by a blazed grating.

Solving this equation for $m = -2$ yields $\theta_i = 67.9°$.

The second wavelength, $\lambda = 266$ nm, is exactly half of $\lambda = 532$ nm. Thus, to keep $m\lambda/a$ fixed, m must be doubled. Hence, for $\lambda = 266$ nm and for $\theta_i = 67.9°$, the negative fourth order ($m = -4$) is blazed. ■

Although blazed gratings are designed for application at a narrow range of wavelengths and at a specified diffraction mode, some devices (e.g., spectrometers) are expected to be used at a wider range of wavelengths. For these devices, the range of applications of blazed gratings must be specified. Typically the grating efficiency - the ratio between the irradiance in the blazed mode and the incident radiance - is expected to exceed 50%. Unfortunately, the grating efficiency also depends on the incident polarization (Loewen 1977). Figure 7.8 presents the variation with incident wavelength of the efficiency of a grating with 150 lines/mm, $\phi = 26.7°$, for incident s and p polarizations. It is evident that the application range of this grating for the s polarization is wider than that for the p polarization. However, as the blaze angle decreases, the application range for the p polarization approaches the range for the s polarization. At $\phi = 5°$, the efficiency is almost independent of polarization. For these gratings, widely used in UV systems, the range of wavelengths at which the efficiency exceeds 50% was found theoretically (Loewen 1977; Loewen, Navière, and Maystre 1977) to be $0.67 < \lambda/\lambda_b < 1.8$, where λ_b is the blaze wavelength.

One important application of gratings is to select the emission wavelength of tunable lasers. Figure 7.9 illustrates a typical configuration of such a tunable

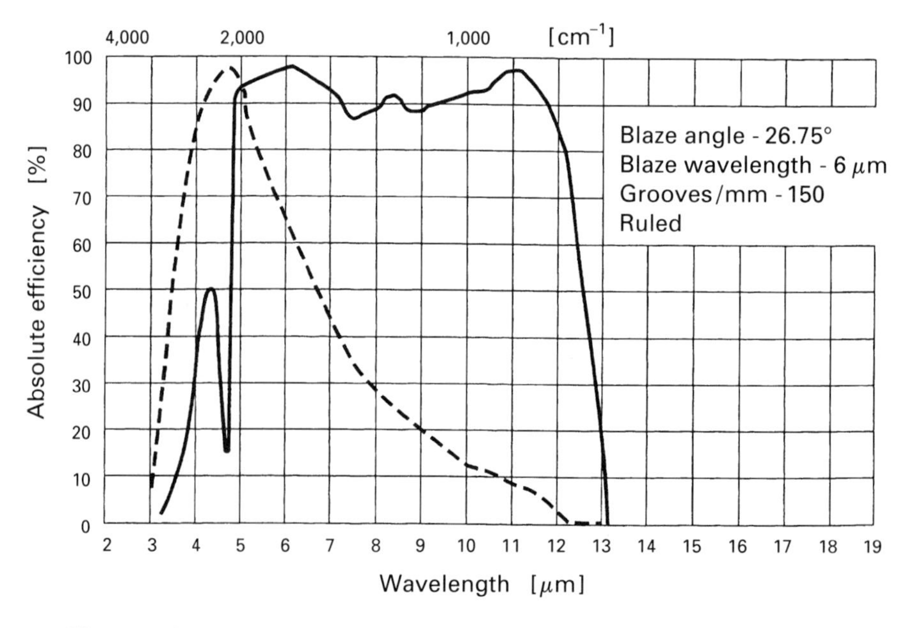

Figure 7.8 Spectral response of a typical grating. Dashed line, p polarization (parallel to the grating grooves); solid line, s polarization (perpendicular to the grating grooves). [Milton Roy Catalog, © Milton Roy Co., Rochester, NY]

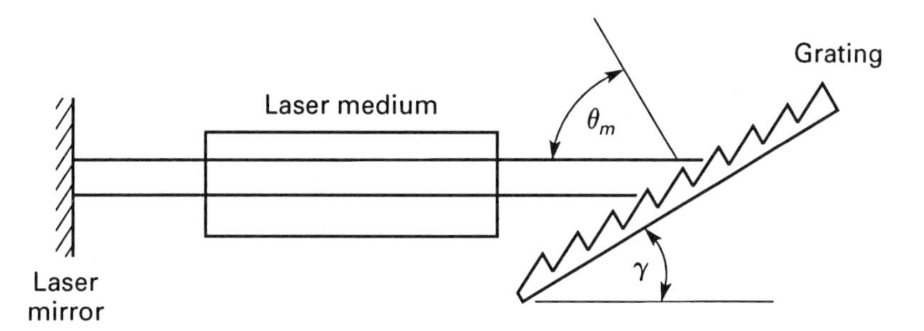

Figure 7.9 Cavity of a tunable laser using a grating in the Littrow configuration as the back mirror.

laser. Although this is a simplified schematic, it shows that the laser beam propagates between an output mirror on one side of the laser medium and the tilted grating on the other side. In each transition through the laser medium, the beam gains energy, part of which is emitted through the output mirror while the remainder is reflected back into the laser medium for additional gain. To complete a round trip, the beam must also be reflected by the grating in the direction of the laser axis. However, owing to its dispersion properties, the grating can reflect only one frequency component along this line. This corresponds to $\theta_i = -\theta_m$ (cf. Figure 7.5); this is known as the *Littrow configuration*. Thus, in this case

$$\sin\theta_m = m\lambda/2a. \tag{7.16}$$

However, for the laser in Figure 7.9, $\theta_m = 90° - \gamma$, where γ is the angle between the grating plane and the laser axis. By varying γ, the wavelength that is reflected back into the laser medium for further amplification can be selected. For efficient operation, the blaze angle of such gratings is selected for operation at high-order modes, typically $m = 5$.

The analysis in this section focused on planar gratings with linear, equally spaced lines. However, other gratings are also available. For example, the grating surface may be curved, either cylindrically or spherically, to focus a diffracted beam. Such gratings are often used in specialized spectrometers. Alternatively, a grating may consist of a transparent flat medium marked with concentric lines. Depending on the pitch (which may vary radially) and the width of these lines, focusing of the transmitted radiation can be obtained. Such circular gratings are called *Fresnel zones* (Das 1991). Unlike Fresnel lenses (Section 2.6), the focusing by a Fresnel zone is strictly by diffraction. Detailed discussion of these special gratings is beyond the scope of this book.

7.4 Moiré Interferometry

Moiré interferometry is an optical technique widely used for visualization and measurement of the planar distribution of two-dimensional strains, out-of-plane distortions of loaded objects, and variations of the index of refraction in

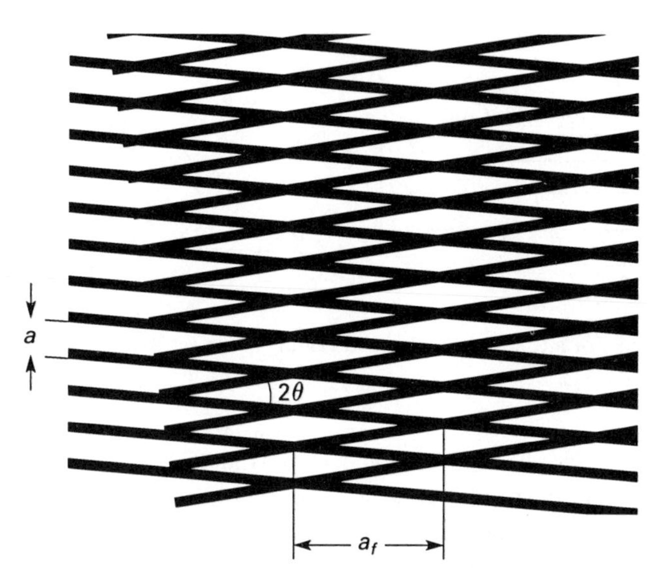

Figure 7.10 Moiré pattern obtained by overlapping two Ronchi gratings.

gas flows or combustion. Although not as sensitive as alternative optical techniques, it enables the imaging of strain fields, gas density variations, and other parameters using relatively simple equipment and with very few limitations.

Despite its misleading name, moiré interferometry is not the result of the interference between two electromagnetic beams. It can be best understood using principles of geometrical optics. To obtain a moiré interference pattern, two (usually coarse) gratings are overlapped as in Figure 7.10. The regions where dark lines of one grating overlap the dark lines of the other grating combine to form a continuous geometrical pattern that visually appears as an interference fringe. The interference pattern observed by this overlap is merely a geometric phenomenon, but it is very sensitive to motion or distortion of any of the interacting gratings. When one of the gratings is translated transversely to its lines, the moiré fringes translate transversely as well. This translation is visually perceptible even when the motion of the grating itself is not; depending on the configurations, translation of any one grating by a few micrometers can induce fringe translation of a few millimeters. Although the study of this technique requires only geometric considerations, diffraction effects induced by any of the interacting gratings may limit its range of use.

Figure 7.11 is a close-up view of the interaction between the lines of two gratings. In most moiré applications the interaction is between two *Ronchi gratings,* which consist of a succession of evenly spaced, alternating opaque and transparent stripes. To simplify the identification of the relevant geometrical parameters, the opaque stripes of the Ronchi grating are depicted by lines in Figure 7.11. Comparison with Figure 7.10 shows that, for an intersection between the opaque lines at an angle of 2θ, bright fringes form along the short diagonals (e.g. BD or CE) of the rhombus patterns (e.g. $ABCD$) enclosed by

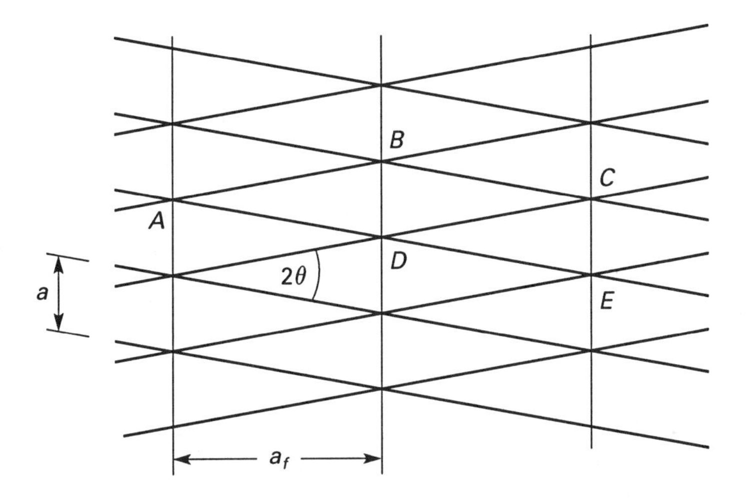

Figure 7.11 Microscopic view of the moiré pattern obtained by overlapping two Ronchi gratings.

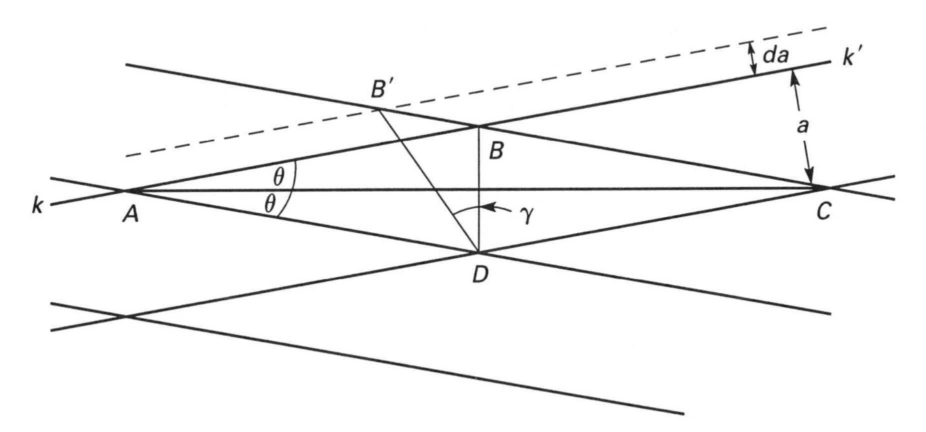

Figure 7.12 Microscopic view of the interaction between a reference and a strained Ronchi grating.

the lines of the gratings. This geometrical configuration is strikingly similar to that of Figure 6.3, where two sets of overlapping plane wavefronts were depicted. Thus, using similar geometrical analysis, for the distance a_f between two adjacent fringes we have

$$a_f = a/(2\sin\theta). \tag{7.17}$$

This result is identical to (6.11) with the exception that the grating pitch a has replaced the wavelength between the interacting wavefronts.

To illustrate the effect of the distortion of one grating on the moiré fringe pattern, assume that one grating is stretched uniformly by normal strain (i.e., the line spacing is increased by da) while the second grating is left undisturbed. This is illustrated in Figure 7.12, where line kk' is replaced by the dashed line.

As a result, the moiré fringes pass through the diagonals of the new rhombi (e.g. $B'D$) and appear to be inclined by an angle γ relative to the previous fringes. Using simple geometrical analysis, it can be shown that

$$\epsilon = 2/(\cot\theta \cot\gamma - 1),$$

where $\epsilon = da/a$ is the normal strain associated with the uniform stretch of the grating. No approximations were necessary to derive this equation. However, in most applications, both θ and γ are sufficiently small that $\cot\theta \gg 1$ and $\cot\gamma \gg 1$. Hence the following simplified equation can be used:

$$\tan\gamma = \epsilon/(2\tan\theta). \tag{7.18}$$

Simply by measuring γ, the normal strain at every point of the imaged field can be determined. However, these measurements require that the test object be marked by a dense array of lines, either by photolithography or mechanically. This is the *object grating*. The strains experienced by that object can be recorded by imaging it through a second grating that overlaps the imaged field; this second grating is often called the *reference grating*. When the strain field is dominated by shear strain, the distortion is dominated by rotation. Thus, the overlap between the distorted grating and the reference grating results in variation of the fringe density (eqn. 7.17) while the fringe orientation remains unchanged.

From (7.18) it is evident that, as θ decreases, the measurement sensitivity increases – that is, detectable γ can be obtained for smaller ϵ. However, when $\theta \to 0$, (7.18) is singular and γ becomes independent of ϵ. Furthermore, at this angle the spacing between fringes formed by two identical gratings approaches infinity (eqn. 7.17). Thus, the measurement when $\theta = 0$ cannot depend on the measurement of the angle of the moiré fringes. To estimate the limiting resolution of this technique (i.e., the smallest detectable strain), assume that the density of the object grating is 4 lines/mm (this is a very coarse grating). With an angle of 5° between the reference and the object grating, the spacing between the moiré fringes (eqn. 7.17) is $a_f = 1.43$ mm. For an $L = 25$-mm sample, the smallest detectable tilt of a moiré fringe is assumed arbitrarily to occur when one end of the fringe shifts by $0.1a_f$ relative to the other end. For this shift $\tan\gamma = 0.1a_f/L$, and from (7.18) $\epsilon = 1\times10^{-3}$. Note that with these coarse gratings, the spatial and strain resolutions are very limited. However, nearly an order-of-magnitude improvement can be achieved by increasing the line density to 40 lines/mm. Lines at densities exceeding 1,200 lines/mm have been successfully recorded (Post, Han, and Ifju 1994), resulting in strain resolution of $\epsilon \approx 10^{-5}$.

To determine the pattern of the moiré fringes when $\theta = 0$, consider first a case where the two overlapped gratings are identical. If the clear lines of one grating overlap the clear lines of the other, the image of the combined gratings appears bright. Conversely, when the dark lines overlay the clear lines, the image appears dark. Thus, independently of their relative position, the image of the overlapped parallel gratings appears without moiré fringes. On the other hand, when the object grating is uniformly stretched in a direction perpendicular to its lines (e.g. in the y direction), its lines shift relative to the undistorted reference grating. Owing to this relative translation, dark lines of one grating may now overlap clear lines of the other, thereby forming a dark fringe. Similarly, where clear lines of one grating overlap the clear lines of the other grating, a bright fringe appears. When the stretch is uniform, an array of parallel and equally spaced fringes appears at a place where no fringe was previously visible. It can be seen that when the image contains N_y dark fringes in the y direction, the distance U_y traveled by a line on the object grating that belongs to the Nth fringe, relative to a line on the first fringe, is

$$U_y = N_y a = N_y/f,$$

where $f = 1/a$ is the line density (or spatial frequency) of the grating. When $\theta = 0$, the straing ϵ_y can be obtained directly from

$$\epsilon_y = \frac{\partial U_y}{\partial y} = \frac{1}{f}\frac{\partial N_y}{\partial y}. \tag{7.19}$$

The strain can thus be measured by counting the fringes over a set distance y. This measurement is analogous to measurements by the vernier of a caliper. However, unlike a caliper (which provides a single measurement), this moiré pattern provides a planar measurement of the distribution of the strain in the y direction. Note that, unlike the result of (7.18), the sensitivity of this measurement (i.e., the smallest strain that can be resolved) increases with f. Thus, the use of high-density gratings is advantageous in this configuration.

Numerous methods have been developed for the application of moiré interferometry to strain measurements (see Post et al. 1994). Figure 7.13 illustrates one such method. The object grating has been recorded by photolithography with line density in excess of 1,200 lines/mm. To record the strain distribution, a second grating was projected on the surface. However, owing to diffraction, ordinary projection of a grating with similar line density was not possible. Instead, a reference grating was generated by intersecting two laser beams at the surface. The angle between these beams was matched (eqn. 6.11) to form a fringe pattern with the same line density as the undistorted grating. Without distortion, the image of the object appeared without any moiré fringes when the density of the projected reference gratings matched the density of the object grating. However, following mechanical or thermal loading, the image of the object appeared to be covered with highly visible moiré fringes of varying density. At each point, the density of these fringes represented the local mechanical or thermal strain. With this technique, the highest line density of the reference grating that can be projeced on a surface is limited by the wavelength of the incident laser beams; for two beams at $\lambda = 514.5$ nm intersecting at approximately 15°, the fringe spacing (eqn. 6.11) is approximately 1 μm. This density of interference fringes determines that the ultimate density of lines to be etched on the surface will be approximately 1,000 lines/mm. Although the density of the reference grating increases with θ, the intersection angle is often

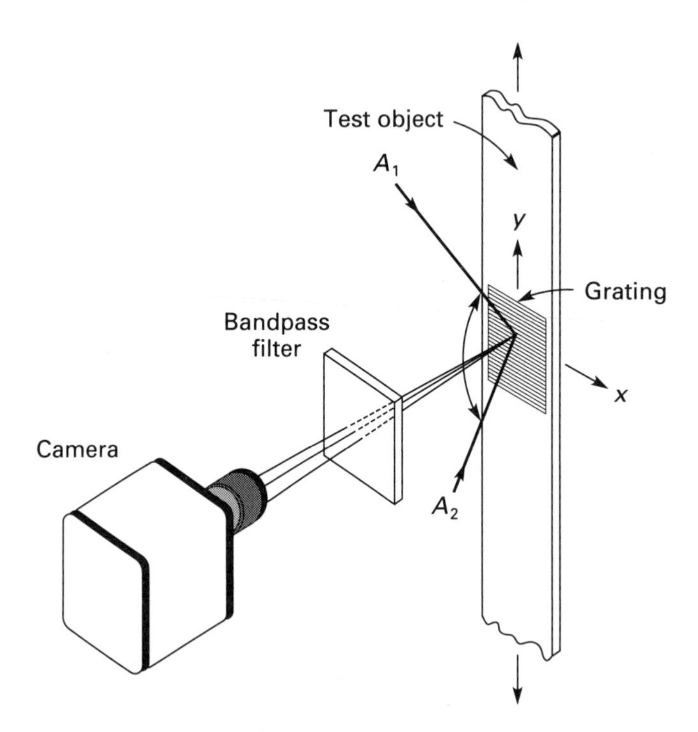

Figure 7.13 Moiré interferometry between a grating etched on the test surface and the fringe pattern obtained by interference between two laser beams (A_1 and A_2). A second pair of laser beams may be used together with a second grating (orthogonal to the first) to obtain an additional strain component.

limited by optical access. The limiting resolution of this technique is measured when the entire image contains only one moiré fringe. For a grating with 1,200 lines/mm and a 25-mm sample, this corresponds to $\epsilon = 3.3 \times 10^{-5}$. Although significantly more sensitive than the alternative moiré technique (where γ is measured), this technique requires very delicate processing of the object. Furthermore, associated with this high sensitivity is also high susceptibility to vibrations. Hence, this moiré technique is mostly used in well-controlled laboratory environments.

In a different application of moiré interferometry, a single grating is used to obtain contoured images of a three-dimensional object. The image is similar to a topographic map, where lines of constant height are drawn. It is obtained by illuminating the object through a flat Ronchi grating that is held in the (x, y) plane (Figure 7.14) and observing the projected shadow through the same grating. If the illumination is by a nominally collimated beam inclined at an angle α relative to the normal to the grating, and if the detection is by a lens with a small numerical aperture and a line of sight that is inclined at an angle β to the normal, then the image of the object is marked by alternately bright and dark fringes that outline lines of constant height, with a height difference of

$$\Delta z = Na/(\tan\alpha + \tan\beta) \tag{7.20}$$

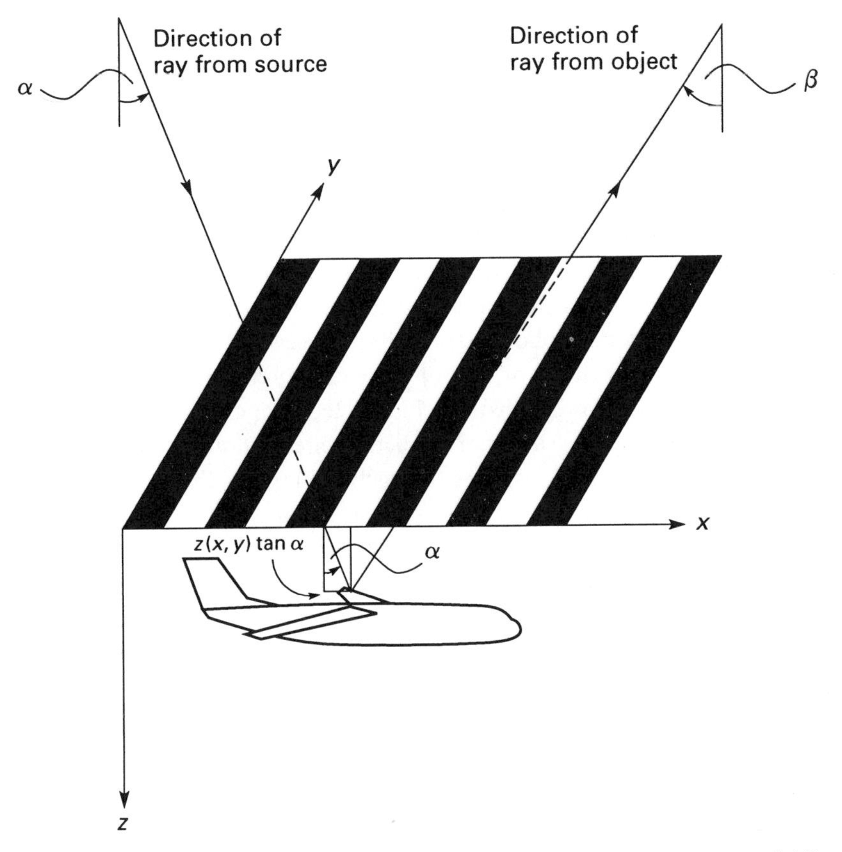

Figure 7.14 Out-of-plane moiré interferometry. [Meadows, Johnson, and Allen 1970, © Optical Society of America, Washington, DC]

between any two points separated by N fringes (Meadows, Johnson, and Allen 1970). Clearly, the sensitivity of the measurement improves by increasing α and β. However, as α increases, reflection at the surface of the grating increases, thereby reducing the light available for illumination. In addition, when β increases, the error associated with the finite collection angle of the lens more than offsets the improved sensitivity (see Research Problem 7.3). Of course, the line density of the grating may also be increased to improve the sensitivity. But, as the pitch decreases, diffraction effects reduce the contrast of the projected grating until they can no longer be resolved. When the lines of the projected grating or its image cannot be resolved, the moiré fringes disappear. However, even before the contrast of the projected grating is reduced, its frequency changes, thereby precluding the possibility of obtaining quantitative measurement from these moiré images. Of course, when the distance R between the grating and the object is well below the Fraunhofer limit (i.e. $R \ll a^2/\lambda$; cf. eqn. 7.3), the projected grating preserves both its contrast and the frequency of the original grating. If further reduction in the pitch is required while R must exceed the Fraunhofer limit, the grating may be placed at distances that are integer multiples of a^2/λ. Although the contrast is diminished at these distances, the original frequency of the grating is recovered in the projection. Thus, if the moiré fringes can be resolved despite the reduction in the contrast of the grating projection, quantitative information can be derived from their images. This restoration of the original grating frequency at determined distances is known as the *Talbot effect* (see Problem 7.4).

Moiré interferometry also offers applications in the fields of combustion or fluid mechanics (Kafri 1980). By placing two Ronchi gratings at the two sides of a flow field, variations in the index of refraction induced by temperature or pressure can be recorded photographically. However, owing to distortions by diffraction effects, the depth of field of these measurements is limited. Furthermore, as with the shadowgraph and schlieren techniques, these moiré measurements are limited by integration of refraction effects along the line of sight.

The few techniques discussed in this section illustrate the potential of moiré interferometry. With the advent of sensitive and high-resolution electronic cameras coupled with fast computers and extensive digital storage capabilities, moiré techniques are expeced to spread into more applications. Moreover, techniques that are now used mainly in laboratory environments will likely be extended to industrial applications such as process or quality control.

7.5 Holography

Holography is one of the most intriguing applications of optics. It depends on the effects of both interferometry and diffraction, so it had to wait for the development of laser sources with long coherence lengths before becoming a practical tool. Owing to the striking visual effects associated with it, holography is often considered to be only an artistic curiosity. Nevertheless, it has far-reaching applications in many unrelated fields. Holograms are now routinely impressed on documents (e.g. credit cards) as a security measure; they are used to display the control panels of aircraft cockpits on their windshield, thereby allowing the pilots to view the various gauges while still monitoring the flight through the window; holograms are used as an archival method for the storage of experimental results; and they are used as a method to detect minute distortions of three-dimensional objects. Although originally invented as a tool for electron microscopy well before the development of lasers (Gabor 1949), holography has never been used for that application.

The name *hologram,* or "total recording," implies that it contains all the information required to reconstruct the image of an object, be it two-dimensional or three-dimensional. Depending on the method of reconstruction, the reconstructed image can appear just as realistic as the original object. In particular, effects (such as parallax and depth) that characterize three-dimensional viewing are preserved by holographic recording. Viewing the image of an object through its hologram is like viewing it through a window. By looking from various angles, different sides of the object can be seen – even objects that are partially hidden behind another object. When part of the hologram is covered, or even broken away, the object can still be seen through the remaining portion. Although to see it the observer may need to peek through the reduced opening and the range of viewing angles may be limited, most of the details of the object are still preserved.

To successfully recreate such a complete image of an object, the "picture" must include - in addition to the intensity distribution - a detailed account of the shape of the wavefronts that emerge from the object and reach the image plane. Thus, unlike ordinary photographs that merely map the distribution (and possibly color) of the irradiance falling on the film, holograms must also contain a record of the distribution of the phase angle ψ of the recorded wavefront. When this information is available, the original wavefront can be faithfully reconstructed. The perception of depth is created when each eye intercepts a different segment of the wavefront. Parallax is seen by changing the line of sight to capture a part of the wavefront that was not seen before. Thus, the perspective is limited only by the size of that wavefront, or ultimately by the aperture of the hologram. Thus, a 360° hologram can afford a view of an object from all its sides.

The principles of complete view of an object are simple. All that is needed is well-reconstructed wavefronts together with the correct variation of amplitude. But recording this information is not that simple! A hologram must include detailed information on the amplitude and phase of the recorded wavefronts, but available photographic media can record only the irradiance at every point; they are sensitive to the distribution of the amplitude of the electromagnetic waves but not to their phases. To compensate for this deficiency, the recording method must include a step in which the local phase information at the image plane is revealed by converting it into a distribution of irradiance. Interferometry is an acceptable method for encoding the phase information of electromagnetic waves into a pattern of varying irradiance.

Figure 7.15(a) illustrates one approach for recording the hologram of a three-dimensional object. The object is illuminated by an expanded laser beam and the scattered radiation, which is marked by a series of wavefronts that mimic its shape, serve as the object beam. The phases of the wavefronts of the object beam can be revealed interferometrically by overlapping it with a second beam - the reference beam. Interference between the two can create an array of bright and dark fringes. The shape of this array is unique to the combination of the two beams. Of course, for interferometry, the illumination must be coherent both longitudinally and transversely. In particular, the difference between the path lengths along any part of the reference beam and any part of the object beam may not exceed the coherence length of the laser. Thus, the coherence length must, at the very least, exceed the dimensions of the object. For example, to record the hologram of a person the coherence length of the laser must exceed ~1 m. For macroscopic objects this requirement limits the sources of illumination exclusively to lasers, which must be used both to illuminate the object and for the reference beam. The optical set-up is designed so that the reference beam combines with radiation scattered from the illuminated

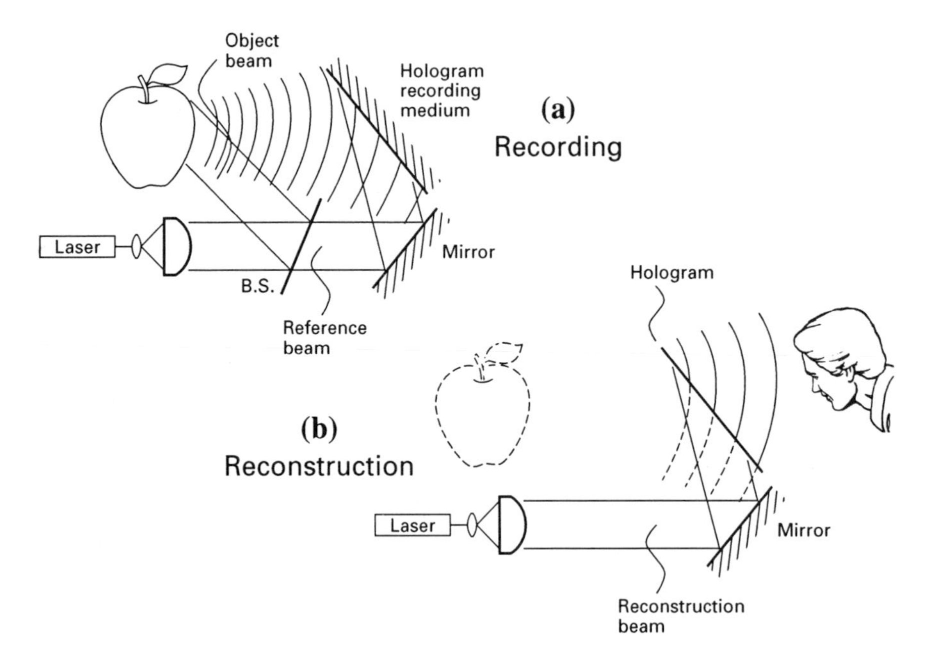

Figure 7.15 Schematic configuration of (**a**) the recording of a hologram and (**b**) the reconstruction of a hologram, where B.S. denotes beam splitter.

object at the plane where the recording medium is placed; for simplicity this is assumed to be the (x, y) plane. The recording medium often consists of a photographic plate. Other media such as photorefractive polymers or inorganic crystals (e.g. lithium niobate, $LiNdO_3$) may also be used and can offer easier recording and faster readout access. Other media may allow overwriting recorded holograms, thereby offering more flexibility than the traditional photographic plates. Because of the inherent need for high recording resolution and low distortion, the use of electronic cameras for holography is very limited. To simplify the discussion of the principles of holography, we will describe here recording by photographic plates only.

The phases of the electromagnetic fields along the object and reference beams depend on their initial phases ϕ_O and ϕ_R and on their optical paths L_O and L_R, respectively. The variation of these phases can be introduced by the following phase terms:

$$\psi_O = \frac{2\pi}{\lambda} L_O = kL_O \quad \text{and} \quad \psi_R = kL_R.$$

At the plane where the hologram in Figure 7.15(a) is recorded, the fields of each of these beams are therefore

$$\begin{aligned} \mathbf{E}_O(x, y) &= E_O(x, y) e^{i\phi_O} e^{i\psi_O(x, y)}, \\ \mathbf{E}_R(x, y) &= E_R(x, y) e^{i\phi_R} e^{i\psi_R(x, y)}. \end{aligned} \tag{7.21}$$

The optical details of the object are encoded by the field amplitude $E_O(x, y)$ and the phase term $\psi_O(x, y)$ of the object beam. The amplitude of the object beam is modulated by the scattering or reflection characteristics of the object, while the phase term varies with its shape. When overlapped with the reference beam, which is described by the amplitude $E_R(x,y)$ and phase $\psi_R(x,y)$, this encoding is revealed by a unique pattern of dense bright and dark interference fringes. The distribution of the irradiance of the overlapped beams is defined by the superposition of their fields (eqn. 6.2) as follows:

$$I(x,y) = \frac{1}{2\eta}[\mathbf{E}_O(x,y)+\mathbf{E}_R(x,y)][\mathbf{E}_O(x,y)+\mathbf{E}_R(x,y)]^*. \tag{7.22}$$

Since the reference beam is coherent with the object beam, $\phi_O = \phi_R$, and since the superposition at the hologram plane can be reduced to reflect only the difference between their optical paths (eqn. 6.3), we have

$$I(x,y) = \frac{1}{2\eta}|\mathbf{E}_R\mathbf{E}_R^* + \mathbf{E}_O\mathbf{E}_O^* + 2E_RE_O\cos[\psi_R - \psi_O]|. \tag{7.23}$$

The (x,y) dependence of the amplitudes of the fields and their phases was omitted for simplicity, but must be assumed. Equation (7.23) describes the shape and the amplitude of the fringes to be recorded. The shape constitutes a record of the distribution of $\psi_O(x,y)$ while the contrast of the fringes (eqn. 6.5) is a record of the distribution of $\mathbf{E}_O(x,y)$ and $\mathbf{E}_R(x,y)$. The first two terms of the equation are simply the irradiance of each of the interacting beams. The third term describes the interference fringe pattern. Note that the field amplitudes of both the object and reference beams, E_O and E_R, influence the amplitude of the interference fringes. Therefore, the contrast of the fringes can be controlled by adjusting the irradiance of the reference beam. To store this information, the photographic medium is exposed to the superimposed fields. The duration of that exposure is monitored carefully to avoid either underexposure or saturation by overexposure. For an ordinary photographic plate, ideal exposure provides a linear dependence between the incident irradiance and the darkening of the plate at any site. After developing and fixing the photographic plate, it appears as a transparency that accurately portrays the original fringe pattern. The contrast reversal associated with recording a negative of that original pattern merely introduces a uniform phase shift of 180° and usually has no significant consequences. With optimal exposure and processing, the transmission constant $t(x,y)$ (see Section 6.5) at any point on the transparency varies linearly with $I(x,y)$. Thus, the transmission constant induced by the interacting fields is obtained by multiplying (7.22) with a proportionality constant β that describes this linear dependence (Goodman 1968, p. 200):

$$t(x,y) = t_R(x,y) + \beta[\mathbf{E}_O\mathbf{E}_O^* + \mathbf{E}_R^*\mathbf{E}_O + \mathbf{E}_R\mathbf{E}_O^*], \tag{7.24}$$

where $t_R(x,y)$ is the transmission induced by irradiance $\mathbf{E}_R\mathbf{E}_R^*$ of the reference beam alone. Often this amplitude is constant and accordingly $t_R(x,y)$ introduces a constant background term.

To reconstruct the phase fronts of the object beam, the processed hologram is illuminated by a beam that is similar to the reference beam – that is, one that is collimated, has the same wavelength, has similarly shaped wavefronts as the reference beam, and is incident on the plane of the hologram at the same angle (Figure 7.15(b)). Although in many applications this may be the reference beam itself, it is now serving a new purpose and is therefore called the *reconstruction beam*.

The field behind the illuminated hologram can be described by diffraction of the incident reconstruction beam through the array of dark and transparent lines. Such diffraction should produce, in addition to the transmitted beam, an array of diffracted modes, where one of them is hopefully the reconstructed image of the object. Unfortunately, detailed calculation of the diffraction by the hologram is too complex. Instead, recall that the transmitted field is proportional to both the incident field $\mathbf{E}_R$ and the transmission constant $t(x, y)$ (see Section 6.5). Therefore, a simple algebraic operation on (7.24) can provide some insight on the properties of the transmitted field. If the reconstructing beam is identical to the reference beam, then its field $\mathbf{E}_R(x, y)$ must be identical to the field of the reference beam. The field transmitted by the developed hologram is therefore

$$\begin{aligned} \mathbf{E}_t &= \mathbf{E}_R t(x, y) \\ &= \mathbf{E}_R t_R + \beta[\mathbf{E}_R \mathbf{E}_O \mathbf{E}_O^* + \mathbf{E}_R \mathbf{E}_R^* \mathbf{E}_O + \mathbf{E}_R \mathbf{E}_R \mathbf{E}_O^*] \end{aligned} \tag{7.25}$$

(Goodman 1968, p. 210). The first term merely represents the propagation of the reconstruction beam behind the hologram. When $t_R(x, y)$ is uniform this propagation continues uninterrupted, with the exception of a uniform attenuation. The second term in (7.25) is also a transmitted version of the incident beam. However, the transmission constant of this component is determined by the irradiance of the object beam, $\mathbf{E}_O \mathbf{E}_O^*$, which may be spatially modulated. Consequently, the transmission of this component may suffer some distortion by diffraction and may have some off-axis scattering. However, if the modulation frequency of $\mathbf{E}_O \mathbf{E}_O^*$ is low, most of the radiation of that component remains along the incident beam axis. The third term is different from the incident beam. Since the irradiance of the reconstructing beam $I_R = \mathbf{E}_R \mathbf{E}_R^*$ is nearly constant, the third term is $\beta I_R \mathbf{E}_O$. This term is identical, to within a scalar constant, to the field of the object beam. The distribution of both its amplitude and its phase are restored. Therefore, this term must represent the reconstructed object beam. Note that the amplitude (and hence the brightness) of the reconstructed object beam depends not just on the brightness of the original object beam but also on the brightness of the reconstruction beam. Finally, the fourth term appears to be proportional to the conjugate of the object beam. It would be exactly proportional to the conjugate of the object beam if the reconstruction were by the conjugate of the reference beam, that is, if the field of the reconstruction beam were $\mathbf{E}_R^*$.

To illustrate the relation between the reconstruction beam and the object beam and its conjugate, consider the following simplified recording and reconstruction configurations (Figure 7.16). The recording of the object is executed by a collimated reference beam that strikes the hologram plate normal to its surface (Figure 7.16(a)). The object is located off axis and its beam forms a

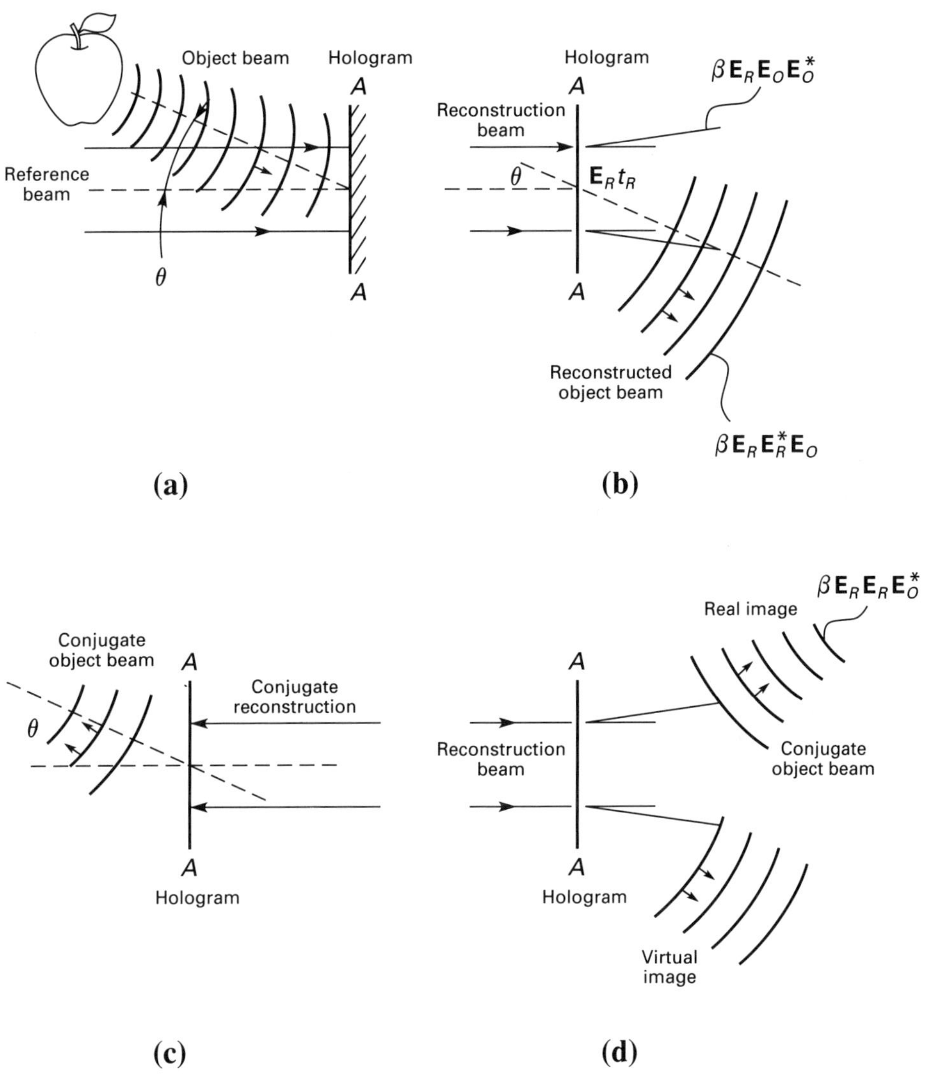

Figure 7.16 Illustration of the relation between the object beam, the reconstruction beam, and their conjugates: **(a)** recording; **(b)** reconstruction by the reference beam; **(c)** reconstruction by the conjugate of the reference beam; **(d)** reconstruction by the reference beam, showing also the conjugate of the object beams.

nominal angle θ with the reference beam; this is the *reference angle*. After developing it, the hologram is placed in its original location and illuminated by the reference beam, which now serves as the reconstruction beam (Figure 7.16(b)). The effect of each of the first three terms of the transmitted field (eqn. 7.25) is shown in the figure; the component of the uniformly attenuated reconstruction beam (first term) remains collimated after the hologram, the second transmitted component of that beam is shown as a slightly diverging beam, and the reconstructed object beam (third term) is represented by a series of wavefronts that appear to continue the propagation of the original object beam along a line that still forms an angle θ with the axis. The object can now be viewed by looking directly into the object beam and appears very realistic. However, the reconstructed object beam does not emanate from a real object. Directly projecting this beam on a screen does not produce an image. Therefore, the object beam corresponds to a virtual image.

To identify the effect of the fourth term in (7.25), assume for a moment that the reconstruction beam is replaced by its conjugate. The conjugate of a collimated beam of plane waves is simply a beam propagating in the opposite direction (Figure 7.16(c)). For illumination by the conjugate of the reference beam, the fourth term in (7.25) produces a reconstructed wave that is exactly proportional to the conjugate of the original object beam. This conjugate is simply the reverse of the original object beam – the wavefronts remain unchanged but are now converging back on the object. Thus, if the object beam consisted of concave wavefronts propagating away from the object, the conjugate beam as illustrated in Figure 7.16(c) consists of identically shaped convex fronts propagating toward the object; Figure 7.16(c) is simply the mirror image of the previous figure. Thus, if the hologram is sufficiently thin, it can be illuminated from the opposite side (i.e., with the original reconstruction beam) and still produce a conjugate of the object beam. Of course, for this illumination the conjugate of the object beam propagates along a line that forms an angle $-\theta$ with the illumination axis. The entire array of beams behind a thin hologram is presented in Figure 7.16(d). It is seen to consist of two beams nominally along the incident beam axis and two object beams: the first, an identical replica of the original object beam; the second, its conjugate. Unlike the object beam, its conjugate can produce an image of the object when projected on a screen and thereby create a real image.

The reconstructed object beam (and possibly its conjugate) may be considered as desirable results of the reconstruction step, but the two beams along the illumination axis are part of a background that needs to be either suppressed or avoided. To separate spatially the object beams from the undesirable background, the reference angle θ must exceed a minimum angle of $\theta_{\min}$. This minimum corresponds to the divergence angle of the object beam, which in turn depends on the expected imaging resolution of the object. If the size of the smallest detail of the object that needs to be resolved by the hologram is r, then the minimum separation angle is

$$\theta_{\min} = \sin^{-1}(\lambda/r) \tag{7.26}$$

(Goodman 1968, p. 214). This equation is consistent with the diffraction angle through a slit (eqn. 7.8), thereby implying that information from the object is carried to the hologram at a nominal divergence angle $\theta_{\min}$. To distinguish between the object beam and the reference beam – and subsequently between the object beam and the reconstruction beam – their angular separation must exceed $\theta_{\min}$. To estimate $\theta_{\min}$, assume that the desired resolution is $r = 0.01$ mm. For an incident wavelength of $\lambda = 514.5$ nm, the reference angle must be $\theta > 3°$ to assure that the object beam can be spatially separated from the reconstruction beam. When a hologram with a $\theta = 3°$ reference angle is reconstructed by a beam with a diameter of 0.1 m, the object and the reconstruction beams overlap each other over an approximate distance of 1.9 m, and will be conveniently separated only past that distance. Typically, the reference angle is significantly larger and the object appears spatially removed from the background beam from much shorter distances.

The reconstruction of a holographic image is similar to transmission by a diffraction grating: the beams along the hologram axis correspond to the 0th-order diffraction mode while the two object beams are the 1st- and −1st-order modes. Thus, by analogy, a diffraction grating may simply be viewed as the holographic image of a plane wave. Of course, the distinction between the reconstructed object beam of a plane wave and its conjugate is trivial. Most diffraction gratings, however, have more than just two diffraction modes. This is due to the sharp contrast between the lines that form these gratings. Thus, by analogy, when the recording of a hologram is not ideal – that is, when $t(x, y)$ is not proportional to the incident fields and the contrast between the lines that form the hologram is abrupt – other diffraction modes, and hence additional images of the object, may be created. These, like the conjugate beam, are false images of the object. As with diffraction through a grating, these higher-order modes are usually much weaker than the first-order modes. Nevertheless, to avoid interference between these false images and the object beam, recording should be designed so that after reconstruction the false images are spatially separated from the object beam.

Numerous holographic recording and reconstruction techniques have been developed to enhance or to suppress one of the object beams, to enhance the reconstruction efficiency, to avoid distortion of the image, or to reduce the requirements for resolution by the recording medium (see e.g. Abramson 1981). The need for high resolution is still the most limiting requirement for holographic recording. To estimate the recording resolution, recall that the hologram is a complex diffraction grating. It is formed by gathering radiation from every point on the object and combining it with the reference beam. Similarly, the reconstructed wavefront is formed by a superposition of numerous wavelets emanating from every point on the hologram that interfere with each other to form the wavefronts of the object beam and its conjugate. Therefore, every point on the reconstructed wavefront is obtained by superposition of radiation from every point on the hologram. This superposition can succeed only if radiation from every point on the hologram is spread by diffraction into a sufficiently wide cone to cover the entire reconstructed wavefront. Thus, if the diffraction angle required to illuminate the wavefront of the object beam is 2α, a diffraction angle of at least 4α is needed to accommodate also the conjugate image (Figure 7.17). For an object beam with a characteristic size of b viewed from a distance L (Figure 7.17), the characteristic width of a fringe d_p for this diffraction angle is

$$4\alpha = \lambda / d_p$$

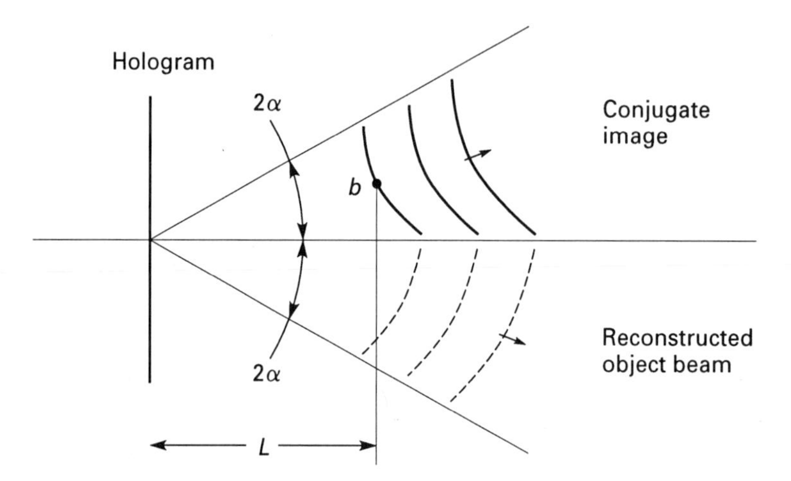

Figure 7.17 The angle of diffraction through a hologram required to accommodate an object and its conjugate.

(cf. eqn. 7.8′), and since

$$\alpha \approx \frac{b/2}{L},$$

the characteristic fringe size is

$$d_p = \frac{\lambda L}{2b}. \tag{7.27}$$

When $L = O(b)$ and illumination is by visible radiation, the characteristic fringe size must be $\sim 1\ \mu$m, thereby requiring a recording density of at least 1,000 lines/mm. Presently, such recording density cannot be achieved by electronic cameras, where typical detector arrays contain approximately 1,000 × 1,000 pixels (unless the size of the imaged object is limited to ~1 mm). Of course, by sacrificing resolution, electronic hologram recording is possible (Jones and Wykes 1989). For photographic recording, such high line density requires photographic emulsions with submicron grain size. Usually, darkening of a grain requires a fixed number of photons; the energy required to darken a grain is only slightly dependent on its size. Therefore, the irradiance required to obtain a desirable contrast in a film with large grain size is smaller than the irradiance for the exposure of a film with fine grains. Consequently, the sensitivity of holographic films is at least 100 times lower than the sensitivity of films used by amateur photographers. To overcome this poor sensitivity, holographic exposures must be longer and must use more intense illumination than ordinary photography. Furthermore, owing to the extremely high resolution and relatively long exposure times needed to record holograms, most holographic recording systems must be vibrationally isolated from their surroundings. Even minute vibrations transmitted through the floor may deteriorate the quality of holograms. Such isolation can be achieved by mounting the entire optical assembly on a heavy tabletop that in turn is "floated" on air cushions.

It is interesting at this point to compare ordinary photographs with a hologram. Figure 7.15(a) illustrates the recording of a hologram. Unlike an ordinary camera, the holographic recording system does not include an imaging lens to project the image of the object on the recording surface. Instead, radiation scattered from any point on the object that has a direct line of sight to the hologram is recorded. Thus, any point on the surface that is geometrically visible from the recording plane can direct radiation to any point on the hologram. The total radiation at any point on the hologram must then be the result of a superposition of all the wavelets from all the visible points on the object. Therefore, each element of the hologram represents a different point of view of the same object, which can produce different perspectives of that object. Furthermore, every part of the hologram contains most of the details of the entire object. Thus, if the hologram is broken, each piece can be used to reconstruct the object. Of course, reducing the accessible size of the hologram comes with a price. The fragmented hologram requires that the object be viewed through a narrower angle, much like looking into a room through a peephole. In addition, the resolution of small details of the object diminishes by fragmenting the hologram (Abramson 1981, p. 49). The idea that a fragmented hologram still describes the entire object - albeit at a reduced resolution - is used to circumvent the resolution limitations of electronic cameras for hologram recording. Certain versions of electronic speckle pattern interferometry (ESPI) are in fact low-resolution holographic measurements of surface displacements (Jones and Wykes 1989).

By contrast, ordinary photographs are recorded through an imaging lens, which inherently can offer only a limited aperture. Therefore, irrespective of the size of the picture, an ordinary photograph is a record of the information that was gathered through that limited aperture of the lens and can include only one angle of view. Attempting to view that photograph from different angles cannot change its appearance. Furthermore, if a piece of the photograph is destroyed, the information contained in that piece is lost.

Although recording of holograms may be complicated and expensive, their high resolution and their extremely large information content make them a useful tool for numerous engineering applications. Holograms can be used to study real-time vibrations (with submicron resolution) of objects that may be as large as a milling machine, or to record instantaneously (for subsequent analysis) the three-dimensional density distribution of supersonic flows.

The application of holography to the interferometric study of minute displacements or distortions of objects is described by Abramson (1981) and Vest (1979). The displacements may be in-plane or out-of-plane, and may be induced by vibrations, stresses, or thermal expansion. In *hologram interferometry,* the holographic image of a vibrating or loaded object is compared to its image when the object is stationary and not loaded. The measurement consists of a comparison between two reconstructed wavefronts: one set of wavefronts corresponds to the reference image of the object, and the second set corresponds to its displaced state. Therefore, displacements that are shorter than the radiation wavelength can be measured. As in ordinary interferometry, two sets of wavefronts are overlapped; if the difference between the two is sufficiently small, they form an array of interference fringes similar to the pattern in Figure 6.3. However, unlike the Michelson–Morley interferometer (where two different surfaces are compared simultaneously) or the Mach–Zender interferometer (where two bodies of gases are compared simultaneously), hologram interferometry allows the surface of an object at one time to be compared to itself at a later time, in particular after it has undergone some changes. Hologram interferometry can be formed in real time by viewing the displaced-object hologram through its reference hologram, or by recording the two holographic images on the same plate and testing the double hologram after processing it. Each of these methods has its own limitations, and the choice depends on the experimental needs.

In real-time hologram interferometry, the tested object is recorded holographically. After developing, the hologram is returned to the holder that retained it during recording. If the hologram is now illuminated with the reference beam, an exact image of the object can be seen as usual. If, in addition, the original object is retained in its place and illuminated by the same laser beam that was used to record its hologram (Figure 7.15(a)), it will be seen from the other side of the hologram and its image will overlap its reconstructed image. Furthermore, if the reconstruction is accurate, wavefronts of the real object and of the reconstructed object beam will overlap each other exactly. However, if the object has been distorted then the wavefront that emanates from its surface is also distorted and – when combined with the original wavefront that is reconstructed by the hologram – forms interference fringes. This pattern is visually apparent and may be recorded photographically. Furthermore, since the hologram represents a permanent record of the original object, the fringe pattern observed when the object is distorted represents the momentary changes in its shape. Thus, when the object is vibrating (owing, e.g., to a momentary acoustical disturbance), each fringe pattern represents an instantaneous record of the shape of its surface.

In an alternative approach to this technique, the same hologram is exposed twice: the object is first recorded in its reference state and then in its distorted state. When developed and illuminated by the reference beam, the hologram reconstructs two object beams, each corresponding to different states of the object. As before, the reconstructed image appears marred by interference fringes that delineate the details of the distortion of that object between exposures. Unlike real-time hologram interferometry, double-exposure hologram interferometry does not impose strict requirements on alignment. In real-time hologram interferometry, the hologram of the reference object may not be distorted and must be returned precisely to its original place. Because the hologram in double-exposure interferometry is not disturbed between the two recording events, these potential difficulties are obviated. On the other hand, the opportunity for continuously observing the distortions of the object is lost.

Holography is also gaining wider acceptance in the field of gas diagnostics. As with the schlieren and shadowgraph techniques, changes in gas density (and hence in its index of refraction) induce changes in the optical path of a beam passing through the flow test section. Because such changes distort the shape of the wavefront, a record of the shape of the wavefront after passage through the test section carries information about the flow during the time of exposure. Although the technique is similar to ordinary interferometry, it offers far more interesting opportunities. The recorded hologram contains the entire information regarding the wavefront that passed through the test section. Therefore, when reconstructed it can be used to produce schlieren images, shadowgraph images, or even interferometric images of the flow. These post-experiment processing steps can be done at a different time or location; furthermore, if more than one hologram of the same test is available, hologram interferometry can be used to compare two sets of data. The primary use of these techniques is for imaging and recording fast transients such as the air density field in the vicinity of supersonically flying projectiles (Havener and Smith 1992), or to record the flow in shock tunnels.

In yet another application of holography, the three-dimensional distribution of the density of certain species, particularly in reacting flows, is recorded (Trolinger and Hess 1992). To obtain these images, the laser wavelength is tuned to the vicinity of an absorption line of the selected species. Owing to anomalous dispersion (eqn. 4.22 and Figure 4.5), the index of refraction changes rapidly near that absorption line. At points where the density of the tested species is relatively high, such changes may be significantly larger than changes induced by variation of the overall flow density. Therefore, the shape of the wavefronts of a beam passing through the reaction zone is controlled primarily by the density distribution of the absorbing species. On the other hand, when the laser is tuned away from that absorbing transition, the only effect on the transmitted beam is induced by the overall variation in the gas density. As with hologram interferometry, the net contribution of the absorbing species can be discerned by overlapping two holograms: one hologram with the laser wavelength near an absorption line and the second with the laser tuned away from that line. The two holograms may be overlapped in real time or by double-exposure recording. Either way, the location and density of the interference fringes delineates the areas where the absorbing species are concentrated. Unlike ordinary absorption spectroscopy, these holographic images describe a three-dimensional distribution of the absorbers.

The discussion here did not include special recording and reconstruction techniques. Some of these techniques, such as the *rainbow hologram* (Abramson 1981, p. 66) that can be reconstructed by white light, broaden the range of applications of holography. However, ultimately the wide use of holography depends on the development of new and more flexible laser sources and advanced recording materials. In particular, new solid-state lasers may simplify the illumination of holograms, while new electronic cameras with large detector arrays may lead to digital recording and replay of three-dimensional images.

References

Abramson, N. (1981), *The Making and Evaluation of Holograms,* London: Academic Press.

Basting, D. (1991), *Industrial Excimer Lasers, Fundamentals, Technology, and Maintenance,* Göttingen: Lambda Physik, pp. 92–3.

Born, M., and Wolf, E. (1975), *Principles of Optics,* 5th ed., Oxford: Pergamon, pp. 401–12.

Cowley, J. M., and Moodie, A. F. (1957), Fourier images: I – the point source, *Proceedings of the Physical Society B* 70: 486–96.

Das, P. (1991), *Lasers and Optical Engineering,* New York: Springer-Verlag, pp. 172–5.

Gabor, D. (1949), Microscopy by reconstructed wave-fronts, *Proceedings of the Royal Society A* 197: 454–87.

Goodman, J. W. (1968), *Introduction to Fourier Optics,* New York: McGraw-Hill.

Havener, G., and Smith, M. S. (1992), Holographic and PLIF measurements of free-flight hypervelocity flows in the AEDC range G facility, AIAA Paper no. 92-3935, American Institute of Aeronautics and Astronautics, Washington, DC.

Hecht, E. (1990), *Optics,* 2nd ed., Reading, MA: Addison-Wesley, pp. 392–515.

Huygens, C. (1690), *Traité de la Lumière,* Leyden; translated by S. P. Thompson (1912), *Treatise on Light,* London: Macmillan.

Iizuka, K. (1983), *Engineering Optics,* Berlin: Springer-Verlag.

Jackson, J. D. (1975), *Classical Electrodynamics,* 2nd ed., New York: Wiley, pp. 67–8, 299–303.

Jones, R., and Wykes, C. (1989), *Holographic and Speckle Interferometry,* Cambridge University Press.

Kafri, O. (1980), Noncoherent method for mapping phase objects, *Optics Letters* 5: 555–7.

Laufer, G. (1984), Instrument for velocity and size measurement of large particles, *Applied Optics* 23: 1284–8.

Loewen, E. G. (1977), Selection rules for diffraction gratings, *Electro-Optical Systems Design* 9(8): 26–9.

Loewen, E. G., Nevière, M., and Maystre, D. (1977), Grating efficiency theory as it applies to blazed and holographic gratings, *Applied Optics* 16: 2711–21.

Meadows, D. M., Johnson, W. O., and Allen, J. B. (1970), Generation of surface contours by moiré patterns, *Applied Optics* 9: 942–7.

Post, D., Han, B., and Ifju, P. (1994), *High Sensitivity Moiré: Experimental Analysis for Mechanics and Materials,* New York: Springer-Verlag.

Sarbach, U., and Kahlert, H.-J. (1993), Excimer laser based micro-structuring using mask projection techniques, *Lambda Highlights,* no. 40, Göttingen: Lambda Physik, pp. 2–4.

Shoshan, I., Danon, N. N., and Oppenheim, U. P. (1977), Narrowband operation of a pulsed dye laser without intracavity beam expansion, *Journal of Applied Physics* 48: 4495–7.

Trolinger, J. D., and Hess, C. F. (1992), Hydroxyl density measurements with resonant holographic interferometry, AIAA Paper no. 92-0582, American Institute of Aeronautics and Astronautics, Washington, DC.

Vest, C. M. (1979), *Holographic Interferometry,* New York: Wiley.

Homework Problems

Problem 7.1

A laser beam is used to measure the distance to the moon by measuring the time of travel of a 10-ns pulse from a large focusing mirror on earth to a retro-reflector on the moon and back to a large collector mirror on earth that gathers the reflected radiation and focuses it on a detector. The distance to the moon is approximately 300,000 km. Find the diameter of the focusing mirror required to project a spot on the moon with a diameter of 10 km. What will be the size of the back-reflected spot after returning to earth if the diameter of the retro-reflector is 20 cm? Also determine the energy intercepted by the collection mirror on earth if the initial pulse energy is 1 J.

Problem 7.2

The pitch of a certain grating is smaller than the incident wavelength. Find the effective pitch that will produce the first-order diffraction mode for $a = \lambda/2$, and find the diffraction angle of that mode for normal incidence.

Problem 7.3

When the readout display of the spectrometer in Figure 7.18 shows that it is tuned to a wavelength of 496 nm, a signal is recorded by the PMT at the exit slit. The only radiation incident on the slit is known to be in the ultraviolet spectral range. Assume that the two mirrors of the spectrometer and its slits are symmetrically located and that the 1,200-lines/mm grating is centered on the line of symmetry.

- **(a)** Assuming that the readout of the spectrometer was recently calibrated, what is the true wavelength of the radiation detected by the PMT?
- **(b)** What is the lowest diffraction order that will make detection of the UV radiation possible?
- **(c)** Find the angle γ through which the grating was turned.
- **(d)** For a slit width of 10 μm, what is the finest spectral feature (in units of cm^{-1}) that can be resolved at this setting of the spectrometer?

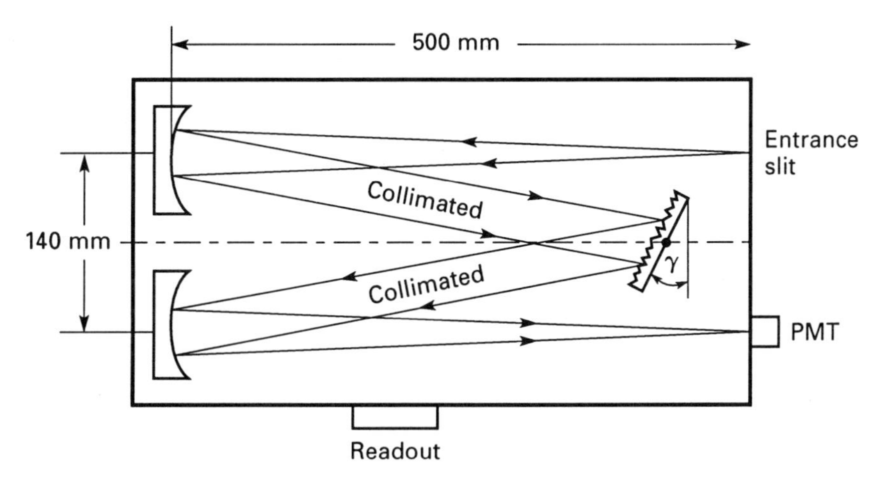

Figure 7.18 Top cutaway view of a typical single monochromator.

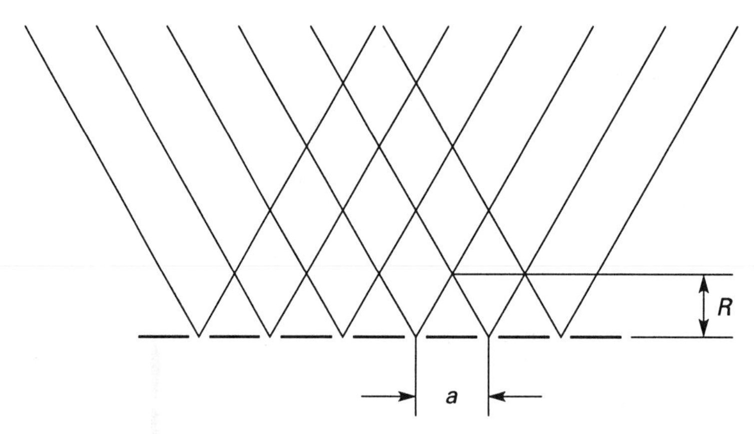

Figure 7.19 Illustration of the Talbot effect.

(e) Can other orders of the grating be used to improve the resolution?
(f) How would a 2,400-lines/mm grating influence the spectral resolution of the radiation of part (a) when the readout is still at 496 nm?

Problem 7.4

The Talbot effect (Cowley and Moodie 1957) is a periodic phenomenon that appears as the reconstruction of the shadow of a Ronchi grating when illuminated by a collimated beam. When the shadow is projected on a screen placed at a distance R from the grating, the period of the shadow equals the period of the grating. The phenomenon occurs because the angular spread $2\theta_{\beta=\pi/2}$ of the diffraction wedge through each of the grating lines is constant for all grating lines. Figure 7.19 illustrates the trajectory of several rays at the edge of each wedge (eqn. 7.8′) emanating from a Ronchi grating with line spacing of a. Assume that the width of each grating line is also a (i.e., assume the dark lines, which are shown in the figure to have finite width, are actually infinitesimally thin). The figure shows that, at planes where the node lines from different slits intersect, the original periodicity of the grating is recovered.

(a) Find the angular spread $2\theta_{\beta=\pi/2}$ of these wedges and show that $R = ma^2/\lambda$, where m is an integer. (Note that when m is an odd integer the periodicity is restored but the image of the grating is shifted laterally by half a period. Only when m is even are both the periodicity and distribution of the irradiance similar to those of the original grating.)
(b) The pattern obtained by the node lines can also be viewed as a pair of overlapped virtual gratings intersecting at an angle $2\theta_{\beta=\pi/2}$. Therefore, the original Ronchi grating can be seen as reproduced at planes where moiré fringes are formed by the two virtual gratings. Use (7.17) to find R.

Problem 7.5

Show that (7.26), which describes the minimum separation angle between the object beam and the reference beam of a hologram, is an alternative statement of the Fraunhofer far-field limit (eqn. 7.3).

Problem 7.6

The visibility function (eqn. 6.5) is a measure of the contrast of the fringes recorded by a holographic plate. Estimate the ratio between the irradiance of the reference beam and that of the object beam that will produce fringe visibility of $V = 0.5$. (*Answer:* $I_R/I_O = 16$.)

Research Problems

Research Problem 7.1

Read the paper by Shoshan, Danon, and Oppenheim (1977).

(a) Verify their equation (1). (Note that the equation for Littrow arrangement is $2a \sin\theta = m\lambda$.)

(b) Verify the calculated estimate of the linewidth of 0.07 Å (p. 4496, col. 2, para. 3).

(c) What UV wavelength can be obtained from this laser cavity using an identical configuration? What will be the linewidth?

Research Problem 7.2

Read the paper by Laufer (1984).

(a) Compare (1) the fringe density obtained by the interference between the first-order diffraction of one grating and the first-order diffraction of the second grating with (2) the fringe density of the moiré pattern obtained by overlapping the two gratings.

(b) Estimate the longest distance from the gratings over which a moiré pattern can be projected.

Research Problem 7.3

Read the paper by Meadows et al. (1970).

(a) Determine which parameters improve the measurement sensitivity.

(b) What is the largest value of α that can be used without excessive loss of signal by the reflection of the incident radiation at the surface?

(c) For a divergence angle of 1 mrad of the incident radiation, find the uncertainty (in percent) of the measurement of Δz for a grating with $a = 1$ mm when $\alpha = \beta = 80°$.

(d) For collection by a lens with NA = 1/32, find the uncertainty (in percent) of the measurement of Δz when $\alpha = 80°$ for $\beta = 80°$ and for $\beta = 45°$.

8 Introduction to the Principles of Quantum Mechanics

Ever splitting the light! How often do they strive to divide that which, despite everything, would always remain single and whole.

Goethe (quoted in Zajonc 1993)

8.1 Introduction

The use of electromagnetic wave theory and geometrical optics to describe propagation of light is satisfactory when there is no exchange of energy between radiation and matter. All observable characteristics – direction of propagation, polarization, diffraction, interference, the energy of radiation, and so forth – are accurately described by one of these so-called classical theories. However, with the exception of absorption by lossy media (eqn. 4.36) or the effects of dispersion (eqn. 4.22), classical theories fail to give reasons for many phenomena that have practical consequences in modern technology and that involve interaction between radiation and matter. Even the partially successful descriptions of dispersion or loss cannot explain the existence of the multitude of spectrally distinct absorption lines; cannot predict the wavelengths for absorption or the wavelengths for anomalous dispersion; cannot account for all the absorbed energy; and, most importantly, cannot predict or describe the effect of optical gain, which is closely related to absorption. Thus, the principles of operation of many important devices of modern optics such as lasers, photodiodes, electronic cameras, and television screens cannot be explained by classical theories. Even the emission by the sun or an incandescent lamp are beyond the scope of these theories. It is evident therefore that a new, modern theory is necessary – one that is either more complete than, or simply complements, the classical theories. This modern theory will be used to explain most of the observations that relate to emission, absorption, energy exchange between electrons, radiation, and more. Although the modern theories to be discussed here can also describe propagation, we prefer to leave these topics to the classical treatment.

The theory of quantum mechanics was developed to answer questions that were outside the scope of electromagnetic wave theory. The need for this new theory emerged with the progress of new and more refined experimental techniques that identified the existence of subatomic particles, such as the electron and the proton, and demonstrated peculiar behavior of these particles when interacting among themselves or with radiation. Several experiments showed that the absorption and emission of radiation is strongly coupled to the behavior of these subatomic particles, as well as to the structure of the atom and of the molecule. Furthermore, some of these experiments seemed to imply that, under certain conditions, subatomic particles behaved as waves while radiation (which was believed to be a wave) behaved as a stream of particles. This latter observation was reinforced by observations that suggested these particles of radiation (or photons) even have momentum. Hence photons, electrons, protons, and neutrons all appeared to have a dual identity, an observation outside the scope of classical theories and, moreover, contrary to fundamental intuition.

Like many theories, quantum mechanics is based on a few plausible postulates, which cannot themselves be proved. These postulates are stated to describe certain simple observations, and their only justification is their inherent simplicity and the successful prediction thereby of other, more complicated phenomena. In this context the postulates of quantum mechanics are similar to the laws of thermodynamics or to Newton's laws of mechanics. Neither could be proved, but all are plausible and so far irrefutable. In this chapter we discuss a few of the experiments that led to the formulation of the postulates of quantum mechanics. Without resorting to the rigorous mathematics of this theory, we will extend these postulates and determine some of their consequences.

The experiments selected for presentation in this chapter are intended to demonstrate that light can be viewed as either an undulatory, wavelike phenomenon or as a corpuscular phenomenon consisting of particles with measurable energy and momentum, and that particles such as electrons can be viewed – similarly to light – as being either waves or particles. The order in which these experiments are presented does not necessarily follow the chronological order of their discovery.

8.2 The Photoelectric Effect

Although numerous experiments can be designed to demonstrate the corpuscular nature of light, the most plausible experiment would attempt to show how one particle of light – a *photon* – is annihilated and replaced by one particle of matter – for example, an electron. Furthermore, if it can be shown that a photon cannot be split then the analogy will be complete. The photoelectric effect was the first such experiment, and it is probably still the simplest one for demonstrating the particle nature of light and for the measurement of the energy of these particles.

Figure 8.1 illustrates a system for the demonstration of this effect. It consists of a pair of metallic electrodes enclosed in a vacuum cell with optically transparent windows. The electrodes are connected to a variable d.c. voltage

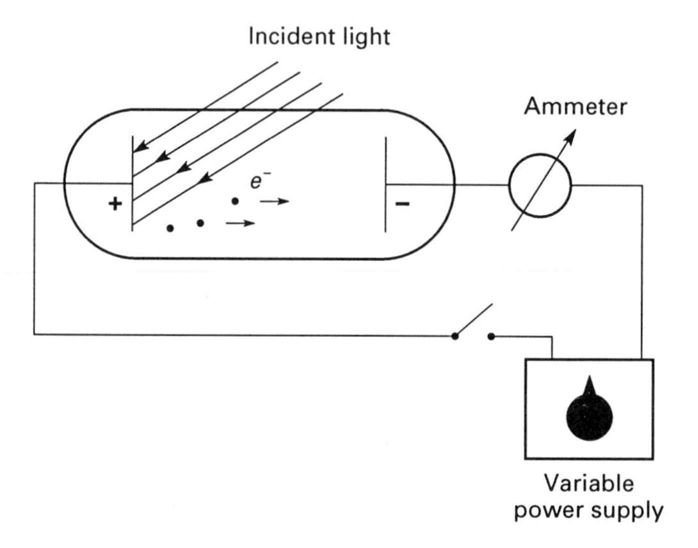

Figure 8.1 Experimental set-up for detection of the photoelectric effect.

source and a sensitive ammeter that measures the current passing in the circuit. Clearly, in the absence of external disturbances (i.e., when no electrons are ejected or absorbed by either electrode), the circuit is open and the current detected by the ammeter approaches zero. Some marginal current that may be detected can be attributed to noise sources. However, when either electrode is illuminated by ultraviolet radiation, current is registered by the ammeter and its extent exhibits strong dependence on some of the illumination characteristics such as irradiance or wavelength. In these experiments, current could be detected even when the anode is illuminated and the voltage across the electrodes, which opposes the motion of electrons, is kept below a certain threshold. The production of electric current by incident radiation is the *photoelectric* effect. Although known since the nineteenth century, this effect was not interpreted correctly until Einstein (1905). The configuration in Figure 8.1 is similar to the photomultiplier tube (PMT) of Section 3.2. However, since the PMT is designed to maximize the current induced by radiation, the illumination is directed toward the cathode rather than the anode and the tube is equipped with several dynodes that are used to amplify the current.

To interpret the effect, Einstein noted that when the voltage and the wavelength of the incident radiation remain constant, the current registered by the ammeter increases linearly with the incident irradiance. However, as the voltage between the anode and the cathode is raised beyond a certain threshold, the current in the circuit could be completely stopped. This is perplexing, because if the voltage difference between the electrodes is considered as a potential barrier then increasing the incident irradiance should restore the current in the circuit. However, results show that the magnitude of the stopping voltage depends only on the frequency ν of the incident radiation; below a certain

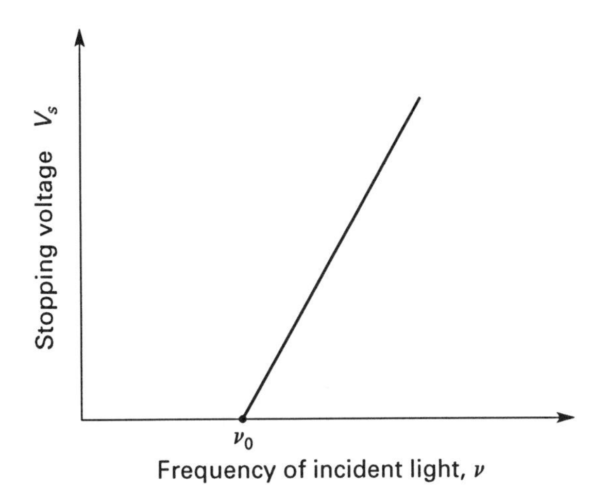

Figure 8.2 Variation of the stopping voltage with the incident light frequency.

frequency ν_0 no voltage is needed to stop the current, while above ν_0 the stopping voltage increases linearly with ν (Figure 8.2).

Similar results can be obtained with a PMT. Owing to its high sensitivity, a PMT may permit observations that were not available to Einstein. When the PMT cathode is illuminated by radiation with a frequency that exceeds a certain threshold, the anode emits current. The variation with time of this current can be recorded by an oscilloscope. As the incident irradiance decreases, the amplitude registered by the oscilloscope trace decreases. By increasing the oscilloscope gain, the signal can still be recorded; however, the trace appears somewhat erratic, as if consisting of many bursts. As the incident irradiance is attenuated further, these bursts appear isolated and all have a similar amplitude. When the incident irradiance is extremely low, a burst may appear on the screen only every few oscilloscope scans. Such behavior can be expected when the radiation is assumed to consist of a stream of particles. When the stream is plentiful, the electron emission is almost continuous; when the stream subsides, the particles appear isolated and the time between them is random. Finally, when the irradiance decreases even further, the time between these particles increases until several oscilloscope scans may be required before one is detected. This is consistent with the hypothesis of Einstein, which explains the photoelectric effect by suggesting that radiation consists of particles, or photons, each carrying a finite quantum of energy that cannot be divided. The indivisibility of these energy quanta is evident from the appearance of isolated bursts on the oscilloscope screen when the incident irradiance is low. Each of these bursts corresponds to a single electron. If photons and electrons could be split, the signal emitted by the PMT at low irradiance would consist of either a low level continuum or a series of low-amplitude bursts. Alternatively, if only photons were divisible, then fragments of photons could be accumulated by the cathode until the energy were sufficient for the release of a single electron. For this hypothesis to be successful, one must assume that all fragments of the photon are incident upon the same site on the surface (which eventually ejects the electron) or, if incident at various sites, then transferred to the site where the electron is to be emitted. This assumption is far less plausible than the assumption that photons are indivisible.

To quantify the results of the photoelectric effect, Einstein hypothesized that the energy of an individual photon is

$$E_p = h\nu \quad \text{or} \quad E_p = \hbar\omega, \tag{8.1}$$

where $h = 6.63 \times 10^{-34}$ J-s is Planck's constant and $\hbar = h/2\pi$. This constant had been used previously by Planck to calculate the radiative energy emitted by hot bodies; he did not relate it to the energy of a quantum of light or a photon. With this interpretation, the effect could be viewed as the ejection of electrons from the surface of the metal. The minimum energy required for this ejection is $h\nu_0$. This minimum energy is usually called the *work function*. It was later shown to be a material property of the metal of the anode, its chemical composition and its crystalline structure. However, once the energy of the incident photon exceeds the work function, the ejected electrons travel with a kinetic energy that can be partly used to overcome the potential barrier across the electrodes. If the potential barrier required to stop these electrons is V_s then their kinetic energy upon ejection had to be eV_s and the photon energy that was required to overcome the work function and to provide this kinetic energy had to be

$$h\nu = eV_s + h\nu_0. \tag{8.2}$$

Equation (8.2) is consistent with the linear relationship presented by Figure 8.2. As the incident irradiance increases without any change in the wavelength, more

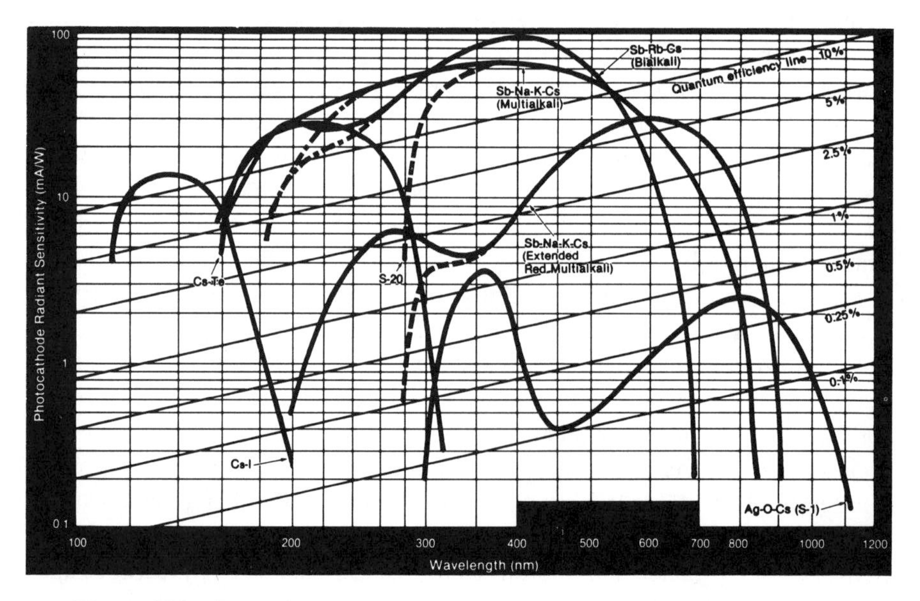

Figure 8.3 Spectral response of several photocathode materials. [Hamamatsu Data Card T82-9-30, © Hamamatsu Corporation, Bridgewater, NJ]

electrons are ejected, thereby inducing higher current. On the other hand, if the incident photon energy falls below the work function then no emission can occur even at high irradiance.

Modern PMTs are constructed with photocathodes that have a low work function. Figure 8.3 presents the spectral response of several metals that are used as photocathodes for commercially available tubes. Despite their relatively low work function, only a fraction of the incident photons can force ejection of electrons. The ratio between the number of emitted electrons and the number of incident photons is the *quantum efficiency*. Depending on the incident wavelength and the photocathode material, the quantum efficiency of standard PMTs varies between 5% and 20%.

8.3 The Momentum of Photons

The results of the photoelectric effect and subsequent experience with PMTs suggest that light consists of indivisible particles carrying prescribed quanta of energy. However, if photons could be shown to have momentum, this would constitute even a more direct proof of their corpuscular nature. Such demonstration is particularly intriguing in that it is well known that photons do not have any mass when at rest. But, by having finite momentum when moving at the speed of light, their relativistic mass may be considered as being nonzero. By having prescribed energy, finite momentum, and nonzero mass, photons have most of the attributes of other elementary particles such as electrons. The only apparent distinction between photons and other elementary particles is their lack of mass at rest and, when compared to electrons and protons, their lack also of an electric charge.

The first experiment to demonstrate that photons possess momentum was performed by Compton (1922; see also Compton 1923). When using a collimated monochromatic X-ray beam to irradiate a target, the wavelength of some of the scattered radiation was observed to increase with the scattering angle. To explain the phenomenon, the scattering was assumed to occur by an exchange of momentum and energy between the incident X-ray photons and the electrons in the target, in a manner consistent with the conservation laws of classical mechanics. If the electrons interacting with the photons are slow then they can be assumed to be at rest relative to the incident photons, and the photon–electron collision and subsequent scattering can be illustrated as in Figure 8.4. Such an exchange of both quantities – momentum and energy – can account for both the scattering of photons and for the variation of the scattered photon frequency with ϕ. However, to conserve momentum, the momentum of a photon must be:

$$\mathbf{p}_p = \hbar\mathbf{k} = \frac{h}{\lambda}\mathbf{e}. \tag{8.3}$$

With this assumption and the expression for photon energy (eqn. 8.1), and by including the classical expression for the momentum of the electron and the

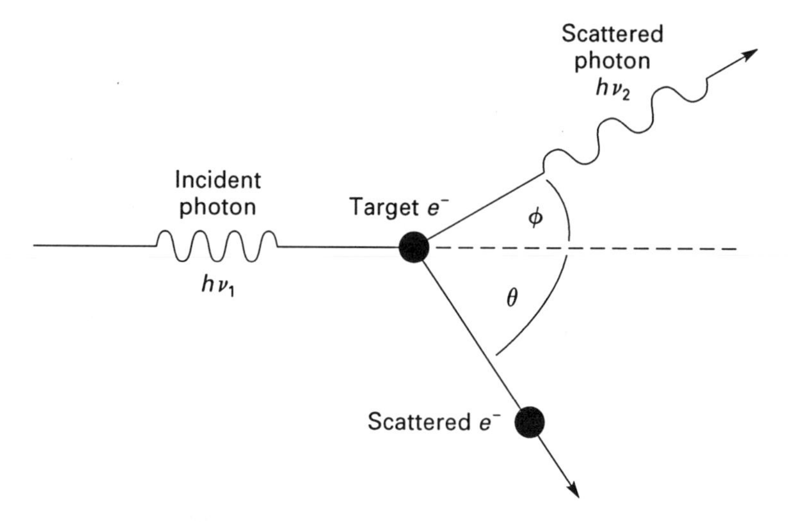

Figure 8.4 Hypothetical experiment for demonstration of the Compton effect.

relativistic term (see Problem 8.1) for its kinetic energy, the following relation between the scattering angle ϕ of the photons and their frequency ν' can be obtained:

$$\frac{1}{\nu'} = \frac{1}{\nu} + \frac{h}{m_e c^2}(1 - \cos\phi), \tag{8.4}$$

where $m_e = 9.11 \times 10^{-31}$ kg is the mass of the electron at rest. Equation (8.4) accurately predicts the results of Compton's experiment, thereby confirming the assumption that photons have a momentum as described by (8.3). The photon momentum points in the direction of the propagation vector **k**, and its magnitude is h/λ. Equation (8.3) can also be obtained using the relativistic kinetic energy of a particle moving at the speed of light and having zero mass when at rest (Problem 8.1).

8.4 Particles as Waves

In the photoelectric effect and in Compton's experiment, photons exchanged energy and momentum with lightweight elementary particles. The first experiment was designed to demonstrate the corpuscular nature of light and to measure the energy of a photon, while the second one allowed the measurement of the photon's momentum. Neither experiment showed any evidence of the undulatory nature of radiation, nor could it be expected to exhibit any undulatory behavior for particles such as electrons. On the other hand, since radiation appears to have a dual character (wavelike and corpuscular), other elementary particles may also be expected to exhibit similar dual behavior. Since the previous experiments were unlikely to show such behavior, an alternative approach is required – an experiment where the diffraction of electrons through a slit or a grid is observed similarly to the diffraction of radiation.

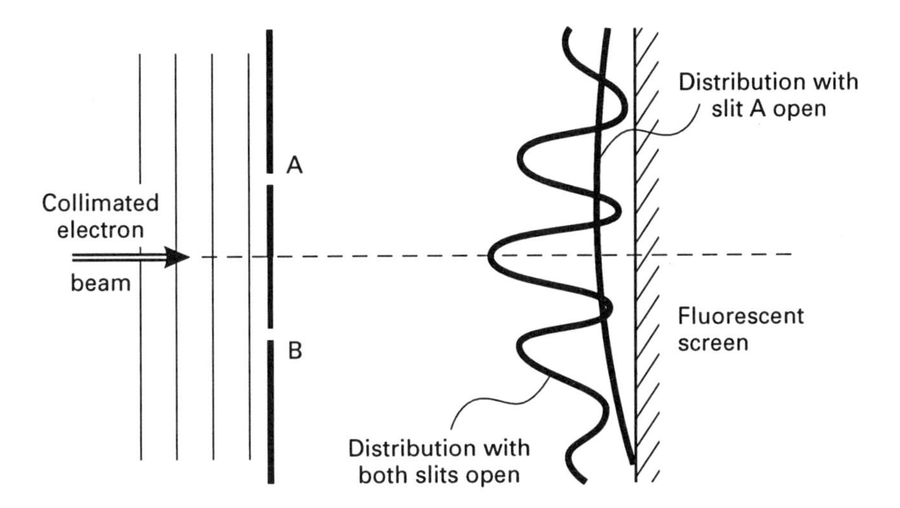

Figure 8.5 Double-slit electron diffraction experiment.

The first experiment suggesting wavelike behavior of elementary particles was reported by Davison and Germer (1927), who showed that if low-energy electrons are scattered from the surface of a crystal then the scattering can be well described using diffraction theory and by assuming that electrons having a momentum p also behave as waves with a wavelength of

$$\lambda = h/p. \tag{8.5}$$

This wavelength of the free electrons was consistent with a previous prediction of de Broglie (1924), and is consequently known as the *de Broglie wavelength*.

An alternative, hypothetical experiment can be designed to further illustrate the consequences of the undulatory behavior of particles. In this experiment (Figure 8.5), a collimated and energetically uniform electron beam is incident on a plate with two narrow but infinitely long slits of width d. The pattern observed on a fluorescent screen placed behind the slits resembles the diffraction pattern obtained behind a double slit illuminated by radiation with the same λ/d ratio. When one of the slits is blocked, the diffraction pattern is consistent with the diffraction of radiation through a single slit. Thus, the diffraction pattern obtained when two slits are open must represent a linear superposition of a certain quantity that depends on the diffraction through each of the individual slits. However, arithmetical point-by-point addition of the diffraction pattern formed on the screen by each slit alone does not reproduce the double-slit diffraction pattern, thereby suggesting that the incident electron flux is not the quantity to be used for the superposition that forms the observed diffraction pattern.

The diffraction pattern obtained through a double-slit illumination by coherent electromagnetic radiation can be used to suggest, through analogy, the properties of the function ψ that best represents the wavelike behavior of elementary particles. That diffraction pattern is calculated by a procedure similar to the derivation of diffraction patterns behind a single slit (e.g., eqn. 7.7); the electric field emerging from each of the slits at the screen is calculated separately, and then the distributions of the two fields are combined to form the total distribution that consists of an interference pattern with nodes and peaks. To obtain the irradiance, a time average of $\mathbf{E}\cdot\mathbf{E}^*$ is calculated (eqn. 4.42). If the analogy between $\mathbf{E}$ and ψ is complete then ψ will also have the necessary properties of linear superposition. Therefore, the diffraction of electrons through each slit may be represented by the distribution at the screen of two complex wavelike functions ψ_1 and ψ_2. Each wave function represents the diffraction through one slit. Assuming that these functions can be superimposed just like the electric field functions, the total function representing the diffraction through both slits is:

$$\Psi = \psi_1 + \psi_2. \tag{8.6}$$

Invoking the analogy between the wave behavior of electromagnetic radiation and the undulatory behavior of particles, the distribution of electrons at the screen (or the probability that an electron is incident at any point at the screen) can be obtained as

$$P = \frac{1}{T}\int \Psi\Psi^*\,dt. \tag{8.7}$$

The amplitude of this function is normalized so that its integral over time and space is unity. With this normalization, P represents a probability function: the probability of finding an electron at a given location or time. However, ψ and Ψ do not by themselves represent intuitive physical quantities. This is in contrast to the undulatory behavior of electromagnetic radiation, where both $\mathbf{E}$ and I represent simple physical quantities.

In order to put the photoelectric effect, Compton's experiment, and the diffraction by photons into a broader perspective, note that in the first two experiments photons exchange energy and momentum with electrons. However, there is no information regarding the exact location of either the photon or the electrons. By contrast, the electron diffraction experiment is not intended to provide any direct information about energy or momentum, but it does permit specifying one of the coordinates of the electron at the moment it passes through the slit. Thus, it may be plausibly argued that an attempt to accurately define the location of a particle results in behavior that is typical of waves, whereas attempts to measure the energy or momentum of particles result in corpuscular behavior. If true, this rule can be used to generalize the dual behavior of elementary particles. However, even more general is the question of what other properties characterize corpuscular and undulatory behaviors. This can be answered by the use of the Hamiltonian, a concept from classical mechanics.

8.5 The Hamiltonian

The preceding results plausibly indicate that if an experiment is designed to measure the momentum of a microscopic particle then its location remains undetermined, and (conversely) if location is measured then momentum is undetermined. These results also imply that, irrespective of their known nature, photons appear as particles in experiments designed to view them as such; conversely, they appear as waves in experiments designed to exhibit their wavelike nature. We may generalize this observation in two steps. In the first step, the measurable quantities that characterize one or the other behavior will be identified. It is expected that such quantities can be grouped into two categories: when a quantity from one category is measured, corpuscular behavior is observed; if a quantity from the other category is measured, undulatory behavior is observed. Furthermore, quantities from one category can be paired with quantities from the other category to form *conjugate pairs*. Thus, the second step will define the consequences of an attempt to measure simultaneously two quantities that form such a conjugate pair. The first step leads to the formulation of the complementarity principle, and the second will result in the uncertainty principle. To determine which quantities belong to these conjugate pairs, or which quantities cannot be accurately measured simultaneously because they are associated with different aspects of the motion, we start by using techniques of classical mechanics for the description of the motion of free particles (Hildebrand 1965, pp. 148–55). These techniques identify the quantities that are necessary to fully define the trajectory and motion of particles on the macroscopic level. However, on the microscopic level they will identify the pairs of conjugate quantities that are also mutually exclusive when measured simultaneously.

Two functions can be used to describe the kinematics and kinetics of free particles: the *Lagrangian* function $L = T - V$ and the *Hamiltonian* $H = T + V$, where T is the kinetic energy and V is the potential energy of the particle. When subjected to conservative forces, the trajectory of a particle is defined by Hamilton's principle:

$$\delta \int_{t_1}^{t_2} L\, dt = 0; \tag{8.8}$$

that is,

> *the trajectory of a particle subjected to conservative forces is such that the difference between the kinetic and potential energies is minimized.*

The Lagrangian function and Hamilton's principle form the basis for Lagrangian mechanics, where n second-order differential equations of motion depend on n independent coordinates $x_1, x_2, \ldots, x_n$ and on n independent velocity components $v_1, v_2, \ldots, v_n$ (Goldstein 1981, p. 55). The harmonic oscillator (4.21) without the second term (dissipation) is an example of such a second-order equation. Solving these equations requires specification of $2n$ initial conditions. This analysis represents n degrees of freedom, and the velocity components are simply time derivatives of the n independent coordinates.

An alternative description of the same physical problem is afforded by Hamiltonian mechanics. Instead of n second-order equations, Hamiltonian mechanics uses $2n$ independent first-order differential equations expressed by $2n$ independent variables. Observations in the previous sections suggest that such independent variables should include the n coordinates that describe the n degrees of freedom and the corresponding n components of the momentum $p_1, p_2, \ldots, p_n$. Transformation from the Lagrangian space (x_n, v_n) to the Hamiltonian space (x_n, p_n) is accomplished by Legendre transformation (see e.g. Callen 1985, p. 137). This transforms the Lagrangian function into the time-dependent Hamiltonian function $H(x_n, p_n, t) = T+V$, which (as required) is described by the n coordinates and n components of the momentum of the system. Accordingly, the linear coordinates and the independent components of the linear momentum form pairs of *canonical* or conjugate variables (x_1, p_1), ..., (x_n, p_n) (Goldstein 1981, p. 339). In addition, H and t are canonical or conjugate variables. Therefore, a system with n degrees of freedom must have $n+1$ conjugate pairs. These Hamiltonian conjugate pairs were identified by two of the fundamental principles of quantum mechanics – the complementarity principle and the uncertainty principle – as the quantities that characterize the corpuscular or the undulatory behavior of particles.

As an illustration, consider the Hamiltonian of a particle in a central force field with a potential energy $V(r)$. The kinetic energy T is expressed by the momentum, and the Hamiltonian in spherical coordinates (r, θ, ϕ) is

$$H(r, \theta, p_r, p_\theta, p_\phi) = \frac{\mathbf{p}\cdot\mathbf{p}}{2m} + V(r) = \frac{1}{2m}\left(p_r^2 + \frac{p_\theta^2}{r^2} + \frac{p_\phi^2}{r^2 \sin^2\theta}\right) + V(r) \qquad (8.9)$$

(Goldstein 1981, p. 345).

8.6 The Complementarity Principle

The Hamiltonian function shows that, in order to fully describe a mechanical system in the classical sense, the coordinates and the corresponding components of the momentum must be fully specified, and the system energy must be specified at any time. However, experience developed in experimentation with microscopic particles suggests that measurements of the momentum and coordinates of such particles are mutually exclusive. Thus, measurement of one quantity at high accuracy precludes the accurate measurement of the other quantity. Furthermore, each quantity represents a characteristic behavior of a particle, either undulatory or corpuscular; one behavior (e.g. wavelike) is observed to the exclusion of the other (e.g. corpuscular). This is a remarkable departure from classical mechanics: the pairs of variables that are required to fully characterize the motion of a classical system are mutually exclusive and cannot be measured simultaneously on a microscopic level. This observation is the basis of Bohr's complementarity principle, which attempts to generalize these simple observations to include all pairs of parameters that are conjugate in the Hamiltonian sense. It also attempts to reconcile the apparent conflict between the conjugate pairs. Bohr (1928) stated the complementarity principle as follows:

Atomic phenomena cannot be described with the completeness demanded by classical dynamics; some of the elements that complement each other to make a complete classical description are actually mutually exclusive, and these complementary elements are all necessary for the description of various aspects of the phenomena.

This statement suggests that the pairs of Hamiltonian conjugate parameters - although necessary for classical description of a system - become mutually exclusive on the microscopic level. In other words, each describes a different aspect of the phenomenon, either wave- or particle-like behavior. On the other hand, both behaviors are part of the same phenomenon. This observation has been restated (Merzbacher 1970) in a more simplified form:

Wave and particle nature are considered complementary aspects of matter. Both are equally essential for a full description, and although they may appear to be mutually inconsistent, they are assumed capable of existence.

With this principle, the apparent paradoxical behavior of light and matter is settled. It determines the quantities that are mutually exclusive and the consequences of the exclusiveness. However, the extent of the exclusiveness of these parameters is defined by the uncertainty principle.

8.7 The Uncertainty Principle

Chronologically, the uncertainty principle (Heisenberg 1927) was stated before the complementarity principle, which was intended to formulate the implications of the uncertainty principle. Thus, whereas the complementarity principle carries primarily conceptual and philosophical weight, the uncertainty principle's importance is as a quantitative measure of the limitations of experiments performed to measure microscopic variables. As will be seen, these limitations exist at all levels of experiments, but their effect is negligible when macroscopic variables are measured. The uncertainty principle states that:

It is impossible to specify precisely and simultaneously the values of both members of pairs of physical variables that describe the behavior of atomic systems and that are conjugate in the Hamiltonian sense. The product of the uncertainties in the knowledge of two such variables must be at least $\hbar$.

By describing the measurement uncertainty of a coordinate by Δx_i, of a linear momentum component by Δp_i, of an angular coordinate by $\Delta \theta_i$, of an angular momentum by ΔJ_i, of the energy by ΔE_p, and of time by Δt, we can restate the uncertainty principle by the following equations:

$$\Delta x_i \cdot \Delta p_i \geq \hbar/2, \tag{8.10a}$$

$$\Delta \theta_i \cdot \Delta J_i \geq \hbar/2, \tag{8.10b}$$

$$\Delta E_p \cdot \Delta t \geq \hbar/2, \tag{8.10c}$$

where all the uncertainties are stated by the root-mean-square (rms) deviation from the average (see e.g. Yariv 1975, p. 13). Each of equations (8.10) suggests that an attempt to measure one of the parameters in the conjugate pair will result in distortion of the value of the other member of the pair. Thus, the first equation suggests that an attempt to measure the coordinate of a microscopic particle, or a photon, must always induce a change in its momentum; if an accurate measurement of the coordinates of a stationary particle is desired, the measurement process itself must induce a motion of the particle, which in turn will represent an uncertainty in the momentum measurement if that was desired also. This statement is not a reflection on the quality of the experiment; it is a law of physics. Approaching a product of uncertainties that is equal to $\hbar$ in a simultaneous measurement of both quantities is analogous to approaching the limit of a reversible process in thermodynamics: although by itself unobtainable, the limit supplies a value against which the quality of an actual measurement can be compared. However, owing to the smallness of $\hbar = 1.055 \times 10^{-34}$ J-s, it is obvious that any measurement of mechanically relevant objects can be achieved at an acceptable level of accuracy. This can also be viewed as a yardstick that can be used to distinguish between problems that can be described by classical approaches and problems (designated here as microscopic) that require quantum mechanical concepts for their description. The following example illustrates the effect of the uncertainty principle on a small system that can still be described by classical theories.

Example 8.1 Find the smallest error that can be expected in the measurement of the momentum of a miniature spherical steel pellet with a diameter of $d = 1\ \mu$m, moving at a velocity of $v_x = 1$ mm/s in the x direction, when its x coordinate is known at an accuracy of $\Delta x/d = 0.1\%$ (the density of steel is $\rho =$ 7,830 kg/m^3).

Solution This pellet is possibly the smallest particle that still has a mechanical significance. Furthermore, owing to its low velocity, its linear momentum is exceptionally small:

$$p_x = \rho \frac{\pi d^3}{6} v_x = 4.1 \times 10^{-18} \text{ kg-m/s}.$$

On the other hand, the uncertainty in the measurement of the momentum (irrespective of the experimental method) is

$$\Delta p_x \geq \frac{\hbar/2}{\Delta x} = \frac{0.528 \times 10^{-34}}{10^{-9}} = 5.28 \times 10^{-26} \text{ kg-m/s}.$$

With these values, $\Delta p_x/p_x \approx 10^{-8}$; that is, the linear momentum of this particle can be measured to at least eight significant digits without violating the uncertainty principle. Furthermore, the coordinate of that particle can be measured with an error of only 1 nm. Thus, although the physical parameters of the particle are exceptionally small and the coordinate is determined within an uncertainty of 1 nm (much shorter than the wavelength of visible light), the momentum can still be measured at an accuracy that far exceeds the accuracy of any practical device. For comparison, the mass of this particle is 4.1×10^{-15} kg, whereas the mass of an electron is $m_e = 9.11 \times 10^{-31}$ kg. Thus, the momentum of an electron will be small relative to the momentum of this pellet even when approaching the speed of light. Hence the effect of the uncertainty principle on the electron will be more pronounced. ■

This example illustrates that, within the macroscopic regime, the effect of the uncertainty principle becomes negligible and so its existence does not create any technical or conceptual difficulty. This is why we can use the principles of Newtonian mechanics. Only at the microscopic level – that is, when the dimensions of particles and their momentum become comparable to the uncertainty predicted by (8.10) – does the effect of the uncertainty principle become noticeable; then the process of measurement inherently influences the measured parameter. To illustrate how the measurement of one parameter can influence its conjugate parameter, consider the γ-ray microscope (Cassidy 1992). This is a Gedankenexperiment (thought experiment) proposed by Heisenberg, where the possible imaging of an electron by a microscope was contemplated. To reduce the blurring of the electron image by diffraction effects, γ-rays (the shortest known wavelength in the electromagnetic spectrum; see Figure 1.3) were proposed as the source of illumination.

Example 8.2 An attempt is made to use a microscope to detect an electron. The microscope can be modeled by a single ideal lens with a focal length f and diameter D and with the electron located at its focal plane (Figure 8.6). A collimated and monochromatic γ-ray beam is used for illumination. Assume that scattering of a single photon is sufficient to image the electron and that the

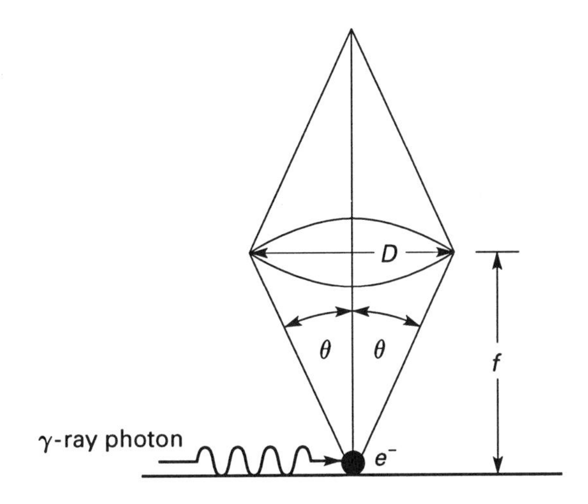

Figure 8.6 Hypothetical experiment for optical imaging of an electron.

wavelength of the scattered photon is λ. Estimate the accuracy of the measurement of the x coordinate of the electron by such imaging, as well as the momentum of the electron immediately after the collision with the photon if its initial momentum is well known.

Solution The uncertainty in microscopic measurements is limited by the imaging resolution, which is ultimately limited by diffraction. The radius of the smallest element that can be resolved by a lens is specified by (7.10). Therefore, the coordinate measurement uncertainty is

$$\Delta x = 1.22(f\lambda/D).$$

To obtain an image, at least one γ-ray photon must collide with the electron and scatter into the collection cone of the lens. Given its wavelength, the magnitude h/λ of the momentum of the scattered photon is well-defined. However, since it can be allowed to scatter anywhere within the lens collection cone, the *direction* of its momentum is ill-defined. Accordingly, the x component of its momentum, p_x (determined by projecting the momentum on the x axis), can be known only approximately. To determine its range of values, Δp_x, we calculate the x component of the momentum for the rays passing through the edge of the lens:

$$\Delta p_x = \pm \frac{h}{\lambda} \frac{D/2}{f}.$$

The new momentum of the electron can be readily determined from the conservation of momentum principle if its initial momentum and p_x for the photon are known. However, since the photon momentum is known with an uncertainty of Δp_x, the momentum of the electron can be known only within a similar uncertainty. Combining the uncertainty associated with imaging of the electron and the subsequent uncertainty in its momentum, for this imaging scheme we have

$$\Delta x \cdot \Delta p_x = 1.22\pi\hbar > \hbar/2.$$

This is consistent with the limits of the uncertainty principle. Although other imaging configurations can be envisioned, all must result in the same inequality. Furthermore, selection of even shorter wavelengths for illumination may improve the resolution but will exacerbate the uncertainty in the momentum measurement. ■

Other experiments can be designed to further confirm the uncertainty principle by attempting to determine simultaneously the coordinate of a particle and its momentum (see e.g. Problem 8.3). However, such experiments usually do not confirm (8.10c), which (as we will see in Section 11.2) is more relevant for gas dynamic measurements. To illustrate how the parameters E_p and t in this equation are paired, we must use the concept of energy on the microscopic scale, the concept of a quantum energy level, and the concept of lifetime (which will be developed in later sections). Alternatively, we can illustrate the concept presented by (8.10c) by evaluating the Fourier transform–limited bandwidth of a short laser pulse and compare the results with the uncertainty principle.

Example 8.3 Using Fourier transformations, determine the minimum bandwidth of a short laser pulse of duration $2\Delta t$ and show that the result is consistent with the uncertainty principle (eqns. 8.10).

Solution Certain laser systems include a device that allows the emission of energy to be condensed into brief pulses of duration ranging from femtoseconds (10^{-15} s) to hundreds of picoseconds. Example 4.2 analyzed the propagation of one such pulse through an optical fiber. There, half of the pulse duration represented the rms deviation from the peak and was used as the limiting parameter in the Fourier analysis (eqn. 4.27). For a stationary observer, the pulse appears as a burst of energy passing by at the speed of light c and with a spread of $\Delta z = c\Delta t$ (Figure 8.7). However, for a pulse of such limited stretch, the minimum bandwidth can be specified by Fourier transformation as follows:

$$\Delta z \Delta k \geq 1/2$$

(eqn. 4.27), where $\Delta k = 2\pi\Delta\nu/c$. Substituting Δt and $\Delta\nu$ for Δz and Δk, respectively, we can rewrite this inequality as

$$\Delta t \Delta \nu \geq \frac{1}{2} \cdot \frac{1}{2\pi}.$$

Multiplying both sides by h, we obtain

$$\Delta E_p \cdot \Delta t \geq \hbar/2,$$

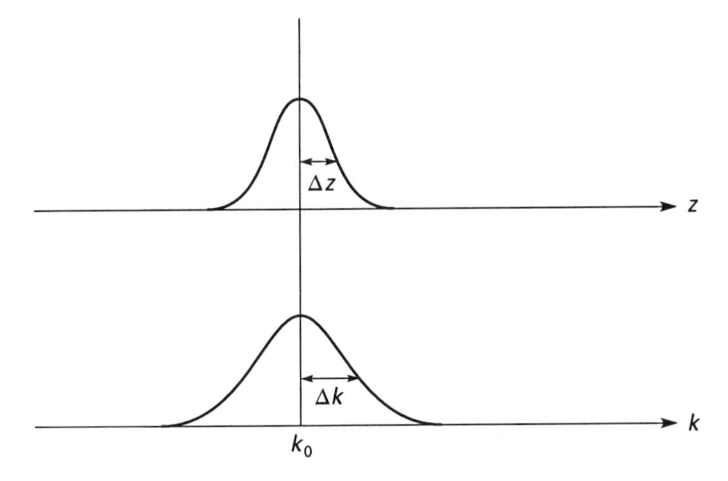

Figure 8.7 Correspondence between the spatial and spectral distributions of the wave function of a photon.

which is identical with the third presentation of the uncertainty principle (eqn. 8.10c). Thus (8.10c) implies that the full bandwidth (typically at half the maximum) of a short pulse with a full duration of $2\Delta t$ is

$$\Delta\nu_{\text{FWHM}} \geq \frac{1}{2\pi\Delta t}. \tag{8.11}$$

■

This result reconfirms our previous observations that radiation can never be made perfectly monochromatic – simply because the time of coherence is limited. Several factors (to be discussed in Chapter 11) reduce the coherence length and thereby broaden the bandwidth of nominally monochromatic radiation. When used for applications where narrowband radiation is required, the impact of all broadening effects must be minimized. Such applications include velocity measurements where Doppler shifts relative to the incident frequency are measured, and high-resolution projection or imaging where dispersion by the imaging optics may limit the resolution. Nevertheless, the limitations associated with the uncertainty principle limit the coherence length and consequently the monochromacity of radiation. When this limit is reached, the bandwidth is said to be *transform limited* or *uncertainty limited.*

8.8 The Correspondence Principle

The primary objective of the theory of quantum mechanics is to explain and predict the phenomena that are observed on an atomic (or microscopic) scale or during the interaction of radiation with atomic or molecular particles. Many of the observations that stem from the scale of the samples cannot be observed when similar experiments are performed on the macroscopic level. Furthermore, apparent conflicts between observations that are made on the microscopic scale are due to our interpretation of the results by means of classical concepts, which in turn are influenced by our experiences on the macroscopic scale. Although the classical theories of mechanics and electromagnetism are valid in most known situations, they fail to predict the results on the microscopic scale and so mislead us when attempting to interpret our observations. Thus, to be successful, the theory of quantum mechanics and the related theory of quantum electrodynamics must be more general than the classical theories. They must predict all the observations made on the microscopic scale in addition to those predicted by classical theories. If successful, then not only will the classical theories be spared but also the sources of conflict (stemming from interpretation of results on the microscopic scale using classical concepts) will be resolved. The principle of correspondence was formulated to express this need. It states:

> *The quantum mechanics theory is a rational generalization of the classical theories that can be recovered at the limit.*

If quantum mechanics is derived correctly then, as the scale of the experiment or the phenomenon increases, effects that can be predicted by classical theories will be also predicted by quantum mechanics. However, the opposite is not necessarily true. Since classical theories are only a specific case of quantum mechanics, they are unable and hence not expected to recover the results of quantum mechanics when the scale is reduced.

8.9 The Wave Function

The three principles stated in the previous sections define the fundamental rules of quantum mechanics. By themselves, these rules do not provide the mathematical tools necessary for modeling the phenomena they describe or for further generalization. Unlike other theories (e.g. thermodynamics) that are derived from a few fundamental postulates (see Callen 1985), the principles of quantum mechanics are not intuitive. Therefore, neither are the mathematical tools to be derived from these principles. Yet, these mathematical tools must in the end reproduce the results of previous experiments and lay the ground for the design of new experiments or applications. Thus, the development of the mathematical functions and equations of quantum mechanics must follow closely the hints provided by the fundamental principles and the related experiments.

The first step in the development of these mathematical tools is to identify a function that describes the motion of the modeled object, its energy, momentum, and its interaction with other objects – all within the limitations imposed by the uncertainty principle. The character of this function must comply with previous observations, on both the microscopic and macroscopic scale. Owing to the projected undulatory nature of the modeled objects, the function to be developed is expected to be similar to the functions of electromagnetic waves (e.g. eqn. 4.11). Consequently, the differential equation that will be used to derive this function should be similar to the wave equation (e.g. eqn. 4.8). Although the full derivation of the equations of quantum mechanics is beyond the scope of this book, we will outline some of the properties of this function. For more information, see Merzbacher (1970).

The double-slit experiment (Section 8.4) provides an excellent insight into the expected properties of the function that describes the electrons diffracted by the slits. The diffraction pattern behind the slits can be explained only if the function that describes each electron is undulatory, just like the equation of the propagation of electromagnetic waves, and if the function that describes the diffraction through one slit can be linearly superimposed with the function that describes diffraction through the other. Such superposition of two wave functions should then produce effects that are analogous to interference. Finally, by analogy with the electromagnetic wave theory, the mean-square value of this function should describe either the distribution of the electrons or their probability density, just as the mean-square value of the electric field describes the energy of the incident radiation. These concepts were already introduced in Section 8.4. Furthermore, it was shown by (8.7) that the probability P of finding an electron at any point on the screen can be determined from the linearly superimposed wave functions. Thus, although the wave function Ψ

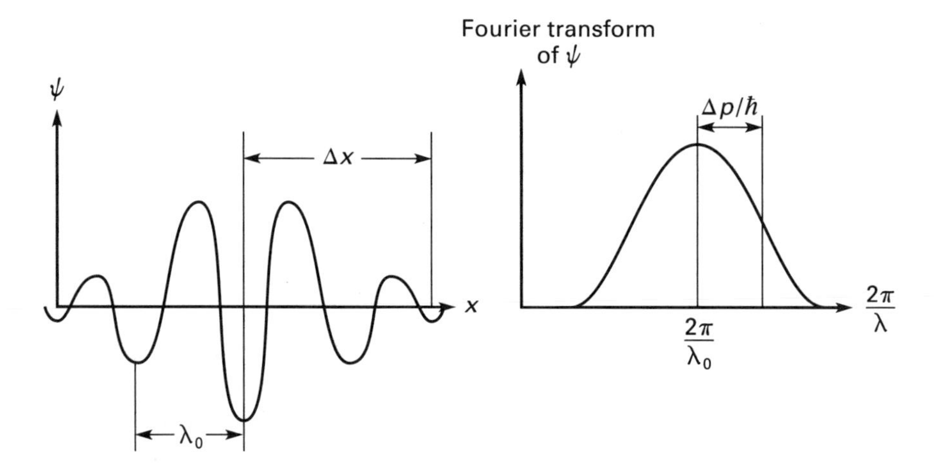

Figure 8.8 Spatial and momentum (or spectral) distribution of the wavepacket of an electron.

does not itself represent an intuitive physical parameter, its mean-square value (eqn. 8.7) represents the probability of finding a particle at a certain coordinate or at a certain time.

Although the concept of probability associated with the wave function is appealing and appears to be compatible with the uncertainty principle, the exact relation between the two is not immediately apparent. However, if an electron is assumed to be described by a wave with a limited extent (i.e., a *wavepacket*) then the nominal location of the electron can be specified by the coordinate of the center of the packet with an uncertainty specified by half of its length (Figure 8.8; cf. Figure 4.8). However, unlike a photon, the Fourier transform of the wavepacket of an electron cannot be expressed by the propagation vector. Instead, the Fourier transform must be represented in the momentum domain (eqn. 8.5), where the momentum is related to **k** by

$$\mathbf{p} = \frac{h}{\lambda}\mathbf{e} = \hbar\mathbf{k}. \tag{8.12}$$

From the Fourier transform of the wavepacket, the momentum uncertainty is

$$\Delta\mathbf{p} = \hbar\Delta\mathbf{k}, \tag{8.13}$$

and accordingly from (4.27) we have

$$\Delta x \cdot \Delta p \geq \hbar/2. \tag{8.10a}$$

Thus, the use of wavepackets for description of particles is consistent with the uncertainty principle. Clearly, as the size of the particle increases, its momentum increases while the wavelength associated with the momentum decreases until – as projected by the correspondence principle – it is no longer necessary to use the wave function to describe the particle and its parameters of motion.

8.10 Quantum States

The wave function of a particle can be obtained by solving the Schrödinger equation (Merzbacher 1970) that describes its Hamiltonian. Although the Schrödinger equation and its assumptions were not described here, suffice it to mention that when properly formulated it is similar to the differential equations describing the propagation of electromagnetic waves, acoustic waves, or the modes of oscillating strings. This formulation is consistent with the complementarity principle where the undulatory behavior of microscopic particles is contemplated. Similarly to these classical problems, Schrödinger's equation can be described either as an initial-value problem (i.e., as a wave propagating from a source to infinity) or as a boundary-value problem (i.e., as a wave oscillating between two boundaries). When the boundary conditions of a wave equation are homogeneous, its solution is expressed by characteristic functions (eigenfunctions), each corresponding to a characteristic value (eigenvalue). Thus, the solution of the transverse displacement of an oscillating string constrained between two walls is expressed by a series of modes, or by their linear superposition. Each solution appears as a sinusoidal wave with an integral number of nodes (Figure 8.9), which are used to specify the characteristic value. The function associated with each node is the characteristic function (see Problem 8.6).

A similar solution describes the oscillation of an electromagnetic wave between two parallel flat mirrors facing each other and separated by a distance L (Problem 8.7). When a mirror is perfectly reflecting, the electric field behind it is zero. Furthermore, when illuminated by normally incident radiation, the

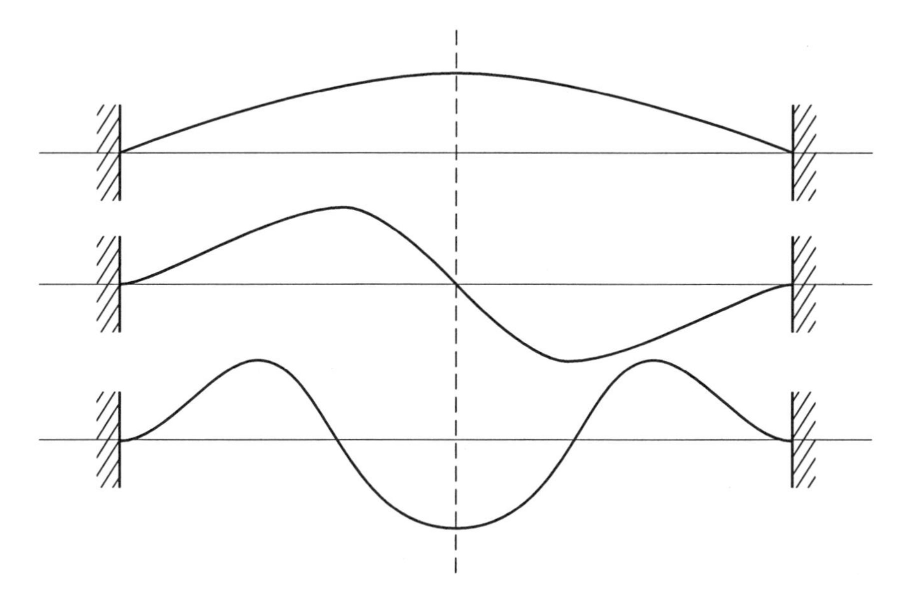

Figure 8.9 Stable oscillation modes of a string constrained between two walls.

electric field vector of the radiation is parallel to the mirror surface. Therefore, from the boundary condition for this field component (eqn. 3.26), the electric field at the front surface of the mirror must also vanish. This establishes homogeneous boundary conditions at both ends of the cavity enclosed by the mirrors. This configuration is identical to a Fabry–Perot etalon with $R = 1$, $\theta = 0°$, and $n = 1$. The wavelengths λ_m of the modes that stably oscillate in the cavity can be obtained directly from the solution of the one-dimensional wave equation with two homogeneous boundary conditions (Problem 8.7) or from the solution for the Fabry–Perot etalon (Section 6.5). The wavelength of the mth-order mode is

$$m\lambda_m = 2L, \quad \text{where } m = 1, 2, 3, \dots. \tag{8.14}$$

The frequency difference between two adjacent modes is

$$\Delta\nu = \frac{1}{\lambda_{n+1}} - \frac{1}{\lambda_n} = \frac{1}{2L} \tag{8.15}$$

(cf. eqn. 6.21), and the energy difference between these modes is $\Delta E_p = h/2L$. Notice that, as the length of the cavity increases, the difference between the energies or wavelengths of two adjacent modes decreases until distinction between individual modes is almost impossible.

The results that describe an oscillating string or an electromagnetic wave oscillating between two parallel mirrors are comprehensible by classical concepts. Therefore, they can be used to suggest the behavioral pattern to be expected of microscopic particles, where guidance by intuition or previous experience may be lacking. If the mathematical methods used to model the motion of elementary particles are similar to the methods used to describe bounded waves, their wave functions – just like the functions of the bounded waves – may also be expected to depend on isolated, or quantized, wavelengths. Accordingly, their energies will be quantized and the modes that are associated with these energies, the *quantum energy states,* will be identified by series of *quantum numbers*.

For illustration, consider a single particle enclosed in a one-dimensional "box" with walls that are infinitely tall and are separated by the distance a (Figure 8.10). To escape the box, the particle must jump over one of the walls. This is possible when the height of the walls is finite. For infinitely high walls, the energy required to escape is infinitely large and escape is impossible. Inside the one-dimensional box, the particle can move only in the x direction with a velocity of v. When $v \ll c$, relativistic effects can be neglected and the kinetic energy of the particle can be expressed as

$$\frac{mv^2}{2} = \frac{p^2}{2m}. \tag{8.16}$$

Because the particle cannot escape the box, the probability of finding it outside the box at $x < 0$ and at $x > a$ is zero. Accordingly, the boundary conditions that must be met by the wave function describing the particle are:

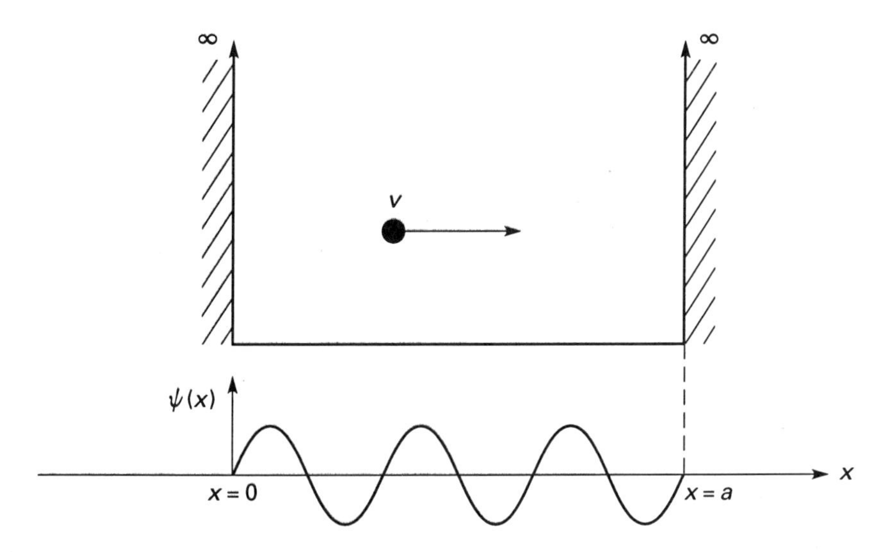

Figure 8.10 Particle in a box and a stable wave function of that particle.

$$\psi(0, t) = \psi(a, t) = 0. \tag{8.17}$$

Assuming that $\psi(x, t)$ is obtained from the solution of a differential equation that is similar to the wave equation, these boundary conditions require that the solution be presented by characteristic functions that are similar to the characteristic functions of the electromagnetic wave between two mirrors. By analogy to (8.14), the wavelength of the wave function describing the nth mode must be $\lambda_n = 2a/n$. This condition guarantees that the phase of the wave function starting at one wall is identical to the phase of the function at that wall after one, or multiple, round trips. Otherwise, the waves of the bouncing particle will destructively interfere with each other. The various wavelengths λ_n are inversely proportional to the momentum, which for the nth mode may be expressed as

$$p_n = \frac{nh}{2a}, \quad \text{where } n = 1, 2, 3, \ldots. \tag{8.18}$$

Since kinetic energy is the only energy mode for this particle, (8.18) and (8.16) can be combined to obtain the energy of the nth mode:

$$E_n = \frac{n^2 h^2}{8ma^2}. \tag{8.19}$$

These are the allowed energies for a microscopic particle in a box with infinitely high walls. Each allowed energy is characterized by an integral quantum number $n = 1, 2, 3, \ldots$. Although incomprehensible by classical concepts, this result shows that a microscopic particle enclosed in a small box can have only isolated energies; it cannot have an energy that is between the allowed states. The energy of each permissible state increases quadratically with the quantum number. By specifying the quantum number of an energy level, its energy can be readily calculated. Since the energy of all states decreases with a, when the width of the box is sufficiently large the states become indistinguishable and merge into a continuum. This convergence toward a continuum is consistent with the correspondence principle and is also consistent with intuition, which dictates that any energy should be allowed for ordinary translation in a closed box.

The variation of the energy of a quantum state with the quantum number that specifies it depends on the type of motion and on the nature of the particle. To illustrate these two concepts, we consider some modes of molecular motion and compare the results obtained for electrons and photons. The energy of certain modes of molecular motion (e.g. vibration) increases linearly with the quantum number (Herzberg 1950), whereas the energy in other modes (e.g. rotation) increases quadratically. When the motion in one mode influences the motion in another mode, higher-order dependence may also occur. For photons oscillating between parallel mirrors, the dependence between the energy of a photon and its quantum number is linear (eqn. 8.15); for the particle in the box, the dependence is quadratic. Although the oscillation of radiation between two mirrors and the motion of a particle in a box are similar, the participating particles are different: the particle in the box has mass when at rest whereas the mass of the photon is zero at rest. This difference accounts for the different functional dependence between the quantum number of a state and its energy (see Problem 8.2).

Several diagrams may be used to describe the relative position of the energy levels and their relative magnitude on the energy scale. Some diagrams simply list the energy and the quantum number or other identifying symbol of the state, and may include markings to designate possible channels of energy exchange between these states. Others present the range of coordinates within which a particle may exist while having that energy; such diagrams are particularly useful to describe molecular vibrations and their energies. For the particle in a box (Figure 8.10), the energy-level diagram (Figure 8.11) is drawn with the abscissa representing the coordinate of the particle and the ordinate representing the energy of the allowed states. Thus, in Figure 8.11 the walls are marked by vertical lines at $x = 0$ and $x = a$ extending to $E_n \rightarrow \infty$. The energy levels are marked by lines that are parallel to the x axis but are limited by the walls. Thus, when in the first energy level, the energy of the particle is constant anywhere in the box. To be promoted from the first to the second energy level, this particle must absorb a quantum of energy of $3h^2/8ma^2$. As the energy increases, the energy gap between adjacent levels appears to increase. Since the energy required to escape the box is $E \rightarrow \infty$, the allowed quantum numbers are $1 < n < \infty$. When at the wall, the velocity of the particle is necessarily zero and its entire energy is in the form of potential energy. Therefore, the boundaries of the box in this diagram can be viewed as a potential curve. A potential curve that has a minimum, such as the box, is often called a *potential well*. Other curves without such a minimum may occur and normally represent an unstable configuration.

Although the solution of the wave function for each of the allowed energy levels of the particle can provide information about its momentum or the likelihood of finding the particle at a certain location in the box, the information

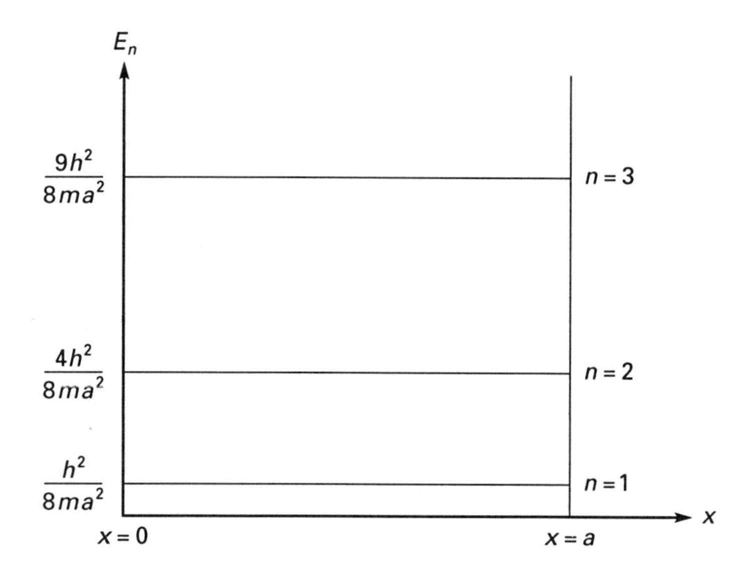

Figure 8.11 Energy-level diagram for the particle in the box of Figure 8.10.

that is primarily relevant for most applications is the energy of each level, the probability that such energy level is occupied by the particle, and the probability that the particle will absorb or emit a quantum of energy while moving from one level to another. Answering these questions is one of the objectives of quantum mechanics. Although (exact or approximate) solutions can be obtained in most cases, most engineering applications require only the spectroscopic parameters without the wave functions themselves. Only when modeling new molecules or new molecular or atomic processes are solutions of the Schrödinger equation required. For other applications, tabulations of necessary data can be found in the literature. Therefore, even such advanced applications as laser-induced fluorescence or Raman spectroscopy can be successfully evaluated by using only the fundamental principles and their primary result. In the next chapters we will introduce the concept of energy quantization as it applies to atomic or molecular systems, and will describe the modes of radiative energy exchange between these states.

References

Ashkin, A. (1980), Applications of laser radiation pressure, *Science* 210: 1081–8.

Bohr, N. (1928), The quantum postulate and the recent development of atomic theory, *Nature* 121: 580–91.

de Broglie, L. (1924), *Thèse,* Paris.

Callen, H. B. (1985), *Thermodynamics and Introduction to Thermostatics,* New York: Wiley.

Cassidy, D. C. (1992), Heisenberg, uncertainty and the quantum revolution, *Scientific American* 266: 106–12.

Compton, A. H. (1922), Secondary radiations produced by X-rays, and some of their applications to physical problems, *Bulletin of the National Research Council* 4(20): 1–54.

Compton, A. H. (1923), The spectrum of scattered X-rays, *Physical Review* (2) 22: 409–13.
Davison, C., and Germer, L. H. (1927), Diffraction of electrons by a crystal of nickel, *Physical Review* 30: 705–40.
Einstein, A. (1905), On a heuristic point of view concerning the generation and transformation of light, *Annalen der Physik* 17: 132–48.
Goldstein, H. (1981), *Classical Mechanics,* 2nd ed., Reading, MA: Addison-Wesley.
Heisenberg, W. von (1927), Über den Anschaulichen Inhalt der Quantentheoretischen Kinematik und Mechanik [On the visualizability content of quantum theoretical kinematics and mechanics], *Zeitschrift für Physik* 43: 172–98.
Herzberg, G. (1950), *Molecular Spectra and Molecular Structure I. Spectra of Diatomic Molecules,* New York: Van Nostrand.
Hildebrand, F. B. (1965), *Methods of Applied Mathematics,* 2nd ed., Englewood Cliffs, NJ: Prentice-Hall, pp. 148–55.
Merzbacher, E. (1970), *Quantum Mechanics,* 2nd ed., New York: Wiley.
Yariv, A. (1975), *Quantum Electronics,* New York: Wiley, pp. 13–17.
Zajonc, A. (1993), *Catching the Light. The Entwined History of Light and Mind,* New York: Oxford University Press, p. 292.

Homework Problems

Problem 8.1

Show that the momentum of a photon is as predicted by (8.3). Use the expression for the photon energy and the following equation for the relativistic kinetic energy:

$$\text{KE} = (m - m_0)c^2,$$

where the relativistic mass is

$$m = \frac{m_0}{\sqrt{1 - v^2/c^2}}$$

and m_0 is the mass at rest.

Problem 8.2

Using the equations of Problem 8.1, determine the relationship between the kinetic energy and the linear momentum of a particle with a mass of m_0 at rest when moving at a velocity $v \ll c$. Compare this result with the relation between the momentum and energy of a photon for which $m_0 = 0$ and $v = c$. Explain the reason for the different functional dependence between these two types of particles.

Problem 8.3

Assume that the position and momentum of a photon are measured by passing the photon through an infinitely long slit with a width of $2\Delta x$, where x is the coordinate transverse to the slit. The radiation can be assumed to be perfectly collimated and monochromatic when entering the slit; that is, upon incidence, $p_x = 0$. Using the results of the diffraction theory for an infinite slit, determine the uncertainty Δp_x in the momentum component in the x direction behind the slit and evaluate the magnitude of $\Delta x \Delta p_x$.

Problem 8.4

A device designed to demonstrate the effect of "light pressure" includes a small turbine consisting of four vanes enclosed in a transparent vacuum chamber (Figure 8.12). Each

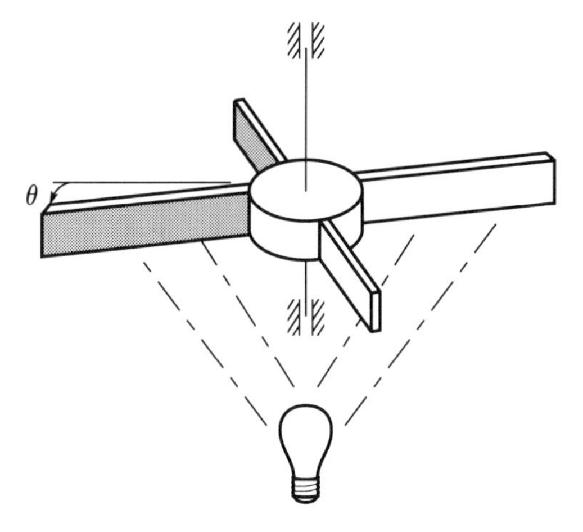

Figure 8.12 Device for demonstration of the effect of "light pressure."

vane is 1 cm wide and 3 cm long; the vanes are painted black on one side but are highly reflecting on the other side. Therefore, when viewed through the cell walls the left vane is black while the right vane is reflecting. Find the magnitude and direction of the torque applied on the turbine by a point source, located 20 cm away from the turbine and emitting 20 W at $\lambda = 500$ nm, when the normal to the vanes facing the source forms an angle θ with the line connecting the point source to the turbine axis. You may assume that the illumination of each vane is uniform and is specified by the illumination at its center.

Problem 8.5

One method for propulsion in long space travels proposes to use solar sails to capture the momentum imparted by solar radiation. Assume that in the vicinity of earth the solar flux is approximately 1 kW/m^2.

(a) Compare the force applied on a 100-m × 100-m sail directly facing the sun when the sail is (1) perfectly absorbing and (2) highly reflective.

(b) Determine the velocity of a 100-kg spaceship after a continuous exposure of 1 hr.

(c) Also find the kinetic energy of the ship and – assuming that its energy was acquired at the expense of the incident photons – find the change in frequency of the average incident photon.

Problem 8.6

The displacement $w(x, t)$ of a vibrating flexible string under a tension T and subjected to a transverse distributed force $F(x, t)$ is

$$\frac{\partial}{\partial x}\left(T\frac{\partial w}{\partial x}\right) - \rho\frac{\partial^2 w}{\partial t^2} + F = 0,$$

where ρ is the linear density of the string in kilograms per meter.

(a) Find $w(x, t)$ for a freely oscillating string (i.e., $F(x, t) = 0$) of length L when constrained at both ends.

(b) Find the possible frequencies of oscillation. (*Answer:* $\omega_n = (n\pi/L)\sqrt{T/\rho}$.)

Problem 8.7

An electromagnetic wave is bouncing between two parallel flat mirrors that face each other and are separated by free space and a distance of L.

(a) Using the wave equation (eqn. 4.8), find the equation of the electric field between these mirrors.

(b) Find the wavelengths and frequencies of the oscillating modes and compare them with the results for the Fabry–Perot etalon (Section 6.5). (*Answer:* $n\lambda = 2L$.)

(c) What is the difference between the energies of two adjacent modes? (*Answer:* $\Delta E_p = hc_0/2L$.)

Research Problems

Research Problem 8.1

Read the paper by Ashkin (1980). Use the results of that paper to solve Research Problem 8.2.

Research Problem 8.2

A glass cylinder with radius of 5 μm and length of 5 μm is located in the center plane of a Gaussian laser beam at the point where the intensity is at half maximum (Figure 8.13). The irradiance distribution in the beam is:

$$I = I_0 \exp\left\{-\frac{2r}{R_0}\right\},$$

where $R_0 = 50\ \mu$m, $I_0 = 10^5$ W/cm^2, and $\lambda = 500$ nm. Although detailed calculation of the distribution of the scattered radiation requires the use of physical optics, a simple approximation can be obtained by tracing several representative rays using geometrical optics. Assume the cylinder can be divided into six longitudinal strips, each $r/3$ wide.

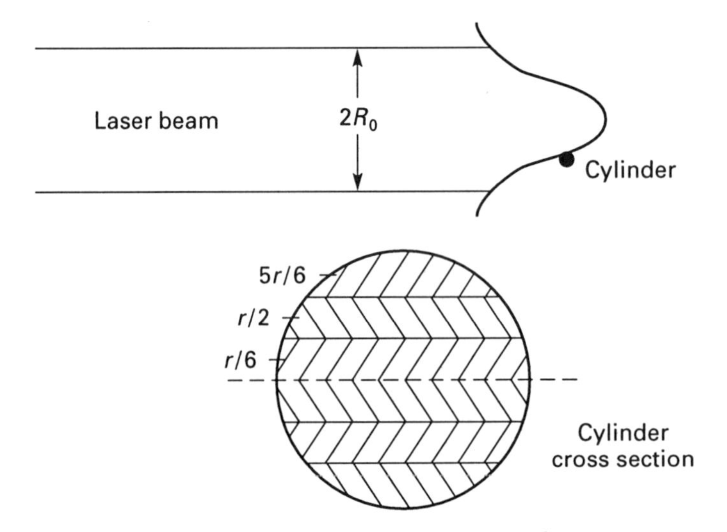

Figure 8.13 Cross section of a glass pellet and is position relative to the laser beam.

Furthermore, assume that the radiation falling on each of these strips can be represented by a single ray passing through their centers at $\pm r/6$, $\pm r/2$, and $\pm 5r/6$.

(a) Calculate the transverse force exerted on the cylinder.

(b) Calculate the axial force exerted on the cylinder.

9 Atomic and Molecular Energy Levels

9.1 Introduction

The emission and absorption of radiation, as well as the conversion of radiation into other modes of energy such as heat or electricity, all involve interaction between electromagnetic waves and atoms, molecules, or free electrons. Such daily phenomena as the radiative emission by the sun, the shielding of earth from harmful UV radiation by the ozone layer, the blue color of the sky, and red sunsets are all – despite their celestial magnitude – generated by microscopic particles. Most lasers depend on emission by excited atoms (e.g. the He–Ne laser), ionized atoms (the Ar^+ laser), molecules (CO or CO_2 lasers), impurities trapped in crystal structures (Nd:YAG or Ti:sapphire lasers), or semiconductors (GaAs diode lasers). Similarly, many scattering processes of interest (e.g., Rayleigh or Mie scattering) result from the exchange of energy and momentum between incident radiation and atomic or molecular species. In the previous chapter we saw that the energy of microscopic particles is quantized: their energy can be acquired, stored, or released only in fixed lumps called *quanta*. The example of the "particle in the box" (eqn. 8.19) illustrated that these energy quanta are specific not only to the particle itself but to the system to which it belongs. Thus, in the box, the energy of the particle is specified by its own mass and by the dimension of the box; in a different box, the same particle will have an entirely different system of energy levels and the quanta will have different magnitudes. Therefore, emission or absorption spectra by such a hypothetical particle depend both on its mass and on the size of the box containing it. By contrast, exchange of energy (or momentum) between two colliding billiard balls, which obey laws of classical mechanics, is determined without the need to consider the dimensions of the pool table.

The example of the particle in the box suggests that any explanation or prediction of radiative processes must include a detailed analysis of the quantum mechanical structure of the atoms or molecules that are part of that process. Although the electromagnetic wave theory simply states that radiation is emitted by oscillating charges, the theory of quantum mechanics requires that the microscopic systems to which these oscillating charges belong also be analyzed. Such microscopic systems include atoms, molecules, or ions to which electrons are attached, as well as solids with encapsulated ionized impurities or streaming free electrons in an electron beam. To calculate the wavelength of the emitted or absorbed radiation, the energy of the levels of the microscopic system must be known. Furthermore, to determine if absorption or emission of radiation are even possible, certain properties of these levels - such as the symmetry or antisymmetry of their wave functions - must also be known. Although a detailed account of the structure and properties of the energy levels of microscopic systems requires extensive theories of quantum mechanics, simple analysis followed by the implementation of measured or analytically derived coefficients is sufficient for quantitative evaluation of most radiative processes.

This chapter presents a brief account of the quantum mechanical structure of certain atomic and molecular species, their typical modes for energy storage, and the conditions for energy exchange between these species and radiation. The discussion will focus primarily on atoms and diatomic molecules that are relevant to engineering applications and on their major features. A reference to the more detailed aspects of their spectroscopy will be given.

The analysis begins with hydrogen, the simplest atom. It contains only one electron and one proton. Because of its simplicity, it was the first atom to be modeled (albeit with limited accuracy). Although the model of the H atom is of little use to the applications described in this book, it serves as an excellent introduction to the reasons for the quantization of energy at the atomic level and its consequences.

9.2 Bohr's Hydrogen Atom

One of the experimental observations in the late nineteenth century that prompted the development of the theory of quantum mechanics was the distinctive emission spectrum of electric discharge through hydrogen. When the glow was analyzed by dispersive elements such as prisms, its spectrum consisted of well-resolved narrowband lines, in contrast to the continuous spectrum of solar radiation or other incandescent sources. The wavelengths of these lines were independent of the discharge parameters and were so well reproduced between experiments that Balmer, Rydberg, and others could group them into series and predict the wavelength of each line of a series by an empirical equation. Each series was assigned an integer $n' = 1, 2, 3, \ldots$ and each line within a series was labeled by another integer $n = 1, 2, 3, \ldots$ where $n \neq n'$; the wavelengths of most observed lines of the H electric discharge were found to vary proportionally with $(1/n^2 - 1/n'^2)$. Despite the success of these empirical formulas, they failed to explain the physical effect causing these lines. The first successful theory was offered by Bohr (1913) at a time when most of the fundamental concepts

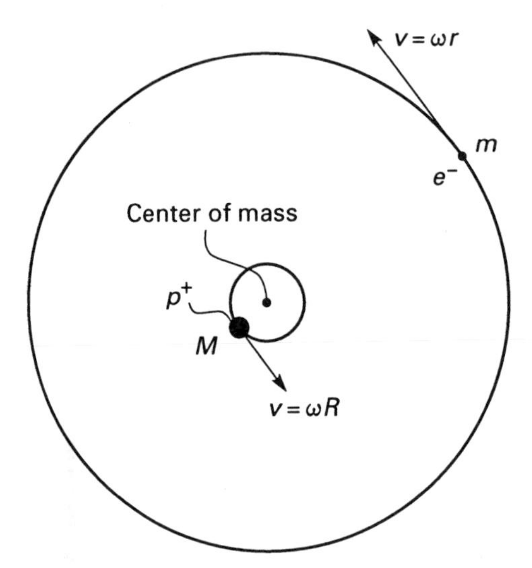

Figure 9.1 Orbits of the electron and proton of a hydrogen atom around the center of its mass.

of quantum mechanics were still unknown. Nevertheless, with few exceptions, it presented a logically plausible model of the hydrogen atom. Subsequent theories of quantum mechanics that were motivated by this model refined it, justified its hypotheses, and put them in a broader context that covered all atoms and molecules. Despite its simplicity, Bohr's general picture of the atom is still accepted as valid and is commonly used to visualize atomic structures.

Bohr's hydrogen atom consists of one electron and one proton, which from a distance appear to be electrostatically neutral. However, for reasons not understood by Bohr, the two charged particles do not collapse together as might be expected from the principles of electrostatics. Instead, they orbit synchronously at a frequency of ω, around a common center, in stable circular orbits (Figure 9.1); the proton in an orbit with a radius R and the electron in another orbit with a radius $r \gg R$. The electrostatic attraction between the positive and negative charges provides the centrifugal force necessary to maintain the stability of these orbits. Since the radius of the electron orbit is much larger than the size of the proton, it is this parameter that ultimately specifies the dimension of the atom; the weight of the atom is determined by the much heavier proton. Viewing the electron as a point particle moving in a stable orbit, almost like a planet in the solar system, is inconsistent with the uncertainty principle. However, at the time this model was offered, this and other concepts such as de Broglie's wavelength were not known. Nevertheless, Bohr hypothesized that, unlike planetary orbits in the solar system, the radii of stable orbits in the hydrogen atom cannot be selected by energy considerations alone; there must be an additional condition that a stable orbit must meet. That additional condition, a critical assumption in Bohr's model, was that the angular momentum of the atom in a stable orbit (i.e., the combined angular momentum of the proton and of the electron) is an integral multiple of $\hbar$. While in this stable orbit, although subject to centripetal acceleration, the electron does not emit radiation and can therefore maintain its motion indefinitely. The lack of radiative emission by the orbiting electron while continuously subject to an acceleration was contradictory to previous observations, whereby accelerating electrons emitted or absorbed radiation. The only mechanism for emission by the atom, as postulated by Bohr's model, involves transition from one stable orbit with an energy E_1 to another with energy E_2. For each transition there is one photon emitted, which for purposes of energy conservation has an energy of $h\nu = E_2 - E_1$. The spectrum associated with transitions between individual orbits must therefore consist of isolated spectral lines.

To calculate the wavelength of the observed spectral lines, the energy of the stable orbits and their radii were calculated by Bohr using principles of classical mechanics and electrostatics together with the new hypotheses (see Problem 9.1). The emission frequency (in units of cm^{-1}) for transitions from a level identified by the integer n' to any other level with an integer n was found to be

$$\bar{\nu} = \frac{me^4}{8\epsilon_0^2 ch^3(1+m/M)}\left(\frac{1}{n'^2} - \frac{1}{n^2}\right) = R_H\left(\frac{1}{n'^2} - \frac{1}{n^2}\right), \tag{9.1}$$

where m and M are the electron and proton masses, respectively. When ϵ_0 is expressed in the MKSA units (see Appendix A), the coefficient R_H in (9.1) is

$$R_H = \frac{me^4}{8\epsilon_0^2 ch^3(1+m/M)} = 109{,}677.576\ \text{cm}^{-1}.$$

This is the *Rydberg constant*. Equation (9.1) was derived by assuming that both the proton and electron are in motion. This is stipulated from Newton's third law of motion (action and reaction). However, by introducing a reduced mass μ to describe the mass of the electron,

$$\mu = \frac{Mm}{M+m}, \tag{9.2}$$

only the relative velocities between the two particles need to be considered (Vincenti and Kruger 1965). Therefore, (9.1) can also be derived by placing the proton at the center of the electronic orbit, replacing m with μ, and calculating the electron motion in a circular orbit around the stationary nucleus (Problem 9.1).

Bohr's model of the hydrogen atom successfully predicts many of the prominent features of the hydrogen spectrum as well as other experimental observations (see Problem 9.2). However, finer details such as splitting of certain lines into two or more closely spaced yet distinguishable lines, or the absence of some lines from the spectrum, could not be explained by this theory. Furthermore, there was no physical justification for the hypothesis that the angular momentum of the atom, when in a stable orbit, is an integral multiple of $\hbar$. When the wavelike nature of microscopic particles was discovered, and in particular after the development of Schrödinger's equation, an accurate solution of the hydrogen atom could be obtained (Merzbacher 1970). The details of this solution are beyond the scope of this book. However, to illustrate how the simple hypotheses of Bohr are related to the detailed solution of Schrödinger's

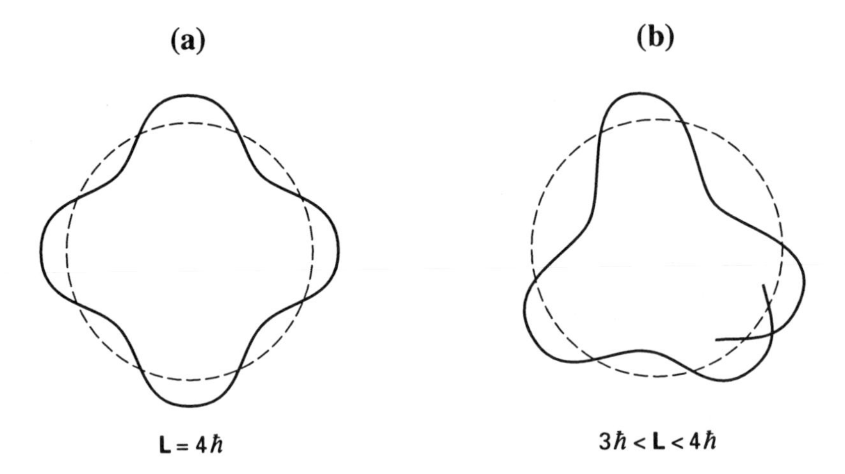

Figure 9.2 Wave functions of the electron of a hydrogen in (**a**) stable and (**b**) unstable orbit.

equation for the hydrogen atom, a heuristic picture of the wavelike behavior of the electron in the hydrogen orbit is presented here.

Assuming that the electron is described by a wave function, its momentum is defined by the wavelength λ of that function. Therefore, the linear momentum is $\mathbf{p} = h/\lambda\mathbf{e}$ (eqn. 8.12) and the angular momentum in a circular orbit is $\mathbf{L} = \mathbf{r} \times \mathbf{p}$. Figure 9.2(a) presents the wave function of an electron in its orbit. The condition that the angular momentum of the electron (when using the reduced-mass presentation) is an integral multiple of $\hbar$ can now be shown to imply that the circumference of the circular orbit is equal to an integral multiple of the wavelength (Figure 9.2(a)), that is, $n\lambda_n = 2\pi r_n$. This serves as the boundary condition for the solution of this wave equation. It is equivalent to the boundary condition for a particle in the box which resulted in wave functions with wavelengths that met the condition $n\lambda_n = 2a$. These conditions imply that the phase of a stable wave function at the beginning of a round trip between the walls of the box, or of a turn around the proton, are matched by the phase after the completion of one such trip.

To further illustrate why this boundary condition is necessary for the establishment of a stable orbit, imagine the wave function as an infinitely long wavy string that follows the orbit around the nucleus. After each turn, a new layer of this imaginary string is laid over a previous layer. If the boundary condition for a stable orbit is not met (i.e., if $n\lambda_n \neq 2\pi r_n$) then the phase of the wave function, which is represented by the wavy string, at the start of a turn cannot match the phase at the end of that turn (Figure 9.2(b)). Thus, as more layers are added, the phase difference increases until the phase of the first layer is matched by an exactly opposite phase of a later layer. Interference causes the cancellation of both. The wave functions of other turns are similarly matched by interfering phases and consequently the entire function is destroyed – that is, the probability of finding the electron in that orbit vanishes. By contrast, when the condition for a stable orbit is met, there can be no destructive interference and the stable wave function is enforced. The solution of the wave function with this boundary condition yields a series of characteristic values and characteristic functions (Section 8.9). Since for a stable orbit $n\lambda_n = 2\pi r_n$, the angular momentum L of that orbit can be obtained directly by substituting for λ_n:

$$L = \mathbf{r}_n \times \mathbf{p} = \frac{r_n h}{2\pi r_n / n}\mathbf{e} = n\hbar\mathbf{e}, \tag{9.3}$$

where n, the principal quantum number, is the integral number of wavelengths that can be accommodated along the orbit circumference and $\mathbf{e}$ is a unit vector pointing perpendicularly to the plane of the orbit. The right-hand side of (9.3) is Bohr's condition for a stable orbit.

9.3 Structure of the Atom

The solutions of Schrödinger's equation for an electron in a circular orbit, together with the boundary condition of (9.3), successfully describe atomic orbits with large radii. These orbits, which are almost perfectly circular, are the *Rydberg states* of these atoms. The spectra for transitions between these states, like the hydrogen spectrum, form regular series similar to that of (9.1). Although they comprise an important group, Rydberg states exist only when atoms are highly excited and hence do not represent the stable states of most atoms. The typical atomic orbit (which is energetically well below the Rydberg state) is elliptical, similar to planetary orbits. As a result, the geometry of these orbits, their energy, and their angular momentum cannot be described by a single quantum number. Instead, the major axis is defined by the principal quantum number n, where

$$n = 1, 2, 3, \ldots, \tag{9.4}$$

and the minor axis is defined by a secondary number l, the azimuthal quantum number. Although the principal quantum number can be any positive integer, the values of the azimuthal quantum number l must be less than n, because l represents the smaller axis of the ellipse. Thus

$$l = 0, 1, 2, \ldots, (n-1) \tag{9.5}$$

(Herzberg 1989, p. 5). Since two quantum numbers are necessary to specify an orbit, the energy must also be specified by two quantum numbers. This is particularly important when the size of the nucleus is relatively large and also for lower orbits (i.e., when the orbit is far from a Rydberg state). The effect of l on the energy cannot be stated by a simple equation, and is computed individually for each state. Exact calculations are available for simple atoms such as He (see e.g. Karplus and Porter 1970). For complex atoms, exact solutions of the complete Schrödinger equation that includes all the electrons and the components of the nucleus are not possible, and approximation methods such as the perturbation theory are used. The results of such calculations and experimental confirmation are widely available in the literature (e.g. Moore 1971).

For orbits in which the nucleus appears as a point charge relative to the radius of the orbit – for example, when $n \gg 1$ – the radius is much larger than the nucleus size and the structure of the atom resembles that of the hydrogen atom. For these orbits, which are called Rydberg states, the energy can be stated by the principal quantum number alone. For these states and when assigning $E = 0$ for an orbit with $n \to \infty$, the energy (in joules) of an orbit with quantum number n is

$$E_n = -\frac{R_H hcZ^2}{n^2}, \tag{9.6}$$

where Z is the atomic number. Thus, for transition from an upper Rydberg level n_1 to a lower level n_2 the emission frequency is:

$$\bar{\nu} = R_H Z^2\left(\frac{1}{n_2^2} - \frac{1}{n_1^2}\right)\ \text{cm}^{-1}. \tag{9.7}$$

When $Z = 1$, the emission frequency of (9.7) is identical to the emission predicted by Bohr's hydrogen atom (eqn. 9.1).

Although the energy associated with the orbit of an electron depends on two quantum numbers, the angular momentum of the orbit is fully specified by l alone. This is unlike Bohr's atom, where the angular momentum was specified by the principal quantum number. For an elliptical orbit, the angular momentum $\mathbf{l}$ is

$$\mathbf{l} = \hbar\sqrt{l(l+1)}. \tag{9.8}$$

Like the classical angular momentum, the quantum mechanical angular momentum is a vector. In most tests the vectorial nature of $\mathbf{l}$ is of no consequence and is not observed. However, when an atom is placed in an external magnetic field, the vector of the angular momentum carries out a precession motion around an axis parallel to the magnetic field vector. This motion of precession draws a cone in space. The precession motion itself is consistent with classical mechanics, but – unlike with classical mechanics – the angle between $\mathbf{l}$ and the precession axis must comply with yet another set of quantization rules. Accordingly, the allowed projections of the vector $\mathbf{l}$ on the precession axis are:

$$\mathbf{m}_l = \hbar m_l, \tag{9.9}$$

where m_l, the magnetic quantum number, can have any integer value between l and $-l$:

$$m_l = l, (l-1), (l-2), \ldots, -(l-1), -l. \tag{9.10}$$

Thus, for every quantum number l, there exist $2l+1$ magnetic quantum numbers. Figure 9.3 illustrates one precession cone, the allowed orientations of the vector $\mathbf{l}$, and their projections when $l = 4$. In the absence of an external magnetic field, the precession disappears and states with different m_l but with the same quantum numbers n and l become indistinguishable. These indistinguishable states are called *degenerate states,* and the number of these degenerate

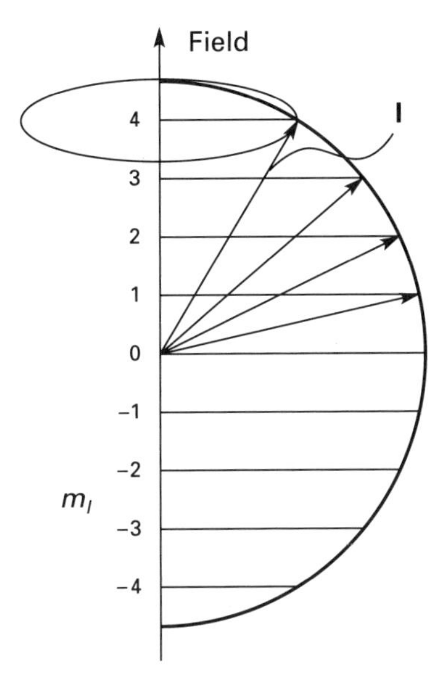

Figure 9.3 Allowed orientations of the angular momentum vector ($l = 4$) in the presence of a magnetic field.

states is the *degeneracy* of the state specified by n and l. States with other n and l may have other degeneracies. The degeneracy is an important parameter that determines the population density of a particular energy level, that is, the number of atoms or molecules within a unit volume that are at the same energy level.

All atoms, with the exception of hydrogen or ionized heavier atoms, have more than one electron that must be assigned into individual quantum states. Since the chemical affinity, electrical characteristics, and optical characteristics of the atom are determined by the position of their electrons, the rules for such assignments must be compatible with macroscopic observations. Although the size of the electron is extremely small relative to the nucleus or the atomic orbit, its spin around its own axis has an important effect on the occupancy of the various states. The angular momentum vector associated with the electron spin is

$$\mathbf{s} = \hbar\sqrt{s(s+1)}, \tag{9.11}$$

where $s = \frac{1}{2}$ is the electron spin quantum number. Similarly to the orbital angular momentum, the spin angular momentum can orient itself with external magnetic fields. The projection of the spin angular momentum takes the following values:

$$\mathbf{m}_s = \hbar m_s, \quad \text{where } m_s = -\tfrac{1}{2} \text{ and } \tfrac{1}{2}. \tag{9.12}$$

Here m_s is the spin magnetic quantum number. In the absence of magnetic fields, the electron in its orbit may point in an arbitrary direction and the two possible states that correspond to m_s are degenerate. Thus, the degeneracy of a single electron in a quantum state is 2. Such doubly degenerate states are called *doublets*.

The four quantum numbers of the atom – n, l, m_l, and m_s – are used for the assignment of electrons into their orbits. Since $s = \frac{1}{2}$, only its orientation – m_s – needs to be considered. Although the electrons are indistinguishable and therefore no particular electron can be assigned to one particular state, the number of electrons in each state is restricted by the following rule:

In one and the same atom, no two electrons can have the same set of values for the four quantum numbers n, l, m_l, and m_s.

This is the *Pauli exclusion principle*. According to this principle, the number of electrons that can be assigned to the level $n = 1$ is determined from the values of other other quantum numbers, which are $l = 0$, $m_l = 0$, and $m_s = \pm\frac{1}{2}$. The spin magnetic quantum number provides the only degeneracy when $n = 1$, and hence the maximum number of electrons that can be assigned to that state is only two. When the number of electrons in an atom exceeds two, the excess must be assigned to a state with $n \geq 2$. When $n = 2$ the allowed azimuthal quantum numbers are $l = 0$ and 1 (eqn. 9.5). The number of electrons allowed into the substate $l = 0$ is limited, as before, to two. However, for $l = 1$, the magnetic quantum number has a degeneracy of 3 ($m_l = -1, 0, 1$) and, according to Pauli's exclusion principle, two electrons with $m_s = -\frac{1}{2}$ and $\frac{1}{2}$ are allowed into each of the m_l states. The total number of electrons in the substate $l = 1$ is therefore six. Together with the two electrons in the substate $l = 0$, the total number of electrons in the level $n = 2$ is eight. Similarly, the total number of electrons allowed into the level $n = 3$ is eighteen. Thus, as the number of electrons in an atom increases, the states with the lowest principal quantum numbers, or *shells*, are filled up before the next shell can accept new electrons. Only when an atom is excited can a shell be populated by an electron before all lower shells are full.

Although the total number of electrons in a shell is restricted, the results obtained by Pauli's exclusion principle merely represent the maximum occupancy of a shell. As the number of electrons in atoms of the periodic table increases, the low-energy orbits are the first to be filled. Thus, for hydrogen in its stable mode, only one vacancy in the shell $n = 1$ is occupied. The next atom in the periodic table, He, has two electrons, filling both vacancies in the $n = 1$ shell. The next atom, Li, has three electrons: the first two in the lower shell $n = 1$ and the third in $n = 2$ and $l = 0$. When a shell is filled, it is unlikely to share its electrons with other atoms and so becomes chemically inert. Thus, only the electrons in external (partially filled) shells must be considered when determining the chemical, electrical, or optical properties of an atom. The energy of each electron in the outer shell is determined by the principal quantum number and by the azimuthal quantum numbers. However, the optical activity – the probability that an electron would absorb or emit a photon while undergoing a transition from one energy level to another – is primarily determined by the total angular momentum, which is obtained by an addition of the azimuthal and spin quantum numbers.

The total angular momentum of an atom is the sum of the angular momenta of each orbit and of each electron. Classically, addition of vectorial quantities is done by using the parallelogram rule. However, in quantum mechanics, vectors can be added only in certain directions. Thus, for two arbitrary angular momentum vectors with quantum numbers A and B the resultant angular momentum is also quantized and can have the following quantum numbers C:

$$C = (A+B), (A+B-1), (A+B-2), \dots, |A-B| \tag{9.13}$$

(Herzberg 1989, p. 25). Thus, when $A = 1$ and $B = 2$, the allowed quantum numbers for their resultant are $C = 3$, 2, and 1.

When i electrons reside in the outer shell of an atom, each must have its own azimuthal quantum number $l_1, l_2, \dots,$ or l_i and its own spin quantum number $s = -\frac{1}{2}$ or $\frac{1}{2}$. Application of the rule outlined by (9.13) for the summation of these numbers strongly depends on the order in which they are added. Usually, when several angular momenta are to be added, it is expected that those that are strongly coupled to each other form a resultant. Subsequently, the resultants of the earlier additions are added to form the total angular momentum. Since the azimuthal angular momenta l_i are strongly coupled to each other, they are added first and their resultant specifies the angular momentum L. For two electrons with azimuthal quantum numbers $l_1 = 1$ and $l_2 = 3$, the possible values of the resultant are $L = 4$, 3, and 2. Usually these quantum numbers are assigned a letter symbol. Accordingly, for $L = 0, 1, 2, 3, 4, \dots$, the assigned letter symbols are S, P, D, F, G, ... Similarly, for $l = 0, 1, 2, 3, 4, \dots$ of an electron, the symbols are s, p, d, f, g, ... After adding the azimuthal quantum numbers, the spin angular momenta are added among themselves to form a resultant spin S. Thus, for two electrons, the total spin quantum numbers are $S = 1$ and 0. Finally, the two resultants are added to form the following total angular momentum J:

$$J = (L+S), (L+S-1), \dots, |L-S|. \tag{9.14}$$

J is an important quantum number that is used to determine the probability that transition between two energy levels will occur while a single photon is emitted or absorbed. On the other hand, the quantum numbers L and S are used to determine the energy of certain electronic configurations. Therefore, the energies of electronic configurations may vary significantly with L quantum numbers or with S numbers. Finally, the degeneracy of the configuration is usually $2S+1$.

Because the relevant quantum numbers determine the energy of each state, its optical characteristics, and its degeneracy, their detailed representation is essential. These parameters are often represented by a compact symbol consisting of letters and numbers. A complete symbol of a state is called a *term*. In a common representation, a term includes the principal quantum number, followed by a letter that represents the angular momentum quantum number L. To the left of this letter, a superscript represents the degeneracy of the state, while to its right a subscript represents its J quantum number. To illustrate, consider the following example.

Example 9.1 Find the terms of an atom with two electrons in its outer shell, where $n = 4$ and the azimuthal quantum number $l_1 = 1$ for one electron and $l_2 = 3$ for the other.

Solution Since the azimuthal quantum numbers are strongly coupled, they are added first. From (9.13), the possible numbers are $L = 4, 3, 2, 1$. For two electrons, each with $s = \frac{1}{2}$, the total spin quantum numbers are $S = 1, 0$. Clearly, when $S = 0$, the values of J are identical with L and the degeneracy is 1. The terms when $S = 0$ are therefore 4^1P_1, 4^1D_2, 4^1F_3, 4^1G_4.

When $S = 1$, the degeneracy is $2S+1 = 3$ and the left superscript is 3. In addition, if $S = 1$ then the J quantum numbers are no longer equal to L. The total quantum numbers that can be formed by (9.14) with $L = 4$ and $S = 1$ are $J = 5, 4, 3$; with $L = 3$ and $S = 1$ they are $J = 4, 3, 2$; and so on with $L = 2$ and $L = 1$. Consistent with the degeneracy, there are three terms for every L, each with a different J in the right subscript. However, these three terms can be written in a concise form, which for $L = 4$ is $4^3G_{5,4,3}$. Similarly, for $L = 3$ the concise form is $4^3F_{4,3,2}$; for $L = 2$, $4^3D_{3,2,1}$; and for $L = 1$, $4^3P_{2,1,0}$. ■

It is evident from this example that as the number of electrons in the outer shell (or as the magnitude of the azimuthal quantum numbers l) increases, the possible combinations of allowed angular momentum quantum numbers increases as well and with it the number of allowed energy states. Usually, the energy levels and their terms are tabulated and the wavelengths for single-photon emission or absorption can be calculated for allowed transitions using the energy difference between the end states. However, optical transitions – those that result in the emission or absorption of a single photon – are limited to certain pairs of states that comply with specific selection rules. Transitions between other pairs of energy levels are optically unallowed; these states do not combine, and the probability for energy exchange between a photon and these states is extremely low. Thus, when two states comply with the selection rules, an electron in the lower energy level may be excited to the upper level while annihilating a photon with energy that matches the energy difference between these states. Conversely, when at the upper state, an electron may release its energy in the form of a photon with energy $h\nu$ that also corresponds to the energy difference between the states. The selection rules that control these optical transitions are complex and may vary from atom to atom or even from state to state within the same atom. However, one general rule appears to apply rigorously to all atoms. It states that optical transitions between two atomic states are allowed when

$$\Delta J = 0, \pm 1 \quad \text{with the added condition that} \quad J = 0 \nrightarrow J = 0. \tag{9.15}$$

Thus, for transitions between two atomic states that involve the emission or absorption of a photon, the change in the total angular momentum must be

0 or ±1, with the added condition that $J = 0$ cannot exchange radiative energy with another $J = 0$ state. In addition to these selection rules, other, more specific rules exist. For example, a selection rule that applies only to small atoms (e.g. sodium) states that:

$$\Delta L = 0 \text{ and } \pm 1 \quad \text{with } \Delta S = 0. \tag{9.16}$$

Graphical presentations of the relative positions of the energy levels of atoms, together with lines connecting states that can be combined by allowed optical transitions, are often used; such diagrams are called *Grotrian diagrams* (Grotrian 1928). Comprehensive Grotrian diagrams are available for numerous atoms and their ions (see e.g. Bashkin and Stoner 1975). Figure 9.4 is a Grotrian diagram of atomic sodium drawn using data of Grotrian (1928) and Bashkin and Stoner (1975). It includes transitions between states in the $n = 3$, 4, and 5 shells. The lowest level in the diagram is the $3\,^2S_{1/2}$ state; this is the ground state of Na. Note that, within the lowest shell, the energy difference between levels with different angular quantum numbers is large. As n increases, the effect of L on the energy decreases until the shells become Rydberg states and the energy difference between states with different L becomes negligible. For the S states (i.e., when $L = 0$) there exists only one J quantum number. For the P states, however, the values for J are $J = \frac{3}{2}$ and $\frac{1}{2}$. With only one electron in the outer shell, all states of Na are doubly degenerate.

The allowed optical transitions of Na are marked by lines connecting the interacting states. They all obey the selection rules of (9.15) and (9.16). Although the wavelengths or frequencies of these allowed transitions can be calculated directly from the energies of the connected states, accurately measured values are included in the diagram. In particular, note the two lines combining the ground state with the $3\,^2P_{1/2}$ and $3\,^2P_{3/2}$ states. These are the sodium D_1 and D_2 lines at $\lambda = 589.6$ nm and $\lambda = 589.0$ nm respectively. These sodium D lines are among the brightest atomic emission lines. The fine splitting between these two lines, the result of coupling between the L and S quantum numbers, is called the *LS fine splitting*. These yellow–orange emission lines are often observed in flames, where small Na impurities are thermally excited and forced to radiate. When this glow is resolved by a spectrometer, two narrowband lines separated by 17.19 cm^{-1} (Moore 1971, vol. I) are observed. The bright emission by these lines is used in efficient Na electric-discharge lamps for illumination of highway intersections. The electric discharge in the lamp excites a large fraction of the Na vapor into the $3\,^2P_{3/2}$ and $3\,^2P_{1/2}$ states, which then decay back into the ground state while emitting the orange–yellow glow. After returning to the ground state, Na atoms may be re-excited to emit again, thereby producing bright emission even at moderate Na densities.

The dominant features of the atomic spectrum are determined by the electrons in the outer shell, with hardly any influence by the significantly heavier nucleus. The most noticeable effect of the nucleus is to finely split some of the atomic energy states. This splitting is much finer than the LS splitting and is called, accordingly, *hyperfine splitting*. Although the hyperfine splitting of Na

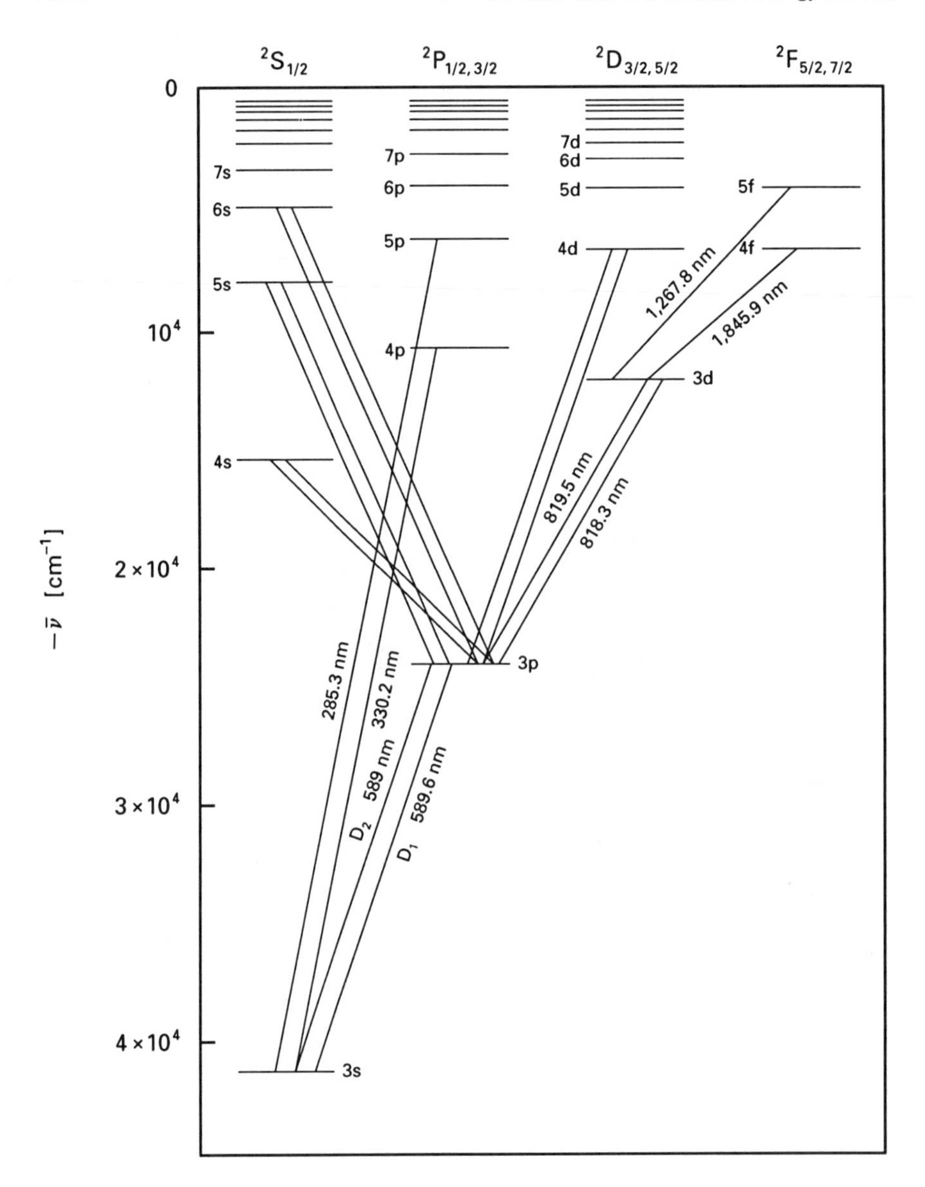

Figure 9.4 Grotrian diagram for atomic sodium (Na). [after Grotrian 1928]

cannot be resolved by ordinary spectrometers, it is resolved by narrowband lasers that can be tuned to excite selected features of the hyperfine structure (Fairbank, Hänsch, and Schawlow 1975).

The hyperfine splitting results from a weak coupling between the nuclear spin I and the total angular momentum J. The rules of addition of the nuclear spin with the total angular momentum are similar to the rules of addition of the other components of the angular momentum (eqn. 9.13). Namely:

$$F = (J+I), (J+I-1), (J+I-2), \ldots, |J-I|. \tag{9.17}$$

The selection rules that apply to the total angular momentum of the atom including the nuclear spin are similar to the rules of (9.15):

$$\Delta F = 0, \pm 1. \tag{9.18}$$

Owing to the weak coupling between I and J, the energy difference between states with different F quantum numbers but with the same J quantum number are much smaller than the fine splitting due to the LS coupling.

To illustrate hyperfine splitting, consider the effect of the nuclear spin of the Na atom, where $I = \frac{3}{2}$, on the D lines. When coupled with I, the $3\,^2P_{3/2}$ state includes four hyperfine states; the $3\,^2P_{1/2}$ state, after coupling with I, includes two hyperfine states. The ground state likewise has one pair of states with different F quantum numbers. Thus, in total there may exist six pairs of hyperfine lines (twelve lines); however, owing to selection rules (eqn. 9.18), only ten lines are observed. Figure 9.5 presents a Grotrian diagram of the hyperfine splitting of these states. The states coupled by allowed transitions are connected with lines. Numbers along these connecting lines indicate the relative intensities of these transitions. The hyperfine splitting between the upper states varies from 15 to 192 MHz. The hyperfine splitting of the ground state is 1,772 MHz. As a result, the spectral separation between lines that include only one of the ground

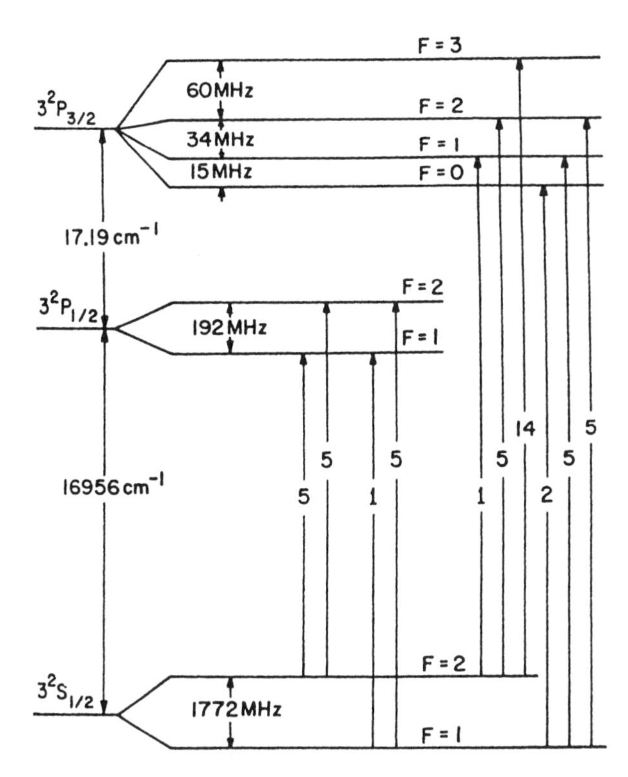

Figure 9.5 Grotrian diagram of the fine splitting of Na (not to scale). [Fairbank, Hänsch, and Schawlow 1975, © Optical Society of America, Washington, DC]

states as the end state ranges from 15 to 200 MHz, while the separation between lines that include both ground states as end states is nominally around 1,772 MHz. For comparison, the difference between the fine-split $3^2P_{3/2}$ and $3^2P_{1/2}$ states is 516 GHz or about 1,000 times larger than the hyperfine splitting. Although the splitting of the upper levels is too fine to be practically resolved, the splitting of the ground state can be readily resolved by transmitting the beam of a tunable narrowband laser through a long cell containing low-density Na vapor and observing the attenuation of the beam when its frequency coincides with the absorption frequency of the laser (Fairbank et al. 1975).

In a more practical application (Zimmermann and Miles 1980), the hyperfine splitting of the Na D lines was used as a frequency marker to determine the Doppler shift introduced by a hypersonic helium flow. In that experiment, the flow was seeded with Na vapor, which was subsequently excited by a tunable laser. Because of the Doppler effect, the excitation frequency of the moving atoms was shifted relative to the excitation frequency of stationary Na atoms. This frequency shift was measured to determine the gas flow velocity. For most gas flows, frequency shifts induced by the Doppler effect are too small and cannot be measured directly by ordinary spectrometers. However, by designing the experiment so that the Doppler shift could be matched against the frequency difference between two lines that are split by the hyperfine structure of the ground state, the absolute value of the Doppler shift and the flow velocity could be determined (see Research Problem 9.1).

9.4 The Diatomic Molecule

The principles that control the structure of atoms can also be applied to the analysis of molecules, their modes of motion, the energy associated with each mode, and the selection rules for single-photon transitions. The analysis of molecules, however, is complicated by the coupling between energy modes of the participating atoms and between the modes of motion that are specific to the molecules themselves; the interaction between the participating atoms perturbs their own states, and this perturbation depends on the number of atoms in the molecule, the type of the bond, and on the molecular motion. Since the complexity of the analysis increases rapidly with the number of atoms in the molecule, *diatomic* molecules (which include only two atoms) are the simplest to model; they are most amenable to rigorous analysis and are usually better characterized than polyatomic molecules. Of primary interest are molecules such as O_2 and N_2, the components of our atmosphere, and such intermediate combustion species as OH, NO, and CO. The first two molecules, O_2 and N_2, are *homonuclear* molecules; that is, they are composed of two identical atoms and are therefore simpler than CO or OH, which are *heteronuclear*. Despite visible differences and large variance in their bond structure, all diatomic molecules share the same modes of motion. Here we describe briefly the principles of their motion and the principal rules that determine the structure of diatomic molecules. For more detailed description see Herzberg (1989). Because of its

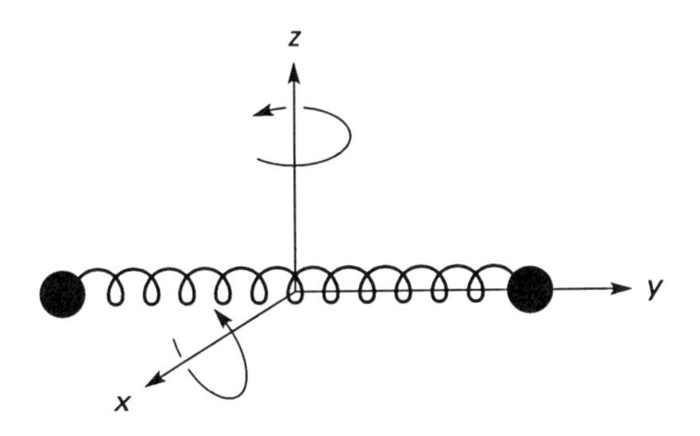

Figure 9.6 A diatomic molecule represented by an imaginary spring-connected dumbbell.

broad scope, a discussion of polyatomic molecules will not be given here; the reader is instead referred to Herzberg (1945).

Although the type of bond between the atoms of a diatomic molecule depends on the electronic structure of their outer shells, it may be viewed simplistically as an elastic oscillating spring. With a pair of atoms attached to its ends, this imaginary spring resembles a dumbbell (Figure 9.6) that oscillates longitudinally. The primary effect of the bond is to restrict the motion of the connected atoms, thereby reducing the number of their degrees of freedom. Since the bond is long relative to the radii of the nuclei of the atoms - where most of the mass is concentrated - each atom can be viewed as a "point particle" that when free can have three degrees of freedom. While the total number of degrees of freedom of two unattached atoms is six, the restriction along the bond reduces this number to five. These include three independent modes of linear translation, and two independent modes of rotation around the x and the z axes of Figure 9.6. Each of these degrees of freedom repesents a mode of energy storage (potential or kinetic) as well as a path for momentum build-up. Although rotation around the y axis in Figure 9.6 is possible, the moment of inertia and hence the rotational momentum and energy associated with it are negligible. Furthermore, since rotation around the x axis is indistinguishable from rotation around the z axis, analysis of only one rotational mode of the diatomic molecule is sufficient and only one quantum number is needed to fully define its rotation. Thus, in the most general case only four degrees of freedom, three translational and one rotational, are needed to characterize the motion of diatomic molecules. Since the bond that links the atoms represents a restriction on their motion, there can be no momentum associated with it; however, the elastic energy stored in the springlike oscillation can be significant and must be considered. The vibration, although not a degree of freedom, is therefore included in the analysis of the molecule. The following sections briefly describe the quantum mechanical aspects of the translation, rotation, and vibration of atomic molecules, starting with the simplest mode of energy storage: linear translation.

Generally, translational motion in the quantum mechanical context is described by the particle in the box with its quantized translational energy (eqn. 8.19). However, in most gaseous samples (e.g. atmospheric air), the linear motion of molecules is not confined; even if it is, the dimensions of macroscopic containers can be considered infinitely large relative to molecular dimensions. Because the energy levels in such a "box" are essentially continuous, the translational energy of molecules in practical applications depends only on the molecular mass and velocity. (The discrete nature of the translational energy does, however, play an important role in the derivation of the entropy of gaseous systems by statistical thermodynamics; see Vincenti and Kruger 1965, p. 92.) Thus, for spectroscopic applications, consideration of the translational energy of individual molecules is rarely evident. With the exception of carefully controlled experiments, the velocity of an individual molecule appears random, and in the absence of directed velocity it depends only on the gas temperature. Although translational motion can be induced or inhibited by the interaction with radiation, the most important result of the random translational motion is broadening of the absorption and emission spectral features by the Doppler shift induced by each absorbing or emitting molecule. A full discussion of this effect is deferred to Section 10.7.

Unlike the translational energy, the rotational energy of molecules is always quantized. This is similar to the quantization of the angular momentum and energy of the electron orbit, the electron spin, or the nuclear spin. To mark the molecular rotation, a new quantum number, J, is introduced. (This is the same symbol used for the total angular momentum of atoms (9.14), but the two should not be confused.) The molecular rotation, similar to rotation in the classical mechanics sense, depends on the moment of inertia and on the frequency of the rotation. The moment of inertia, in turn, depends on the masses of the atoms and the separation between them, which varies owing to the oscillation of the springlike bond. However, as a first approximation, the separation between the atoms is regarded as constant, so we can view the molecule as a *rigid rotator*. The energy of such a hypothetical rigid rotator is

$$E_{\text{rot}} = BJ(J+1), \tag{9.19}$$

(Herzberg 1989), where the rotational constant B is a molecular parameter that describes the rotational energy; it is specific to each molecule and to the energy state of its electrons in the outer shell.

The error introduced by the rigid-rotator approximation is small, but it can impose limitations on certain applications where highly accurate spectroscopic measurements (e.g., of gas temperature) are needed. To correct for this error – that is, to include the effect of the vibrational motion on the moment of inertia – certain parameters of the vibration must first be defined. Therefore, description of the nonrigid rotator will be presented after our discussion of molecular vibration.

The angular momentum of the rigid rotator, like the angular momentum of the electrons in their orbit, is a vector. By analogy to the electron it is therefore expected that, when subjected to a magnetic field, the angular momentum of the rigid rotator will point only in select azimuthal directions that are specified by the magnetic quantum number M. Like the magnetic quantum number of the atom (eqn. 9.10), the values of M are integers that cannot exceed the value of J:

$$M = J, (J-1), (J-2), \ldots, -(J-1), -J. \tag{9.20}$$

Thus, associated with J there are $2J+1$ quantum numbers M. In the absence of magnetic fields, the states specified by the M quantum numbers are indistinguishable and are therefore degenerate. Accordingly, the degeneracy of the J rotational level of the rigid rotator is $2J+1$.

The quantum numbers J and M are the only numbers needed to describe the rigid rotator when the angular momentum around the y axis (Figure 9.6) is negligible. However, occasionally the angular momentum of the electrons in their orbits may have a small component pointing along the y axis. When this component cannot be neglected, the rotator is viewed as a symmetric top spinning around the y axis simultaneously with its spin around one of the perpendicular axes. Although the resultant of these two rotations is still marked by J, it is no longer perpendicular to the internuclear axis. The new component for the rotation around the y axis is marked by a new quantum number Λ. Of course, $J > \Lambda$ and therefore the values for Λ, which are also integers, must be

$$J = \Lambda, (\Lambda+1), (\Lambda+2) + \cdots. \tag{9.21}$$

The vector associated with J for the symmetric top is different from the vector associated with J of the rigid rotator primarily because of their different orientations; the angular momentum of the rigid rotator points perpendicularly to the internuclear axis, whereas the angular momentum of the symmetric top is inclined relative to that axis. A component with a quantum number K that is perpendicular to the internuclear axis may still be defined, but since J and Λ are integers, K may not always be an integer. Although the effect of Λ on the rotational energy is small, it does have an effect on the selection rules for optical transitions and on the transition probability between rotational lines (see Section 9.5).

In its most simplistic approximation, the bond between the atoms of the diatomic molecule can be regarded as an elastic spring with a restoring force of $F = kr$, where k is the spring constant and r denotes internuclear separation. The spring is at equilibrium when the separation between the atoms is r_e. However, when this equilibrium is perturbed, the spring begins to vibrate around its equilibrium position. In order to compute the motion parameters, the spring may be viewed as an undamped harmonic oscillator, which classically is described by the following equation of motion:

$$\mu \frac{d^2(r-r_e)}{dt^2} + k(r-r_e) = 0. \tag{9.22}$$

By using the reduced mass μ (eqn. 9.2), one atom can be viewed as stationary and the entire oscillatory motion can be attributed to the other atom. The solution of (9.22) describes the oscillation of the spring around the equilibrium position r_e at a characteristic frequency of

$$\nu_{\text{osc}} = \frac{1}{2\pi}\sqrt{\frac{k}{\mu}}. \tag{9.23}$$

For the CO molecule this frequency is $\sim 6 \times 10^{13}$ Hz. This is an extremely high frequency, unattainable by any system of macroscopic dimension. On the other hand, it is comparable to the frequency of infrared radiation, thereby suggesting that part of the IR emission by CO molecules is associated with the vibrational motion.

The elastic energy stored by the spring during oscillation can be calculated by integrating the mechanical work done on it. As expected, the stored elastic energy varies with the separation between the atoms. When the spring is fully stretched, the velocity of the oscillating atoms is zero and the entire energy of the vibration is stored by the spring as elastic energy; elsewhere during the oscillation cycle, part of the vibrational energy is stored in the spring as elastic energy while the rest is the kinetic energy. Since energy is conserved, the sum of the potential and kinetic energies is a motion constant that equals the mechanical work done on the spring when fully stretched. Thus, the total vibrational energy can be calculated by the following integral:

$$V = k\int_{r_e}^{r_{\max}} (r - r_e)\,dr = \frac{1}{2}k(r_{\max} - r_e)^2,$$

where $r_{\max}$ is the amplitude of the oscillation. The result of this integral, the variation of the vibrational energy with the oscillation amplitude, can be presented graphically in an energy diagram (Figure 9.7) by a parabola with its vertex at r_e. Classically, as Figure 9.7 suggests, the oscillator can be stretched to any amplitude and can therefore have any energy. Point A on the parabola

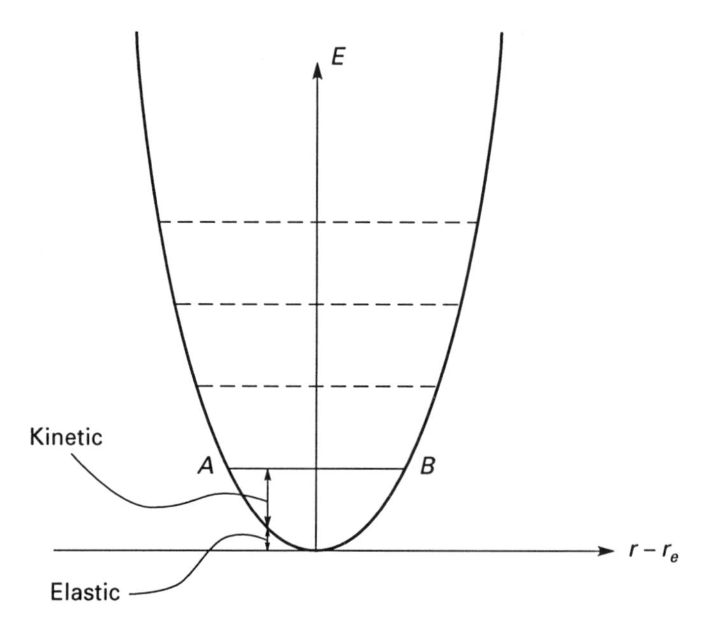

Figure 9.7 Energy-level diagram of the harmonic oscillator.

marks such a point of arbitrary energy. When at point A, the oscillator is fully compressed and its entire energy is elastically stored. However, as it begins to stretch, part of the elastic energy is released and so becomes kinetic energy, until it is fully released when the internuclear separation is r_e. From that point, as stretching continues, the kinetic energy is reclaimed by the spring until at point B the energy is again entirely elastic. Since energy is conserved throughout the entire motion, the horizontal line connecting points A and B represents the entire vibrational energy. Thus, at any point during the oscillation, the kinetic energy can be represented by the vertical line connecting that point on line AB with the parabola underneath; the elastic energy is the remainder.

Unlike the energy of classical oscillators, quantum mechanically the vibrational energy is quantized and therefore only certain oscillation amplitudes are allowed. Heuristically the parabola in Figure 9.7 can be viewed as a potential well, which is similar to the potential well presented by the box in Figure 8.11. In other words, the boundary conditions of the harmonic oscillator, like those of the particle in the box (eqn. 8.17), represent a constraint on the motion of the oscillator, which cannot extend beyond set limits. Thus, by analogy to the particle in the box, the solution of the characteristic value problem of the harmonic oscillator generates a series of characteristic functions, each with the following vibrational energy:

$$E(v) = h\nu_{\text{osc}}\left(v + \frac{1}{2}\right) \tag{9.24}$$

(Herzberg 1989), where $v = 0, 1, 2, \ldots$ are the vibrational quantum numbers and ν_{osc} is the characteristic frequency of the oscillation. Conveniently, ν_{osc} can also be obtained from the classical solution of the harmonic oscillator (eqn. 9.23). The quantized vibrational energy levels of the harmonic oscillator are presented in the energy-level diagram of Figure 9.7 by the dashed horizontal lines. Each level, like level AB, represents the exchange between the kinetic energy and the elastic energy while the total energy of that vibrational mode is conserved.

The numerical value of the emission frequency associated with vibrational transitions is usually quite large when expressed in hertz. For a more convenient presentation, wavenumber units [cm^{-1}] are used. To simplify the calculation of the emission frequency, the energy of the various molecular states are also presented in these units by

$$G(v) = \frac{E(v)}{hc} = \omega_e\left(v + \frac{1}{2}\right), \tag{9.25}$$

where $G(v)$ is called the *vibrational* term value and ω_e is the oscillation frequency (in cm^{-1}) of the harmonic oscillator. The term value can be viewed as the frequency (in cm^{-1}) of a hypothetical emission from that state to a zero energy state. Apart from a constant hc, $G(v)$ represents the energy of the vibrational state.

From (9.25), or by inspecting Figure 9.7, it can be seen that the energy difference $\Delta G(v)$ between any pair of adjacent vibrational levels is constant and is proportional to the oscillation frequency:

$$\Delta G(v) = \omega_e.$$

Thus, the emission frequency for transitions from any vibrational state to an adjacent lower state is ω_e, and all transitions involving $\Delta v = \pm 1$ are spectroscopically indistinguishable. This is a unique property of the harmonic oscillator that is *never* observed in measured vibrational spectra of diatomic molecules, where the energy difference between adjacent vibrational levels is the largest for transitions between $v = 0$ and $v = 1$ and decreases as v increases. Thus, although the harmonic oscillator model is sufficient to predict such important characteristics of molecular vibration as its characteristic frequency or the quantization of the vibrational energy, it fails to predict other important details.

Before introducing corrections to the harmonic oscillator model, consider another discrepancy between this model and the actual diatomic molecule. The parabola in Figure 9.7 describes the variation with the vibrational energy of the oscillation amplitude around r_e. Since the parabola is unbounded, the amplitude of the oscillator when stretched is unbounded as well. However, realistic molecules dissociate when stretched too far: when the vibrational energy exceeds the energy of dissociation, the separation between the molecules approaches infinity without any addition of energy. Similarly, when compressed, the energy of molecules approaches infinity as $r \to 0$ owing to electrostatic repulsion between the nuclei. A physically plausible potential well must account for both effects by including two asymptotes, a horizontal asymptote when $r \to \infty$ and a vertical asymptote for $r \to 0$. Two types of potential curves are consistent with these asymptotes (Figure 9.8): one curve, like the parabola in Figure 9.7, includes a potential well; the second curve shows energy decreasing montonically

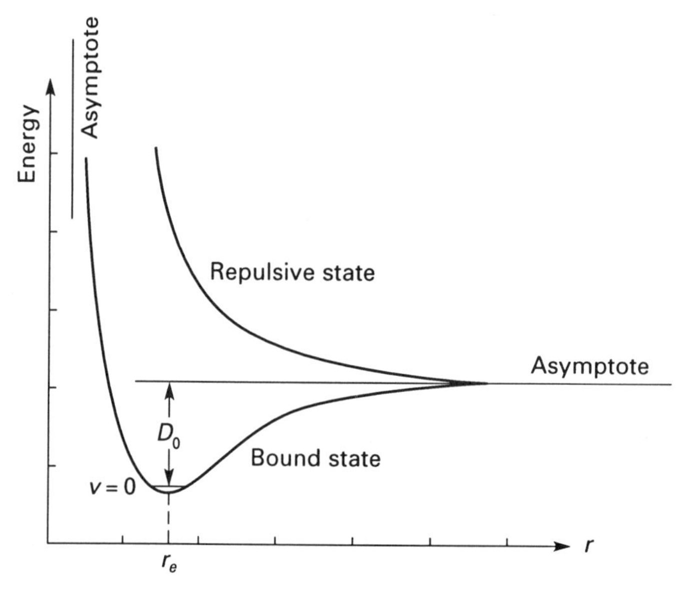

Figure 9.8 Energy-level diagram of an anharmonic oscillator showing bound and repulsive states.

with r. Both curves are physically plausible. Energy levels of stable molecules are described by curves with a potential well. These molecules dissociate only when their energy exceeds the dissociation energy D_0, where D_0 is measured from the ground vibrational level $v = 0$. Thus, vibrational and rotational levels with energy below D_0 are bound states. On the other hand, potential curves where the energy decreases monotonically with r represent unstable molecules – for example, virtual molecules that are temporarily formed during collisions between atoms of inert gases. In such a state, a molecule dissociates withcut completing even a single vibrational cycle. These states are called *repulsive* or *predissociated* states.

To model bound-state oscillations we first note that, near the bottom, the potential well resembles the parabola of Figure 9.7. Thus, it can be modeled by the equation of motion of the harmonic oscillator (eqn. 9.22) with the addition of higher-order terms to account for the deviation of the curve from a parabola as it departs from its bottom. These higher-order correction terms represent the anharmonicity of the oscillator. Near the bottom these correcting terms (and hence the anharmonicity) are small, but they increase as the vibrational energy and its amplitude increase. The term value of the vibrational levels of a bound anharmonic oscillator is

$$G(v) = \omega_e\left(v+\frac{1}{2}\right) - \omega_e x_e\left(v+\frac{1}{2}\right)^2 + \omega_e y_e\left(v+\frac{1}{2}\right)^3 + \cdots \tag{9.26}$$

(Herzberg 1989). The first term in this equation is the solution of the harmonic oscillator with ω_e its oscillation frequency, while the subsequent terms represent the higher-order correections with x_e and y_e their coefficients. When $x_e \to 0$ and $y_e \to 0$, the results of the harmonic oscillator (eqn. 9.25) are recovered.

An important distinction between the harmonic and the anharmonic oscillators is the variation with v of the energy difference between two adjacent levels. For a transition from $v+1$ to v, the emission frequency is:

$$\Delta G(v) = G(v+1) - G(v) \approx \omega_e - 2\omega_e x_e(v+1) + \cdots. \tag{9.27}$$

Clearly, as v increases the energy separation $\Delta G(v)$ between adjacent levels decreases. When these levels can no longer be resolved they are said to *converge into a continuum*. For low vibrational quantum numbers, the variation of the vibrational energy with v can be observed spectroscopically by measuring the emission, or absorption, frequencies for transitions from several vibrational states to adjacent states. By spectroscopically measuring the frequencies of two different vibrational transitions, the magnitudes of ω_e and x_e can be determined experimentally. To determine the value of higher-order terms (e.g. y_e), additional, highly resolved spectroscopic measurements are needed. Although higher-order terms can be added to (9.26) and (9.27), most applications require the use of only ω_e, x_e, and y_e. A comprehensive listing of these coefficients is presented by Herzberg (1989). Table 9.1 presents recently determined coefficients for several diatomic molecules.

Table 9.1 *Spectroscopic constants for vibrational and rotational energy levels of selected diatomic molecules in their ground and one excited electronic states*

Molecule	T_e	ω_e	$\omega_e x_e$	$\omega_e y_e$	B_e	α_e	D_0
O_2 $X\,^3\Sigma_g^-$ [a]	0	1,580.3	12.073	0.0546	1.4457	0.0157	41,260
O_2 $B\,^3\Sigma_u^-$ [b]	49,792	709.56	10.92	−0.0176	0.8187	0.0127	57,128
N_2 $X\,^1\Sigma_g^+$ [c]	0	2,358.6	14.32	−0.0023	1.9982	0.0172	78,714
N_2 $A\,^3\Sigma_u^+$ [c]	50,204	1,460.6	13.87	−0.0103	1.4546	0.0180	28,959
OH $X\,^2\Pi$ [d]	0	3,735.2	82.21		18.871	−0.714	35,419 [e]
OH $A\,^2\Sigma$ [d]	32,682	3,184.3	97.84		17.355	−0.807	51,287 [e]
NO $X\,^2\Pi$ [f]	0	1,904.0	13.97	−0.0012	1.7046	0.0178	52,427
NO $B\,^2\Pi$ [f]	45,919	1,036.9	7.460		1.076	0.0116	
CO $X\,^1\Sigma^+$ [g]	0	2,169	13.3	0.0308	1.93	0.0175	90,450
CO $A\,^1\Pi^+$ [g]	64,982	1,516	17.3		1.61	0.0223	25,532

Note: Values are given in units of cm^{-1}.
[a] Hébert, Innanen, and Nicholls (1967); see also Veseth and Lofthus (1974).
[b] Cheung et al. (1986).
[c] Lofthus and Krupenie (1977).
[d] Dieke and Crosswhite (1962).
[e] Carlone and Dalby (1969).
[f] Suchard (1975).
[g] O'Neil and Schaefer (1970).

As noted in our discussion of the rigid rotator, the vibrational and rotational motions of diatomic molecules are coupled. As the bond connecting the atoms is stretched, the moment of inertia is changed and with it the rotational energy. Similarly, the rotation affects the vibrational motion through the centrifugal force. Thus, even though (9.19) and (9.26) accurately describe independent rotational and vibrational motions, additional terms are needed to correct for the coupling between them. Of course, owing to the high frequency of oscillation (which may approach 10^{14} Hz), the correction need only account for time-averaged effects. Both corrections are small and are often neglected, but for high-resolution spectroscopy, which is used for accurate thermodynamic measurements of gas flows, they must be included. The effect of time-averaged variation on the moment of inertia induced by the vibrational motion is to modify the rotational constant. Thus, for the *vibrating rotator,* the rotational constant is

$$B_v = B_e - \alpha_e\left(v+\frac{1}{2}\right)+\cdots \tag{9.28}$$

(Herzberg 1989). To be consistent with the vibrational term value, B_v is expressed in units of cm^{-1}. To account for the effect of the centrifugal force on the vibrational motion, a new coefficient D_v (which depends on the quantum number v) is introduced:

$$D_v = \frac{4B_e^3}{\omega_e^2}+\beta_e\left(v+\frac{1}{2}\right) \tag{9.29}$$

(Herzberg 1989), where B_e, α_e, and β_e are constants that are reported in the literature for most molecules of practical interest.

The total energy of a diatomic molecule in a state prescribed by the quantum numbers v and J can be obtained by adding the vibrational and rotational term values. The total represents the energy of a diatomic molecule in its ground electronic state. It does not include the effect of electronic excitation, which must be added separately. By modifying the rotational energy equation (eqn. 9.19) by including corrections for the vibrating rotator and *rotating oscillator,* the following term value for a state that can be identified by the (v, J) quantum numbers is obtained:

$$G(v,J) = \omega_e\left(v+\frac{1}{2}\right) - \omega_e x_e\left(v+\frac{1}{2}\right)^2 + \cdots + B_v J(J+1) - D_v J^2(J+1)^2 + \cdots. \tag{9.30}$$

The discussion of the various energy modes of diatomic molecules has until now ignored the effect of the energy state of the electrons. Only modes that depend on the motion of the molecule itself were considered. Of course, the electrons in their orbits around the atom can store energy, even when in their ground state, the lowest and most stable shell. However, by measuring the molecular energy relative to the electronic ground state, term values of that electronic state can be expressed without reference to the electron energy (eqn. 9.30). When one (or more) of the electrons is excited, the entire molecular energy is increased. Furthermore, electronic excitation modifies the chemical properties of the atom and thereby modifies the strength of the bond, the equilibrium separation r_e between the atoms, the vibrational parameters, the rotational parameters, and the dissociation energy. Thus, the potential well of an electronically excited molecule is located above the potential well of the ground electronic state, its vertex is shifted laterally (to the left or to the right of the vertex of the ground state), and its shape changes relative to the shape in its ground electronic state.

Figure 9.9 is the energy-level diagram of three of the potential curves of N_2. This is a simplified version of a more complete diagram (Lofthus and Krupenie 1977) of the electronic energy levels of N_2, where potential curves for numerous electronic excitations of N_2 and N_2^+ were presented. Two of the curves in Figure 9.9 illustrate bound states, while the third is a repulsive predissociated state. If all the electrons of the molecule are in their lowest possible state, then the energy is described by the lower curve, which is identified by the term symbol $X\,^1\Sigma_g^+$. This symbol is unique, and only one electronic state in that molecule can have it. Like the term symbols for atomic levels, it describes parity and symmetry properties of the state. The capital roman letter identifies the electronic state, while the capital greek letter represents the total angular momentum of the electrons in their orbit – that is, the orbital angular momentum. The superscript to the left of the greek letter is usually the multiplicity of the state, which depends on the total spin quantum number S. The superscript to the right of the greek letter may be either + or − and the subscript may be either g or u; they identify symmetry properties of the state and are needed to determine the selection rules for transitions between electronic states (Section 9.5). Further discussion of this labeling is beyond the scope of this book.

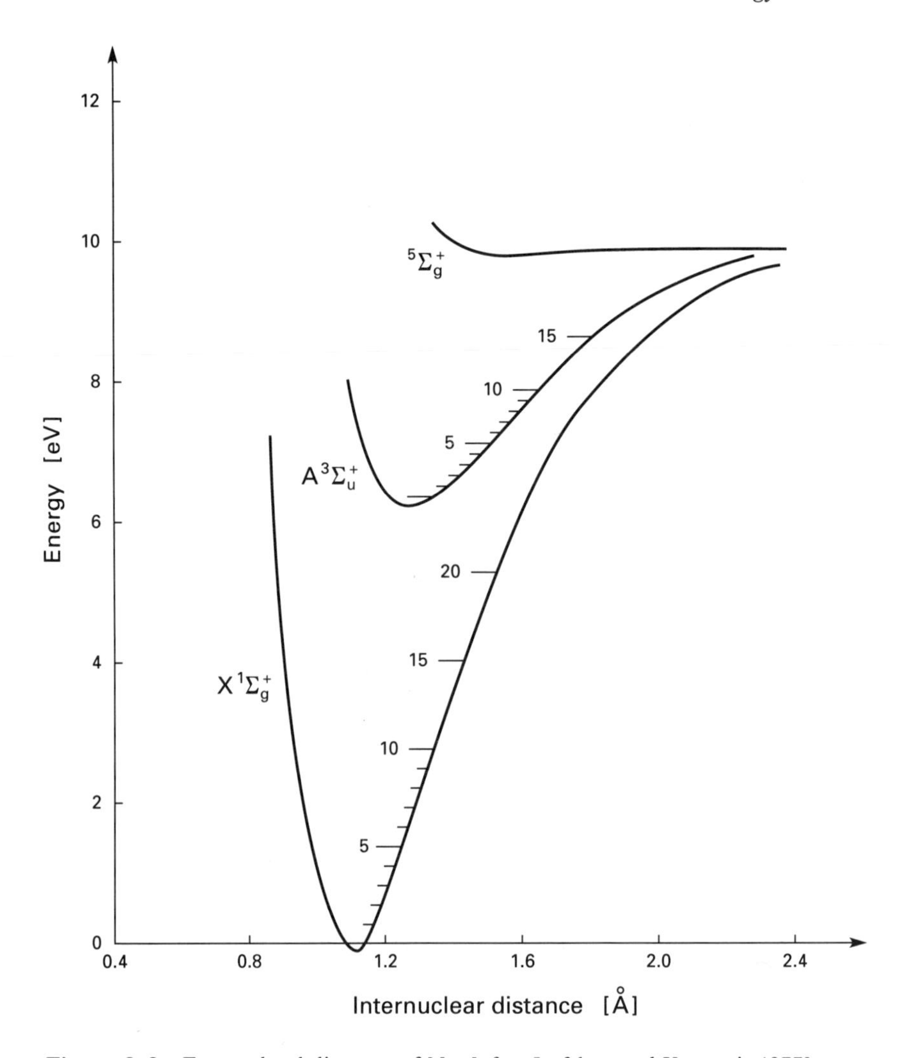

Figure 9.9 Energy-level diagram of N_2. [after Lofthus and Krupenie 1977]

Although the excited electronic state $A^3\Sigma_u^+$ in Figure 9.9 is above the ground electronic state, it is still a bound state, albeit shallower than the potential well of the ground electronic state. The energy difference [cm^{-1}] between the vertex of the excited state and the vertex of the ground state is the electronic term value. Thus, for the $A^3\Sigma_u^+$ state, $T_e = 50{,}204\ cm^{-1}$. Values of T_e for excited electronic states of several molecules are presented in Table 9.1. Although the use of T_e as a measure of the electron energy is unambiguous, it depends on locating the vertex of the potential energy curve. However, that energy point is inaccessible even when $v = 0$ (eqn. 9.30). Thus, instead of T_e, electronic states may be characterized by T_0, which is measured from $v'' = 0$ in the ground electronic state to $v' = 0$ in the upper electronic state. This is similar to D_0, which is also measured relative to the energy of $v = 0$. To convert T_0 to T_e and vice versa, the vibrational term values $G(v'=0)$ and $G(v''=0)$ are added or subtracted.

The term value for an electronically excited *ro-vibronic state* (i.e., rotational-vibrational and electronic) is:

$$G(T_e, v, J) = T_e + \omega_e\left(v+\frac{1}{2}\right) - \omega_e x_e\left(v+\frac{1}{2}\right)^2 + \cdots + B_v J(J+1) - D_v J^2(J+1)^2 + \cdots. \quad (9.31)$$

For the ground electronic state when $T_e = 0$ cm^{-1}, (9.31) is identical to (9.30).

The third electronic state in Figure 9.9, $^5\Sigma_g^+$, is a repulsive state. Without a potential well, electronic excitation to this state results in dissociation of the molecule. As the separation between the atoms increases, the energy of the molecule asymptotically approaches the dissociation energy of N_2. The difference between the energy at any point on this curve and D_0 is added after the dissociation to the translational energy of the molecular fragments. Owing to the large number of repulsive (or predissociative) states and their short lifetime, many are poorly characterized and some may not even have a roman letter assignment.

Inspection of Figure 9.9 reveals other features of the potential well of the excited electron. Most notable is the larger equilibrium separation, r_e, between the nuclei. Although this separation often increases with electronic excitation, it may also decrease. Therefore, curves for excited electronic states are shifted laterally - to the left or to the right - relative to the ground state. As will be shown in Section 9.5, this has a significant effect on the probability for optical transitions between excited electronic states. It is also evident from Figure 9.9 that the potential well of the excited bound state is shallower than the ground state, thereby implying that dissociation of an electronically excited molecule requires the addition of less energy than dissociation of the ground state. Finally, the spacing between the vibrational and rotational states varies, and so the various coefficients vary, with electronic excitation. This can be seen in Table 9.1, where the oscillation frequency of N_2 in the ground state is $\omega_e = 2{,}358.6$ cm^{-1} but only $\omega_e = 1{,}460.6$ cm^{-1} in the excited $A^3\Sigma_u^+$ state.

Energy-level diagrams are available in the literature for many molecules of practical interest, and may be used to estimate some of their optical characteristics. Parameters such as emission wavelength can be obtained (within a limited accuracy) by simple geometrical measurement. This is illustrated by the following example.

Example 9.2 Excimer lasers consist of a mixture of noble gas (such as Kr, Ar, or Xe) and a halogen (such as F_2 or Cl_2), diluted in a buffer gas (usually He). Because of their chemically inert properties, noble gases do not react with halogens when both gases are unexcited. Thus, although collisions between the halogen and the noble atoms occur, their interaction is limited in time and scope. This is represented by the repulsive energy curve, $X\,^2\Sigma^+$, in Figure 9.10, where the dissociation energy after collision between Kr and F approaches zero.

When an electrical discharge is applied to the mixture, one or both of the mixture components is excited. While excited, the noble gas is no longer inert

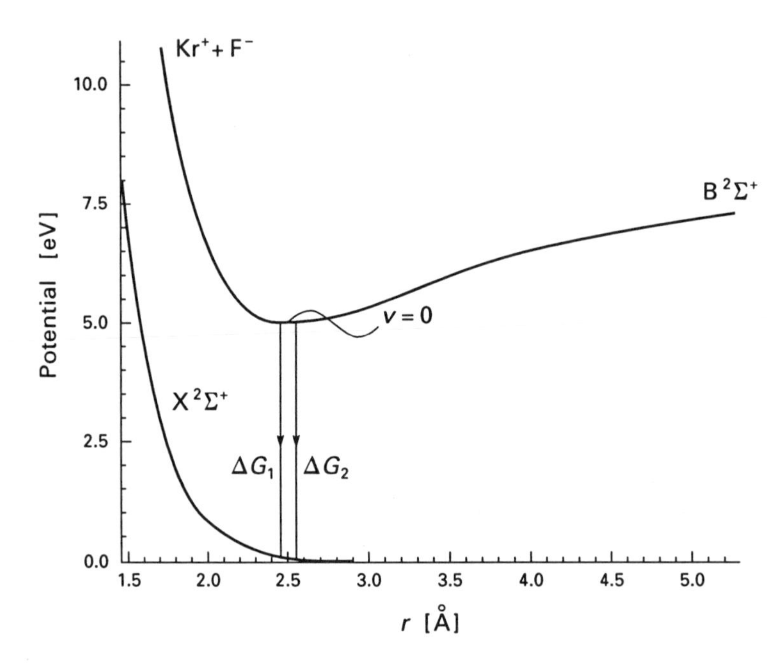

Figure 9.10 Energy-level diagram of a KrF molecule. [prepared by the author using data presented by Hay and Dunning 1977]

and can form a stable bond with the halogen atom. One of these states for the KrF molecule is $B^2\Sigma^+$, which combines Kr^+ with F^- (Hay and Dunning 1977). This state with its potential well (Figure 9.10) includes bound vibrational levels with $\omega_e = 340\ \text{cm}^{-1}$. However, when the energy of this state is released – for example, by optical emission at nominally $\lambda = 248$ nm – the molecule returns to the $X^2\Sigma^+$ state. Such transition may occur at any time during the vibrational motion, which for $v = 0$ is depicted by a horizontal line at the bottom of the well. Owing to the curvature of the $X^2\Sigma^+$ state, the energy difference between the excited state and the ground state depends on the separation between the nuclei. Thus, the emission wavelength depends on the phase of the vibration at the time of emission. The extreme value of the emitted photon energies is marked by the vertical lines ΔG_1 and ΔG_2.

Using this graphical presentation, find the range of emission wavelengths of the KrF molecule. Compare this result with calculation of the extremes of the emission range using the potentials presented in Table 9.2 for three different nuclear separations of both states (Hay and Dunning 1977).

Solution Although the emission range of the KrF molecule can be obtained by direct graphical measurements, an error of at least ~1% is introduced by the limited accuracy of drafting devices. The reader is encouraged to find the emission range from such measurements. Here we use the data of Table 9.2 to calculate that range. The point at $r = 2.51$ Å is at the vertex of the bound potential curve. Therefore, the term value of the first vibrational state ($v = 0$) is:

Table 9.2 *Potentials for selected points on the potential curve of the KrF molecule*

r [Å]	$X\,^2\Sigma^+$ [cm^{-1}]	$B^2\Sigma^+$ [cm^{-1}]
2.38	1,305	41,341
2.51	765	40,858
2.65	443	41,083

$$G(T_e, v) \approx T_e + \frac{\omega_e}{2} = 40{,}858 + 170 = 41{,}028 \text{ cm}^{-1}.$$

These are also the term values G_1' and G_2' at both ends of the vibrational cycle. By interpolation of the data in Table 9.2 for the $B^2\Sigma^+$ state, the internuclear separations at maximum amplitude are found to be $r_1 = 2.46$ Å and $r_2 = 2.62$ Å. At these separations, the potentials of the repulsive state $X\,^2\Sigma^+$ are $G_1'' = 955$ cm^{-1} and $G_2'' = 522$ cm^{-1}. The extremes of the emission frequencies are therefore:

$$\Delta G_1 = G_1' - G_1'' = 41{,}028 - 955 = 40{,}073 \text{ cm}^{-1};$$

$$\Delta G_2 = G_2' - G_2'' = 41{,}028 - 522 = 40{,}506 \text{ cm}^{-1}.$$

For comparison, the experimentally measured emission by KrF lasers ranges from 40,180 to 40,360 cm^{-1}. ■

To accurately calculate the emission wavelengths of ro-vibronic transitions (i.e., transitions that involve simultaneous change of electronic, vibrational, and rotational quantum numbers), the term values of the end states are calculated using (9.31). The spectroscopic constants for these calculations are widely available (see e.g. Herzberg 1989). Table 9.1 presents such constants for several molecules at their ground electronic and one excited electronic states. The dissociation energy D_0, measured from $v = 0$ of these electronic molecular states, is also included. However, the value of β_e (eqn. 9.29) is extremely small and was therefore omitted. Thus, when $J \gg 1$ and when correction for the rotating oscillator is needed, only the first and the larger term in (9.29) should be used.

With all the energy modes of the diatomic molecule now identified, it is instructive to note their relative magnitude. Although the following comparison does not apply to all molecules and to all their energy levels, it is applicable to many molecules at or near their ground state. As will be seen, the least energetic mode is the rotation, followed by translation, vibration, and by the most energetic, the electron orbital energy.

Example 9.3 Find the energy (in cm^{-1}) required to excite the O_2 molecule from its ground ro-vibronic state, $X\,^3\Sigma_g^-$ ($v''=0$, $J''=1$), to $B\,^3\Sigma_u^-$ ($v'=1$, $J''=3$). Evaluate separately the contributions of the electronic excitation, the vibrational, and the rotational excitations, and compare them to the translational energy when $T = 300$ K. Assume that the kinetic energy of a molecule can be calculated from the expression for the average translational energy of perfectly elastic particles $e_{tr} = \frac{3}{2}kT$.

Solution The problem specifies an excitation in which the electron energy, the vibrational energy, and the rotational energy of the molecule are increased simultaneously by the absorption of a single photon. However, the same result could be obtained in three steps, raising the energy of each mode separately. The total energy of excitation is then obtained by summing the individual contributions. Thus, the change in the vibrational energy can be obtained using the first part of (9.30) as follows:

$$\Delta G(v) \approx \left[\omega_e'\left(v'+\frac{1}{2}\right)-\omega_e'x_e'\left(v'+\frac{1}{2}\right)^2\right]-\left[\omega_e''\left(v''+\frac{1}{2}\right)-\omega_e''x_e''\left(v''+\frac{1}{2}\right)^2\right],$$

where v', ω_e', and x_e' represent the parameters of the upper state and v'', ω_e'', and x_e'' those of the lower state. Using the spectroscopic constants in Table 9.1 we obtain for $\Delta G(v)$, the change in the vibrational energy,

$$\Delta G(v) = 295.74 \text{ cm}^{-1}.$$

Similarly, from the second part of (9.30) and the coefficients of Table 9.1 we have for $\Delta G(J)$, the change in the rotational energy,

$$\Delta G(J) = G(3') - G(1'') = 9.60 - 2.86 = 6.74 \text{ cm}^{-1}.$$

The effect of the centrifugal forces $D_v \approx 10^{-7}$ cm^{-1} is negligible for the parameters of this problem. The last component is the electron excitation energy,

$$T_e = 49{,}792 \text{ cm}^{-1},$$

which is obtained directly from Table 9.1.

For comparison, the translational energy is

$$e_{tr} = \frac{3}{2}\frac{kT}{hc_0} = \frac{3}{2}\frac{(1.38\times10^{-23})300}{(6.63\times10^{-34})(3\times10^{10})} = 312.22 \text{ cm}^{-1}.$$

To convert it to wavenumber units [cm^{-1}], e_{tr} was divided by $h = 6.63\times10^{-34}$ J-s and $c_0 = 3\times10^{10}$ cm/s.

Since the translational energy is not affected by optical excitation, the total excitation energy $\Delta G(T_e, v, J)$ is obtained by summing the electronic, vibrational, and rotational energies:

$$\Delta G(T_e, v, J) = 49{,}792 + 295.74 + 6.74 = 50{,}094.48 \text{ cm}^{-1}.$$

The wavelength of a photon with this frequency is

$$\lambda = \frac{10^7}{50{,}094.5} = 199.62 \text{ nm};$$

this is an ultraviolet radiation.

Note that the rotational energy is significantly smaller than either the vibrational energy or the electronic excitation energy. Accordingly, the spacing between adjacent levels is the smallest for rotational states, larger for vibrational states, and largest for electronic states. Because of their high density, rotational states are rarely shown in energy-level diagrams such as Figure 9.9, where their presence is only implied.

Although the translational energy is relatively large, it cannot be excited directly by optical methods. Usually, energy from other energy modes is funneled to the translational mode through intermolecular collisions. Therefore, in most spectroscopic studies the effect of the translational energy (with the exception of its effects on line broadening) is not considered. ■

Although the molecular term value (eqn. 9.31) is a significant refinement of the simplified term values of the rigid rotator or the harmonic oscillator, it still neglects important parameters that influence the energy and degeneracy of molecular states and the probability of transitions between them. One important consideration that will be mentioned here only briefly is the coupling between the angular momenta of the molecule. There exist rules that determine the coupling order of the nuclear angular momentum, the orbital quantum momentum, the spin angular momentum, and the molecular angular momentum. These are similar to the coupling rules between the angular momenta of atoms (the LS coupling of eqn. 9.14) and may vary from molecule to molecule or even between energy levels of the same molecule. For illustration, the strongest coupling for O_2 in its ground state is between the component of the orbital angular momentum Λ along the y axis and the perpendicular component of the nuclear angular momentum K. Therefore, Λ is added to K to form a total molecular angular momentum. From (9.21), the resultant should be labeled with a J quantum number. However, since the correct coupling sequence for this molecule requires that the total electron spin S also be added, the resultant of the first addition is assigned a new quantum number N; that is, $N = K + \Lambda$. The total quantum number is then $J = N + S$. Other diatomic molecules (e.g. OH) obey different coupling sequences. Furthermore, the coupling sequence may change with electronic excitation. These coupling methods were first suggested by Hund and are known as *Hund's cases* (Herzberg 1989, p. 219). The coupling of the angular momenta of O_2 is consistent with Hund's case (b). Thus, as for atoms, for each N there is (eqn. 9.14) a multiplicity of $2S+1$, which often takes the values 1, 2, and 3. Because of the electric and magnetic fields induced by the atoms, this multiplicity results in split energy levels: when $S = 1$, each rotational level with a select quantum number N is split into three closely spaced lines. The primary use of the various Hund cases is to calculate the fine structure of molecular states and to specify the allowed transitions between them. An illustration of the effect of the Hund coupling on the rotational energy levels of O_2 and on its electronic transitions will be presented (Figure 9.12) in the next section, where the selection rules for molecular transitions are discussed.

9.5 Selection Rules for Single-Photon Transitions in Diatomic Molecules

In Section 9.3 it was shown that the probability for optical transitions between certain atomic states is limited by selection rules (eqns. 9.15 and 9.16).

Similar rules can be found to identify the allowed single-photon transitions between molecular states. However, owing to the complexity of molecular structure relative to atomic structure, the selection rules for single-photon molecular transitions are more involved and must cover additional criteria for rotational, vibrational, and electronic transitions. To further complicate matters, rules for pure transitions (rotational–rotational, vibrational–vibrational, etc.) are different from the rules for mixed transitions (e.g., ro-vibrational transitions that involve simultaneous change in the rotational and vibrational quantum numbers). We will present here a brief compilation of these rules; for a comprehensive review, see Herzberg (1989).

Single-photon transitions that do not involve excitation of an electron are responsible for the infrared spectrum of diatomic (and also polyatomic) molecules. These are typically transitions that occur between vibrational and rotational levels in the ground electronic state of molecules. Although infrared emission may also occur in transitions between electronic states, the infrared spectrum of diatomic and polyatomic molecules is usually associated with transitions within the same electronic state. In most vibrational transitions, the change in the vibrational quantum number is $\Delta v = \pm 1$. To estimate the wavelength of such a transition consider NO, a diatomic molecule with typical molecular constants. In its ground electronic state, the emission frequency for vibrational transition of NO is approximately 1,900 cm^{-1} – corresponding to $\lambda =$ 5.26 μm – well in the IR range of the spectrum. For pure rotational transitions in NO the emission is at only $\sim$10 cm^{-1}. When combined, mixed vibrational-rotational transitions are thus also in the infrared range. Similarly, the greenhouse effect in the earth's atmosphere is the result of the infrared absorption spectrum of CO_2. These triatomic molecules are transparent to the visible portion of the solar spectrum but absorb parts of the infrared radiation emitted by the warm surface of the earth. Thus, when the concentration of CO_2 in the atmosphere increases, heating of the surface of the earth by solar radiation remains unimpeded but a larger fraction of the earth's IR emission is trapped by the increased CO_2 concentration, thereby preventing efficient cooling of the surface and the atmosphere. Similarly, H_2O vapor is a strong IR absorber. Unlike triatomic molecules or diatomic molecules such as NO, there exist a large group of diatomic molecules (e.g. O_2 and N_2) that do not have such an IR spectrum. Thus, a fundamental rule that identifies such molecules must be presented before formulating the specific rules that identify allowed transitions.

To emit (or absorb) electromagnetic radiation, a source must have a dipole moment (Section 3.6). Without the oscillation of such a dipole moment, there can be no electromagnetic wave (eqn. 4.21). This is evident from classical considerations, but can also be derived from quantum mechanical concepts. A dipole moment may be permanent or induced. Permanent dipoles are formed by structural asymmetries that distort electronic orbitals and tend to raise the concentration of electronic charges at one side of the molecule relative to the other. Permanent dipoles occur in heteronuclear molecules (NO, OH, CO, etc.), where uneven sizes of the two nuclei influence the electronic orbits and their distribution around the molecule. By contrast, homonuclear molecules (O_2, N_2, etc.) are perfectly symmetric, and accordingly their charges – positive and negative – are symmetrically distributed. Thus, homonuclear molecules in their ground electronic state do not have a dipole moment. Consequently, the probability of single-photon emission or dipole radiation between vibrational and rotational levels of homonuclear molecules in their ground state approaches zero, and infrared spectra for these molecules are either absent or extremely weak. On the other hand, when subjected to an intense magnetic or electrostatic field, a dipole moment can be induced in these otherwise unpolarized molecules, and a weak infrared spectrum may occur. Alternatively, when electronically excited or when ionized, homonuclear molecules have a dipole moment and so dipole radiation or dipole absorption become possible.

The selection rules that control the infrared spectra of heteronuclear molecules are dominated by their ideal configurations; that is, the selection rules for harmonic oscillators and rigid rotators determine the overall behavior of these molecules in absorption and emission of a single photon. The corrections for anharmonicity or the vibrating rotator have only a secondary effect on the probability for transition. For the harmonic oscillator, selection rules require that for a single-photon (i.e. dipole-allowed) interaction, the change in the vibrational quantum numbers between the upper state v' and the lower state v'' is

$$\Delta v = v' - v'' = \pm 1. \tag{9.32}$$

Thus, from $v''=0$ a heteronuclear molecule (or an electronically excited homonuclear molecule) can only absorb a photon when its energy is raised within the same electronic state by one vibrational quantum number. The frequency of such a transition, in the harmonic oscillator approximation, is ω_e (eqn. 9.27). At higher vibrational levels, both an increase or decrease by one vibrational quantum number are possible by absorption or emission, respectively. This rule applies also (though less rigorously) to anharmonic oscillators and to rotating oscillators. Therefore, transitions with $\Delta v = \pm 2$ or $\Delta v = \pm 3$, which are the "overtones" of the fundamental transition of $\Delta v = \pm 1$, are possible, albeit at an extremely low probability.

For dipole-allowed transitions between the rotational levels of the rigid rotator, the change in the rotational quantum numbers between the upper state J' and the lower state J'' is

$$\Delta J = J' - J'' = \pm 1. \tag{9.33}$$

This rule applies also (and rigorously) to vibrating rotators. However, when effects of the symmetric top are introduced, this selection rule is modified to read:

$$\begin{aligned} &\Delta J = \pm 1 && \text{when } \Lambda = 0; \\ &\Delta J = 0, \pm 1 && \text{when } \Lambda \neq 0. \end{aligned} \tag{9.34}$$

The energy difference between adjacent vibrational levels is generally much larger than the energy difference between adjacent rotational states. Therefore, in any vibrational state, a molecule can be at any one of numerous rotational levels and yet remain energetically below the next vibrational state. When a vibrational transition takes place, it may also be accompanied by a change in the rotational quantum number. Of course, the energy change for vibrational transitions accompanied by a change of $\Delta J = 1$ is larger than the energy change when $\Delta J = 0$ or $\Delta J = -1$. Therefore, vibrational spectra of diatomic molecules can be divided into three groups of lines, or *branches,* that are associated with $\Delta J = J' - J'' = -1$, 0, and 1; these are labeled the P, Q, and R branches, respectively. However, for most diatomic molecules $\Lambda = 0$, so according to (9.34) the Q branch (i.e. $\Delta J = 0$) is missing. The NO molecule is one of the few cases where a Q branch is observed. The following example illustrates the grouping of the rotational lines in the vibrational spectrum of NO into three branches.

Example 9.4 Using the data in Table 9.1, calculate the frequencies of the absorption lines between two vibrational levels of gaseous NO. Although at room temperature most of the rotational levels and some of the vibrational states may be occupied by a large number of molecules, limit the calculation to absorption by molecules that occupy the $J'' = 0$–10 levels in the $v'' = 0$ state. Present your results graphically by assigning a vertical line of fixed height on the frequency axis for each absorption transition.

Solution The selection rules of the harmonic as well as the anharmonic oscillator restrict excitation by absorption to $\Delta v = 1$, so the frequency for pure vibrational excitation is

$$\nu_0 = G(v''+1) - G(v'') \approx \omega_e - 2\omega_e x_e(v''+1)$$

(eqn. 9.27). Substituting data from Table 9.1 for the $X\,^2\Pi$ state, we find that the absorption frequency of NO in the ground electronic state and without any change in the rotational energy is $\nu_0 = 1{,}876.06\ \text{cm}^{-1}$.

For the P branch, the frequency shift relative to ν_0 induced by a change in the rotational quantum number is calculated using the rotational component of (9.30) and the rotational constants of the vibrating rotator (eqn. 9.28):

$$\Delta\nu_P = B'_v J'(J'+1) - B''_v J''(J''+1) = B'_v(J''-1)J'' - B''_v J''(J''+1).$$

Similarly, for the Q branch the frequency shift relative to ν_0 due to the rotational transition is

$$\Delta\nu_Q = (B'_v - B''_v)J'' + (B'_v - B''_v)J''^2,$$

and for the R branch the frequency shift relative to ν_0 due to the rotational transition is

$$\Delta\nu_R = B'_v(J''+1)(J''+2) - B''_v J''(J''+1).$$

The effects of the centrifugal force associated with D_v (eqn. 9.29) have been neglected relative to the other terms. The frequency of a line in the P branch is $\nu_P = \nu_0 + \Delta\nu_P$; in the Q branch, $\nu_Q = \nu_0 + \Delta\nu_Q$; and in the R branch, $\nu_R = \nu_0 + \Delta\nu_R$. However, for the sake of simplicity, only the shifts of the rotational

Table 9.3 *Absorption frequencies relative to ν_0 for several rotational levels of NO in the* $X\,^2\Pi$ ($v''=0$) *state*

J''	$P(J'')$	$Q(J'')$	$R(J'')$
0		0	3.356
1	−3.391	−0.0356	6.676
3	−10.28	−0.2136	13.21
5	−17.31	−0.534	19.60
7	−24.49	−0.997	25.85
10	−35.51	−1.96	34.96

Note: Values are given in units of cm^{-1}.

branches relative to ν_0 were calculated, and are presented in Table 9.3 for several J'' quantum numbers.

Note that the transitions are identified by the quantum number of the lower rotational state J''. Although this choice is particularly convenient for absorption spectroscopy (where all transitions originate at the lower level), it has been adopted also for other spectroscopic applications where emissions from excited states occur. The only rotational line that preserves the fundamental vibrational frequency is Q(0). This holds for all diatomic molecules, even when the effect of the centrifugal force is included. Thus, for molecules without a Q branch (i.e., when $\Lambda = 0$), there are no pure vibrational transitions. The remaining lines of the Q branch are densely clustered next to the Q(0) line. The P branch occupies the low-frequency side of the spectrum, while the R branch is at the high-frequency side. Since the separation between most of the lines of the Q branch is smaller than their bandwidth, they overlap each other and their absorption (or emission) are superimposed. This results in a stronger absorption (or emission) by the Q branch even if all other parameters, such as the population of the absorbing states, are identical to the other branches.

The bar chart (Figure 9.11) illustrates the rotational structure of the vibrational absorption of NO. In reality, the height and width of each line depend on the thermodynamic and spectroscopic properties of the gas sample. Because we have not yet considered such properties, each transition was represented by a single thin line with fixed height. However, to depict the effect of clustering of transitions in the Q branch, lines that are separated by less than 0.3 cm^{-1} were stacked one above the other, thereby making them appear taller than lines in the other two branches.

For comparison, an absorption spectrum of NO is also presented (Gillette and Eyster 1939). Owing to its strong absorption, the Q branch in this graph was recorded when the partial pressure of NO was lower than when the other branches were recorded. However, the shape of the Q branch and the location of the rotational lines of the R branch are matched well by the bar chart. The splitting of the rotational lines in the P branch is caused by overlapping transitions from an adjacent electronic state. ■

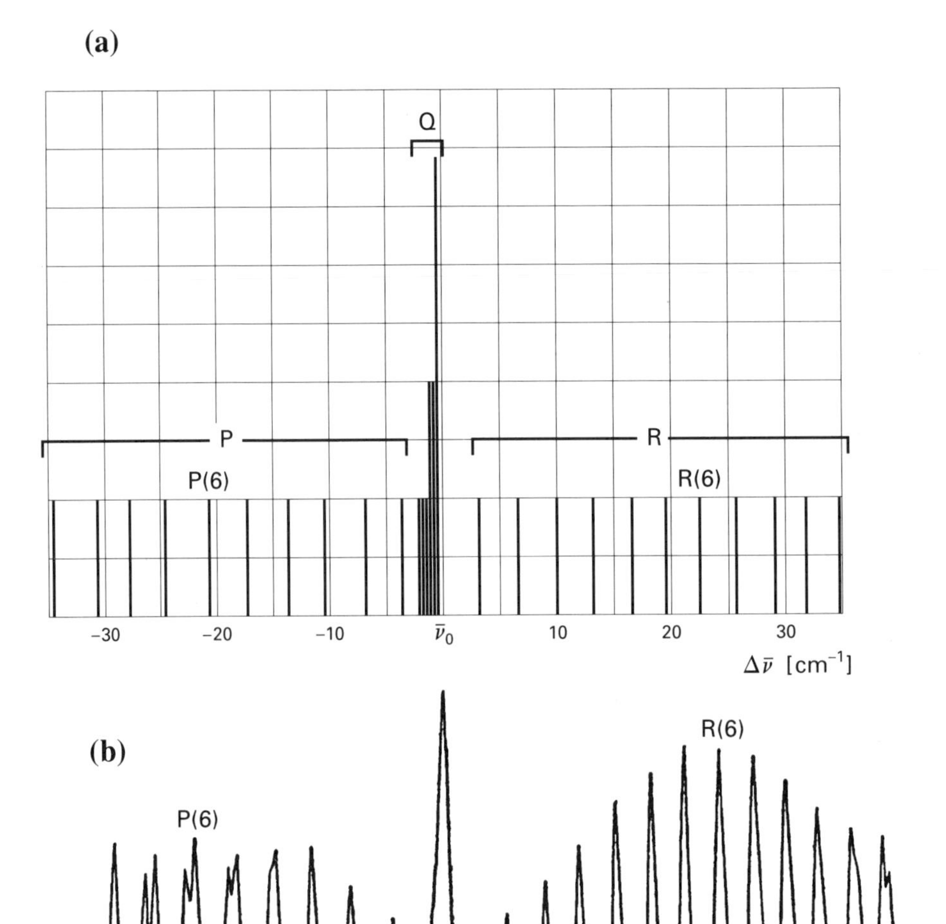

Figure 9.11 Comparison between (**a**) bar chart of the rotational spectrum of NO and (**b**) the measured spectrum. [Gillette and Eyster 1939, © American Physical Society, College Park, MD] Since the rotational term values for NO are slightly different from those of (9.30), there may be a discrepancy of up to 2 cm^{-1} between the results of this example and those of Gillette and Eyster.

Although the discussion here of the properties of molecular energy levels did not require the use of molecular wave functions, it is instructive to note that the selection rules are determined by symmetry properties of these functions. Usually, the symmetry of the initial and final states determine if a transition involving certain modes of energy exchange is allowed. Thus, the symmetry properties of the end states that are required for single-photon transitions are different from the symmetry properties for transitions involving multi-photon exchange or collisional energy exchange between molecules. Of particular interest with regard to symmetry are the homonuclear molecules with $\Lambda = 0$. Because of their exceptionally high structural symmetry, these molecules have wave functions that are either symmetric or antisymmetric; that is, inversion of a homonuclear molecule around any axis may at most result in a sign change of its wave function. For these molecules, the symmetry of the wave functions of the rotational states alternates between symmetric and antisymmetric. Thus, if states with even J quantum numbers are symmetric then states with odd J quantum numbers are antisymmetric, and vice versa. For these molecules, a general selection rule is that states with symmetric wave functions cannot undergo transitions to states with antisymmetric functions, that is,

symmetric $\nleftrightarrow$ antisymmetric.

This is a strict prohibition that holds for transitions caused by either collisions or radiation. Thus, such transitions as single-photon excitation, where the additional requirement that $\Delta J = \pm 1$ (eqn. 9.34) must be met, are restricted. On the other hand, nondipole transitions - where $\Delta J = \pm 2$ is allowed - preserve the symmetry of the wave function and can therefore be induced. Accordingly, single-photon rotational transitions are not observed in the spectra of homonuclear molecules, whereas multi-photon transitions (e.g. Raman scattering) are observed (see Section 11.5).

The symmetry properties of the rotational states of homonuclear molecules and the corresponding selection rule are responsible for another unique feature of their rotational spectra. It is evident from the previous discussion that once a molecule is created with a symmetric rotational wave function, it cannot be transformed into an antisymmetric wave function merely by undergoing a transition to such a state. Thus, rotationally symmetric molecules are segregated from rotationally antisymmetric molecules; if one group is empty, molecules from the other group cannot be transformed to make up for molecules in the empty group. Particularly notable in this regard are O_2 molecules. Owing to their electronic structure, the wave functions of O_2 molecules in the ground electronic state satisfy the symmetry characteristics of states with odd J values only. Therefore, only these levels can be populated, while states with even J quantum numbers remain empty. This is visible in all rotational spectra of O_2, where transitions from the ground electronic state show lines with odd J quantum numbers while lines for transitions from levels with even J quantum numbers - whether by Raman scattering, dipole radiation, or any other multi-photon process - are missing. Because collisional processes and other energy transfer mechanisms cannot fill the unpopulated states, the structure of alternatively missing lines is permanent.

The symmetry properties of the end states also control the dipole-allowed transitions between electronic states of molecules. Owing to their important role, the symmetry properties are represented in the term symbol of each electronic state. Generally, electronic wave functions possess two types of symmetry properties; one type of symmetry is marked by a superscript + or − to the right of the capital greek letter that symbolizes the total angular momentum, while the other symmetry property is marked by a right subscript of g or u. Although most selection rules for dipole-allowed transitions depend on the Hund coupling case that prevails for the interacting electronic states, the following (somewhat general) rules for single-photon transitions between electronic levels of homonuclear molecules can be stated:

$$g \leftrightarrow u, \quad g \not\leftrightarrow g, \quad u \not\leftrightarrow u. \tag{9.35}$$

Thus, for O_2, the strong Schumann–Runge band in which a single photon can be absorbed by transition from $X\,^3\Sigma_g^-$ to $B\,^3\Sigma_u^-$ is allowed. Although the right superscript is unchanged in this transition, the rule implied by it, $\Sigma^- \leftrightarrow \Sigma^-$, is not general and holds only for Hund cases (a) and (b). Similarly, rules that specify the change in the orbital angular momentum - that is, dipole-allowed transitions between states with different greek letters - are not general. On the other hand, the Schumann–Runge band is compatible with yet another general rule:

$$\Delta S = 0. \tag{9.36}$$

This rule implies that only states with the same multiplicity can combine with each other. Thus, for nitrogen, the $A^3\Sigma_u^+$–$X\,^1\Sigma_g^+$ transition is dipole-unallowed because the multiplicity of the states is changed - even though it includes states with u and g subscripts and identical superscripts. A more detailed discussion of the selection rules for electronic transitions of various molecules and Hund cases is presented by Herzberg (1989, p. 240).

The selection rules for dipole-allowed transitions between molecular rotational levels that belong to different electronic states also depend on the Hund cases that apply to these states. Furthermore, when two states with two different Hund cases are coupled by a dipole transition, the selection rules may need to be specified separately. To illustrate one possible group of rules, consider Hund's case (b) describing the O_2 molecule. The selection rules for rotational transitions between electronic states that comply with this case are:

$$\Delta N = 0, \pm 1 \quad \text{but} \quad \Delta N \neq 0 \text{ for } \Sigma \leftrightarrow \Sigma \text{ transitions.} \tag{9.37}$$

Here $N = K + \Lambda$ is the sum of the components of the angular momentum perpendicular and parallel (respectively) to the internuclear axis. (With O_2, for example, $N = K$ because $\Lambda = 0$.) Because $\Delta N = 0$ is unallowed for the $\Sigma \leftrightarrow \Sigma$ dipole transitions, the Q branch is missing from the Schumann–Runge band and only the P and R branches are observed. In addition, although coupling between N and S (or between K and S for the $X\,^3\Sigma_g^-$ and $B\,^3\Sigma_u^-$ states of O_2) is weak, they can be added to form a total quantum number J. For O_2 (where $S = 1$), for every N quantum number there exist three quantum numbers $J = N+1$, N, and $N-1$ corresponding to three hyperfine split energy states. Transitions between these hyperfine split states occur primarily when $\Delta J = \Delta N$. Although transitions with $\Delta J \neq \Delta N$ do occur, they are much weaker than the $\Delta J = \Delta K$ transitions and are therefore called *satellite lines*. A diagram illustrating the hyperfine splitting and the primary allowed transitions of the Schumann–Runge band of O_2 is presented in Figure 9.12.

Selection rules for dipole-allowed transitions between the vibrational levels of different electronic states (i.e., for vibronic transitions) are less rigorous than

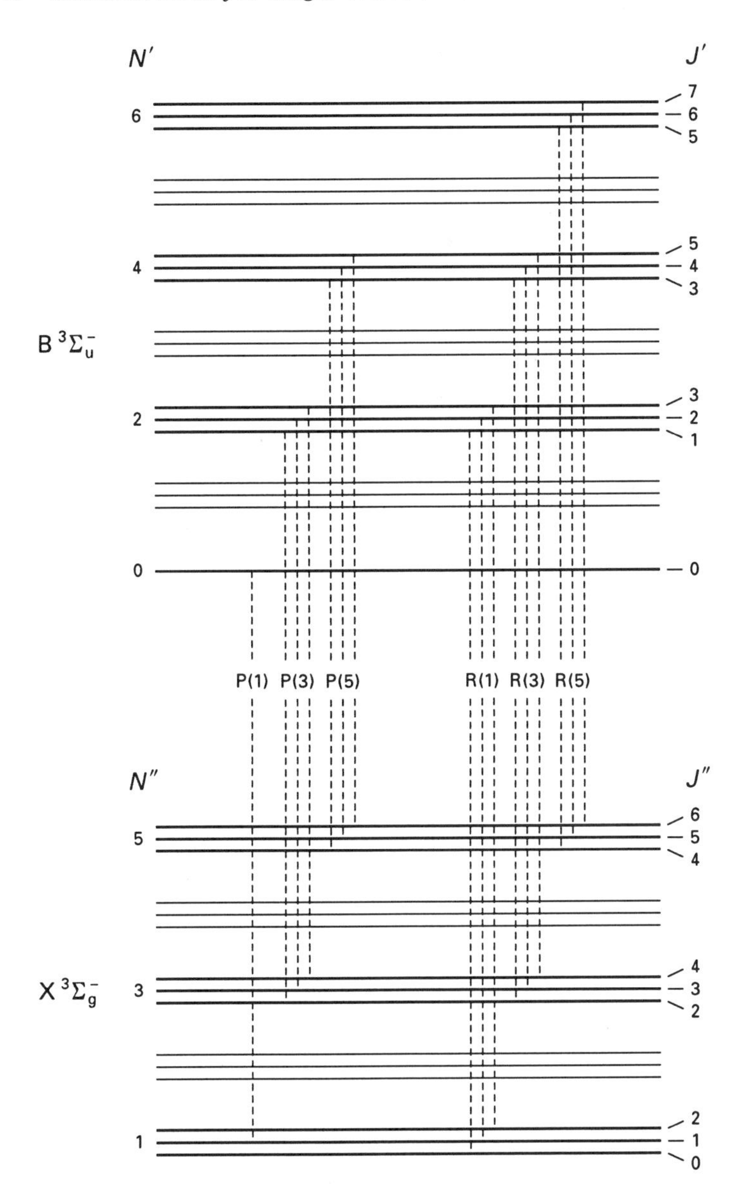

Figure 9.12 Diagram illustrating allowed rotational transitions in the Schumann–Runge band of O_2.

the rules for vibrational transitions within the same electronic state. Assuming that the symmetry requirements of the electronic and rotational wave functions are satisfied, transitions between the vibrational states are limited primarily by mechanical constraints. Thus, owing to their finite inertia, the velocity of the nuclei cannot change abruptly during a transition. Similarly, the separation between the nuclei after the transitions must remain comparable to their separation before it. Because of the first restriction, transitions occur mostly when molecules are near the ends of their oscillation cycles, when the velocity of the nuclei approaches zero. The second restriction favors transition between vibrational states with similar internuclear separation at one end of the oscillation cycle. This is the *Franck–Condon principle*, which can be stated as follows (Herzberg 1989):

> *The electron "jump" in a molecule takes place so rapidly relative to the vibrational motion that immediately afterwards the nuclei still have nearly the same relative position and velocity as before the "jump".*

According to this principle, transitions between any vibrational level in one electronic state to any other vibrational level in another electronic state may take place as long as the electronic or rotational transitions are allowed. However, some vibronic transitions are more likely to occur than others. To illustrate how the Franck–Condon principle affects ro-vibronic transitions, consider the O_2 molecule in the $B\,^3\Sigma_u^-$ ($v'=4$) state (Figure 9.13). This vibronic state can be excited from the ground vibronic state, $X\,^3\Sigma_g^-$ ($v''=0$), by single-photon absorption of an ArF laser beam at nominally 193 nm. Once excited, the molecule can relax back to the ground electronic state while emitting a photon. Although the excitation took the molecule from $v''=0$ to $v'=4$, this is not the most probable transition. A more likely transition (though at a different wavelength) combines $v'=4$ with vibrational states that – at any of their oscillation endpoints – have nearly the same internuclear separation as one of the oscillation endpoints of $v'=4$. This is illustrated by the vertical line connecting the

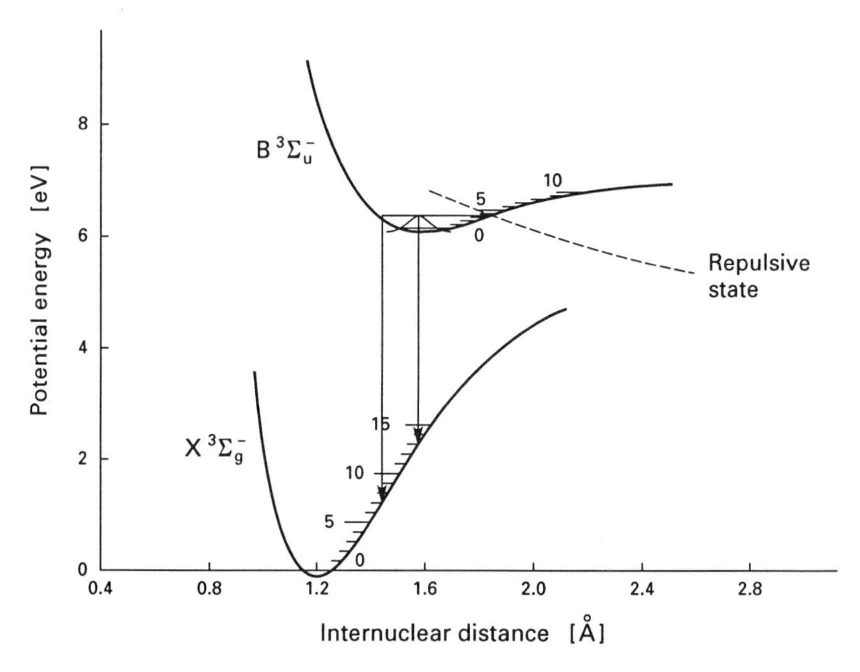

Figure 9.13 Energy-level diagram of O_2 in the $X\,^3\Sigma_g^-$ and the $B\,^3\Sigma_u^-$ electronic states, showing vibronic transitions from $v'=0$ and 4. [after Krupenie 1972]

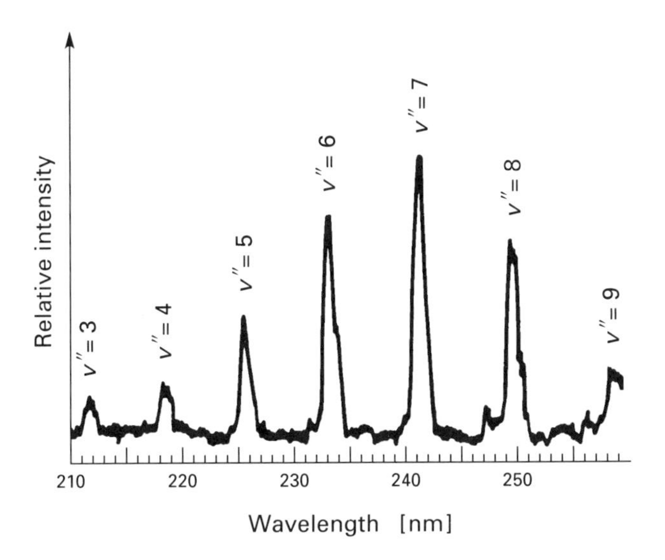

Figure 9.14 Dispersion spectrum of O_2 following an excitation of the $B\,^3\Sigma_u^-$ state. [Laufer, McKenzie, and Huo 1988, © Optical Society of America, Washington, DC]

left endpoint of $B\,^3\Sigma_u^-$ $(v'=4)$ with $X\,^3\Sigma_g^-$ $(v''=7)$. Thus, in the transition from $v'=4$ to $v''=7$, the velocity of the nuclei remains at or near zero while the separation between the nuclei is only slightly changed. A similarly likely transition is from $v'=4$ to $v''=6$. The original $B\,^3\Sigma_u^-$ $(v'=4) \leftarrow X\,^3\Sigma_g^-$ $(v''=0)$ excitation, although obviously possible, is approximately 10^4 times less probable than either the $v'(4) \rightarrow v''(6)$ or $v'(4) \rightarrow v''(7)$ transition (Krupenie 1972).

The variation in the transition probabilities as implied by the Franck–Condon principle can be observed in the dispersion spectra of excited molecules. These spectra are obtained by collecting the radiation emitted by excited molecules and separating it into its spectral components. Figure 9.14 illustrates the dispersion spectrum of O_2 following an excitation of the $B\,^3\Sigma_u^-$ $(v'=4)$ state. The spectrum includes transitions to $v''=3$, 4, 5, 6, 7, 8, and 9. As expected, the substantial overlap between the $v'=4$ state and the $v''=6$ and $v''=7$ states results in the strongest transitions.

The extent of the overlap between the wave functions of pairs of vibronic states, and hence the probability for vibronic transitions, can be calculated. The results are represented by parameters, named Franck–Condon factors, that are specific to the interacting levels. Comprehensive lists of Franck–Condon factors are available for numerous molecules (see e.g. Krupenie 1972 for O_2 or Lofthus and Krupenie 1977 for N_2). Alternatively, the target states for the most probable vibronic transitions can be identified graphically; one or both endpoints in the oscillation span of these target states are the nearest to the vertical lines drawn from the endpoints of the original state. Such graphical inspection may be used for all transitions when $v' \neq 0$ and $v'' \neq 0$. When $v=0$, the probability function that describes the molecular oscillation has a maximum at the equilibrium point – that is, when $r = r_e$. An approximate distribution of the probability function is shown for $v' = 0$ in Figure 9.13. Owing to this exceptional property of the $v = 0$ state, transitions from the center of its span to the endpoint of other v levels are most likely. This is illustrated in Figure 9.13 by the vertical line connecting $v' = 0$ with $v'' = 13$. For the manifold of lines originating from $v' = 0$, this is the transition with the largest Franck–Condon factor (Krupenie 1972).

The Franck–Condon principle can also be applied to determine the probability for transitions between bound and repulsive states, and hence the probability that an excited molecule dissociates. Consider the $B\,^3\Sigma_u^-$ electronic state in Figure 9.13. It is crossed by a repulsive state that intersects the potential curve near $v' = 4$. The energy required for transition from $v' = 4$ to the repulsive state is extremely small, and the change in the separation between the nuclei during such a transition is also small. Therefore, the transition can occur spontaneously, that is, without collisions between the excited molecule and a neighboring molecule. Once a predissociated transition occurs it cannot be reversed, and the molecule proceeds along the "slope" of the repulsive state until it dissociates. Indeed, excitation of $v' = 4$ is a significantly more efficient path for the dissociation of O_2 than the excitation of the adjacent $v' = 0, 1, 2, 3$ or $v' = 5, 6, 7, \ldots$ states.

The discussion in this chapter showed that, on the atomic and molecular levels, energy is stored in quantized states. These states represent the energy associated with the motion of electrons around the nucleus, their spin, the spin of the nucleus, or the rotation and vibration of molecules. Rules that identify dipole-allowed transitions were presented. However, these were only selection rules; they could neither determine the probability of a transition occurring nor describe the modes of optical transitions. Thus, additional principles that govern the rate of certain transitions and their probability will be presented in the next chapter. These principles will not only reveal the brightness of emission lines and the extent of absorption, but will also allow us to determine the population density in various energy states under conditions of equilibrium or optical excitation.

References

Bashkin, S., and Stoner, J. O., Jr. (1975), *Atomic Energy Levels and Grotrian Diagrams,* Amsterdam: North-Holland.

Bohr, N. (1913), On the constitution of atoms and molecules, *Philosophical Magazine and Journal of Science* 26(151): 1–25.

Carlone, C., and Dalby, F. W. (1969), Spectrum of the hydroxyl radical, *Canadian Journal of Physics* 47: 1945–57.

Cheung, A. S.-C., Yoshino, K., Parkinson, W. H., and Freeman, D. E (1986), Molecular spectroscopic constants of $O_2(B\,^3\Sigma_u^-)$: the upper state of the Schumann–Runge bands, *Journal of Molecular Spectroscopy* 119: 1–10.

Dieke, G. H., and Crosswhite, H. M. (1962), The ultraviolet band of OH, *Journal of Quantitative Spectroscopy and Radiative Transfer* 2: 97–199.

Fairbank, W. M., Jr., Hänsch, T. W., and Schawlow, A. L. (1975), Absolute measurement of very low sodium-vapor densities using laser resonance fluorescence, *Journal of the Optical Society of America* 65: 199–204.

Gillette, R. H., and Eyster, E. H. (1939), The fundamental rotation–vibration band of nitric oxide, *Physical Review* 56: 1113–19.

Grotrian, W. (1928), *Graphische Darstellung der Spektren von Atomen und Ionen mit ein, zwei und drei Valenzelektronen,* Berlin: Springer-Verlag, pp. 20–1.

Hay, J. P., and Dunning, T. H. Jr. (1977), The electronic states of KrF, *Journal of Chemical Physics* 66: 1306–16.

Hébert, G. R., Innanen, S. H., and Nicholls, R. W. (1967), The O_2 $B^3\Sigma_u^-$–$X^3\Sigma_g^-$ Schumann–Runge system. In *Identification Atlas of Molecular Spectra,* vol. 4, Toronto: York University Report Centre for Research in Experimental Space Science and Department of Physics.

Herzberg, G. (1945), *Infrared and Raman Spectra of Polyatomic Molecules,* New York: Van Nostrand.

Herzberg, G. (1989), *Molecular Spectra and Molecular Structure I. Spectra of Diatomic Molecules,* Malabar, FL: Krieger.

Karplus, M., and Porter, R. N. (1970), *Atoms and Molecules: An Introduction for Students of Physical Chemistry,* Menlo Park, CA: Benjamin.

Krupenie, P. H. (1972), The spectrum of molecular oxygen, *Journal of Physical and Chemical Reference Data* 1: 423–534.

Laufer, G., Krauss, R. H., and Grinstead, J. H. (1991), Multiphoton ionization by N_2 by the third harmonic of a Nd:YAG laser: a new avenue for air diagnostics, *Optics Letters* 16: 1037–9.

Laufer, G., McKenzie, R. L., and Huo, W. M. (1988), Radiative processes in air excited by an ArF laser, *Optics Letters* 13: 99–101.

Lofthus, A., and Krupenie, P. H. (1977), The spectrum of molecular nitrogen, *Journal of Physical and Chemical Reference Data* 6: 113–307.

Merzbacher, E. (1970), *Quantum Mechanics,* 2nd ed., New York: Wiley.

Moore, C. E. (1971), *Atomic Energy Levels as Derived from the Analysis of Optical Spectra* (National Standard Reference Data Series NSRDS-NBS 35, vols. I–III), Washington, DC: National Bureau of Standards.

O'Neil, S. V., and Schaefer, H. F. III (1970), Valence-excited states of carbon monoxide, *Journal of Chemical Physics* 53: 3994–4004.

Suchard, S. N. (1975), *Spectroscopic Data,* part B, New York: Plenum.

Veseth, L., and Lofthus, A. (1974), Fine structure and centrifugal distortion in the electronic and microwave spectra of O_2 and SO, *Molecular Physics* 27: 511–19.

Vincenti, W. G., and Kruger, C. H., Jr. (1965), *Introduction to Physical Gas Dynamics,* Malabar, FL: Krieger, pp. 89–93.

Zimmermann, M., and Miles, R. B. (1980), Hypersonic-helium-flow-field measurements with the resonant Doppler velocimeter, *Applied Physics Letters* 37: 885–7.

Homework Problems

Problem 9.1

Using principles of classical mechanics and electrostatics to describe the orbital energy and angular momentum of the electron–proton pair in the hydrogen atom, calculate the energy and the radius of the nth orbit of the hydrogen atom. Assume that both the electron and the proton are orbiting around a common center (Figure 9.1). Also show that, by using the reduced mass $mM/(m+M)$ for the mass of the electron, the same results can be obtained by considering the proton to be stationary with the electron orbiting around it at a radius $r_n = 4\pi\epsilon_0 n^2\hbar^2/me^2$.

Problem 9.2

In his original paper, Bohr (1913) hypothesized that, in order to observe emission from a highly excited energy state of the hydrogen atom, the mean distance between atoms in the gas sample must exceed the diameter of the excited atom. Otherwise, the volume of the outer shell would exceed the volume accessible to that atom. Assume therefore that the mean diameter of a highly excited atom is $d_m = 1/N^{1/3}$, where N is the number of atoms in a unit volume.

- **(a)** Find the maximum number of lines that can be observed for emission from the highest possible excited states to the $n = 1$ state of a Bohr hydrogen atom in a discharge tube at room temperature, where $P = 7$ mm Hg. (*Hint:* Use the Loschmidt number from Appendix B to determine the discharge density.)
- **(b)** Determine the pressure of a celestial gas at $T = 30$ K if 33 lines, all corresponding to a final state of $n = 1$, are observed.

Problem 9.3

Review the Grotrian diagram (Figure 9.5) of the hyperfine structure of sodium. Try to predict the emission spectrum of the D_1 and D_2 lines of Na vapor by assuming that the upper levels are uniformly populated and that the excited population is not depleted by the emission process. Mark each allowed transition by a vertical line, and separate the lines by distances that are proportional to their frequency differences. Assign a height to each line that is proportional to the number marked by the arrow in Figure 9.5 connecting the end levels of the transition. When two or more adjacent lines are separated by less than 200 MHz, regard them as spectrally unresolved and replace them by a single line with a height that is the sum of their heights.

Problem 9.4

Calculate the range of emission wavelengths of the KrF excimer laser for hypothetical transitions from $v = 1$. Use the data in Table 9.2 for your calculations.

Problem 9.5

Calculate the electronic term value T_0 for O_2 in the $B\,^3\Sigma_u^-$ state, and for N_2 in the $A\,^3\Sigma_u^+$ state, using the parameters of Table 9.1. (*Answer:* 49,356.9 cm^{-1} and 49,754.9 cm^{-1}.)

Problem 9.6

An ArF laser that can be tuned from 192.9 to 193.7 nm is to be used for certain measurements. It is necessary to transmit the laser beam through a long path in atmospheric air. Unfortunately, there exist certain rotational levels in the ground vibrational and electronic state of O_2 that may absorb the laser radiation while undergoing a $B\,^3\Sigma_u^- \leftarrow X\,^3\Sigma_g^-$ transition in the Schumann–Runge band. Such inadvertent attenuation of the beam may be avoided by tuning the laser away from such rotational transitions. To identify these transitions:

- **(a)** find the final vibrational state v' for excitation of $v'' = 0$ by the ArF laser;
- **(b)** find the rotational levels within the $v'' = 0$ state that can be excited, and identify the target rotational states for the P and R branches and calculate the absorption wavelengths of these lines.

Problem 9.7

Repeat Problem 9.6 for excitation of the $v''=1$ state. Assuming that the number of molecules occupying the $v''=0$ state is the same as the number of molecules at the $v''=1$ state (this assumption implies that the sample is not in equilibrium), determine which of these vibronic transitions is stronger. Use for this evaluation the energy-level diagram in Figure 9.13.

Research Problems

Research Problem 9.1

Read the paper by Zimmermann and Miles (1980). In this paper, a tunable laser is used to measure the Doppler shift of Na vapor seeded in a hypersonic He flow. The laser beam was passed through the test section at an angle to the flow axis and then reflected back on itself (Figure 9.15). Thus, by propagating in the forward direction a Doppler downshift could be detected, and when retro-reflected a Doppler upshift was detected. Figure 2 in that paper presents two spectra of the hyperfine structure obtained in these experiments; one spectrum is associated with the Doppler upshift and the second with the Doppler downshift. Owing to limitations of the experiment, only the ground-state hyperfine splitting of the D_2 line was resolved. However, the frequency difference due to this splitting was known and could be used as a frequency marker to determine the sum of the Doppler upshift and downshift.

(a) Using the Grotrian diagram in Figure 9.5, determine the frequencies of each of the lines in the D_2 structure relative to the absorption line from $3^2S_{1/2}$ $(F=2)$ to $3^2P_{3/2}$ $(F=3)$. Draw an approximate spectrum by assigning to each line a height that is proportional to the numbers marking the lines of the allowed transitions. Compare your result with the spectrum in the paper or with Figure 11.8.

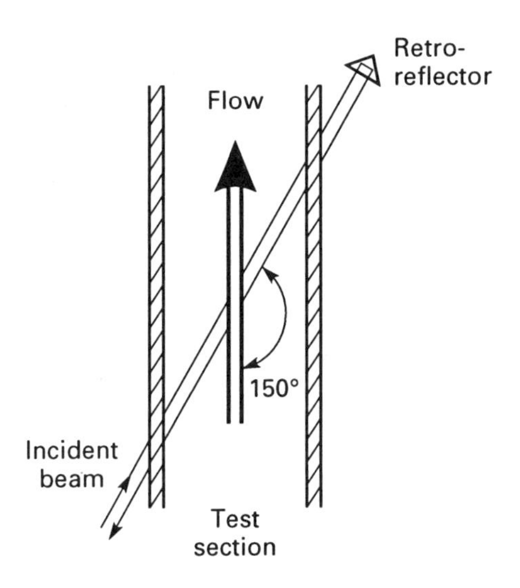

Figure 9.15 Experimental set-up for velocity measurements using the Doppler shift of Na excitation.

(b) Assume that the incident laser beam forms an angle of 150° with the flow axis. Determine the Doppler shift for this and the retro-reflected beams when the flow velocity is 1,000 m/s. Draw the Doppler upshifted and downshifted spectra.

(c) Assume that the hyperfine splitting of the upper states cannot be resolved and that it appears as one broad line. Determine the minimum flow velocity required to separate the incident beam spectrum from the reflected beam spectrum.

(d) Estimate the minimum measurable velocity when the incident angle is 135°.

(e) Estimate the velocity measurement uncertainty when the incident angle is 150° and the width of each of the spectral lines (at FWHM) is 700 MHz.

Research Problem 9.2

Read the paper by Laufer, Krauss, and Grinstead (1991). The paper discusses a process in which six photons are stacked together to ionize N_2. The ion is formed in an excited electronic and vibrational level $B^2\Sigma_u^+$ ($v'=8$). From this level, the ionized molecules cascade through several vibrational states within the same electronic state. At each vibrational level, some molecules radiate while decaying to one of the vibrational states of the lower electronic level of the ion $X^2\Sigma_g^+$. This process is demonstrated by the dispersion spectrum of the emission in figure 1 of the paper.

(a) Using the known energy of the ionizing photons, calculate the energy required [eV] to produce an ion at the $B^2\Sigma_u^+$ ($v'=8$) level.

(b) From the parameters in the spectrum of figure 1, estimate the energy difference between the $v''=0$ state and the $v''=1$ state in the $X^2\Sigma_g^+$ electronic level of the ion. (*Hint:* Find two emission lines that originate at the same vibrational level in the $B^2\Sigma_u^+$ state.)

(c) From the parameters in the spectrum of figure 1, estimate the energy difference between the $v'=0$ state and the $v'=1$ states in the $B^2\Sigma_u^+$ electronic level of the ion.

(d) Neglecting anharmonicity of the excited electronic state, use the results of parts (a), (b), and (c) to estimate the ionization energy of N_2.

10 Radiative Transfer between Quantum States

10.1 Introduction

Until now, our discussion of the interaction between radiation and matter has concentrated only on the spectral aspects of radiation. The results could determine the wavelengths for absorption and emission or the selection rules for such transitions, but could not be used to determine the actual extent of emission or absorption. These too are important considerations which are needed to fully quantify radiative energy transfer. Unfortunately, none of the classical theories can predict the extent of emission from an excited medium, or even the extent of absorption. Although the discussion in Section 4.9 (on the propagation of electromagnetic waves through lossy media) touched briefly on the concept of attenuation by absorption, it failed to show the reasons for the spectral properties of the absorption or to accurately predict its extent. We will see later that the classical results are useful only as a benchmark against which the actual absorber is compared. The objective of this chapter is therefore to present an introduction to quantum mechanical processes that control the emission and absorption by microscopic systems consisting of atoms and molecules. The results will then be used to predict the extent of emission by media when excited by an external energy source and to evaluate the absorption of incident radiation.

It is now well recognized that all emission or absorption processes are the result of transitions between quantum mechanical energy levels. Even when an emission (e.g. solar radiation) is spectrally continuous, it is caused by transitions between discrete quantum mechanical states that are either densely spaced or have extremely wide bandwidths so that one line merges into the other. Although most radiative processes are complex and involve simultaneous absorbing and emitting transitions between several energy levels, their fundamental principles can be explained by considering the interaction between pairs of isolated states. When such simplifications are justified, interaction between other, unrelated energy levels can be ignored and a radiating system may be viewed as a two-level system where emission takes place from the upper, or *excited* state and absorption from the lower, or *ground* state.

Emission and absorption by a microscopic system are rarely isolated events. Most media of engineering significance contain vast numbers of atoms or molecules; some of these molecules may be in the ground state while others are in an excited state. The net observable effect of such large ensembles is an integrated outcome of all their contributions. Thus, although the initial step in evaluating the extent of energy transfer requires characterization of the emitting system, the net radiative transfer depends also on the statistical properties of the medium: the number of atoms or molecules that occupy each of the interacting levels. The actual number of the occupants of each energy level can be accurately determined when the medium is at thermodynamic equilibrium or, in rare cases, when the departure from equilibrium is well characterized. A complete analysis of radiative transfer thus requires consideration of the thermodynamic properties of the medium together with the quantum mechanical properties of the atoms and molecules that compose it. This is normally accomplished using *rate equations,* which combine the time-dependent variation of the population density in one energy level with the population density of other levels by using the transition rates between these states. Input to these equations includes terms that describe the rates of each possible transition and information regarding the thermodynamic constraints that influence the population of these states. Many optical effects are well described by the inclusion of only two levels in the rate equations. However, in complex systems such as laser media, three, four, or more states may need to be coupled together in order to accurately characterize the process. Radiative transfer in these systems requires calculation of the transition rates between all participating levels. Such systems will be discussed in Chapter 12.

The discussion of radiative energy transfer begins by considering spontaneous emission and absorption by two-level systems. At low levels of radiation, or when the system is at thermodynamic equilibrium, these are the only processes that are detected. When a two-level system is in its ground state, a photon at an energy that exactly matches the difference between the states can be absorbed, with its energy used for the excitation of the system to the upper state. Conversely, when the system is at an excited state, it may dispose of its energy by spontaneously emitting a photon with energy that matches the energy difference. These processes are appropriately called *absorption* and *spontaneous emission*. However, there also exists a third energy exchange process: *stimulated emission*. Like absorption, stimulated emission occurs when a two-level system encounters a photon at an energy that exactly matches the energy difference between the levels. However, unlike absorption, it can occur only when the system is at its excited state, and results in the emission of a photon that is an identical twin of the incident photon. Since stimulated emission occurs only when an excited atom or molecule encounters a photon, it is a rare event under most circumstances. Only when the population of the excited state is sufficiently large or when it exceeds the population of the lower state can stimulated emission have an appreciable effect. However, complete analysis of radiative transfer requires the inclusion of all three processes.

10.2 Spontaneous Emission

When a two-level system is excited, it has several alternatives for the disposal of its excess energy: it may give it up by collisions, either with neighboring molecular or atomic systems or with the walls of its container; if the energy is sufficiently high, it may shed an electron and thereby become ionized; or it may collide with a photon and so be forced to release its energy by stimulated emission. In all these interactions, energy exchange occurs instantaneously at the time of the collision. Thus, although the affinity of the two-level system for such transitions controls the process, the rate of successful energy exchange events must also depend on the affinity of the collisional partners and on the rate of such collisions. However, there exists one process – spontaneous emission – that allows the system to release its energy independently of its surroundings. In the absence of such a process, systems would not be able to seek their lowest possible energy without the assistance of an external agent. Owing to spontaneous emission, any microscopic system – even if isolated – can eventually reach its lowest energy state by emitting a photon that matches the energy difference between the upper and the lower states.

Because spontaneous emission does not require an external agent, it cannot be accelerated or slowed and is fully described by only one parameter: the characteristic time τ_{21} required for the emission to occur. This characteristic time is a measure of the time that the excited state (level 2 in Figure 10.1) can survive; it is considered to be the *natural lifetime* of that state. This lifetime is a unique property of each excited state, whereas the lifetime at the ground state of any two-level system is $\tau_{12} \to \infty$. The latter is intuitively obvious; since promotion to an excited state requires an energy input, a two-level system cannot hop spontaneously from the ground state to an excited state. Thus, for a two-level system,

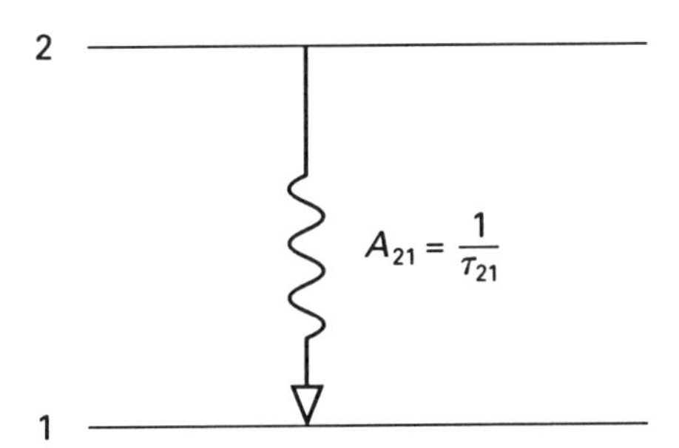

Figure 10.1 Transition due to spontaneous emission between two quantum mechanical states.

only the lifetime τ_{21} of the excited state needs to be specified. Nevertheless, it represents the coupling between both states of the system. When an atomic system is at an excited state, it may often elect to relax to one of several lower energy states, and each possible transition may have a different lifetime that is unique to the coupled states. To help distinguish between them, each lifetime is marked by two subscripts, the first for the upper state and the second for the lower state.

The natural lifetime is a good parameter for characterization of transients or for comparison of a spontaneous emission transition with other short-duration effects. However, to calculate the time-dependent population density of the excited state, the following rate constant is more convenient:

$$A_{21} = \frac{1}{\tau_{21}}\ \text{s}^{-1}. \tag{10.1}$$

This is the Einstein A coefficient, the rate constant for depletion of the excited state by spontaneous emission. Along with the wavelength for optical transitions, each pair of quantum states must also be characterized by their Einstein A coefficient.

In the absence of competing depletion processes, the rate of transitions from an upper to a lower state must increase proportionally both with A_{21} and with the population density of the excited state. This plausible statement is represented mathematically by the following rate equation:

$$\frac{dN_2}{dt} = -A_{21}N_2. \tag{10.2}$$

Usually N_2 is defined as the number of excited atomic systems within a unit volume of the sample. Depletion of N_2 is understood from the negative sign. Although this rate equation is plausible, the reader is cautioned that there are numerous processes (e.g. two-body collisions) where the depletion of the excited state varies nonlinearly with N_2.

Ideally, for each atomic particle that was removed from level 2, one particle must be added to level 1. Therefore the total number of two-level particles, N_0, is conserved:

$$N_0 = N_1 + N_2 = \text{constant}.$$

With this conservation condition, (10.2) may also be written as a rate equation describing the replenishment of level 1:

$$\frac{dN_1}{dt} = A_{21}N_2. \tag{10.3}$$

Equation (10.2) is a first-order differential equation. Thus, when a sample is prepared with an initial excited population density of N_2^0 and then left unperturbed, the time-dependent excited-state population density decays as:

$$N_2 = N_2^0 \exp\{-A_{21}t\} = N_2^0 \exp\left\{-\frac{t}{\tau_{21}}\right\}. \tag{10.4}$$

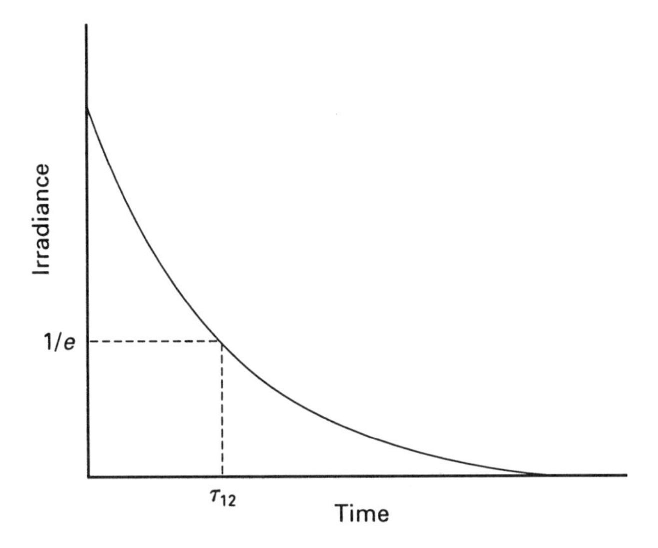

Figure 10.2 Variation with time of the radiative spontaneous decay of an excited state following a pulsed excitation.

In this solution, the natural lifetime τ_{21} appears as the time required to deplete the population of the excited state to $1/e$ of its initial value. The same exponential decay behavior also describes the irradiance due to the spontaneous emission (Figure 10.2). Thus, if a quantum state is excited by a narrowband tunable light source within a time that is short relative to τ_{21}, and if the emission is detected by a fast detector, then the output appears as an exponentially decaying pulse, where τ_{21} is the time measured between the onset of the pulse and the time of its decay to $1/e$ of its peak amplitude. As will be shown in more detail in the next section, owing to the short duration of the spontaneous emission, the emitted radiation can never be perfectly monochromatic. Similarly, owing to the limited lifetime of the upper state, the energy can be specified only within the limitations of the uncertainty principle. Thus, instead of a sharply defined energy, the upper state is a band of finite width centered around a nominal energy. The spectral width of the emission matches the width of this energy band, just as the nominal photon frequency matches the nominal energy difference between the end states of the atomic system. This apparent broadening of the emission spectrum and of the energy distribution in the upper state has important consequences in certain thermodynamic measurements of gas flow properties.

10.3 Natural Lifetime and Natural Broadening

Natural lifetime is a parameter that is specific to each pair of interacting states. Therefore, values for natural lifetime can vary widely between atoms and molecules, or even between states of an individual atom or molecule. On one end, metastable or long-lived states (e.g., the $v'=1$ level in the ground electronic state of O_2) may have a lifetime in excess of several milliseconds. On the other end, the lifetime may be as short as 10 ps (e.g., the lifetime of the predissociated $B^3\Sigma_u^-$ ($v'=4$) state of O_2). The lifetimes of excited states of other atomic-scale particles usually vary between these extremes and therefore, consistent with the uncertainty principle, the energy of these states cannot be specified exactly. By viewing the natural lifetime as the uncertainty associated with the measurement of the time of excitation, $\Delta t = \tau_{21}$, the *full* extent of the energy uncertainty of an excited state is

$$\Delta E_2 \geq \frac{h}{2\pi\tau_{21}} \tag{10.5}$$

(cf. eqn. 8.11). Because of its limited lifetime, the upper state cannot be viewed as a sharply defined energy level. Instead, it appears as a band with a *full* width ΔE_2 centered around E_2. Consequently, when excited, a particle may have an energy within the range specified by this band, which can be viewed as the probability distribution of finding the particle at or near E_2. It is most likely to have its energy in the vicinity of E_2, but it may have (albeit at diminishing probabilities) any other energy. This probability is described by a distribution function with its maximum at E_2 and with a full width at half maximum (FWHM) of ΔE_2 as defined by (10.5). This function associates, with each possible value of energy near the excited state, a probability of finding the excited system with that energy. Since the overall probability of finding the excited system at the excited state is unity, the peak probability must be adjusted to assure that the integral of the probability function over the entire energy range is unity – that is, the probability function must be normalized.

At the ground state the natural lifetime approaches infinity, so when left alone the system will remain at that state indefinitely. Therefore, the probability distribution of that state is a δ function. This too is a normalized probability function, but with a FWHM approaching zero. When a transition takes place from an excited state to the ground state, it can originate from any point on the probability distribution curve of the excited state to the ground-state δ function. Thus, the range of photon energies (and, accordingly, the linewidth) of the emission is influenced only by the FWHM of the excited-state probability distribution function. Combining (10.1) and (10.5), the emission spectrum appears as a line with a FWHM of $\Delta\nu_{21}$, where:

$$\Delta\nu_{21} = \frac{A_{21}}{2\pi} = \frac{1}{2\pi\tau_{21}}. \tag{10.6}$$

(Note that the inequality sign of (10.5) was replaced here by an equal sign; this will be justified in what follows.) This spectral broadening, which is specific to the transition between levels 2 and 1, does not depend on the motion of the system or the interaction between the system and other systems, and is therefore called *natural broadening*. Other broadening mechanisms will be discussed in Section 10.7.

Equation (10.6) successfully predicts the FWHM of the spontaneous emission line, but fails to describe its exact shape. The uncertainty principle, used to estimate the FWHM of the probability distribution of the excited state, is not sufficiently detailed to predict the exact structure of that distribution function, which eventually determines the shape of the emission spectral lines. However, the ability to predict and measure the shape of a spectral line is an essential component in many thermodynamically related spectroscopic measurements. The shape of a naturally broadened line can be determined using other concepts of quantum mechanics or by classical concepts, for example, Fourier transformation of the transient electromagnetic field. Since the latter is plausible and simple, it will be used here to derive directly the mathematical shape of spontaneous emission lines.

To obtain the Fourier transform of spontaneous emission, consider it as a time-dependent phenomenon that follows an instantaneous excitation. Therefore, it is an exponentially decaying pulse of radiation, where the point $1/e$ of its peak amplitude is reached in a time of τ_{21} (Figure 10.2). The Fourier transformation of the electromagnetic field of this pulse to the frequency domain must be described by a spectral line with a finite width (see Section 4.7). Owing to the quadratic dependence between radiative energy and its electromagnetic field (eqn. 4.42), the time constant for the exponentially decaying field is $2\tau_{21}$. The nominal frequency of this field, $\omega_0 = (E_2 - E_1)/\hbar$, depends on the energy difference between the end states. Thus, the time-dependent component of the electromagnetic field of the exponentially decaying irradiance is

$$\mathbf{E}(t) = \mathbf{E}_0(e^{-i\omega_0 t} + e^{i\omega_0 t})e^{-A_{21}t/2} \tag{10.7}$$

(cf. eqn. 4.14), where $A_{21}/2$ stands for $1/2\tau_{21}$ in the exponential term. The frequency distribution of the electromagnetic field $\mathbf{E}(\omega)$ is obtained by Fourier transformation from the time domain to the frequency domain (see Problem 10.1). However, the spectral distribution of the electric field itself is of little interest in spectroscopic applications. Therefore, using the relation between the electric field and the irradiance (eqn. 4.42), the following frequency distribution of the irradiance $I(\omega)$ is obtained from the Fourier transform of the field $\mathbf{E}(\omega)$:

$$I(\omega) = I_0 \frac{A_{21}/2\pi}{(\omega - \omega_0)^2 + (A_{21}/2)^2} = I_0 g_N(\omega). \tag{10.8}$$

This distribution is the irradiance $I(\omega)$ per unit frequency in the frequency range of ω and $\omega + d\omega$. It includes two components: the amplitude I_0, which determines the energy content of the emission line; and the *lineshape factor* $g_N(\omega)$, which represents the spectral content of the emission. Thus, the height of the spectral lines is determined primarily by I_0, while their frequency at the line center and their bandwidth are included in the lineshape factor. For spontaneous emission, the lineshape factor is

$$g_N(\omega) = \frac{A_{21}/2\pi}{(\omega - \omega_0)^2 + (A_{21}/2)^2}. \tag{10.9}$$

This lineshape factor describes the natural broadening and has a Lorentzian shape; see Figure 10.3(a). It is centered at ω_0 with a FWHM of $\Delta\omega = A_{21}$ rad/s.

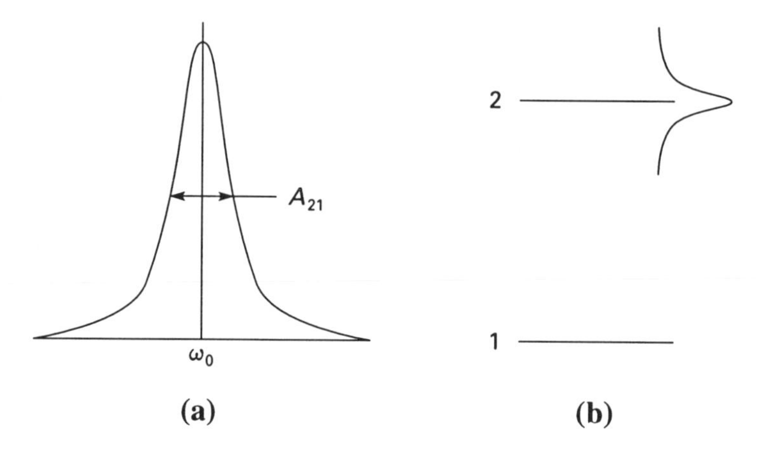

Figure 10.3 (**a**) The lineshape associated with spontaneous emission compared with (**b**) the probability distribution function of the excited state at or near E_2.

In units of hertz this bandwidth is $\Delta\nu_{21} = A_{21}/2\pi$, as predicted by (10.6). This frequency distribution, apart from a factor of h, is identical to the probability distribution function of the excited state at or near E_2 – that is, it is proportional to the probability of finding the excited system at or near E_2 (Figure 10.3(b)). Since the probability distribution function is normalized (i.e., the integral of the probability function must be unity), the lineshape factor must also be normalized and so must meet the condition

$$\int_{-\infty}^{\infty} g_N(\omega)\,d\omega = 1. \tag{10.10}$$

Equation (10.9) has already been normalized.

This condition of normalization can be applied to the computation of the total irradiance carried by a spectral line of spontaneous emission. Therefore, the total irradiance is

$$I = \int_{-\infty}^{\infty} I(\omega)\,d\omega = I_0 \int_{-\infty}^{\infty} g_N(\omega)\,d\omega. \tag{10.11}$$

However, since $g_N(\omega)$ is normalized, the spectrally integrated irradiance is I_0. In most spectrally unresolved measurements the results are presented by the irradiance I_0, whereas most spectroscopic measurements are presented by $I(\omega)$. To distinguish between these quantities, $I(\omega)$ is often called the *spectral irradiance*.

Although the discussion in this chapter is limited to two-level systems, in reality very few radiative interactions occur between two isolated levels. Under almost any excited state there exist numerous energy levels. Although many of them cannot be reached from that excited level, selection rules permit transitions to many others. Thus, although each allowed transition couples two levels, there can be more than just one allowed transition from one excited state. Of course, one excited system can choose only one transition, but within a large group of excited systems there can be many different transitions, all originating at the same excited state. The rate of each of those transitions is determined by the Einstein A coefficient specific to the coupled states. Barring competing effects (such as collisions), the relative heights of the spontaneous emission lines are directly proportional to the relative magnitudes of the corresponding Einstein A coefficients. This is confirmed by comparing the heights of the lines (in the dispersion spectrum in Figure 9.14) of transitions from the $B\,^3\Sigma_u^-$ $(v'=4)$ state of O_2 to several vibrational levels in the $X\,^3\Sigma_g^-$ state with their Einstein A coefficients (see Example 10.2). Although similar dispersion spectra may be used to determine the relative magnitudes of the A coefficients, the absolute values of these coefficients must be either calculated or determined by other measurements. The dispersion spectrum was obtained after the simultaneous excitation of several rotational levels in the $B\,^3\Sigma_u^-$ $(v'=4)$ state. The resultant spectrum does not resolve the individual rotational transitions, but does show clearly the lines for vibrational transitions from $v'=4$. Such a group of spectral lines formed by different transitions from the same state is called a *band.*

Although the transition rate of each of the lines of a band depends on its own A coefficient, the rate of loss of excited-state population (or the probability for transition when a single particle is analyzed) depends on all the allowed transitions of that band. This is illustrated in Figure 10.4, where three allowed transitions are shown from level 4 to lower levels 1, 2, and 3 with Einstein A coefficients A_{41}, A_{42}, and A_{43}, respectively. The rate of population loss in the excited state is controlled now by an effective rate A_{eff}, which is determined by the following sum:

$$A_{\text{eff}} = \sum_{i=1,2,3} A_{4i}.$$

Obviously, the rate of each individual transition is slower than the combined rate and so the fraction of all the excited systems that undergo transition to a selected level i (where $i = 1$, 2, or 3) is

$$S = \frac{A_{4i}}{A_{\text{eff}}}. \tag{10.12}$$

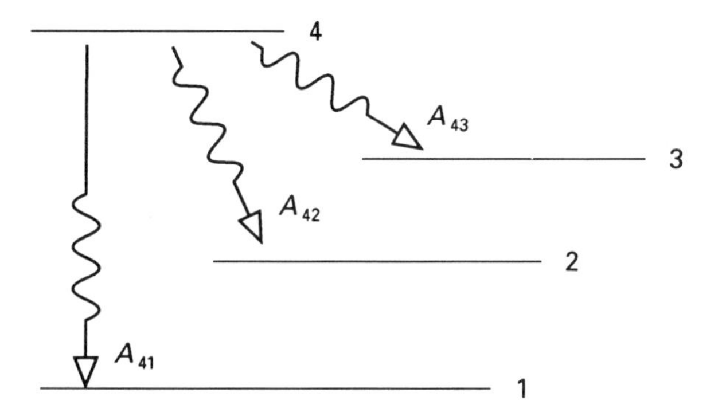

Figure 10.4 Spontaneous emission from one excited state to a manifold of lower levels.

Table 10.1 *Einstein A coefficients for ro-vibronic transitions from $v'=0$, $N'=2$, $J'=\frac{5}{2}$ in the $A^2\Sigma \rightarrow X^2\Pi$ system of OH*

	Branch					
Band	$R_1(1)$	$Q_1(2)$	$P_1(3)$	$Q_{12}(2)$	$P_{12}(3)$	$O_{12}(3)$
$0\rightarrow 0$	8.071×10^4	5.167×10^5	4.960×10^5	1.129×10^5	1.880×10^5	7.457×10^4
$0\rightarrow 1$	2.834×10^2	1.541×10^3	1.277×10^3	3.578×10^2	5.441×10^2	2.041×10^2
$0\rightarrow 2$	2.212×10^1	1.397×10^2	1.333×10^2	3.178×10^1	5.356×10^1	2.142×10^1
$0\rightarrow 3$	2.047	1.272×10^1	1.199×10^1	2.940	4.989	2.008

Note: Values are given in units of s^{-1}.

The lifetime of the excited state is now determined by $1/A_{\text{eff}}$ and the bandwidth is determined by $\Delta\nu_{\text{eff}} = A_{\text{eff}}/2\pi$ (eqn. 10.6). The following example illustrates the effect of competing transitions on the broadening of all the transitions from one excited state.

Example 10.1 The Einstein A coefficients for transitions from the $v'=0$, $N'=2$, $J'=\frac{5}{2}$ state in the $A^2\Sigma \rightarrow X^2\Pi$ band of OH are listed in Table 10.1 (Dimpfl and Kinsey 1979). Find the linewidth of the transitions from this state and calculate the percentage of the transitions in the $0\rightarrow 0$ $P_1(3)$ line.

Solution Table 10.1 lists the coefficients for three primary branches (P_1, Q_1, and R_1) and three satellite branches (Q_{12}, P_{12}, and O_{12}) for four vibronic transitions. The subscripts adjacent to the branch symbol distinguish transitions from the $J'=N'+\frac{1}{2}$ state that are listed here from transitions originating at a nearby $J'=N'-\frac{1}{2}$ state.

The Einstein A coefficients, or the rates, are the highest for the $0\rightarrow 0$ vibronic transitions and decline progressively for transitions to states with higher v'' quantum numbers. Thus, although transitions to vibrational levels $v''>3$ are possible, their rates are extremely low and so were not included. Also note that the transition rates of the primary rotational branches are higher than the rates of the satellite branches.

The lifetime of this state is determined from the inverse of the effective Einstein coefficient, which (after neglecting the effect of transitions to states with $v''>3$) is

$$A_{\text{eff}} = \sum A_{00} + \sum A_{01} + \sum A_{02} + \sum A_{03}$$
$$= 1.467\times 10^6 + 4.207\times 10^3 + 4.019\times 10^2 + 3.669\times 10^1 = 1.473\times 10^6 \text{ s}^{-1}.$$

Each summation in this equation was carried out over all the rotational lines within each vibrational band and is therefore the sum of all the elements in the corresponding row of Table 10.1. The linewidth of all the transitions from this state, $\Delta\nu$, which is controlled by A_{eff}, is

$$\Delta\nu = \frac{1.473\times10^6}{2\pi} = 0.234 \text{ MHz}$$

(cf. eqn. 10.6). This is an extremely narrow line, too narrow to be resolved by most instruments. Furthermore, other physical effects such as collisions or molecular motion cause the linewidth to increase significantly (see Section 10.7). Therefore, measurement of this width requires not only high-resolution instruments but also trapping the molecule and isolating it from its surroundings.

The percentage of excited OH molecules that radiate in the $0\to0$ $P_1(3)$ line, as calculated using (10.12), is

$$S[P_1(3)] = \left(\frac{4.960\times10^5}{1.473\times10^6}\right)100 = 33.7\%.$$

Of the six available rotational branches in the $0\to0$ band, the $P_1(3)$ and the $Q_1(2)$ are the strongest, with a comparable percentage of molecules emitting in each of these lines. Thus, approximately two thirds of all the excited molecules emit through spontaneous transitions in one of these two lines, while the remaining third of the excited molecules emit in all the other allowed transitions. Therefore, the rate of transitions from the excited state is dominated by $0\to0$ ro-vibronic transitions. ■

The A coefficients in Table 10.1 strongly depend on the target vibrational level, but they are also strongly influenced by the rotational branch of the transition. The variation with the target vibrational state was attributed to the Franck–Condon principle in Section 9.5. The dependence on rotational transition is described by the *Hönl–London factor.* Unlike vibrational transitions, rotational transitions are limited to a few branches by strict selection rules. The Hönl–London factor, which is specific to each electronic transition, assigns a probability to each rotational branch. At larger rotational quantum numbers J, the sum of these factors for each rotational state approaches $2J+1$ – the degeneracy of these states. The A coefficient for a specific ro-vibronic transition is obtained by multiplying the probability of a vibronic transition by the corresponding Hönl–London Factor. The coefficients in Table 10.1 already include these factors. For the Schumann–Runge band of O_2, these coefficients are reported by Tatum (1966).

The results of Example 10.1 demonstrate that although the net number of molecules selecting a specific path for spontaneous emission is controlled by the A coefficient for that transition, the linewidth and the effective lifetime of the excited state are controlled by the sum of the A coefficients of all transitions. Other mechanisms that are specific to the molecular structure can also diminish the lifetime of an excited state. Although these alternative mechanisms compete with the emission process and may not bring the system back to its ground state, they are part of the natural spontaneous process of the molecule and may therefore influence its natural lifetime. One such effect is dissociation of the molecule by transition from an excited state to an unbound predissociated state. Figure 9.13 illustrates the dissociation of O_2 in the $B\,^3\Sigma_u^-$ $(v'=4)$ state by such transition. Although the transition to the unbound state is not radiative, its rate P can be specified and, together with the A coefficients, determines the lifetime of the excited state and the linewidth of the emission to the ground electronic state. The *fluorescence yield* – the fraction of molecules that reach a specific vibrational state $v''=i$ from the excited state $v'=j$ – is also influenced by the predissociation rate P. This yield may be expressed as

$$S_i = \frac{A_{ji}}{\Sigma_i A_{ji} + P}. \tag{10.13}$$

Thus, when $P \gg A_{\text{eff}}$, most of the excited molecules are lost by dissociation well before reaching their natural lifetime for emission. The few spontaneously emitting transitions that do occur before dissociation are broadened by the short lifetime. Although the predissociative transition is not accompanied by emission, its rate can be determined by measuring the linewidth of one of the allowed radiative transitions. To illustrate the effect of predissociation on spontaneous emission, consider the predissociated transitions of O_2 as follows.

Example 10.2 The calculated contribution of predissociative transitions (i.e., transitions that lead to dissociation) of O_2 from the $B\,^3\Sigma_u^-$ $(v'=4)$ state to the linewidth of spontaneous emissions from that state is 3.693 cm^{-1} (Cheung et al. 1993). The calculated Einstein A coefficients (in s^{-1}) for the strongest vibronic transitions from this state are:

$$A_{43} = 3.523\times10^5,\ A_{44} = 1.132\times10^6,\ A_{45} = 2.516\times10^6,\ A_{46} = 3.876\times10^6,$$

$$A_{47} = 3.964\times10^6,\ A_{48} = 2.338\times10^6,\ \text{and}\ A_{49} = 4.461\times10^5$$

(Allison, Dalgarno, and Pasachoff 1971). Find what percentage of the excited molecules can reach the $v''=7$ level by spontaneous emission.

Solution Separation of the contribution of the predissociation to the natural linewidth from the contribution of the Einstein A coefficients is experimentally impossible. The observed natural linewidth is always proportional to the sum of all the rates of natural transfer from the excited state. Reporting the contribution of the predissociation to the linewidth, which by itself is not a measurable quantity, can nonetheless serve as an estimate of the rate P of predissociation, which is calculated from the reported contribution to the linewidth using (10.6):

$$P = 2\pi\Delta\nu = (2\pi\cdot3.693)(3\times10^{10}) = 6.96\times10^{11}\ \text{s}^{-1},$$

where unit conversion (from cm^{-1} to s^{-1}) required the introduction of $c_0 = 3\times10^{10}$ cm/s. The fraction of molecules that reach $v''=7$ is therefore

$$S_7 = \frac{A_{47}}{A_{\text{eff}}+P}\times100 = \frac{3.964\times10^6}{(1.46\times10^7)+(6.96\times10^{11})}\times100 = (5.7\times10^{-4})\%,$$

where $A_{\text{eff}} = \sum A_{4i}$. Of all the allowed transitions from $v' = 4$, the rate of decay to $v'' = 7$ is the highest. Therefore, the remarkably low percentage of molecules reaching that state is evidence that predissociation dominates the lifetime (as well as the natural linewidth) of the excited state. Less than 0.001% of excited molecules radiate back to the bround state. However, those few spontaneous transitions occur at a time that is short relative to the predissociation lifetime, which in turn is short relative to the time between collisions. Thus, the effects of collisions on the fluorescence emissions (which will be discussed in Chapter 11) are negligible (Laufer, McKenzie, and Fletcher 1990). Other molecules, such as OH, present similar effects of strong predissociation on the spontaneous emission (Andresen et al. 1988). ■

10.4 Absorption and Stimulated Emission

Our discussion of spontaneous emission highlighted radiative transitions that occur without any interaction of the emitting two-level system with other particles and that are independent of external fields. For these transitions, the rate is defined by characteristics of the system itself and cannot be accelerated or slowed. However, there exist two radiative processes that depend not only on the system properties but also on the properties of the incident radiation. One such process, absorption, is intuitively understood. The counterpart of absorption, stimulated emission, was first suggested by Einstein as a symmetric complement to the absorption process; namely, the same radiative mechanism that forces a two-level system into an excited state could also force it back down to the ground state. In any macroscopic medium, both processes occur simultaneously, so whenever there is absorption there is also stimulated emission. Furthermore, the two compete with each other, and the net outcome of any radiative transfer can be fully quantified only by accounting for these two processes and for spontaneous emission.

Figure 10.5 describes schematically a typical absorption experiment. A cell containing molecules or atoms that can be considered as two-level systems is illuminated uniformly from the left by "white light" (radiation with uniform spectral distribution). Thus, the spectral irradiance $I_0(\nu)$ of this source, measured per unit frequency, is constant over a certain frequency range $\Delta\nu$, where the frequency ν is measured in hertz. The atomic-scale particles in the cell are uniformly distributed at a density of N particles per cubic centimeter (the abbreviated notation for density units is thus cm^{-3}). The nominal energy of the upper state is E_2; that is, the Lorentzian distribution of the occupational probability of the upper state is centered around E_2. The energy of the lower state is E_1. The center frequency ν_0 of the incident radiation coincides with photon frequency associated with the energy difference: $h\nu_0 = E_2 - E_1$. After passing through the entire length L of the cell, part of the radiation is absorbed and the new spectral distribution (showing an absorption line) is depicted to the right of the cell. For naturally broadened absorbers and when the absorption is

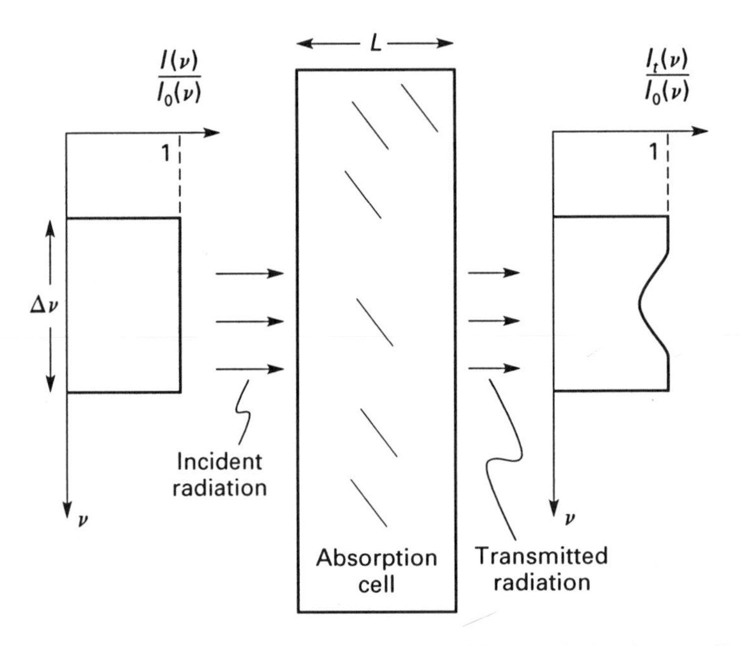

Figure 10.5 Absorption of spectrally uniform radiation by a uniformly thick and homogeneous absorber having a single absorption line.

weak, the shape of this line resembles the Lorentzian of (10.9). Thus, although the upper state was unoccupied before absorption took place, the spectral shape of the absorption spectrum, like the shape of the spontaneous emission, is determined by the probability distribution of that state – the same distribution that determined the lineshape of spontaneous emission. When the absorption is strong – that is, when a significant portion of the irradiance at the line center is absorbed, the shape of the absorption line may deviate from the Lorentzian of natural broadening. For such absorptions, which are said to be *saturated,* the center of the line is typically flattened and the width of the line increases.

As predicted by (4.36), the attentuation increases exponentially with the distance propagated through the cell, with the density of the absorbing medium, and with its absorption properties. The absorption is spectrally non-uniform; it peaks at ν_0 and tapers off away from that frequency. To account for this spectral variation we introduce a new parameter, the absorption coefficient κ_ν. With this frequency-dependent coefficient, the transmitted spectral irradiance at any frequency ν is

$$I_t(\nu) = I_0(\nu)e^{-\kappa_\nu L}. \tag{10.14}$$

Thus, κ_ν is largest at the absorption line center, while sufficiently away from ν_0 the attenuation is negligible and the spectral irradiance remains unchanged at $I_0(\nu)$.

The energy given up by the radiation is not lost; it is absorbed by the cell's atomic particles, which as a result are excited to their upper state. The absorbers are believed to be independent of each other, so absorption by one particle does not influence the absorption by its neighbor. The rate of absorption and hence the rate of excitation of two-level systems certainly depend on characteristics of the incident radiation. However, since the absorbers are independent of each other, the rate of excitation is also proportional to their density N_1 in the ground state. Therefore, the absorption rate equation is:

$$\frac{dN_1}{dt} = -W_{12}N_1, \tag{10.15}$$

where W_{12} is the absorption rate coefficient measured in units of 1/seconds.

Although the units of this rate constant and its role in the rate equation are identical to those of the Einstein A coefficient, the two are physically different. Whereas the A coefficient is an inherent property of the system and is independent of any outside factor, W_{12} depends on the frequency of the incident radiation and on its irradiance. Thus, without incident radiation, the particles in the cell remain unexcited and $W_{12} \to 0$. Furthermore, even when the cell is illuminated, if the frequency of the incident radiation is away from ν_0 then the rate of excitation still can approach zero. For the experiment described by Figure 10.5, W_{12} is the rate of absorption induced by a spectrally uniform broadband radiation. Other values of W_{12} are expected when the incident irradiance is narrowband.

To find the dependence of W_{12} on the properties of the incident radiation, the interaction between the radiation and the absorbers can be viewed as simple two-body collisions between atomic particles and photons. Some of these collisions result in elastic scattering, where the colliders exchange momentum but preserve their own energy. In such collisions, the incident photons bounce off the particles without changing their wavelength. However, other collisions result in absorption, where the colliding photon is annihilated and both its momentum and energy are taken up by the absorber. In most applications the loss of momentum is inconsequential and only the energy exchange is noted. The rate of absorption, as defined by (10.15), must then be proportional to the rate of collisions between the interacting particles and must therefore be proportional to both the density N_1 of the absorbers and the density of photons with a frequency ν. Since photons cannot be accumulated, defining their density may prove an elusive task. However, if the collimated beam is viewed as a stream of photons, then the spectral irradiance $I(\nu)$ is the energy flux of radiation at a frequency ν crossing a unit area at a velocity c:

$$I(\nu) = \rho(\nu)c, \tag{10.16}$$

where $\rho(\nu)$ is the radiative energy within a unit volume, per unit frequency, at the frequency ν. A similar expression for the intensity can be derived when the radiation is propagating away from a point source. Equation (10.16) is analogous to the continuity equation of fluid mechanics, where the flux of mass q of a fluid at a density ρ flowing at a velocity u is $q = \rho u$. When $\rho(\nu)$ is divided by the photon energy $h\nu$, the result is the number of photons per unit frequency enclosed at a given moment by a unit volume. Thus, the probability for absorption must be proportional to both the density of the absorbers and the density of the photons. Although a proportionality coefficient for the interaction between the photons and the absorbers can be defined, the conventional definition (Einstein 1917) uses an energy density representation. With this convention, the rate of absorption transitions induced by a spectrally uniform broadband beam is

$$\frac{dN_1}{dt} = -B_{12}\rho(\nu)N_1, \tag{10.17}$$

and the absorption rate constant is

$$W_{12} = B_{12}\rho(\nu) = B_{12}I(\nu)/c. \tag{10.18}$$

Another relation between the absorption rate constant and parameters of the incident radiation is found by comparing the number of photons lost by the incident radiation (which can be calculated directly from eqn. 10.14) with the loss of absorbers dN_1/dt (eqn. 10.15). Since every lost photon removes one absorber from the ground state, constants that control the loss of photons must be matched by the constant that controls the absorption rate. The first derivative of (10.14) (see Problem 10.3) is used to calculate the spectral irradiance, $\Delta I(\nu)$, lost by passage through a thin layer of absorbers Δx:

$$\Delta I(\nu) = I_0(\nu) - I_t(\nu) = I_0(\nu)\kappa_\nu \Delta x. \tag{10.19}$$

Such linear dependence between the absorbed irradiance and the absorption path is valid only for optically thin absorbers, that is, when $\Delta I(\nu)/I_0(\nu)$ is small. Equation (10.19), if divided by the photon energy $h\nu$, yields the rate of loss experienced by the group of photons with frequency ν after passing a distance Δx through the absorber. On the other hand, the loss of absorber population in the ground state depends on the absorption of photons of all frequencies within the absorption band. To calculate the total irradiance absorbed by the layer, the spectral distribution of the incident radiation must be defined. For simplicity, assume again that the incident radiation is spectrally uniform with a bandwidth of $\Delta\nu$ and spectral irradiance $I_0(\nu)$. The irradiance absorbed by the thin layer of absorbers is then obtained by the following integration of $\Delta I(\nu)$ over the entire frequency spectrum:

$$\Delta I = \int [I_0(\nu) - I_t(\nu)]\, d\nu = I_0(\nu)\Delta x \int \kappa_\nu \, d\nu. \tag{10.20}$$

Far from ν_0, where $\kappa_\nu \to 0$, the irradiance remains constant at $I_0(\nu)$ and the integrand approaches zero. In other words, spectral irradiance outside the absorption line does not contribute to the excitation of the absorbers and is not attenuated. It is therefore sufficient to calculate the integral only in the vicinity of ν_0, where absorption is nonnegligible.

The absorbed irradiance is used by the absorbers to raise their energy from level 1 to level 2. For each absorbed photon at the nominal energy of $h\nu_0$, one system is excited to the upper state. The rate of absorbing transitions per unit volume, $-dN_1/dt$, must therefore equal the number of photons of all frequencies that were absorbed per unit volume:

$$\frac{\Delta I}{h\nu_0 \Delta x} = -\frac{dN_1}{dt}.$$

The link between W_{12} and the total absorption by a medium containing N_1 absorbers per cubic centimeter can now be found by introducing (10.15) and (10.20) into this equation as follows:

$$\int \kappa_\nu \, d\nu = \frac{W_{12}}{I_0(\nu)} N_1 h\nu_0. \tag{10.21}$$

After multiplying and dividing by c and using (10.18) to replace W_{12} with the Einstein B coefficient, we have:

$$\int \kappa_\nu \, d\nu = \left(\frac{W_{12}c}{I_0(\nu)}\right)\frac{N_1 h\nu_0}{c} = B_{12}\frac{N_1 h\nu_0}{c}.$$

Finally, the Einstein B coefficient for an absorption from level 1 to level 2 of a two-level system expressed by the absorption coefficient is

$$B_{12} = \frac{c}{N_1 h\nu_0}\int \kappa_\nu \, d\nu. \tag{10.21$'$}$$

The absorption coefficient κ_ν is a macroscopic property of the absorbing medium in the cell and as such increases proportionally with its density. However, after division by N_1, the right-hand side of (10.21′) describes the property of an individual absorber. Therefore, B_{12}, as described by (10.21′), depends only on the properties of the individual absorber. Furthermore, we have shown that B_{12} represents the absorption integrated over the entire absorption line of spectrally uniform radiation.

The units of B_{12} can be readily determined from (10.21′). In the MKSA system they are $m^3/J\text{-}s^2$. In an alternative derivation, where the frequency is defined in units of cm^{-1}, B_{12} is defined in units of $m^2/J\text{-}s$. To convert (10.21′) to these alternative units, B_{12} is divided by $c_0 = 3\times10^8$ m/s. Other definitions of B_{12}, and consequently other units, are often given in the literature (see e.g. Gray and Farrow 1991).

Equations (10.18) and (10.21′) demonstrate that B_{12} is a molecular or atomic constant that is specific only to the absorbing transition. It is unlike W_{12}, which can assume different values for different parameters of the incident radiation. It has different values when the incident source is broadband, or narrowband tuned to the absorption line center or away from it. Therefore, before using (10.15) to calculate rates of absorption, W_{12} must be re-evaluated to account for the characteristics of the incident source. For spectrally uniform radiation with a bandwidth of $\Delta\nu$, the radiative spectral energy density $I_0(\nu)$ is

$$I_0(\nu) = I/\Delta\nu;$$

otherwise, for spectrally non-uniform radiation, the spectral energy density is

$$I(\nu) = I g_L(\nu), \tag{10.22}$$

where $g_L(\nu)$ is the normalized lineshape factor of the incident radiation (cf. eqn. 10.9). This lineshape factor should be distinguished from the lineshape function of the absorbers $g(\nu)$. For naturally broadened spontaneous emission, $g_N(\nu)$ is defined by (10.9). (Although it was there defined as $g_N(\omega)$, those units can be readily converted to 1/hertz.) Other effects that may influence the absorption lineshape factor are discussed in Section 10.7. To obtain the absorption rate constant for spectrally non-uniform radiation, the uniform spectral irradiance term in (10.20) must be replaced with $I(\nu)$, which is then kept inside the integral where it is convolved with the absorption lineshape. Of course, such modification of (10.20) affects only W_{12} (see eqn. 10.24). It cannot affect B_{12}, which has already been defined as a molecular parameter using the absorption of spectrally uniform radiation.

The absorption of radiation by a two-level system may be presented in an alternative, graphical way. Assume that the absorber can be viewed as a sphere with an effective cross section σ_{12}, which is not necessarily identical to the geometrical cross section. Instead, this hypothetical cross section delineates a region around the absorber that serves as a photon "trap." Whenever a photon pierces through the boundaries of this region, it is annihilated and the absorber is excited to the upper state. In this picture, the boundaries of the trap around a weak absorber are close to its core and its absorption cross section is small. Conversely, a strong absorber has a large absorption cross section. The combined cross section of all absorbers is $N_1\sigma_{12}$. An incident beam with spectrally uniform irradiance at nominally ν_0 can be viewed (eqn. 10.16) as a collimated stream of photons with an approximate flux (photons per unit time and unit area) of $I/h\nu_0$. The rate at which this stream is piercing through the combined cross section must, by the definition of σ_{12}, equal the product of the flux and the combined cross section, which in turn equals the rate of transitions from level 1:

$$\frac{dN_1}{dt} = -\frac{I}{h\nu_0}\sigma_{12}N_1. \tag{10.23}$$

Comparing with (10.15), we find that

$$W_{12} = \frac{I}{h\nu_0}\sigma_{12}.$$

When the linewidth $\Delta\nu$ is broad relative to the absorption linewidth and when the spectral irradiance is uniform, the absorption cross section can also be expressed using (10.18) in terms of the Einstein B coefficient:

$$\sigma_{12} = \frac{B_{12}h\nu_0}{c\Delta\nu}.$$

As $\Delta\nu$ increases, σ_{12} decreases, which implies that much of the radiation of a spectrally broad source is not coincident with the absorption line and so its losses due to absorption are small.

With this definition of the absorption cross section, its magnitude depends on the spectral distribution of the source and is therefore of little use. Instead, we derive a cross section for absorption of radiation with a bandwidth that is narrow relative to the absorption bandwidth. This cross section will depend on the properties of the absorber, and the only independent parameter will be the incident wavelength. To distinguish this highly monochromatic irradiance from broadband irradiance, and to indicate that its nominal frequency is ν, we denote it by $I_\nu = I(\nu)\,d\nu$. The absorption of such narrowband radiation depends on its frequency relative to the nominal absorption frequency ν_0; it is most pronounced at ν_0 and diminishes away from the line center.

The frequency dependence of the absorption has already been described by (10.14), which applies also to the absorption of narrowband radiation. Therefore, the frequency dependence of the cross section can be derived from that equation. Assuming that the frequency dependence of the absorption coefficient is characterized by the absorption lineshape $g(\nu)$ and by a parameter κ_0 that depends on the absorption at the line center,

$$\kappa_\nu = \kappa_0 g(\nu).$$

The irradiance lost by a beam when passing through a layer at a depth of Δx (eqn. 10.20) must match the energy gained by the absorbers (eqn. 10.15). But now the incident irradiance is narrowband and so the integration of (10.20) can be omitted. Comparing (10.15) and (10.20), we obtain for the absorbed irradiance

$$\Delta I_\nu = W_{12} N_1 h\nu \Delta x = I_\nu \kappa_0 g(\nu) \Delta x. \tag{10.24}$$

It is readily seen that when the incident radiation is narrowband, W_{12} depends on the absorption lineshape factor $g(\nu)$ and so varies with the incident frequency. Before this relation can be further reduced, κ_0 must be replaced by B_{12} using (10.21′) and the condition $\int g(\nu)\,d\nu = 1$. With this substitution, the absorption cross section for narrowband irradiance at frequency ν is

$$\sigma_{12}(\nu) = \frac{B_{12} h\nu_0}{c} g(\nu) \tag{10.25}$$

(see Problem 10.4). Equation (10.25) can be used to calculate $\sigma_{12}(\nu)$ when B_{12} is expressed in units of $m^3/J\text{-}s^2$. For an alternative equation for the absorption cross section, see Example 10.3. This cross section may vary as the absorption lineshape varies. Nevertheless, it is now a measure of the absorption properties of an absorber at prescribed physical conditions. Finally, with this definition of the absorption cross section and using (10.21′), the absorption coefficient can now be replaced with $\sigma_{12}(\nu)$ to yield

$$\kappa_\nu = N_1 \sigma_{12}(\nu),$$

and the attenuation of a narrowband beam when passing a distance L through a cell containing absorbers at a density of N_1 is

$$I_\nu(L) = I_\nu(0) e^{-N_1 \sigma_{12}(\nu) L}. \tag{10.26}$$

Although the cross section presented in (10.25) is often comparable to the geometrical cross section of the absorber, it may be either significantly larger or smaller. This is illustrated by the following comparison of the absorption cross section of two transitions of the OH molecule.

Example 10.3 Find the absorption cross section of the $A^2\Sigma\ (v'=0) \leftarrow X^2\Pi\ (v''=0)\ Q_1(2)$ and of the $A^2\Sigma\ (v'=3) \leftarrow X^2\Pi\ (v''=0)\ Q_1(2)$ transitions of OH. The Einstein B coefficients for these transitions are $B_{00} = 3.024 \times 10^9$ $m^2/J\text{-}s$ and $B_{03} = 1.682 \times 10^7\ m^2/J\text{-}s$, and their nominal wavelengths are 308.08 nm (Dieke and Crosswhite 1962) and 244.99 nm (Quagliaroli 1993), respectively. The lifetime of the $v'=0$, $N'=2$, $J'=\frac{5}{2}$ state can be determined from the results of Example 10.1. The $v'=3$, $N'=3$, $J'=\frac{5}{2}$ state is strongly predissociated; its predissociation lifetime is 312 ps (Heard et al. 1992).

Solution When the Einstein B coefficient is expressed in units of $m^3/J\text{-}s^2$, the absorption cross section is calculated directly by (10.25). However, the B coefficients in the present problem are expressed in units of $m^2/J\text{-}s$. These units can be converted to the units required in (10.25) by multiplication by c_0. Conversely, if (10.25) is multiplied by c_0 then the units of the B coefficients as stated in the present problem can be used directly. Thus, for the $0 \rightarrow 0$ transition,

$$\sigma_{00}(\nu) = B_{00} h\nu_{00} g(\nu) \approx \frac{B_{00} h\nu_{00}}{\Delta\nu_{00}}.$$

Although $g(\nu)$ is a Lorentzian and is well defined, calculations of the cross section are often simplified by replacing $g(\nu)$ with $1/\Delta\nu_{00}$ for the $0 \leftarrow 0$ band, where $\Delta\nu_{00} = 7.8 \times 10^{-6}\ cm^{-1}$ is the absorption linewidth calculation in Example 10.1 using the effective Einstein A coefficient. The frequency of this transition can now be stated as

$$\nu_{00} = \frac{10^7}{\lambda\ [nm]} = \frac{10^7}{308.08} = 32{,}459\ cm^{-1},$$

and the absorption cross section is

$$\sigma_{00}(\nu) \approx \frac{(3.024 \times 10^9)(6.63 \times 10^{-34})(32{,}459)}{7.8 \times 10^{-6}} = 8.34 \times 10^{-15}\ m^2.$$

Although by its definition $\sigma_{00}(\nu)$ must depend on the frequency of the incident radiation, the present approximation of $g(\nu) \approx 1/\Delta\nu_{00}$ renders it uniform throughout the entire absorption band. Similarly, the cross section for the $3 \leftarrow 0$ transition is

$$\sigma_{03}(\nu) \approx \frac{(1.682 \times 10^7)(6.63 \times 10^{-34})(40{,}818)}{0.0170} = 2.67 \times 10^{-20}\ m^2.$$

For comparison, the effective geometrical cross section of a molecule is determined by the cross section of the imaginary sphere traced out by its rotational motion. Therefore, with a distance of approximately 1 Å (Carlone and Dalby 1969) between the OH nuclei, the area of the geometrical cross section $\sigma_{geo} = 1.57 \times 10^{-20}\ m^2$ is comparable to σ_{03} but is significantly smaller than σ_{00}. On the other hand, weaker absorptions may result with cross sections that are significantly smaller than the geometrical cross section. Nevertheless, the geometrical cross section can serve as a yardstick for comparison with cross sections of absorption or other molecular and atomic interactions. ■

Stimulated emission is the counterpart of absorption. Whereas absorption is an interaction between photons at the nominal frequency ν_0 and two-level systems at the ground state, stimulated emission is an interaction between the same photons and two-level systems in their excited state. Like absorption, the photons force the system out of its present state. However, since the only other available state is the lower state, each successful interaction results in an emission of a new photon while demoting the system to the lower state. The newly emitted photon is identical to the stimulating photon and combines with it to form a *coherent pair.* Figure 10.6 compares this emission process with the previously identified processes of spontaneous emission and absorption. The left side of the figure shows a two-level system and its radiative environment just prior to the transition, while the right side shows the outcome of each transition. The first two interactions in the figure are intuitively plausible. However, the third interaction - stimulated emission - is unique, particularly because it is the only process that results in the amplification of radiative fields. Although spontaneous emission can contribute to existing fields, the spontaneously emitted photons are unrelated; that is, they are incoherent with each other or with the existing field. On the other hand, stimulated emission reproduces all the characteristics of the incident field. Therefore, when the incident

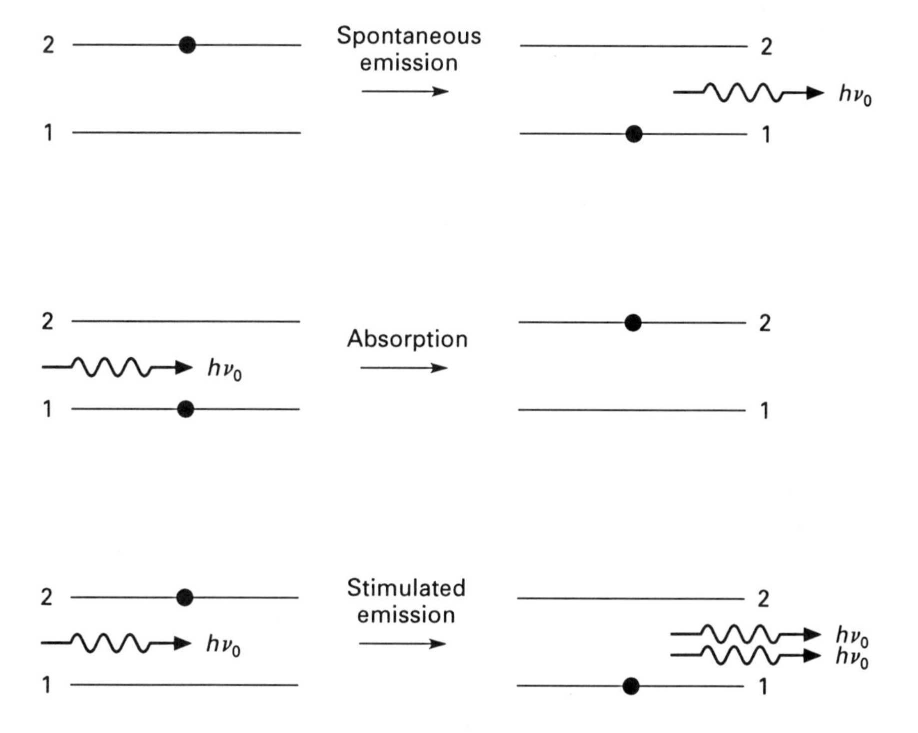

Figure 10.6 Schematic comparison between spontaneous emission (top), absorption (center), and stimulated emission (bottom).

field is coherent, all emitted photons are coherent with the field as well as with each other. This emission is the source of laser radiation.

To model the stimulated emission process, assume that the contents of the cell in Figure 10.5 are replaced by two-level systems, all in the excited state. Experimentally, it is nearly impossible to realize this objective. However, if it were possible, the spectral irradiance around ν_0 would be amplified by passing it through the cell, with the strongest amplification at ν_0. This amplification may be viewed as an inverse absorption (i.e., as a negative loss). Therefore, the time-dependent population N_2 of the excited state can be described by a similar rate equation:

$$\frac{dN_2}{dt} = -W_{21}N_2, \tag{10.27}$$

where W_{21} is the rate constant for stimulated emission. Like W_{12}, this rate constant depends on the properties of the incident radiation as well as on the properties of the emitter. These properties of the emitter are described by a new Einstein B coefficient (Einstein 1917), B_{21}. For incident radiation with spectral irradiance of $I(\nu)$, B_{21} is defined similarly to B_{12} (eqn. 10.18) as follows:

$$B_{21} = \frac{W_{21}c}{I(\nu)}. \tag{10.28}$$

Equations (10.18) and (10.28), together with (10.1), specify three coefficients (A_{21}, B_{12}, and B_{21}) that characterize the rate of radiative energy transfer between any two quantum mechanical states. Each pair of coupled quantum mechanical energy levels has its own Einstein coefficients, which can be derived from the quantum mechanical properties of these states and can be shown to be related to each other. It is thus sufficient to evaluate, either experimentally or theoretically, only one coefficient; the other two coefficients can then be calculated from the known one. The relationship between these coefficients must be such that conditions of thermodynamic equilibrium of any medium are met both for the energy stored by the medium itself and for the radiative energy field that interacts with it. When this equilibrium is established, the role of the three Einstein coefficients is to balance the emission and absorption between the various energy states with the density of their population. Thus, the relationship between the Einstein coefficients can be derived from two statistical distributions that describe thermodynamic equilibrium. The first is the distribution of the population density among all the available energy levels when a medium is at thermodynamic equilibrium; the second depicts the radiative field associated with the equilibrated medium. The next section describes these distributions and through them provides the link between B_{12}, B_{21}, and A_{12}.

10.5 The Statistics of Thermodynamic Equilibrium in Microscopic Systems

All media – whether gaseous, liquid, or solid – contain a vast number of atomic-scale particles, and each of these particles has numerous quantized energy levels. In the gas phase, these particles move about and collide with each other and with the walls of their container. During these collisions, energy can be exchanged, thereby raising one particle to a higher energy state while lowering its collisional partner to a lower level. In the liquid and solid phases, these particles vibrate and bounce against their immediate neighbor while exchanging energy. Energy exchange also occurs by absorption and emission of photons; some of the particles in the excited states emit (mostly spontaneously), whereas others absorb stray photons. Therefore, the photons that are continuously generated by emission and annihilated by absorption coexist with the atomic particles.

When such media are excited by a sudden input of energy such as a flash of light, an electric discharge, or a shock wave, many of these particles are excited to higher energy levels. However, once the disturbance disappears, the energy that was acquired by these particles is redistributed among the particles and among the coexistent photons by collisions, radiative emissions, and absorptions until a new state of equilibrium is reached. At this equilibrium the medium appears on the macroscopic scale to have again reached a steady state, and will remain in this state indefinitely without outside intervention. This is compatible with the classical definition of equilibrium in the thermodynamic sense. However, on the microscopic scale the activity remains unceasing: particles exchange energy with each other, with their external boundaries, and with the photon population surrounding them. Thus, although the energy of an individual particle may change numerous times every second, the established equilibrium implies that the number of particles occupying any of the available energy levels remains independent of time and accordingly the population density of one level relative to another is also fixed. The population density of any state of a species depends on the overall density of that species, but the relative population of two states of the same species is independent of the density and is a measure of only one macroscopic thermodynamic property: the temperature. Similarly, the distribution of the photon population (i.e., the number of photons at each frequency) that coexist within this medium depends on the thermodynamic conditions. Since these photons are continuously emitted and absorbed by the individual transitions, their net distribution at any time is an image of the population distribution among the various quantum states. And, like the population distribution among the energy states, the distribution of the photon population at thermodynamic equilibrium is a function of the temperature alone. Thus, when therodynamic equilibrium is fully established, although the state of any one particle may remain undefined, the statistical distributions among the various states and among the photons that are generated and consumed by the various transitions are well defined. Ideally, these distributions depend solely on such macroscopic thermodynamic properties as temperature, mole fraction, and density.

The analysis of the interaction of microscopic particles among themselves and with radiative fields is part of the science of statistical mechanics (see e.g. Vincenti and Kruger 1965). Its results are used to study such energy transfer modes as heat by conduction or convection, species transfer by diffusion, momentum transfer by viscosity, and (of course) radiative transfer. Theories of statistical mechanics are also used to derive macroscopic thermodynamic properties from microscopic considerations, or to study material properties. For the objective of this book, only the distribution of the population of atomic-scale particles among their various energy states and the spectral distribution of the radiative field surrounding them is described. These results will then be used to derive the links between the three Einstein coefficients. In addition, since temperature influences the statistical distribution of the energy among the atomic-scale particles of all media and consequently the spectral distribution of the radiation they emit, we will identify methods for the measurement of temperature from the measurement of either of these distributions.

The statistical distribution among the available quantum mechanical states of atomic-scale particles in thermodynamically equilibrated media is described by the *Boltzmann distribution*. When the intrinsic properties of a medium cannot be changed without an external intervention, the population density – that is, the number of particles in a unit volume, N_i, that occupy the ith energy level – depends only on the temperature T and is described by

$$N_i = Ae^{-E_i/kT}, \tag{10.29}$$

where A is a coefficient yet to be determined, E_i is the energy of that state, and $k = 1.38 \times 10^{-23}$ J/K is the Boltzmann constant. Figure 10.7 shows the variation of the relative population density with energy in a gas of a hypothetical molecule. The population density is shown relative to the population density in the zero energy state for three temperatures: T_0, $2T_0$, and $3T_0$. The relative population densities of four levels of that molecule (1, 2, 3, and 4) are marked by horizontal bars.

As shown by the Boltzmann distribution, the largest number of particles is at the lowest possible energy state. As the energy of a state increases, the number of particles occupying it decreases exponentially. As temperature increases, the population of higher energy states increases at the expense of the population of lower states. However, no matter how high the temperature, the population of an upper state remains smaller than the population of a lower energy state. Even when the temperature approaches infinity, the upper-state population can at most equal the population of any lower state. Thus, *population inversion* – the condition where the population of an upper state exceeds the population of a lower state – is incompatible with thermodynamic equilbrium. Of course, population inversion can be induced by numerous methods such as a shock wave, a flame, an electric discharge, or an optical excitation. However, once the disturbance is removed, the population among the various energy states is redistributed and, following a finite transient period, the distribution invariably returns to the pattern predicted by (10.29). During the time the external disturbance is applied and during the transient redistribution period, temperature *cannot be defined*. Conversely, when equilibrium is established, temperature is fully defined by the Boltzmann population distribution. This

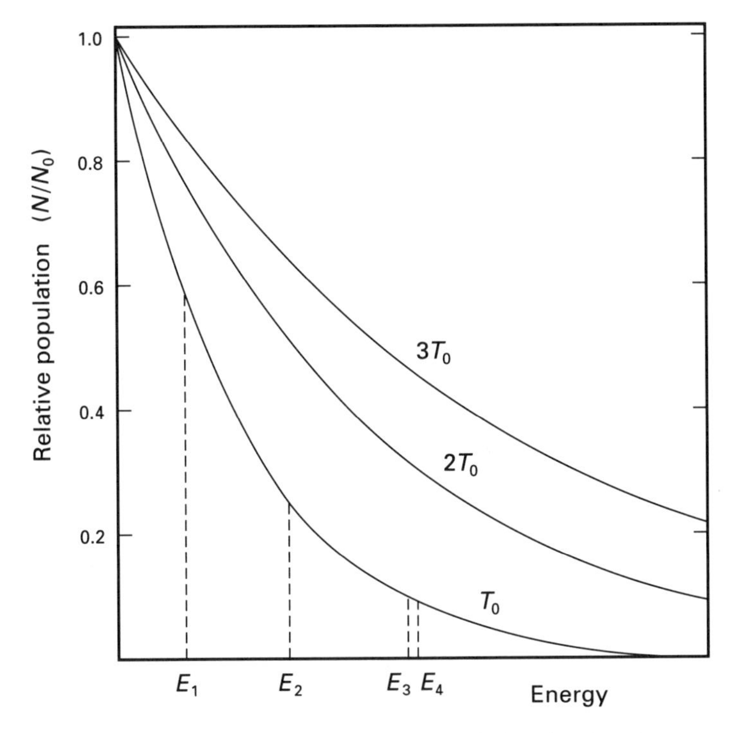

Figure 10.7 Variation of the relative population density of quantum states with energy (Boltzmann's distribution).

means that, at equilibrium, measurement of the relative population of two or more known energy levels can serve as a measure of temperature.

The distribution of (10.29) describes well the relative population of discrete states. However, energy levels are often degenerate: more than one state can have exactly the same energy. Because all degenerate states must (according to eqn. 10.29) have the same population density, the total population occupying the degenerate states must increase with their degeneracy. To illustrate this effect, consider levels 3 and 4 in Figure 10.7. These distinguishable levels are energetically close to each other and accordingly have nearly the same population. Thus, the total population of these two states is approximately twice the population of each state separately. Even as the energy gap between them narrows, so long as they remain two separate levels the sum of their populations will remain twice the population of each. By extension, when the degeneracy is g_i, the total population of all degenerate states is

$$N_i = A g_i e^{-E_i/kT}. \tag{10.30}$$

The distribution described by (10.30), apart from the unspecified coefficient A, is complete. Therefore, the population density of energy level i relative to level j is obtained by the following ratio:

$$\frac{N_i}{N_j} = \frac{g_i}{g_j} \exp\left\{-\frac{E_i - E_j}{kT}\right\} = \frac{g_i}{g_j} \exp\left\{-\frac{\Delta E}{kT}\right\}. \tag{10.31}$$

This ratio is independent of the coefficient A, and thus may be used to determine either the relative population density of two states (when the temperature is known) or the temperature (when the population distribution is known).

Another parameter that can be derived from (10.31) is the threshold temperature beyond which the population of a certain state is considered to be significant. Such a threshold temperature is used to estimate the degree of excitation of quantum mechanical energy levels. Assume that a vibrational level is considered to be excited when its population density is $1/e$ of the population of the ground vibrational state. For such a criterion, the threshold temperature Θ_v required to vibrationally populate a molecular species is found by setting the exponent of (10.31) to unity. Using the difference $\Delta G(v)$ between the term values of the two lowest vibrational states (eqn. 9.27), this temperature, or the *vibrational temperature* of the molecule, is

$$\Theta_v = \frac{hc_0 \Delta G(v)}{k} = \frac{hc_0}{k}(\omega_e - 2\omega_e x_e).$$

For O_2 in the ground electronic state, $\Theta_v = 2{,}243$ K. For comparison, the rotational temperature of O_2 in its ground electronic state is $\Theta_r = 20.8$ K. Thus, at a temperature of only ~21 K, the population of the $N=3$ level is already $1/e$ of the population of the $N=1$ level. (Recall that, for O_2, rotational levels with even N quantum numbers are unpopulated.) As indicated by these results, at atmospheric conditions and at thermodynamic equilibrium, the population of O_2 is concentrated mostly in the $X\,^3\Sigma_g^-$ $(v=0)$ vibronic state while remaining negligible at any other vibrational or electronic state. On the other hand, owing to their close spacing, many of the rotational states in the ground vibronic state are well populated and measurements of the relative population of two or more rotational states can be a sensitive method of determining the temperature of cold gases. The use of the vibrational level population for such measurements becomes practical when the temperature approaches or exceeds Θ_v.

Example 10.4 Using a spectroscopic technique, the vibrational population of the $v=1$ state of O_2 was measured relative to the population of the $v=0$ state. Find the temperature of the gas and the temperature measurement uncertainty $\Delta T/T$ when the population density ratio is $R = N_{(v=1)}/N_{(v=0)} = 0.1 \pm 0.01$.

Solution The problem does not specify whether the rotational structure was resolved in this measurement. We therefore assume that, in the course of this spectroscopic measurement, the contribution of all the rotational lines was integrated and the population density ratio R includes all the rotational levels within each vibrational state. From (10.31), the explicit temperature in terms of the ratio R is

$$T = -\frac{hc_0 \Delta G(v)}{k \ln R} = -\frac{hc_0(\omega_e - 2\omega_e x_e)}{k \ln R};$$

using the data of Table 9.1 and Table B.1, its numerical value is

$$T = -\frac{1.4413(1{,}580.3 - 2\cdot 12.073)}{\ln(0.1)} = 974 \text{ K}.$$

The accuracy of this temperature measurement depends both on the accuracy with which the spectroscopic coefficients are specified and on the error ΔR introduced by the measurement of the population density ratio. Assuming that the contribution of the uncertainty in the spectroscopic coefficients is negligible, the effect of ΔR on the temperature measurement uncertainty can be obtained from the following exact differential of (10.31):

$$\frac{\Delta R}{R} = \frac{hc_0 \Delta G(v)}{kT^2} \Delta T.$$

Thus, the temperature measurement uncertainty is

$$\frac{\Delta T}{T} = \frac{\Delta R}{R}\left[\frac{hc_0 \Delta G(v)}{kT}\right]^{-1}$$

(Grinstead, Laufer, and McDaniel 1993). The term in brackets is the *sensitivity* of this measurement. With high sensitivity, the temperature measurement uncertainty can remain small even when the uncertainty in the spectroscopic measurement is high. Thus, selection for these measurements of two levels with the largest possible energy difference can reduce the temperature measurement uncertainty. For the present experiment, the sensitivity is 2.33 and for $\Delta R/R =$ 10%, the temperature measurement uncertainty is

$$\frac{\Delta T}{T} = 0.43 \frac{\Delta R}{R} = 4.3\%.$$

■

In order to obtain the population density of an individual state of a certain species, and not just the ratio of the population densities of two states, the proportionality coefficient A of (10.29) must be determined. It can be conveniently associated with the total density N of that species, which is a thermodynamic property that may be independent of the temperature. Thus, by summation of the population in all the energy states of the species,

$$N = \sum_i A g_i e^{-E_i/kT}.$$

In terms of the total density, the proportionality factor of (10.29) is

$$A = \frac{N}{\sum_i g_i e^{-E_i/kT}}$$

and with this the Boltzmann distribution becomes

$$N_j = N \frac{g_j e^{-E_j/kT}}{Q}, \tag{10.33}$$

where

$$Q = \sum_i g_i e^{-E_i/kT} \tag{10.34}$$

is the *partition function*. By this definition, the partition function is obtained from the summation of the population of the entire manifold of energy levels of each species. However, from a practical standpoint such an approach is not feasible. Instead, the summation is carried out over a specific group of states. For example, at atmospheric conditions and at thermodynamic equilibrium, the overwhelming majority of the O_2 molecules of air are in their ro-vibronic ground state and so the partition function need only include summation over the rotational states in the $X\,^3\Sigma_g^-$ ($v=0$) level. This specific function, the *rotational partition function,* is thus obtained as

$$Q_r = \sum_J (2J+1)\exp\left\{-\frac{hc_0}{kT}B_v J(J+1)\right\} \quad \text{or} \quad Q_r \approx \frac{kT}{hc_0 B}. \tag{10.35}$$

The second formula was obtained by replacing the summation with an integration with respect to J that was then carried out from 0 to ∞ (Herzberg 1989, p. 125). For homonuclear molecules such as O_2, where states with alternating J quantum numbers are empty, Q_r must be divided by 2 (Fast 1962, pp. 275–80).

The vibrational partition function, Q_v, must be evaluated (see e.g. Vincenti and Kruger 1965, p. 135) only when the population density of higher vibrational states is significant (i.e., when $T \geq \Theta_v$). Because each vibrational state includes its own manifold of rotational states, the combined partition function for both rotational and vibrational states is obtained by multiplying the two separate partition functions. Thus the population density of a given ($v; J$) state in the ground electronic state is

$$N(v;J) = N\frac{(2J+1)\exp\left\{-\dfrac{hc_0\omega_e(v+1/2)}{kT}\right\}\exp\left\{-\dfrac{hc_0 B_v J(J+1)}{kT}\right\}}{Q_v Q_r}, \tag{10.36}$$

where only the first-order terms of the term values (eqn. 9.30) were included. Similarly, when the partition function of the electronic states, Q_e, must be included, the combined partition function is obtained by the product $Q_r Q_v Q_e$.

Boltzmann's distribution predicts that the population density of an individual state – not accounting for the degeneracy – is lower than the population of a lower state. However, with degeneracy included (eqn. 10.31), the combined population of the degenerate states of one energy level may exceed the combined population of the degenerate states of a lower energy level. This excess, which typically occurs among the rotational levels of molecules, does not represent a departure from thermodynamic equilibrium or an inversion. Because of their ever-increasing degeneracy, higher rotational states become highly populated as temperature increases. These multiply degenerate states act as "sinks" to the molecules of lower states. Thus, although individual "sub"-states at an upper J level are always less populated than "sub"-states in a lower J level, their combined population can exceed that of the lower state. It is thus possible to find, in the manifold of rotational states, one state with a quantum number

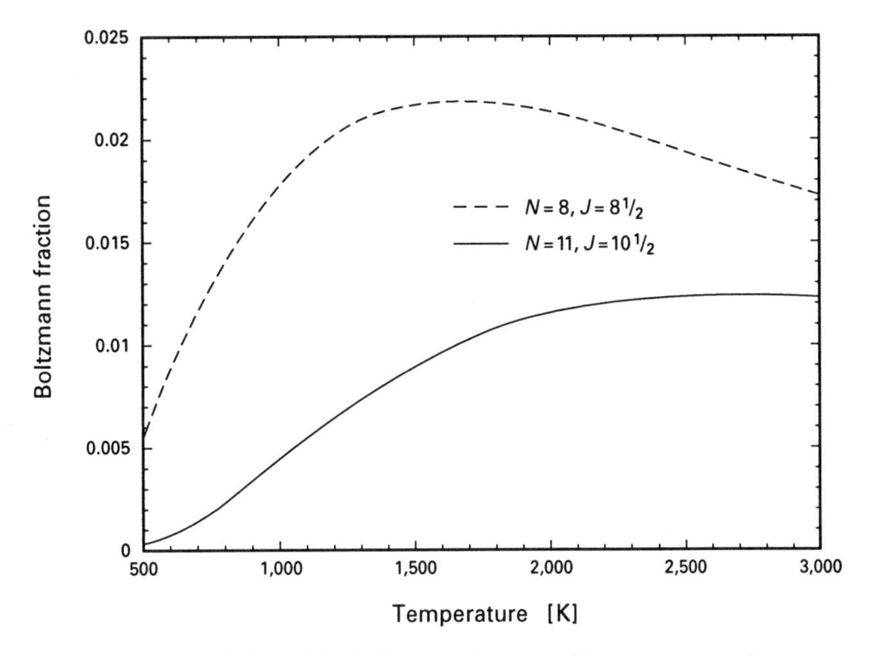

Figure 10.8 Variation of the Boltzmann fraction with temperature of two rotational states of OH in the $X\,^2\Pi$ ($v=0$) state.

J_{max} that has a combined population exceeding the population of any state above or below. This unique temperature-dependent quantum number, which can be obtained from the derivative of (10.36) with respect to J, is:

$$J_{max} = \sqrt{\frac{kT}{2hc_0 B_v}} - \frac{1}{2} \tag{10.37}$$

(Herzberg 1989, p. 124). Thus, as temperature rises, the population density of a rotational state increases until it reaches a maximum. As temperature increases even further, the population of that state declines, as the need to feed the population of upper states (with higher degeneracy) increases. Figure 10.8 illustrates this phenomenon by describing the variation of the population density with temperature of two rotational states of OH in the ground vibronic state. The population densities are represented by the Boltzmann fraction,

$$\beta_i = \frac{g_i}{Q} \exp\left\{-\frac{\Delta E_i}{kT}\right\}. \tag{10.38}$$

This is the fraction of all OH molecules that occupy that state. Clearly, the population density of the $N=8$ state increases until a well-defined maximum is reached. Beyond that maximum, the population of the $N=8$ state declines as temperature increases. For this molecule, the state with the maximum population at 3,000 K is $N_{max}=10$. Therefore, the maximum of the $N=11$ level is reached beyond that temperature and cannot be observed in the figure.

In the course of certain chemical reactions, in rapid combustion processes, or in electrical discharges, a large number of highly excited molecules may be formed. The excitation may include a large population in upper electronic and vibrational states. Generally, these excited species are not in thermodynamic equilibrium and the distribution among these states does not comply with (10.33). As thermodynamic equilibrium is being recovered, closely spaced states become equilibrated well before the entire population is. Thus, it may become possible for the distribution among the rotational states of certain vibrational levels - either in the ground electronic state, or even in an excited electronic state - to approach equilibrium while the medium itself remains unequilibrated. Analysis by a high-resolution monochromator of the emission spectrum of flames can be used to confirm such partial equilibrium among rotational states of an otherwise unequilibrated medium. The results of such spectroscopic analysis may be used to measure the intermediate temperature. If the rotational emission spectrum is well resolved, the emission by individual rotational lines is proportional to the population of the rotational states. Thus, if the Einstein A coefficient of the J rotational line is A_J and the population of the emitting state is N_J, then the signal recorded for this emission line is

$$S_J = \eta \frac{N_0}{Q_r} A_J g_J \exp\left\{-\frac{E_J}{kT}\right\},$$

where N_0 is the population of the vibronic state and η is the collection and detection efficiency of the monochromator system. If the Einstein A coefficients and the energies of all the emitting states are known, equilibrium can be confirmed by drawing a Boltzmann plot showing the variation of $\log[S_J/g_J A_J]$ of each of the resolved lines with E_J/k. When all the recorded states are in thermodynamic equilibrium, the Boltzmann plot forms a straight line with a slope of $-1/T$. The other parameters of the experiment, $\eta N_0/Q_r$, usually do not vary from line to line and so their contribution merely offsets the plot without affecting the temperature dependence. Similar plots may be obtained for other energy-level manifolds. However, a straight line with a $-1/T$ slope can be expected only at equilibrium.

Example 10.5 The emission from an oxyacetylene flame has been analyzed spectroscopically and the emission of OH lines in the $A^2\Sigma\ (v'=0) \rightarrow X\,^2\Pi\ (v''=0)$ band was measured (Dieke and Crosswhite 1962). Table 10.2 lists the relative measured signal S_N of several rotational lines in the R_2 branch together with their relative transition probabilities (including the degeneracy) $A_N g_N$ and their energies E_N (in cm^{-1}). Determine if this rotational manifold of OH is at thermodynamic equilibrium; if it is, calculate the intermediate flame temperature.

Solution Rotational equilibrium can be confirmed graphically by a Boltzmann plot, where the relative signal of the emission lines is related to the energies of the emitting states by:

$$\ln \frac{S_N}{A_N g_N} = \text{constant} - \frac{hc_0 E_N}{kT},$$

Table 10.2 *Spectroscopic parameters and measured irradiance for rotational lines in the* R_2 *branch of the* $A^2\Sigma$ $(v'=0) \rightarrow X^2\Pi$ $(v''=0)$ *band of OH*

N	S_N	$A_N g_N$	E_N [cm^{-1}]
1	68	2.7	32,474
4	273	12.8	32,778
7	378	24.8	33,383
10	352	37	34,280
13	255	49.1	35,459

Note: S_N denotes relative measured signal; $A_N g_N$ denotes relative transition probabilities including degeneracy; E_N denotes emitting-state energy.

where the constant represents the experimental parameters of $\eta N_0/Q_r$. Figure 10.9 (overleaf) presents the Boltzmann plot for this experiment. This is certainly a straight line, and the temperature represented by the inverse of the negative slope is 2,727 K. Thus, although the flame front itself is most likely not in equilibrium, the temperature of the rotational states is defined. Following the rotational manifold, equilibrium is established by the vibrational manifold and then by the electronic states. It is difficult to predict the relationship between the rotational, vibrational, and eventually the real temperatures of the gas. The result depends on the extent of energy stored by the nonequilibrated states. Nevertheless, intermediate measurements like the one here are useful as an indicator of the expected temperature after equilibrium is established. ■

The Boltzmann distribution as described by (10.33) can be viewed as a steady-state phenomenon. In the absence of external disturbances, this distribution of the population of energy states remains unchanged indefinitely. However, when viewed microscopically, the medium appears very lively. The microscopic particles that make up this medium are excited or de-excited by collisional energy exchange with other particles; they translate and vibrate; and, just as important, they emit or absorb radiation. Thus, when the population distribution among the energy states of particles inside a closed and insulated container remains constant, the distribution of the photons in that vessel must remain unchanged as well. Since these photons are the net result of the unceasing emission and absorption transitions between the energy states of the particles, their distribution must be temperature-dependent, just like the distribution of the population density. Unfortunately, the radiation inside the container and its spectral distribution cannot be easily detected to verify this hypothesis, because opening the container to sample the radiation may disrupt its thermodynamic equilibrium. However, an alternative equilibrium configuration may be used whereby a perfectly absorbing surface is illuminated by an outside source. If the incident irradiance exactly balances any surface emission, the object remains

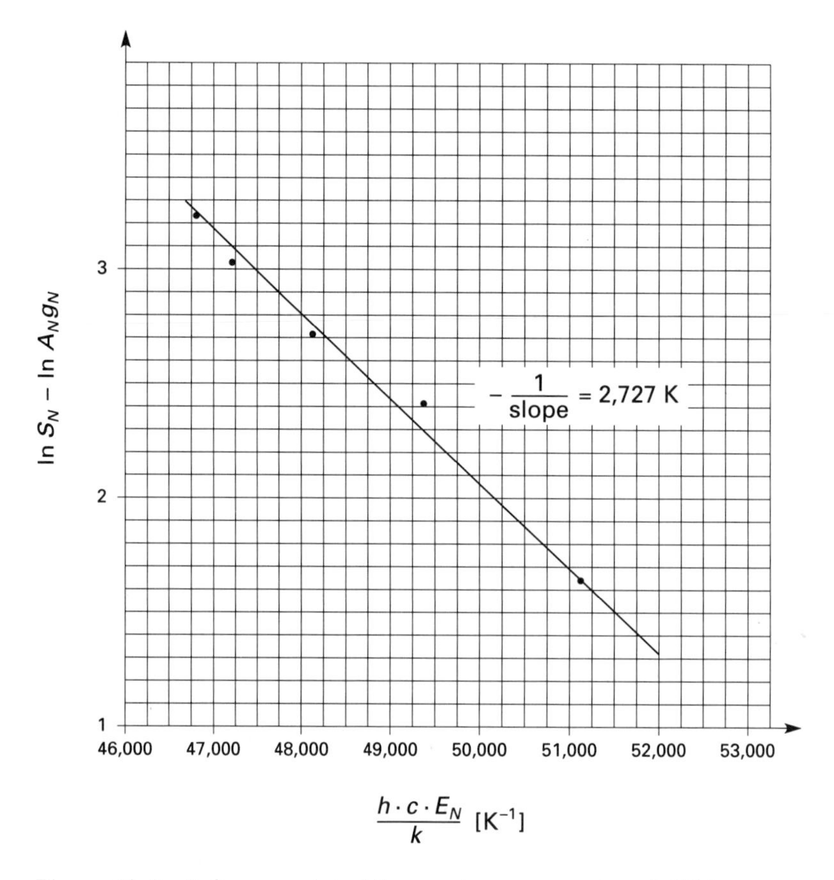

Figure 10.9 Boltzmann plot of the spontaneous emission of OH in an oxyacetylene flame.

in equilibrium - both internally and with its surroundings. Since objects that perfectly absorb all visible radiation appear black, the emission by such objects is called *blackbody radiation*. Figure 10.10 presents the spectral distribution of the emission of an ideal blackbody at various temperatures. It can be seen that, as the temperature of the radiator increases, the spectral irradiance increases and the peak of the spectral distribution shifts toward shorter wavelengths. Thus, objects at a temperature of 1,000 K emit mostly in the infrared, whereas the emission of the sun (with surface temperature of 5,800 K) peaks in the visible.

This phenomenon was first modeled correctly by Planck, who considered the radiation inside the insulated container as consisting of particles with an energy of $h\nu$. The statistical principles of photons are fundamentally different from those of atomic particles. While the number of atoms in a closed volume is fixed, the number of photons is changing and may be a function of the energy content of that volume. By correctly predicting the unique statistical behavior of photons and their energy distribution, Planck found the spectral

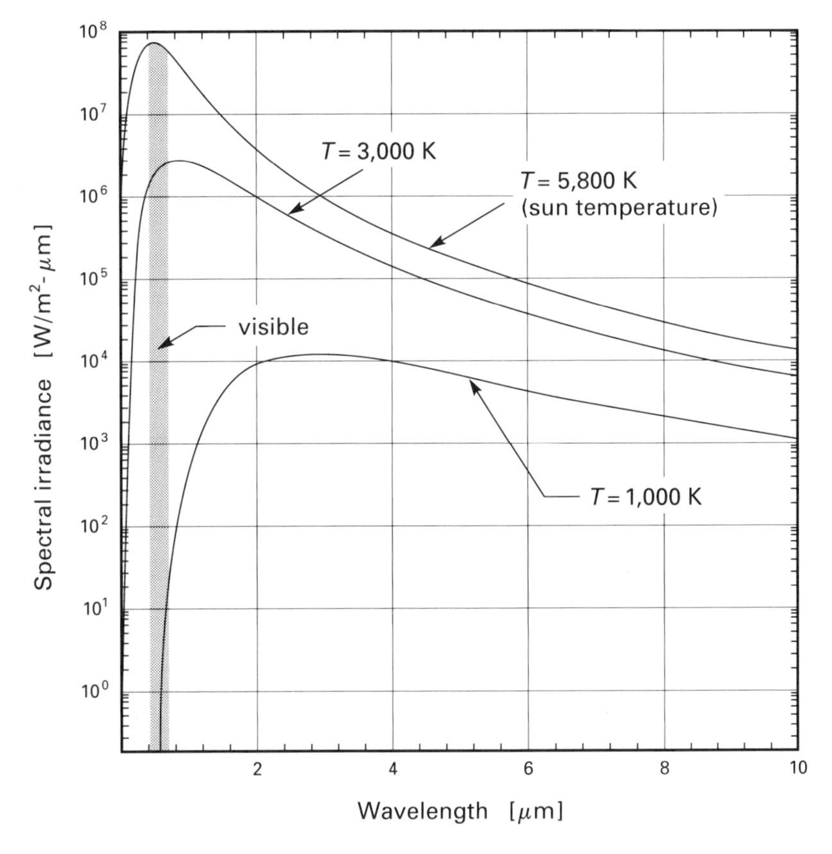

Figure 10.10 Spectral distribution of the emission of an ideal blackbody at various temperatures.

energy density per unit frequency in the frequency range of ν and $\nu+d\nu$ of this radiation to be

$$\rho(\nu)=\frac{8\pi n^3 h\nu^3}{c_0^3(e^{h\nu/kT}-1)}, \tag{10.39}$$

or, when expressed by the wavelength (see Problem 10.8),

$$\rho(\lambda)=\frac{8\pi n^3 hc_0}{\lambda^5(e^{hc_0/\lambda kT}-1)}.$$

The photons within this volume travel in all directions at the speed of light c_0. Thus, when coming off a surface at equilibrium, there are as many photons traveling inward as there are traveling outward. To determine the energy flux emerging from a hot surface, each element is enclosed by a hemisphere (Vincenti and Kruger 1965, p. 440) and the radiation passing through such envelope is obtained by integration over the entire hemisphere. Thus, the hemispherical spectral *emissive power* – the spectral irradiance $I(\nu)$ emerging from each area element of a surface at temperature T – is

$$I(\nu)=\frac{2\pi n^3 h\nu^3}{c_0^2(e^{h\nu/kT}-1)} \quad \text{or} \quad I(\lambda)=\frac{2\pi n^3 hc_0^2}{\lambda^5(e^{hc_0/\lambda kT}-1)}. \tag{10.39'}$$

The distribution of the spectral irradiance in Figure 10.10 is obtained by the second of these equations.

The irradiance emitted by an ideal blackbody shows a strong temperature dependence. But unlike the temperature dependence of the Boltzmann distribution, the temperature dependence of blackbody radiation involves an increase in the total number of photons with temperature as well as an increase in their total energy. Thus, both the total irradiance and the frequency of the distribution peak vary with temperature. The variation with temperature of the wavelength of the peak irradiance λ_{max} is obtained by (10.39). With the temperature expressed in degrees Kelvin and the wavelength in nanometers, λ_{max} is

$$T\lambda_{max}=2.898\times10^6 \text{ nm-K}. \tag{10.40}$$

This is *Wien's displacement law.* Hence, for an approximate temperature of 5,800 K at the surface of the sun (Noyes 1982), the peak of the solar emission spectrum is at $\lambda_{max}=500$ nm.

The total irradiance emitted by a surface element at all wavelengths is obtained by integrating (10.39′) over the entire spectrum:

$$I=\int_0^\infty I(\nu)\,d\nu=\sigma T^4. \tag{10.41}$$

This is *Stefan–Boltzmann's law,* where $\sigma=5.6697\times10^{-8}$ W/m^2-K^4 is the Stefan–Boltzmann constant. Owing to the strong temperature dependence of the emission by a blackbody, radiative emission is a major source of energy loss by hot bodies. A notable example is the sun, where almost the entire energy generated at the much hotter core is disposed of by radiating at the surface. Thus, despite the significant subsurface temperature gradient fed by the solar fusion energy, the surface is kept at its observed temperature only by blackbody emission. However, the following illustrates that, as solar radiation propagates away from the surface, it is spread out and so irradiance collected on a spherical envelope away from the sun is much lower than the surface irradiance.

Example 10.6 Find the solar irradiance at the edge of the earth's atmosphere.

Solution The irradiance at the surface of the sun can be calculated from (10.41) using the known temperature of 5,800 K (in fact, the temperature was determined by measuring the sun's spectral distribution). In the absence of absorption or scattering losses in space, the emitted energy is conserved as it propagates away from the sun. Nevertheless, it diminishes due to the expanding spherical envelope that it intercepts. Therefore, the irradiance at the edge of the earth's atmosphere is determined by the spherical envelope that includes earth's orbit around the sun. Since the radius of earth's orbit is $R_e=214R_s$, where R_s is the radius of the sun, the irradiance at the edge of the earth's atmosphere is

$$I_e = \sigma T^4 \left(\frac{R_s}{R_e}\right)^2 = \frac{(5.6697 \times 10^{-8})(5{,}800^4)}{214^2} = 1.4\ \text{kW/m}^2.$$

This irradiance includes the visible, the infrared, and the harmful UV components. Thanks to the screening properties of the atmosphere, most of the UV and part of the IR components are absorbed before reaching the earth's surface. Thus, at the surface of the earth, only 60% of the incident radiation can be collected. This irradiance is an important parameter used in the design of solar collectors. ■

The strong temperature dependence of blackbody radiation presents attractive opportunities for thermal imaging and temperature measurements. Although measurement of the total irradiance may be the simplest technique, it is often hindered by imperfections of the emitter. Depending on the surface structure and its composition, the emission characteristics may be different from those of an ideal blackbody. Since the strongest absorption and consequently the strongest emission are exhibited by ideal blackbodies, all other surfaces may have emission characteristics that are only a fraction of the blackbody emission. This fraction between the measured emission and the emission of the ideal blackbody is described by the *emissivity,* $0 < \epsilon \le 1$. The emissivity may be constant through the entire spectral range or it may vary with wavelength or temperature. Thus, for accurate temperature measurement based on total irradiance, the emissivity and its spectral characteristics must be specified.

To avoid the dependence on emissivity, measurements of the radiation's spectral distribution are often made. Such measurements can be simplified by measuring the irradiance at two different wavelengths. The ratio of these two measurements is strictly temperature-dependent when the emissivity is the same at these two wavelengths. However, for surfaces with spectrally varying emissivity, such measurements may be confounded by this spectral dependence. Other spectral techniques include measurement of the emission color, for example, by matching the color of the hot surface against the color of a heated filament at known temperature.

10.6 Consolidation of Einstein Coefficients and the Oscillator Strength

The spectral distribution of blackbody radiation was derived by Planck, who correctly assumed that the emission represents the statistical distribution of photons emitted by hot bodies and is part of the equilibrium between the radiation and the radiator. Thus, the details of the Boltzmann distribution (eqn. 10.33) and the details of blackbody radiation (eqn. 10.39) are coexistent. In fact, one is the result of the other. When the Boltzmann distribution within a medium is disturbed, the blackbody radiation associated with that medium is disturbed as well. Conversely, disturbance of the balance between incident and emitted radiation of a blackbody radiator results in a redistribution of its atomic particles among their energy states. The interaction between these two distributions, both of which are part of the thermodynamic equilibrium, is controlled by the three radiative processes of spontaneous emission, absorption, and stimulated emission. Thus, relations among the three Einstein coefficients A_{21}, B_{12}, and B_{21} can be derived (Einstein 1917) using the correspondence between these two distributions.

Derivation of the relation among the three Einstein coefficients can be simplified by analyzing the energy transfer between only two isolated states. However, the distribution of the atomic particles of a medium at thermodynamic equilibrium among their energy states is not maintained by transitions between pairs of states. Instead, it involves complex energy exchanges among all energy levels of all the particles in the medium. To preserve the equilibrium distribution in the face of this activity, any contribution to the population of a state (e.g., state 1) from another state (state 2) must be balanced by a separate process that drives any excess population back to the contributing state. Otherwise, the equilibrium distribution between states 1 and 2, as well as between state 1 and any other state, will be disrupted. Owing to the large selection of transitions allowed from any state, the restoring process is unlikely to be directly from state 1 back to state 2. Nevertheless, its end effect must be just that. The restoration of any disruption of the equilibrium is controlled by the *principle of detailed balance,* which states:

> *For every process that tends to alter an equilibrium distribution, there exists an inverse process. The process and its inverse occur at the same frequency, thereby offsetting each other.*

Although the principle does not specify any procedure for the restoration of a perturbed equilibrium, neither does it preclude any physically allowed route. One can conceivably find two states that maintain their relative equilibrium population by radiative processes among themselves. Other processes (which most probably exist) between these two states, and between these and any other states, are balanced out separately and are therefore independent of the radiative processes that couple these two states directly. Thus, although such a pair of states is definitely not isolated from the other states, the analysis of the radiative transfer between them can be carried out as if they were isolated. With this assumption, the three rate equations that described each radiative process independently (eqns. 10.2, 10.17, and 10.27) can now be combined to describe the relation between the population densities of any pair of states at equilibrium. Since these two states are now regarded as isolated, radiative transitions from one state must be added to the population of the other and vice versa, and at equilibrium each state's population must remain constant. Therefore, only one rate equation is needed to describe the interaction between the three radiative processes and the population of these states. The equation for one of these states at equilibrium – for example, the upper level (level 2) – is

$$\frac{dN_2}{dt} = -A_{21}N_2 + B_{12}\rho(\nu)N_1 - B_{21}\rho(\nu)N_2 = 0, \tag{10.42}$$

where N_1 and N_2 represent the entire population, including degeneracy, of the lower and upper states respectively. Requiring as in (10.42) that the net transition rate from level 2 be zero merely implies that the population density of that level is at steady state. This requirement does not by itself imply that the population of that level is at thermodynamic equilibrium. Therefore, additional conditions that describe equilibrium must be stated.

When at thermodynamic equilibrium, the ratio N_1/N_2 of the population densities of the two levels is determined by the Boltzmann distribution. Using (10.31), both N_1 and N_2 can be eliminated from (10.42); after re-arrangement, the following expression for the spectral energy density $\rho(\nu)$ associated with the radiative interaction between these states can be obtained:

$$\rho(\nu) = \frac{A_{21}}{B_{12}(g_1/g_2)e^{h\nu_0/kT} - B_{21}},$$

where $h\nu_0$ is the photon energy for transitions between these levels. However, when thermodynamic equilibrium between all energy levels of the medium prevails, $\rho(\nu)$ is the spectral energy density of a blackbody radiator at the temperature of the medium and (10.39) can be used to describe it. Comparing the two expressions for $\rho(\nu)$ at thermodynamical equilibrium, we can obtain the following condition relating the three Einstein coefficients to each other:

$$\rho(\nu) = \frac{8\pi n^3 h\nu_0^3}{c_0^3(e^{h\nu_0/kT} - 1)} = \frac{A_{21}}{B_{12}(g_1/g_2)e^{h\nu_0/kT} - B_{21}}.$$

Because all three Einstein coefficients are temperature-independent spectroscopic properties, the following condition must also be met to eliminate the potential temperature dependence implied by the previous equation:

$$\frac{g_1}{g_2}B_{12} = B_{21}. \tag{10.43}$$

With this condition, the temperature-dependent exponential term is canceled on both sides of the equation, thereby providing a second relation between the Einstein A and B coefficients:

$$\frac{A_{21}}{B_{12}} = \frac{g_1}{g_2}\frac{8\pi n^3 h\nu_0^3}{c_0^3} \quad \text{or} \quad \frac{A_{21}}{B_{21}} = \frac{8\pi n^3 h\nu_0^3}{c_0^3}. \tag{10.44}$$

The first of these relations (eqn. 10.43) is most surprising. It states that when the degeneracies of the interacting states are equal, the Einstein B coefficients for absorption and stimulated emission are equal. Stated differently, this condition implies that when the populations of the upper and lower states are equal, absorption and stimulated emission balance each other almost exactly; the only imbalance is due to spontaneous emission. The other condition (eqn. 10.44) and (10.43) together form a complete set, leaving only one Einstein coefficient to be specified independently. Thus, the entire radiative transfer process between any two states can be fully characterized by only one Einstein coefficient.

The unification of the three radiative processes can be carried one step further by comparing them with the classical electromagnetic theory, where absorption is attributed to the dissipating term of dipole oscillations (eqn. 4.21). A hypothetical medium consisting of $\bar{N}$ ideal, elastically bound electrons with charge e and mass m_e can provide, through the classical theory, a yardstick against which all other absorptions are measured. The spectrally integrated absorption of these oscillating electrons is exceptionally strong and so the absorption of all other media can be expressed as a fraction of this absorption. For this hypothetical medium, the integrated absorption coefficient (Mitchell and Zemansky 1961, p. 96 and its appendix) is

$$\int \kappa_\nu \, d\nu = \frac{\pi e^2}{m_e c(4\pi\epsilon_0)} \bar{N}, \tag{10.45}$$

where κ_ν, the absorption coefficient of the ideal oscillators, is as defined by (10.14). In ordinary media, κ_ν increases with the density N_1 of absorbers at their lower state. By introducing the *oscillator strength* f_{12}, which is the ratio $\bar{N}/N_1$ used to calculate the density of the ideal oscillators required to produce the same absorption effect as N_1 actual absorbers, the absorption by any medium can be presented as

$$\int \kappa_\nu \, d\nu = \frac{\pi e^2}{m_e c(4\pi\epsilon_0)} N_1 f_{12}. \tag{10.46}$$

For most absorbers, $0 < f_{12} \leq 1$. Equations (10.45) and (10.46) are expressed in the MKSA unit system.

To be useful, f_{12} must be expressed either in terms of one of the Einstein B coefficients or in terms of A_{21}. Using (10.21′) to substitute for the integrated absorption coefficient, the oscillator strength expressed by B_{12} is

$$f_{12} = \frac{m_e h}{\pi e^2}(4\pi\epsilon_0)\nu_0 B_{12}. \tag{10.47}$$

To express the oscillator strength in terms of A_{21}, the first of equations (10.44) is used to replace B_{12}. Therefore,

$$f_{12} = \frac{m_e c^3}{8\pi^2 e^2}(4\pi\epsilon_0)\frac{g_2}{g_1} \cdot \frac{A_{21}}{\nu_0^2} \quad \text{or} \quad f_{12} = 1.50\left(\frac{g_2}{g_1} \cdot \frac{A_{21}}{\bar{\nu}_0^2}\right). \tag{10.48}$$

The coefficient of the second of these equations was evaluated for frequency, $\bar{\nu}_0$ [cm^{-1}] and A_{21} [s^{-1}]. Finally, there also exists an oscillator strength for the $2 \to 1$ transitions. From the relation between B_{12} and B_{21} (eqn. 10.43), the following relation between f_{12} and f_{21} is readily deduced:

$$g_1 f_{12} = g_2 f_{21}. \tag{10.49}$$

The following example illustrates calculation of the oscillator strength.

Example 10.7 The Einstein A_{21} coefficient for the $A^2\Sigma$ $(v'=1) \rightarrow X^2\Pi$ $(v''=0)$ $P_2(8)$ transition of OH is 1.22×10^5 s^{-1} and for absorption the B coefficient is $B_{12} = 1.50 \times 10^{17}$ m^3/s^2-J (Dimpfl and Kinsey 1979). Find the oscillator strengths for both transitions and compare them.

Solution The frequency for this transition (Dieke and Crosswhite 1962) is $\bar{\nu}_0 = 34{,}957.52$ cm^{-1}. Combining (10.48) and (10.49), the oscillator strength for emission is

$$f_{21} = 1.50 \frac{A_{21}}{\bar{\nu}_0^2} = \frac{(1.50)(1.22 \times 10^5)}{(34{,}957.52)^2} = 1.498 \times 10^{-4}.$$

The result is dimensionless because the oscillator strength is merely a measure of the strength of this transition relative to an ideal transition.

The oscillator strength for absorption is obtained directly by substitution into (10.47) as follows:

$$f_{12} = \frac{m_e h}{\pi e^2}(1.113 \times 10^{-10})(1.048 \times 10^{15})(1.50 \times 10^{17}) = 1.31 \times 10^{-4}.$$

The MKSA values for m_e, h, and e were taken from Tables A.1 and B.1.

To compare these two results, the degeneracy of the excited state and that of the ground state must be calculated. For this transition, $J = N - \frac{1}{2}$. Thus, the degeneracy of the excited state, where $N' = N'' - 1$, is

$$g_2 = 2J' + 1 = 2(6.5) + 1 = 14,$$

and for the ground state the degeneracy is $g_1 = 16$. Using (10.49) and these degeneracies, the oscillator strength for absorption can be calculated from that for emission:

$$\frac{g_2}{g_1} f_{21} = \frac{14}{16}(1.498 \times 10^{-4}) = 1.31 \times 10^{-4} = f_{12}.$$

This matches the previous result for f_{12}.

The oscillator strength for this transition is much less than unity, which is not unusual for molecular transitions. The strongest transitions normally occur between atomic levels. For comparison, the oscillator strength of strong atomic transitions approaches unity – for example, the $6P_{3/2} \rightarrow 6P_{1/2}$ transition of Cs, where $f_{12} = 0.814$ (Stone 1962). There exist a few atomic transitions with $f > 1$, for example, $7P_{3/2} \rightarrow 7S_{1/2}$ for Cs, where $f_{21} = 1.115$. ■

10.7 Pressure- and Temperature-Broadening Effects

The analysis of the previous sections treated each of the participating atomic particles as if isolated from its surroundings. With that assumption, all two-level systems of a species have the same nominal absorption or emission frequency and the same linewidth, independently of its thermodynamic properties. This may be true on rare occasions where atomic particles are kept isolated and motionless. Otherwise, atomic particles move about when in the gas phase while colliding with each other, or oscillate about their confining crystalline structure when in the solid phase. As a result of their motion and mutual collisions, the spectral properties of these particles are no longer uniform, and neither are they independent of the thermodynamic conditions of the medium. Owing to collisions, the transition linewidth is broadened and its frequency is shifted; owing to motion, the absorption or emission frequencies are Doppler-shifted, which also results in broadening of spectral lines. Effects of collisional or motion-induced line broadening are observed even at moderate temperatures and pressures. We will discuss the effects of line broadening associated with molecular or atomic collisions and with the motion of the absorbing and emitting species.

Collisions between atomic particles can be classified into two categories, elastic and inelastic. Inelastic collisions involve significant energy exchange, energy that is sufficient to excite or de-excite a system from its present quantum state. Elastic collisions, on the other hand, do not involve any energy exchange; elastic collisions can only alter the direction of the translational velocity. The effect of elastic collisions on emission and absorption is therefore more subtle than the effect of inelastic collisions. Although elastic collisions do not remove the colliding particles from their present quantum states, they do perturb their wave function. Figure 10.11 illustrates the wave function of an energy state undergoing elastic collisions. The effect of two of these collisions – separated by a characteristic time τ_c – is shown.

Although the frequency of the wave function is not changed by the collisions, abrupt phase changes are evident in the figure. Their effect on the energy level is similar to the effect of the natural lifetime τ_{21}; owing to the limited duration of a single train of the wave function, the energy uncertainty associated with it is extended. This is another manifestation of the uncertainty principle.

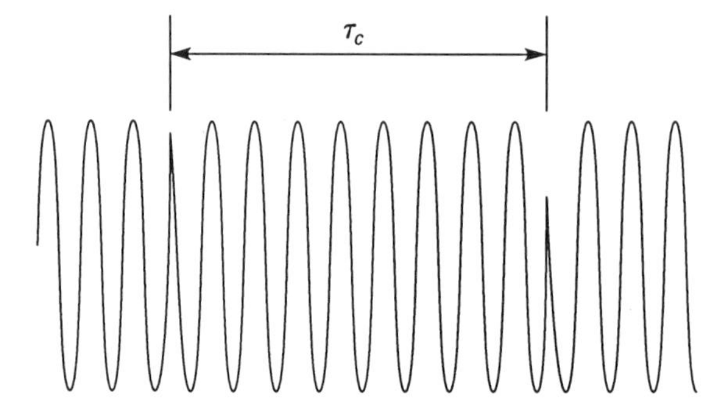

Figure 10.11 Wave function of an energy state undergoing elastic collisions, showing the effect of two collisions separated by a characteristic time τ_c.

Therefore, the energy uncertainty increases when τ_c decreases and optical transitions from that state are broadened. But unlike natural broadening, where the lifetime of only the excited state is finite, collisions affect both ground and excited states equally, so the total collisional broadening $\Delta\nu_C$ is twice the broadening that would result from the perturbation of a single state. For a rate of collision $Z_C = 1/\tau_c$ of an atomic particle with any other perturbing particle, the following collisional broadening is obtained by comparing this process with natural broadening (eqn. 10.6):

$$\Delta\nu_C = Z_C/\pi. \tag{10.50}$$

Collisional broadening is another manifestation of the uncertainty principle, so the lineshape factor for collisional broadening, $g_C(\nu)$, like the lineshape factor for natural broadening, is Lorentzian. With the transition frequency ν [Hz] and with $\Delta\nu_C$ replacing ω and A_{21} in (10.9), the Lorentzian lineshape factor for collisional broadening is

$$g_C(\nu) = \frac{\Delta\nu_C/2\pi}{(\nu-\nu_0)^2+(\Delta\nu_C/2)^2}. \tag{10.51}$$

Note that $g_C(\nu)$ is specified in units of 1/hertz, unlike $g_N(\omega)$ which was specified (eqn. 10.9) in seconds per radian.

Although the effect of collisional broadening was described for a single particle, it is certainly a process that depends on the thermodynamic properties of the medium surrounding that particle. Dephasing events may occur both in condensed phases as well as in the gas phase. In all phases the rate of these events depends on temperature, but in the gas phase pressure dependence is more dominant. To illustrate this, consider a gaseous mixture with two molecular species, A and B. The rate of collisions of one molecule of species A with any molecule of species B (Vincenti and Kruger 1965, p. 52) is

$$Z_{CB} = N_B\sigma_{AB}\left(\frac{8\pi kT}{\mu_{AB}}\right)^{1/2}, \tag{10.52}$$

where N_B is the density of species B, σ_{AB} is defined (Mitchell and Zemansky 1961, p. 170) as the effective cross section for collisions between A and B, and μ_{AB} is the reduced mass of the two species (eqn. 9.2). In single-component gases, the collisions are between molecules of the same species (collisions of A with B are indistinguishable from collisions of B with A). The rate of these collisions per molecule (Vincenti and Kruger 1965, p. 52) is

$$Z_{CA} = \frac{N_A\sigma_{AA}}{2}\left(\frac{16\pi kT}{m_A}\right)^{1/2}, \tag{10.53}$$

where N_A is the density of species A, σ_{AA} is the effective cross section for self-collisions, and $m_A = 2\mu_{AA}$ is the molecular mass of species A.

The combined rate of collisions between a molecule of species A and any other molecule in the mixture containing i species is obtained by adding the rates of the collisions between species A and all the other species:

$$Z_C = Z_{CA} + Z_{CB} + \cdots = \sum_i Z_{Ci}. \tag{10.54}$$

To associate Z_C with macroscopic thermodynamic properties, assume that the gas is ideal. In other words, assume that the density of each species can be calculated from its partial pressure and its temperature (see e.g. Wark 1988, p. 408), thereby allowing the density N_i of species i to be replaced with macroscopic thermodynamic properties such its partial pressure P_i and temperature T. For illustration, the rate of collisions in a single-component gas when the specific volume is kept constant is

$$Z_{CA} = P\frac{\sigma_{AA}}{2}\left(\frac{16\pi}{m_A kT}\right)^{1/2} \approx \frac{P}{T^{1/2}}, \tag{10.55}$$

where (from the ideal gas law) $N_A = PN_{Av}/R_uT = P/kT$ (Vincenti and Kruger 1965, p. 7), and where $R_u = 8.314$ kJ/kmole-K is the universal gas constant and $N_{Av} = 6.023 \times 10^{26}$ kmole^{-1} is Avogadro's number. In terms of the collisional broadening $\Delta\nu_{C0}$ at a known reference thermodynamic state specified by the pressure P_0 and temperature T_0, the collisional broadening at any other thermodynamic state is:

$$\Delta\nu_C = \Delta\nu_{C0}\frac{P}{P_0}\left(\frac{T_0}{T}\right)^{1/2}. \tag{10.56}$$

Collisional broadening increases linearly with pressure. Even though it is also temperature-dependent, the effect of pressure is so dominant that the entire mechanism is called *pressure broadening*. Since all molecules in the gas collide at the same nominal rate, this broadening is uniform for all molecules whenever the pressure and temperature are uniform. Accordingly, it is often called *homogeneous broadening*.

The equations (10.52 and 10.53) that describe the rates of collisions were obtained under the assumption that the cross section for collisional broadening is independent of temperature. While this may often be true, reports show that for certain molecules the cross section may vary with temperature. Thus, for O_2 in air, the cross section for collisional broadening varies as $T^{-0.2}$ (Cann et al. 1979). For this variation, (10.56) for the pressure broadening of the Schumann–Runge band of O_2 in air is

$$\Delta\nu_C = 0.3P\left(\frac{273.2}{T}\right)^{0.7} \text{cm}^{-1}. \tag{10.57}$$

This semi-empirical equation was adjusted to yield the broadening in units of cm^{-1} when the pressure is given in atmospheres and the temperature in degrees Kelvin.

When a spectral line is analyzed, its width is influenced both by natural and by collisional broadening. Mathematically, the resultant lineshape is obtained by the convolution integral of two Lorentzian lineshapes (eqns. 10.9 and 10.51), which is also a Lorentzian with a FWHM that is the sum of the widths of the two separate lineshape factors. This observation is physically plausible.

The effective lineshape is the result of two processes, one that diminishes the population density of a state at a rate of A_{21} and the other that perturbs the states at a rate of $2Z_C$. Therefore, the density of unperturbed states diminishes at a rate of $A_{21}+2Z_C$ and the combined linewidth, which is proportional to the sum of these rates, is

$$\Delta\nu_{NC} = \Delta\nu_{21} + \Delta\nu_C. \tag{10.58}$$

The third line-broadening effect to be discussed here, after natural and pressure broadening, is induced by the random motion of atomic particles while in the gas phase. In the absence of a net gas flow, the velocity vector of each particle (atom or molecule) points randomly in any direction and its magnitude may vary widely, with the most probable velocity exceeding the speed of sound. Thus, radiation incident from an arbitrary direction encounters absorbers with velocity components pointing toward and away from its propagation vector; an observer moving with the absorber must experience a Doppler shift (eqn. 6.24) that depends on the magnitude of the velocity and its direction relative to the propagation vector. Thus, even if the radiation is not exactly resonant with the nominal absorption frequency ν_0, there are still absorbers that – owing to the Doppler shift – sense the incident frequency as being at ν_0. Because of this shift, the absorption coefficient κ_ν at any incident frequency ν_1 depends not only on the absorption characteristics of the absorber at ν_1 and its density, but also on the number of absorbers that had their absorption frequency shifted to coincide with ν_1. Similarly, the emission by moving emitters appears to a stationary observer as being Doppler-shifted. The extent of emission at any frequency ν_1 depends not only on the characteristics of individual emitters but also on the number of emitters whose emission frequency was shifted to ν_1 by their motion relative to the observer.

Figure 10.12 depicts the lineshape factor $g_D(\nu)$ associated with the motion of the absorbers. The group of absorbers having sufficient velocity to shift the absorption frequency from ν_0 to ν_1 (i.e., to allow absorption at ν_1 by absorbers with a resonance of ν_0) is marked in the figure by a bar with a length that is proportional to the density of these absorbers. The infinitesimally narrow width of that bar suggests that this group is capable of absorption in a narrow range of frequencies; that is, the combined magnitude of the natural and pressure broadening of its members is negligible. Similar rectangles can be assigned to other velocity groups, at frequencies that depend on the velocities of their members. Clearly, when the absorbers are free to move, there exist an infinite number of such bars, which combine to form an absorption lineshape that is no longer defined by the absorption characteristics of a single absorber. Instead, it is defined by the outcome of the velocity distribution of the microscopic gas particles, where each particle interacts with the incident radiation independently of the others. Accordingly, this broadening mechanism is termed *inhomogeneous* or *Doppler broadening*.

The lineshape factor for Doppler broadening is derived from the velocity distribution of the absorbers (or emitters). By analogy with the Boltzmann

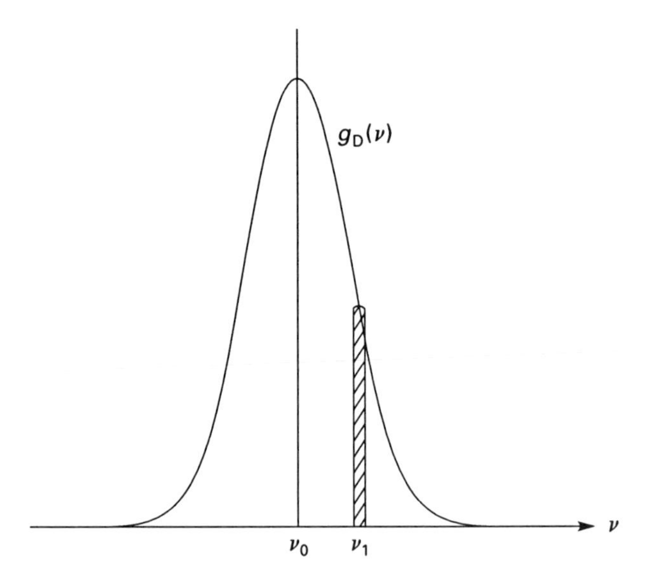

Figure 10.12 The lineshape factor $g_D(\nu)$ associated with the random motion of absorbers.

distribution for other energy modes, the distribution associated with the kinetic energy of moving particles with a velocity component $\mathbf{V}_1$ is

$$f(\mathbf{V}_1) = \sqrt{\frac{m}{2\pi kT}} \exp\left\{-\frac{m\mathbf{V}_1^2}{2kT}\right\}. \tag{10.59}$$

(Vincenti and Kruger 1965, p. 44). Equation (10.59) is only part of the Maxwellian distribution in which all three velocity components of the kinetic energy are included. To transform this distribution into a distribution of the probability that the absorption by a particle is resonant with an arbitrary incident frequency ν, the relation between the frequency shift $\nu - \nu_0$ and the velocity $\mathbf{V}_1$ (eqn. 6.24) is incorporated into (10.59) to form the following lineshape factor:

$$g_D(\nu) = \frac{c}{\nu_0} \sqrt{\frac{m}{2\pi kT}} \exp\left\{-4 \ln 2 \frac{(\nu - \nu_0)^2}{\Delta\nu_D^2}\right\}, \tag{10.60}$$

where $\Delta\nu_D$, the Doppler linewidth as defined by the FWHM of $g_D(\nu)$, is

$$\Delta\nu_D = \frac{2\nu_0}{c} \sqrt{\frac{2 \ln 2 \cdot kT}{m}}. \tag{10.61}$$

Owing to its associated exponential function, this lineshape factor is called *Gaussian* after the similar statistical distribution. The only thermodynamic property that enters into this lineshape factor is temperature. This is to be expected, since the velocity of the absorbers depends on temperature only. Unlike pressure broadening, where both pressure and temperature must be known in order to determine the lineshape function, Doppler (or *temperature*) broadening can be determined even when only the temperature is known.

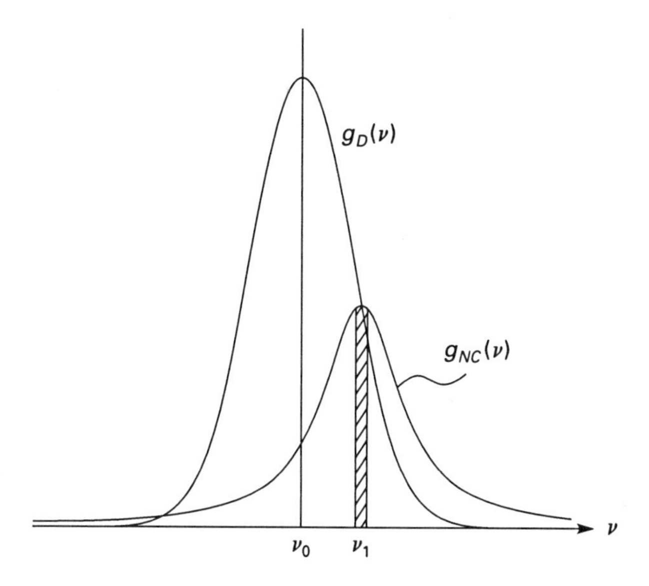

Figure 10.13 The ideal Doppler lineshape factor of Figure 10.12 overlapped with the Lorentzian lineshape factor of one absorbing group.

The Doppler lineshape factor (eqn. 10.60) was derived for particles with negligible natural and pressure broadenings, $\Delta\nu_{NC} \ll \Delta\nu_D$. However, unless temperature is exceptionally high or pressure is extremely low, the effect of these broadening mechanisms cannot be ignored. Therefore, the velocity group of particles that are resonant with an incident frequency ν_1 may also absorb, owing to pressure or natural broadening, at other frequencies in the vicinity of ν_1. Similarly, velocity groups that are not exactly resonant with ν_1 may still contribute to the absorption at ν_1. This is illustrated in Figure 10.13, where the ideal Doppler lineshape factor of Figure 10.12 is again presented. However, the infinitely thin bar that was previously used to depict the velocity group resonant with ν_1 is now replaced by a Lorentzian lineshape, centered at that frequency. Clearly, absorption by particles in this velocity group is possible at frequencies other than ν_1 whenever the overlap between the two curves in the figure is non-negligible. Similarly, absorption by particles from other groups whose Lorentzian "wings" extend to ν_1 may absorb at that frequency as well. The effective lineshape that incorporates all three broadening mechanisms is obtained by the convolution of two functions – a Lorentzian and a Gaussian.

Unlike the previous convolution, which combined natural broadening with pressure broadening, the convolution between the first two Lorentzian broadening mechanisms and the Doppler broadening is complicated by the Gaussian functional shape of $g_D(\nu)$. The lineshape factor that combines natural and pressure broadening with Doppler broadening is obtained by the following convolution integral of the Lorentzian distribution with the Gaussian distribution (Armstrong 1967):

$$g(D,B) = \frac{1}{\Delta\nu_D}\sqrt{\frac{\ln 2}{\pi}}K(D,B), \quad \text{where } K(D,B) = \frac{B}{\pi}\int_{-\infty}^{\infty}\frac{e^{-y^2}}{B^2+(D-y)^2}\,dy \tag{10.62}$$

and where the broadening parameter B is given by

$$B = \sqrt{\ln 2}\,\frac{\Delta\nu_{NC}}{\Delta\nu_D} \tag{10.63}$$

and the detuning parameter D by

$$D = \sqrt{\ln 2}\,\frac{\nu-\nu_0}{\Delta\nu_D}. \tag{10.64}$$

The combined lineshape factor is the *Voigt profile*. It needs to be calculated whenever the linewidth of the Lorentzian component of the broadening is comparable to the Gaussian component. Because of its complexity, the integral cannot be solved. Alternatively, approximate solutions (Armstrong 1967) and tables (Young 1965) were developed to evaluate it. However, at high temperature, when the broadening is dominated by the motion of the particles ($B \ll 1$), the Voigt profile approaches the Doppler broadening lineshape and can be approximated by the Gaussian lineshape function. Similarly, at a low temperature but at a high pressure, $g(D,B)$ can be approximated by the Lorentzian lineshape factor. The following example illustrates a case where the effect of Doppler broadening can be neglected for most practical applications.

Example 10.8 The predissociation linewidth of a particular transition from the $B^3\Sigma_u^-$ ($v'=4$, $N'=10$) state of O_2 is 3.624 cm^{-1}. The frequency of that transition is $\nu_0 = 51{,}800$ cm^{-1}. Find the temperature at which Doppler broadening will dominate the lineshape of this transition.

Solution As was shown in Example 10.2, the short lifetime associated with this predissociation dominates the other natural transition rates and hence must be the major component of natural broadening. In addition, it is likely to be large relative to pressure broadening, as will be shown using (10.57). Since neither temperature nor pressure were given in the problem, they must be estimated. The temperature at which Doppler broadening becomes dominant is expected to be very high. Therefore, the effect of pressure broadening on the lineshape will be assessed at a similarly high temperature, for example, 10,000 K. The pressure will be selected at 5 atmospheres but (depending on the results) may be reassessed. With these thermodynamic parameters, the contribution of pressure broadening to the linewidth is:

$$\Delta\nu_C = (0.3)5\sqrt{\frac{273.2}{10{,}000}} = 0.25 \text{ cm}^{-1}.$$

This is the width of a Lorentzian lineshape function and may therefore be added arithmetically to the natural linewidth:

$$\Delta\nu_{NC} = \Delta\nu_N + \Delta\nu_C = 3.874 \text{ cm}^{-1}.$$

For the Voigt profile to be mostly Gaussian, the broadening parameter B (eqn. 10.62) must be less than unity. To evaluate B, $\Delta\nu_D$ must be evaluated with T as a parameter:

$$\Delta\nu_D = \frac{2(51{,}800)}{3\times 10^8}\sqrt{\frac{(2\ln 2)(1.38\times 10^{-23})T}{32(1.67\times 10^{-27})}} = (6.53\times 10^{-3})T^{1/2}\ \text{cm}^{-1}.$$

With the condition that $B<1$, the temperature must satisfy

$$(6.53\times 10^{-3})T^{1/2} > 3.874\sqrt{\ln 2} \quad \text{or} \quad T > 243{,}960\ \text{K}.$$

This temperature is exceptionally high. For comparison, the energy $E=kT$ that corresponds to this temperature is in excess of 21 eV (the reader should try to confirm this). The dissociation energy of O_2 is only 5 eV. Thus, at this temperature less than 1% of the entire oxygen population will remain undissociated, so the notion of molecular spectrum at that temperature is meaningless. ■

The concepts developed here and in Chapter 9 supply the foundation for most of the work involving interaction of radiation with matter. Applications of relevance to engineers, such as laser-induced fluorescence or Raman scattering, can be modeled by using the concepts of two-level systems as presented here. Some of these applications will be discussed in Chapter 11. However, the theory of light amplification and the generation of laser beams require the use of three- and four-level systems, which are certainly more complicated than the two-level system of this chapter. Discussion of such systems is deferred to Chapter 12.

References

Allison, A. C., Dalgarno, A., and Pasachoff, N. W. (1971), Absorption by vibrationally excited molecular oxygen in the Schumann–Runge continuum, *Planetary and Space Science* 19: 1463–73.

Andresen, P., Bath, A., Gröger, W., Lülf, H. W., Meijer, G., and ter Meulen, J. J. (1988), Laser-induced fluorescence with tunable excimer lasers as a possible method for instantaneous temperature field measurements at high pressures: checks with atmospheric flame, *Applied Optics* 27: 365–78.

Armstrong, B. H. (1967), Spectrum line profiles: the Voigt function, *Journal of Quantitative Spectroscopy and Radiative Transfer* 7: 61–88.

Cann, M. W. P., Nicholls, R. W., Evans, W. F. J., Kohl, J. L., Kurucz, R., Parkinson, W. H., and Reeves, E. M. (1979), High resolution atmospheric transmission calculations down to 28.7 km in the 200–243-nm spectral range, *Applied Optics* 18: 964–76.

Carlone, C., and Dalby, F. W. (1969), Spectrum of the hydroxyl radical, *Canadian Journal of Physics* 47: 1945–57.

Cheung, A. S.-C., Mok, D. K.-W., Jamieson, M. J., Finch, M., Yoshino, K., Dalgarno, A., and Parkinson, W. H. (1993), Rotational dependence of the predissociation linewidths of the Schumann–Runge bands of O_2, *Journal of Chemical Physics* 99: 1086–92.

Dieke, G. H., and Crosswhite, H. M. (1962), The ultraviolet bands of OH, *Journal of Quantitative Spectroscopy and Radiative Transfer* 2: 97–199.

Dimpfl, W. L., and Kinsey, J. L. (1979), Tables of Einstein *A*-coefficients and *B*-coefficients for the $A^2\Sigma$–$X\,^2\Pi$ system of OH and OD, *Journal of Quantitative Spectroscopy and Radiative Transfer* 21: 233–41.

Einstein, A. (1917), On the quantum theory of radiation, *Physikalische Zeitschrift* 18: 121–8; translation in ter Haar, D. (1967), *The Old Quantum Theory,* Elmsford, NY: Pergamon, pp. 167–83.

Fast, J. D. (1962), *Entropy,* New York: McGraw-Hill, pp. 275–80.

Garmire, E. M., and Yariv, A. (1967), Laser mode-locking with saturable absorbers, *IEEE Journal of Quantum Electronics* 3: 222–6.

Gray, J. A., and Farrow, R. L. (1991), Predissociation lifetimes of OH $A^2\Sigma^+$ ($v'=3$) obtained from optical-optical double-resonance linewidth measurements, *Journal of Chemical Physics* 95: 7054–60.

Grinstead, J. H., Laufer, G., and McDaniel, J. C., Jr. (1993), Rotational temperature measurement in high-temperature air using KrF laser-induced fluorescence, *Applied Physics B* 57: 393–6.

Heard, D. E., Crosley, D. R., Jeffries, J. B., Smith, G. P., and Hirano, A. (1992), Rotational level dependence of predissociation in the $v'=3$ level of OH $A^2\Sigma^+$, *Journal of Chemical Physics* 96: 4366–71.

Herzberg, G. (1989), *Molecular Spectra and Molecular Structure I. Spectra of Diatomic Molecules,* Malabar, FL: Krieger.

Laufer, G., McKenzie, R. L., and Fletcher, D. G. (1990), Method for measuring temperatures and densities in hypersonic wind tunnel air flows using laser-induced O_2 fluorescence, *Applied Optics* 29: 4873–83.

Mitchell, A. C. G., and Zemansky, M. W. (1961), *Resonance Radiation and Excited Atoms,* Cambridge University Press.

Noyes, R. W. (1982), *The Sun, Our Star,* Cambridge, MA: Harvard University Press.

Quagliaroli, M. T. (1933), Modeling the laser-induced fluorescence of the OH molecule, M.Sc. Thesis, University of Virginia, Charlottesville.

Stone, P. M. (1962), Cesium oscillator strengths, *Physical Review* 127: 1151–6.

Tatum, J. B. (1966), Hönl–London factors for $^3\Sigma^\pm$–$^3\Sigma^\pm$ transitions, *Canadian Journal of Physics* 44: 2944–6.

Vincenti, W. S., and Kruger, C. H., Jr. (1965), *Introduction to Physical Gas Dynamics,* Malabar, FL: Krieger.

Wark, K., Jr. (1988), *Thermodynamics,* 5th ed., New York: McGraw-Hill.

Young, C. (1965), Tables for calculating the Voigt profile, Technical Report no. 05863-7-T, University of Michigan, Ann Arbor.

Homework Problems

Problem 10.1

Using the Fourier transformation of the time-dependent field associated with spontaneous emission, show that the spectral distribution of the radiation is a Lorentzian as described by (10.8). Also show that $g_N(\omega)$ is normalized.

Problem 10.2

Measure the relative heights of the O_2 fluorescence lines in Figure 9.14 for transitions from the $B\,^3\Sigma_u^-$ ($v'=4$) state to the $v'=3,4,5,6,7,8,9$ levels in the $X\,^3\Sigma_g^-$ ground electronic state. Compare the relative heights with the calculated Einstein A coefficients reported in Example 10.2. Estimate the error associated with this measurement.

Problem 10.3

Derive the absorption equation (eqn. 10.14) by integrating (10.19) over the entire length of the cell shown in Figure 10.5. Also derive (10.20).

Problem 10.4

Assume that a two-level absorber consisting of molecules at a density of N_1 is illuminated by spectrally narrow radiation with a spectral irradiance $I(\nu)$ and linewidth $d\nu$ that is narrow relative to the absorption linewidth. The absorption lineshape of the two-level system is a Lorentzian with a linewidth of A_{12} rad/s. Express W_{12} in terms of $I(\nu)$ and B_{12} and derive an expression for the frequency-dependent absorption cross section (eqn. 10.25).

Problem 10.5

Certain dyes are used as saturable absorbers to control laser cavities (Garmire and Yariv 1967). These dyes act as shutters by absorbing incident radiation of low irradiance. But when the irradiance increases, dye molecules are forced from the ground state to the excited state, thereby depleting the ground state. Once the population of the excited state relative to the ground state is sufficiently high, stimulated emission offsets absorption, the transmission losses become small, and the dye becomes effectively transparent.

(a) Assuming that such an absorber is a two-level system, write the rate equation for the population of the ground state N_1 for transmission of an incident beam with an irradiance I through a thin dye cell.

(b) Describe the loss of irradiance of a collimated beam passing a distance dx through the dye by writing an expression for dI/dx in terms of the population densities N_1 and N_2 of the two states. What is the requirement (in terms of N_2/N_1) that will turn the dye transparent?

Problem 10.6

Derive the partition function Q_v for the vibrational levels of a molecule and find the population density of a rotational level J in the $v = 0$ vibrational state when $T \ll T_v$.

Problem 10.7

Verify the results of Figure 10.8 by calculating the Boltzmann fraction of the $N = 8$ state at $T = 1{,}500$ K. (*Hint:* Although not shown by (10.36), the electronic partition function must also be included. Owing to the electronic degeneracy of the $X\,^2\Pi$ state and the fine splitting, the partition function of the electronic state is $Q_e = 4$.)

Problem 10.8

The energy emitted by a blackbody radiator in the frequency range of ν to $\nu + d\nu$ is $I(\nu)\,d\nu$, where $I(\nu)$ is defined by (10.39′). Derive the term for $I(\lambda)$ so that $I(\lambda)\,d\lambda$ will represent the energy in the wavelength range of λ to $\lambda + d\lambda$.

Problem 10.9

Find the surface temperature of an emitter when the irradiance at 2 μm is 10% of the irradiance at 1 μm. (*Answer:* 9,100 K.)

Problem 10.10

The solar irradiance at the edge of the earth's atmosphere was calculated in Example 10.4. About 40% of the incident radiation is absorbed (or scattered) by the atmosphere.

(a) Assume you intend to design a perfect planar solar collector to be used for heating water. Name all the loss mechanisms that you may expect to control and those you may not control in your quest for a perfect collector. What is the highest water temperature you may expect to obtain from your perfect collector when placed at sea level?

(b) Assume you intend to design a perfect concentrating collector. Recognizing that the second law of thermodynamics prevents the flow of heat from a cooler body to a hotter body, determine the maximum concentration ratio, [flux incident on the collector]/[flux at the collector's focus], that may be expected at the edge of the atmosphere and at sea level.

(c) What is the maximum temperature that can be obtained by this collector at the edge of the atmosphere and at sea level?

(d) What type of collector would you recommend for home heating and what type for power generation? Justify your recommendations.

Problem 10.11

To estimate the results of the greenhouse effect in earth's atmosphere, assume that the blackbody emission by its surface is expected to maintain an average temperature of 300 K. Owing to an increase in the concentration of CO_2 molecules, the absorption of earth's blackbody emission by the atmosphere increases in the spectral range of $10 \pm 1\ \mu m$ by 10%. Find what the temperature of earth's surface must be in order to rid itself of the additional energy trapped by the atmosphere. (*Answer:* 301 K.)

Problem 10.12

The oscillator strength of the $6\,^2S_{1/2} \rightarrow 7\,^2P_{3/2}$ transition of Cs is 1.74×10^{-2} (Stone 1962). The transition wavelength is 455.53 nm. Find the absorption cross section when the broadening is 0.1 cm^{-1}. (*Answer:* $6.14 \times 10^{-18}\ m^2$.)

Problem 10.13

Using the ideal gas law and the physical constants given in Appendix B, calculate the Loschmidt number.

Problem 10.14

Calculate the pressure broadening of O_2 in atmospheric air at $P = 1$ atm and $T = 1{,}000$ K relative to its broadening at $P = 1$ atm and $T = 273.2$ K. Assume that:

(a) the cross section for collisional broadening is temperature-independent; and

(b) the cross section for collisional broadening varies with temperature as prescribed by (10.57).

Research Problems

Research Problem 10.1

Read the paper by Cann et al. (1979).

(a) How is the concept of optical depth defined by this paper?

(b) Using the data in that paper, calculate the optical depth of the ozone layer at $\lambda = 230$ nm from the top of the atmosphere to altitudes of 28.65 km and 40 km.

(c) Calculate the same for O_2.

(d) Determine the attenuation of solar radiation at that wavelength by the atmosphere from its top to these altitudes, and calculate the solar spectral irradiance at that wavelength that can be detected there.

(e) Assume that the ozone is depleted to 10% of its original density in a 5-km layer starting at 30 km above sea level. Describe how the depletion will affect the solar irradiance at $\lambda = 230$ nm at an altitude of 28.65 km.

Research Problem 10.2

Read the paper by Dieke and Crosswhite (1962). This paper describes the use of the emission spectra of OH in an oxyacetylene flame for the measurement of flame temperature. The technique depends on rapid equilibration of the rotational population of excited OH molecules. It is therefore assumed that in flames where OH is formed in an excited state by chemical reactions, the population distribution among the rotational states reaches an equilibrium even in the absence of vibrational or electronic equilibrium. The emission of a ro-vibronic line depends therefore on the Franck–Condon factor of the appropriate vibrational level, as well as on the rotational transition probability. However, the relative brightness of rotational lines that are part of one vibronic band (e.g., the $0 \rightarrow 0$ transition) depends only on the rotational transition probability. The probabilities A_k for all the branches of the $^2\Sigma \rightarrow {}^2\Pi$ band are listed in table 4 of the paper. The relative intensities of these lines are listed in table 13 or table 14.

(a) Draw a Boltzmann plot for $0 \rightarrow 0$ transitions in the P_1 branch. Calculate the temperature and compare it with the reported temperature.

(b) Calculate J_{max} at $T = 1{,}000$ K and 3,000 K for the upper electronic state and compare the result with the results of figure 3 of the paper.

(c) Calculate which rotational state will have at $T = 2{,}000$ K the same emission intensity as the emission from a rotational level with $K = 3$. Determine the same when $T = 1{,}000$ K and 3,000 K and compare the result with the results of the paper.

(d) Assuming that the temperature is determined by the iso-intensity method, determine the measurement resolution if only one pair of rotational transitions is used for that measurement.

11 Spectroscopic Techniques for Thermodynamic Measurements

When you can express in numbers that of which you speak, you have the beginning of a science. Until that time your knowledge is meager and unsatisfactory.

Lord Kelvin

11.1 Introduction

Spectroscopic techniques are an excellent tool for thermodynamic measurements. Typically they do not require a direct contact with the tested medium and are therefore considered nonintrusive. This is a significant advantage over measurements that require physical probes, particularly when the environment at the point of measurement is hostile owing to such extreme conditions as high temperature, aggressive chemical reaction, or fast flows. In addition, many spectroscopic techniques can be adapted to image the two- or three-dimensional distribution of select properties. With the advent of electronic cameras, images can be digitized and stored almost instantaneously by computers. Exposure times of as short as 5 ns are possible with the use of CCD (charge-couple device) intensified cameras. Recording rates of up to 12 kHz can be achieved with image converter cameras, and storage of hundreds of images is possible on most desktop computers. These device characteristics enable such fast phenomena as explosions or detonation waves to be recorded electronically; depending on the selected spectroscopic technique, images of the distribution of thermodynamic parameters of such events can be obtained.

Spectroscopic techniques can be classified in several ways. A simplistic but useful classification includes only two groups: *passive techniques,* which depend on spontaneous emission by the tested object; and *active techniques,* which require an excitation of the tested object to induce detectable emission. Some of the techniques of the first group have been discussed previously, for example, the emission by a blackbody radiator (Section 10.5) that depends on the temperature of the emitting object. Analysis of this radiation can be used to determine the temperature of an object without physical contact. Similarly, spectroscopic analysis of the emission of OH in a flame (Section 10.5) was used to determine the flame temperature and its state of equilibrium. Such measurements

are desirable when the tested medium is inaccessible (e.g., in astronomical measurements or remote sensing). In other remote measurements, the gathered radiation from a remote star can even by analyzed to determine its velocity relative to the observer.

On the other hand, in active spectrsocopic techniques the tested medium is illuminated by an external source. The thermodynamic properties are then determined from an analysis of the absorbed or the scattered radiation. Techniques such as laser-induced fluorescence (LIF), planar laser-induced fluorescence (PLIF), and Raman scattering are active. These techniques are more selective than the passive techniques and allow measurements when the tested medium does not radiate spontaneously. Hence these techniques enable measurements of the properties of cold objects, or analysis of nonradiating species in hot media. The use of an external light source permits stroboscopic illumination along with synchronized recording; this allows one to time the measurement with a desirable event, or to avoid the interference of competing phenomena such as undesirable glow or background radiation.

Active spectroscopic techniques can be further divided into resonant and nonresonant techniques. With *resonant* techniques, the illuminating source is usually tunable and its wavelength can be adjusted until it coincides with a selected transition of the tested medium, thereby forcing it to undergo an absorption or stimulated emission. These resonant techniques include absorption spectroscopy, resonantly enhanced Raman techniques, and more. For cases where the illumination does not match any transition of the tested species, active *nonresonant* techniques - such as Raman or Rayleigh scattering - can be used as an alternative. Resonant techniques typically yield a much stronger signal than do nonresonant techniques, but the difficulty of matching the wavelength of the incident source with a specific transition of the tested species is a serious limitation.

The use of spectroscopic techniques for thermodynamic diagnostics usually depends on the availability of bright and monochromatic sources. Other desirable properties of the illuminating source may include tunability, accurate control of the illumination duration and its timing, or predictable distribution of the radiative energy within the beam. Although noncoherent sources may have some of these characteristics, most can be obtained only from laser sources. Therefore, most active techniques have evolved along with the development of new lasers. Active spectroscopic techniques are available for measurements of temperature, pressure, velocity, and species concentration in gases and liquids, as well as for measurement of surface temperature, pressure, and adsorbates of solids. Techniques can be tailored for synchronized instantaneous imaging or for high-sensitivity measurement. However, despite their success, most techniques are not widely available. Many require expensive equipment and a high level of specialization. Nevertheless, in many applications they offer the only existing method of diagnostics. Here we describe several resonant and nonresonant spectroscopic techniques. These are presently among the most advanced techniques and are gaining wider acceptance in engineering applications.

11.2 Laser-Induced Fluorescence

Laser-induced fluorescence (LIF) is one of the most developed resonant techniques. As its name implies, LIF requires excitation of the tested species by a light source - typically a laser - that is resonant with a selected transition. It can therefore be viewed as a derivative of absorption spectroscopy. The objective of LIF measurements is to determine the population density of selected quantum states and consequently some or all of the thermodynamical properties of the tested medium. For these measurements, a narrowband laser is tuned to excite one or more transitions in the tested atom or molecule. Using a spectrally narrow bandwidth, individual rotational transitions that originate from certain vibrational and electronic levels can be selectively excited. For atomic spectroscopy, the fine and even the hyperfine structure can be resolved, whereupon the excited species may fluoresce. In the absence of competing energy loss mechanisms, the population of the excited state and hence the emitted fluorescence are proportional to the population of the ground state. At thermodynamic equilibrium, the population of an individual rotational state is proportional to the total density of the probed species and to its temperature (eqn. 10.36). Therefore, when the density is known, a single LIF measurement can be used to determine the gas temperature. Otherwise, with two independent LIF measurements (i.e., with the excitation of two different states), both the density of the probed species and its temperature can be determined. Alternatively, the Doppler shift experienced by the absorbing species when moving relative to the source can be measured by LIF and analyzed to determine the flow velocity of gases or liquids.

Although in many ways LIF is similar to absorption spectroscopy, it presents several unique advantages. The measured LIF signal is emitted by the excited species. Therefore, optical detection can be designed to collect fluorescence from a selected sample volume along the exciting laser beam. Depending on the application, either single-point or multiple-point measurements are possible. Furthermore, by expanding the laser beam into a "light sheet" and using a camera to detect the fluorescence, imaging of the molecular properties within the laser light sheet is possible. On the other hand, the need to selectively excite a specific transition limits LIF to particular species. Systems designed to measure the properties of one species can rarely be used to measure other species. Nevertheless, for diagnostics in hostile combusting environments and for high-speed flows, LIF offers unique opportunities that overcome many of its disadvantages. Many of the best-developed LIF techniques are aimed at measuring species that occur naturally in such environments: O_2, OH, NO, the CH bond of hydrocarbons, et cetera. Despite their inherently different designs, the principles of most LIF techniques are similar.

Figure 11.1 is a diagram of the energy levels that participate in a typical LIF experiment. A narrowband laser beam is tuned until its photon energy $h\nu_{ge}$ matches the energy difference between the ground state g and the excited (upper)

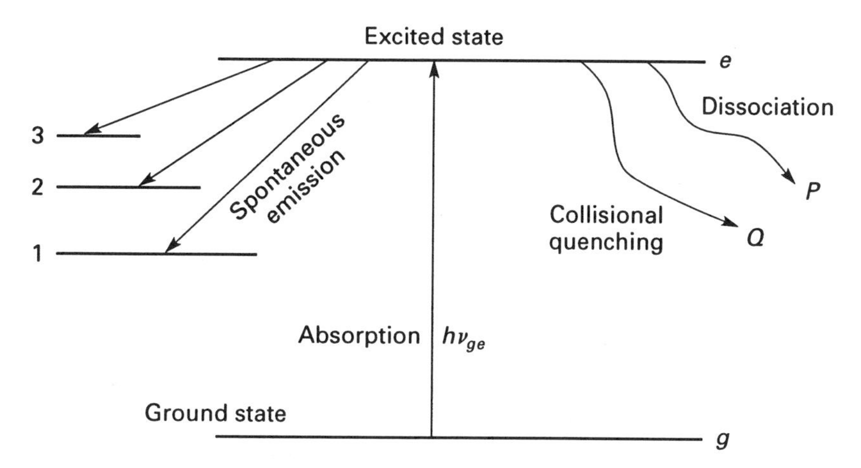

Figure 11.1 Possible modes of energy transfer from an excited state.

state e. If single-photon transitions between these states are allowed, part of the population in the ground state is excited. Once in the upper state, the excited species may relax spontaneously to any one of the lower states - which in this figure are presented schematically by levels 1, 2, or 3 - while emitting a photon. Alternatively, it may give up its energy by dissociation or ionization, or may be removed from this state by collisionally exchanging energy with another object. The latter is known as *collisional quenching,* and includes primarily collisions with other molecules and atoms. The diagram shows spontaneous emission to three energy levels and in some complex molecules the entire LIF process may involve tens or even hundreds of energy levels, but each of these spontaneous transitions is decoupled from the other transitions. The entire LIF process can therefore be modeled as isolated radiative interactions consisting of separate two-level systems. Of course, stimulated emission may also be present, but it does not contribute to the emission detected off the laser beam axis.

For the excitation of a transition with a spectrally integrated absorption cross section σ_{ge} by an incident beam of irradiance I, the rate of excitation is

$$\frac{dN_e}{dt} = \frac{I}{h\nu_{ge}}\sigma_{ge}N_g \tag{11.1}$$

(cf. eqn. 10.23). Implicit in this equation is the assumption that the bandwidth of the laser beam exactly matches the absorption bandwidth. This may be achieved by estimating the absorption linewidth and matching the laser bandwidth accordingly. In some applications, however, the laser bandwidth exceeds that of the absorbing species. For these applications I represents the fraction of the incident irradiance that coincides spectrally with the absorption line. Alternatively, the convolution between the spectral distribution of the laser beam and the spectral dependence of the absorption cross section (eqn. 10.25) must be calculated. At thermodynamic equilibrium, the density of the ground-state population N_g is determined by the Boltzmann distribution (eqn. 10.33):

$$N_g = N_0 \frac{g_g e^{-E_g/kT}}{Q} = N_0 \beta_g(T), \tag{11.2}$$

where

$$\beta_g = \frac{g_g e^{-E_g/kT}}{Q} \tag{11.3}$$

(cf. 10.38) is the Boltzmann fraction – that is, the fraction N_g/N_0 of the tested species that occupies the probed state.

Although (11.1) and (11.2) are sufficient to describe the excitation of the tested species by a laser, the time and spatial variation of the beam irradiance may introduce unnecessary complexity into the calculation. Instead, for pulsed excitation with a duration Δt, the excited population may be expressed by the total incident energy, which is obtained by temporal and spatial integration of the irradiance. Often Δt is determined by the duration of the incident laser pulse; it might instead be the exposure time of the detector used for the fluorescence measurement. When both the laser and the detection are pulsed, Δt is the shortest of the two durations. The total laser energy E_L incident during an LIF experiment is then

$$E_L = \int_A I\,dA\Delta t, \tag{11.4}$$

where A is the cross-section area of the laser beam. Therefore, with this total incident energy, the total number of photons lost by the incident beam by absorption while traversing a sample element of length Δl is

$$n_A = \frac{dE_L}{h\nu_{ge}} = \frac{E_L}{h\nu_{ge}} N_0 \beta_g \sigma_{ge} \Delta l. \tag{11.5}$$

This number of photons is equal to the number of molecules excited along a slice Δl of the beam.

From its excited state, a molecule may return spontaneously back to its original ground state while emitting a photon at the same wavelength as the incident radiation. Unfortunately, other elastic scattering processes (e.g. Mie scattering) may interfere with detection. Thus, for better discrimination, other transitions (e.g., transitions to level 3 in Figure 11.1) are selected for detection. By placing a bandpass filter between the excited sample and the detector, a specific emission can be detected while detection of the emission occurring via other transitions is blocked. However, the net number of molecules emitting in the desired band is only a fraction of the total number of excited molecules. In the absence of energy loss mechanisms, this fraction is defined by (10.13). Most often, such loss mechanisms cannot be neglected. In most applications of LIF, an excited molecule may be collisionally quenched by colliding with another molecule, exchanging energy with it at a rate of Q_C. Whether the exchange involves energy loss or gain, the excited molecule is removed from the state from which it was expected to radiate and is therefore lost to the fluorescence process. Similarly, predissociation of excited molecules at a rate P may deplete the excited state. Finally, stimulated emission by the intense incident laser beam may further deplete the excited state at a rate S. Therefore, by extension of (10.13), the fraction of molecules that emit spontaneously by a transition to a selected state (e.g., level 3 in Figure 11.1) is

$$\mathrm{SV} = \frac{A_{e3}}{\sum_i A_{ei} + P + Q_C + S}. \tag{11.6}$$

This is the *Stern–Volmer coefficient* (Stern and Volmer 1919). When other mechanisms (e.g. ionization) effectively deplete the excited-state population, their rate may also be added to the denominator of SV. The Stern–Volmer coefficient is a measure of the yield of the LIF process. The number of emitted photons is obtained by multiplying the number of excited molecules by SV. Thus, the total number of photons emitted by the volume enclosed by a slice of Δl along the laser beam is

$$n_e = n_A \times \mathrm{SV} = \frac{E_L}{h\nu_{ge}} N_0 \beta_g \sigma_{ge} \Delta l \frac{A_{e3}}{\sum_i A_{ei} + P + Q_C + S}. \tag{11.7}$$

Of these emitted photons, only a small fraction can be collected and analyzed by the detection system. Most are lost by the limited aperture size of the collection lens or telescope, by the bandpass filter, and by the limited quantum efficiency of the detector. Evaluation of the expected signal requires introduction of parameters that specify these components.

Figure 11.2 presents a typical LIF set-up designed for gas-phase measurements using a tunable pulsed laser. The incident laser beam is split into two

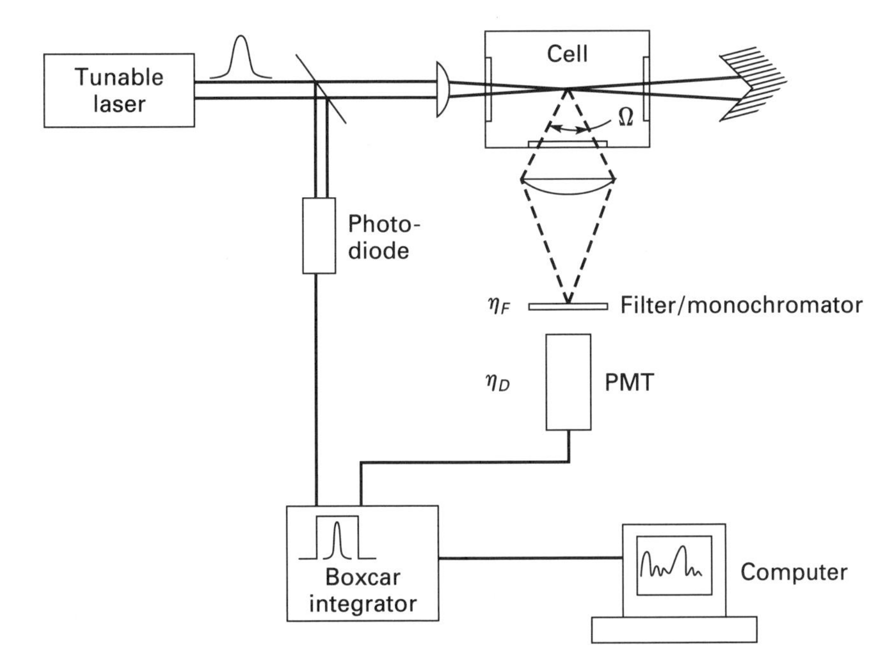

Figure 11.2 Experimental layout for LIF measurements.

parts: one part (consisting of most of the energy) is directed to the sample cell, while a small fraction of the beam is directed toward a photodiode where the energy of each pulse is measured. The fluorescence induced at the focal volume of the beam is collected by a telescope with a collection solid angle Ω. The collected radiation is then passed through a bandpass filter that transmits only the desired radiation at a transmission efficiency of $\eta_F < 1$. The transmitted radiation is then detected by a photomultiplier tube (PMT), where the incident photons are converted into photoelectrons. The efficiency of this conversion is limited by the transmission efficiency of the PMT windows, the structure of the photocathode and the dynodes, and other factors. However, the conversion ultimately is limited by the quantum efficiency η_D. This is a quantum mechanical property of the photocathode that specifies what fraction of incident photons are converted by the photoelectric effect into electrons. The number of photoelectrons n_{pe} emitted by the photocathode is the measurable signal of the LIF process. The detectable signal, expressed in photoelectrons, is therefore

$$\begin{aligned} n_{pe} &= \left(\frac{\Omega}{4\pi}\eta_F\eta_D\right)n_e \\ &= \left(\frac{\Omega}{4\pi}\eta_F\eta_D\right)\frac{E_L}{h\nu_{ge}}N_0\beta_g\sigma_{ge}\Delta l\frac{A_{e3}}{\sum_i A_{ei}+P+Q_C+S}. \end{aligned} \tag{11.8}$$

This signal is amplified internally by the dynodes of the PMT (Section 3.2) and the output is processed electronically.

Although the fluorescence is separated from background radiation by the bandpass filter, there may be a significant level of continuously emitted background radiation that is spectrally coincident with the LIF signal. Furthermore, the detector and the electronic system are likely to generate noise that interferes with the measurement. On the other hand, owing to the short duration of the laser pulse, the LIF signal is present only briefly (and synchronously) with the laser pulse. Therefore, for increased rejection of any background signal or noise, the detection can also be pulsed (or gated) and synchronized electronically with the laser pulse.

Equation (11.8) includes most of the parameters that influence pointwise LIF measurements. It does not, however, account for laser beam attenuation by an optically thick medium along the incident beam path, nor for reabsorption of the LIF emission by a large density of molecules in the ground state. These effects may be considerable in certain applications and must be accounted for (Quagliaroli et al. 1993). Other effects, such as depletion of the ground state by an intense laser beam, may also introduce a considerable error (Laufer, McKenzie, and Fletcher 1990). On the other hand, saturation effects associated with stimulated emission from the excited state are included in the Stern–Volmer coefficient by the rate term S (Daily 1977).

Planar imaging of the distribution of thermodynamic properties of gas or samples is an important application of LIF (see e.g. Kychakoff et al. 1982 and Lee et al. 1992). By expanding the laser beam into a light sheet and judiciously

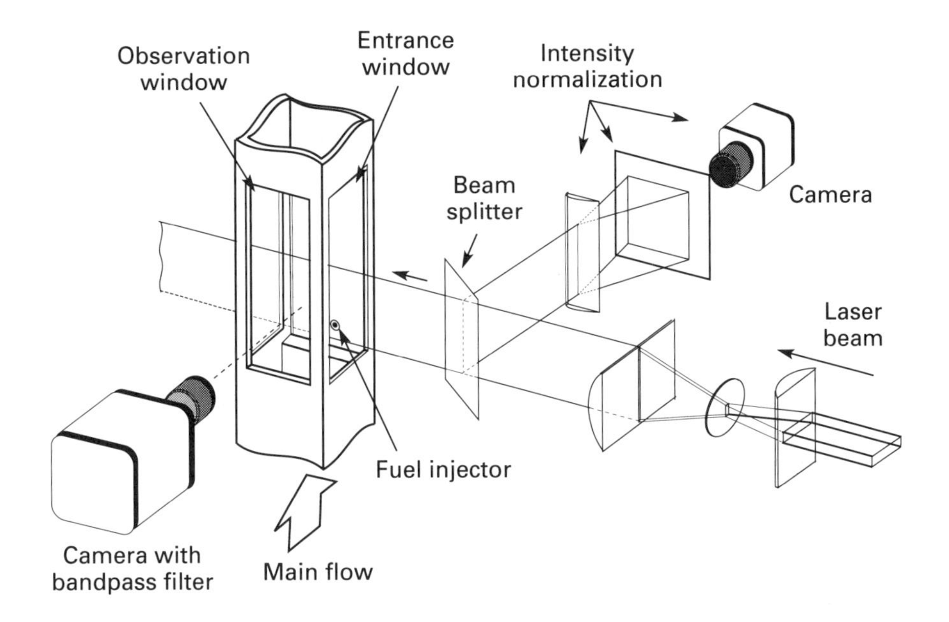

Figure 11.3 Experimental layout for PLIF measurements.

passing it through the tested sample, selected sections can be imaged. Figure 11.3 illustrates a typical set-up of a PLIF experiment. The primary difference between the LIF and the PLIF techniques is the use of a laser light sheet for illumination. Therefore, instead of a PMT, a camera equipped with a bandpass filter is used for the fluorescence detection. A second camera may be needed for normalization of laser energy fluctuations and to account for spatial variations of the energy distribution. When LIF is imaged in hot gases, background luminosity may overwhelm the signal. Therefore, as with point measurements, gated detection may be necessary; this is achieved by gating intensified CCD cameras, which may provide gates as brief as 10 ns together with synchronization of the laser pulse.

Owing to the extended size of the laser beam needed for planar imaging, the signal available from each volume element in most PLIF experiments is low relative to the signal available in ordinary pointwise LIF. That signal is usually recorded by an electronic camera consisting of an array of pixels with m_p columns and n_p rows. Since each pixel images the fluorescence from a specific sample element, the laser light sheet can also be viewed as being divided into $m_p \times n_p$ elements (Figure 11.4). When the energy distribution transverse to the laser beam is uniform (or else corrected by normalization), each row of pixels in the object plane is illuminated by an energy of E_L/n_p for every laser pulse. Similarly, the length of each element imaged by the camera is $\Delta l = L/m_p$, where L is the length of the section along the light sheet that is imaged by the camera. With these two modifications, the signal (in terms of photoelectrons) at each of the camera pixels becomes

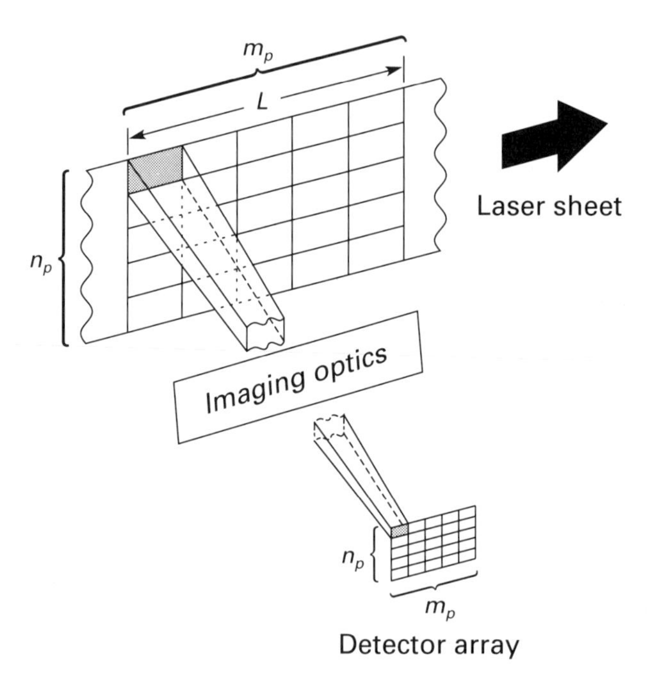

Figure 11.4 Imaging of a section of a laser light sheet of length L on an array with $m_p \times n_p$ pixels.

$$n_{pe} = \left(\frac{\Omega}{4\pi}\eta_F\eta_D\right)\frac{E_L}{h\nu_{ge}n_p}N_0\beta_g\sigma_{ge}\frac{L}{m_p}\frac{A_{e3}}{\sum_i A_{ei}+P+Q_C+S}. \tag{11.9}$$

Equation (11.9) may be used to estimate the expected signal when the medium properties are known, or to determine those properties from the measured PLIF signal. The following example illustrates such a calculation for PLIF obtained in an optically thin medium, where absorption and trapping are negligible.

Example 11.1 Calculate the number of photoelectrons to be generated at one pixel of a CCD camera with a quantum efficiency of $\eta_D = 0.1$ and a collection efficiency of $\Omega/4\pi = 10^{-2}$ in a PLIF measurement of OH at a density of $N_{\mathrm{OH}} = 5\times10^{21}$ m^{-3}. The laser bandwidth matches the $(3\leftarrow 0)$ $P_1(8)$ transition and the collected fluorescence is from the $(3\rightarrow 2)$ $Q_1(7)$ transition. The Boltzmann fraction of the probed state at the temperature of this test is $\beta_g = 0.023$. The incident laser energy of 240 mJ/pulse at 40,296.3 cm^{-1} (Andresen et al. 1988) is spread into a light sheet that corresponds to 100 pixel rows in the image plane. Each of these rows is further divided by the detector array into pixels, where each pixel captures radiation from a 1-mm section. The fluorescence passes through a filter with a peak transmission efficiency of $\eta_F = 0.1$. The absorption linewidth, including Doppler broadening, is estimated at $\Delta\nu = 18$ GHz. For the absorption, the Einstein B coefficient is $B_{03} = 3.461\times10^{15}$ m^3-s^{-2}-J^{-1} and, for the $Q_1(7)$ emission, $A_{32} = 3.24\times10^5$ s^{-1} (Dimpfl and Kinsey 1979). The excited state is strongly predissociated, with a predissociation rate of $P = 0.57\times10^{10}$ s^{-1} (Heard et al. 1992).

Solution The cross section for absorption by this transition is

$$\sigma_{03} \approx \frac{B_{03} h\nu_{03}}{\Delta\nu c} = 5.137 \times 10^{-22}\ \text{m}^2$$

(cf. eqn. 10.25). Although ν_{03} and $\Delta\nu$ were specified in units of cm^{-1} and GHz, respectively, they must be expressed in the same units when calculating the cross section.

The population density of the $v' = 3$ state immediately after the excitation is strongly controlled by the rate of predissociation. This rate exceeds the total rate of spontaneous emission by approximately four orders of magnitude. Therefore, most of the excited molecules dissociate before they emit, thereby reducing the fluorescence efficiency. On the other hand, since the predissociation rate under these experimental conditions is about ten times faster than the rate of collisions, effects of collisional quenching can be neglected. With these assumptions, the effective Stern–Volmer coefficient for this transition is

$$\text{SV} \approx \frac{A_{32}}{P} = \frac{3.24 \times 10^5}{0.57 \times 10^{10}} = 5.68 \times 10^{-5}.$$

The signal in terms of photoelectrons per pixel is obtained by introducing these results into (11.9) as follows:

$$\begin{aligned} n_{pe} &= \left(\frac{\Omega}{4\pi}\eta_F \eta_D\right) \frac{E_L}{h\nu_{03} n_p} N_{\text{OH}} \beta \sigma_{03} \frac{L}{m_p} \times \text{SV} \\ &= (0.01 \times 0.1 \times 0.1) \frac{0.240}{(6.63 \times 10^{-34})(1.2089 \times 10^{15})100} \\ &\quad \times (5 \times 10^{21}) 0.023 (5.137 \times 10^{-22}) 0.001 (5.68 \times 10^{-5}) \\ &= 1{,}004 \text{ photoelectrons/pixel.} \end{aligned}$$

■

Of course, the signal predicted by this example is observed only when the illumination as well as the density of OH are uniformly distributed. Typically, both are non-uniform and large variations in the signal are possible. Figure 11.5 shows an image of the distribution of OH molecules in a supersonic H_2–air combustion flow obtained under conditions similar to those of Example 11.1 and using the configuration illustrated in Figure 11.3 (Quagliaroli et al. 1995). The flow of heated supersonic air at a Mach number $M = 2$ enters the image plane from below and the injection of H_2 is perpendicular to that plane. The raw PLIF signal in this image was converted to represent the distribution of N_{OH} using the results of a calibration measurement in an environment where the density of OH was accurately known.

The success of PLIF imaging depends on the signal level. Therefore, all system designs include optimization of the illumination, signal collection, and signal processing systems. However, the outcome of the measurement ultimately depends on the absorption Einstein B coefficient, the Einstein A coefficient of the emitting transition, and the competing energy loss mechanisms that determine the magnitude of SV (eqn. 11.6). Thus, based on signal considerations

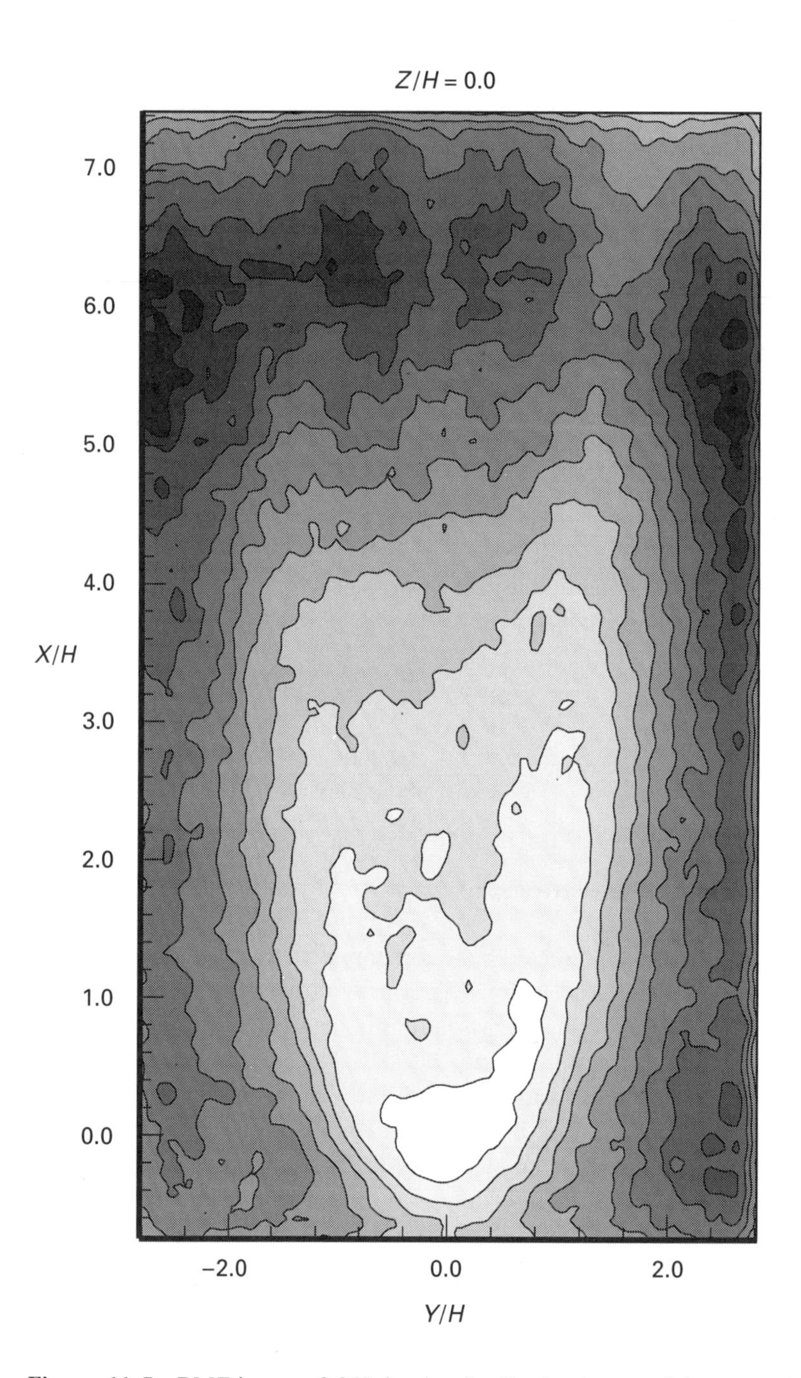

Figure 11.5 PLIF image of OH density distribution in a model supersonic hydrogen–air combustor. Hydrogen is injected transversely (out of the page, along the Z axis, where Z is measured from the step face) behind an $H = 5.1$-mm step into the air flow, which is from the bottom up.

alone, LIF transitions should include strong absorptions to enhance the excitation, a strongly emitting transition for the detection, and (if possible) a non-predissociated state. However, when the absorbing transition is too strong, the laser beam may be significantly attenuated even when propagating through a moderately thick medium. In addition, if the emitting transition terminates at a level where the naturally occurring population density is high, reabsorption of the emitted photons (i.e., *radiation trapping*) may further complicate the measurements. Therefore, the selection of an excitation path must include considerations not only of the expected fluorescence signal but also of the effect of the absorption on the incident laser beam and of the emitted fluorescence as they propagate through the sample.

Figure 11.6 illustrates the calculated effect of absorption on an incident laser beam by OH at 2,000 K (Quagliaroil 1993). Two laser systems were considered for the excitation; a frequency-doubled tunable dye laser system that can be tuned around nominally $\lambda = 283$ nm to excite transitions in the strongly absorbing $1 \leftarrow 0$ band, and a tunable KrF laser beam that can be tuned around nominally $\lambda = 248$ nm to excite transitions in the weakly absorbing $3 \leftarrow 0$ band. To account for both the density and the propagation path lengths through the medium, the line integral of the product of $N_{\mathrm{OH}}\,dL$ along the distance traveled

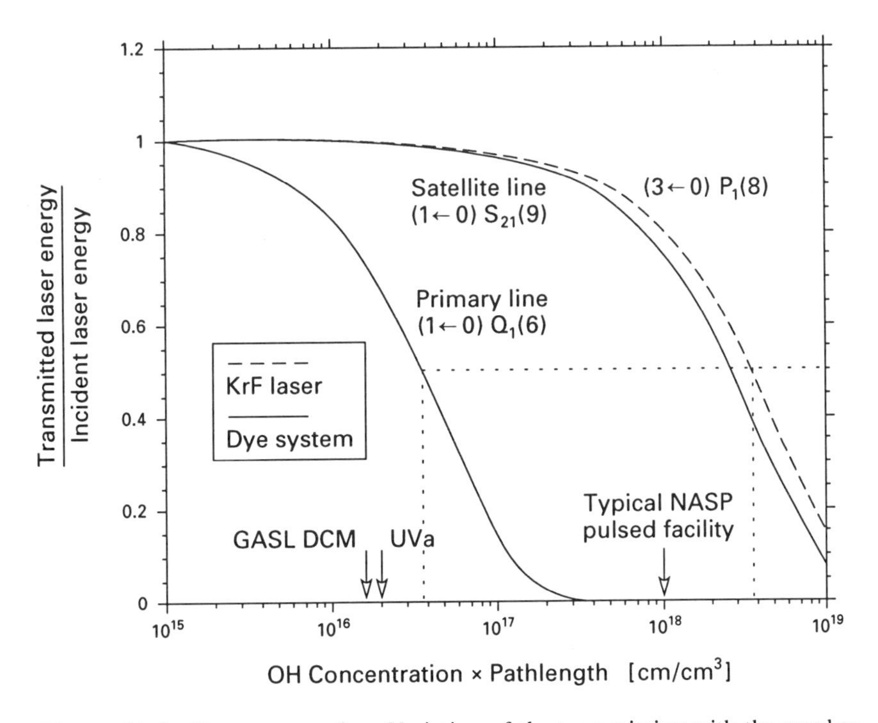

Figure 11.6 Beam attenuation: Variation of the transmission with the number of OH molecules intercepted per unit area of a doubled dye (solid line) or tunable KrF (dashed line) laser beam passing through reacting flow at static temperature of 2,000 K. Typical absorption depths in three supersonic combustion test facilities are indicated.

by the beam is presented as the independent variable. To further facilitate the interpretation of these results, the conditions of three different test facilities are marked. Figure 11.6 shows that when the $1 \leftarrow 0$ transition is excited, the beam is attenuated by approximately 15% after propagating only 2 cm through the medium, even when the concentration of OH is at a moderate level of 5×10^{15} cm^{-3}. Without correcting for this attenuation, an error of at least its magnitude is expected when LIF is used for density measurements. For temperature measurements, the error may be either larger or smaller than this uncertainty (see Example 10.4 of Section 10.5). In large facilities, where both N_{OH} and the propagation distances are much larger, attenuation of the beam may preclude LIF measurements. Alternatively, a satellite transition in the $1 \leftarrow 0$ band or a transition in the $3 \leftarrow 0$ band may be used.

A related effect that may distort LIF measurements in hot gases is radiation trapping. When the radiative transition terminates at a low quantum state and when the temperature is high, the density of the population in the end state may be sufficient to reabsorb, or trap, a large portion of the LIF emission. The extent of radiative trapping depends on the integrated concentration and temperature of OH along the detection optical path. Therefore, in complex three-dimensional flows, corrections for this distorting effect may not be possible.

Figure 11.7 illustrates the effect of trapping on the LIF of OH by presenting the variation of the normalized transmission of the emitted fluorescence with the number of OH molecules intercepted, per unit area, along the detection path. As before, the dye laser system is represented by excitations in the $1 \leftarrow 0$ $Q_1(6)$ and the satellite $1 \leftarrow 0$ $S_{21}(9)$ bands. Two humps are apparent in each of these curves. The upper hump is due to fluorescence trapping in the $0 \rightarrow 0$ band. The lower hump, due to the fluorescence trapping in the $1 \rightarrow 1$ band, becomes apparent only after the radiation emitted by the $0 \rightarrow 0$ band is depleted. Despite the low attenuation of the incident beam when it is tuned to the satellite band (Figure 11.6), the effect of trapping is similar to that experienced when the primary band is excited. In applications involving small facilities, the effect of such trapping is moderate. But even there, with a larger concentration of OH, trapping becomes significant and may cause a non-uniform signal loss of as much as 70%. Clearly, the signal generated by the doubled dye–laser system in large facilities is likely to be fully trapped. Thus, even if the beam can penetrate the flow when the laser is tuned to the satellite transition, no signal will be detected when the optical path length is too long.

For contrast, Figure 11.7 also shows the effects of fluorescence trapping for excitations by the tunable KrF laser. Two cases are presented: attenuation by trapping for radiative transitions in the $X\,^2\Pi$ ($v''=2$) level, and attenuation by trapping for radiative transitions to the $X\,^2\Pi$ ($v''=3$) level. As expected, the trapping by the $v''=2$ level in the ground electronic state of fluorescence in the $3 \rightarrow 2$ band is larger than the trapping by the $v''=3$ level. However, in most applications the trapping by both transitions is sufficiently small, and the selection may be dominated by considerations of signal level and discrimination

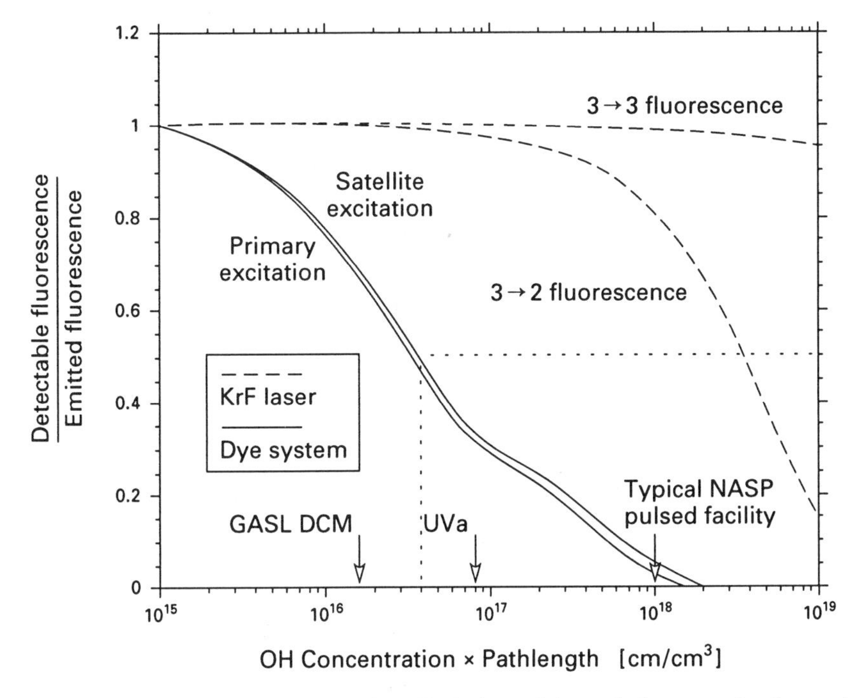

Figure 11.7 Fluorescence trapping: Variation of the ratio between the detected and emitted fluorescence, with the number of OH molecules intercepted per unit area, along the detection path for fluorescence induced by a doubled dye (solid line) or tunable KrF (dashed line) laser beam in reacting flow at static temperature of 2,000 K. Typical absorption depths in three supersonic combustion test facilities are indicated.

of the laser-induced OH fluorescence against chemiluminescence or competing LIF from other combustion species.

These considerations illustrate the analysis required for the design of an effective LIF system. However, even when considerations of signal beam attenuation and radiation trapping are successfully resolved, collisional quenching may introduce significant uncertainties. When the lifetime of the excited state is long relative to the time between collisions, a large number of excited molecules may have the opportunity to collide with other molecules while exchanging energy. Whether energy is gained or lost by such collisions, the excited species is removed from the radiating state before emission occurs, thereby reducing the available fluorescence. Consequently, the emission depends not only on the population density at the ground state, but also on the rate of quenching collisions (Crosley 1981).

The effect of collisions on the emission bandwidth was discussed in Section 10.7. Similar parameters also describe collisional quenching. For collisions between two different species A and B, the probability for the removal of species A from its excited state is controlled by the quenching cross section σ_{AB}^{Q}. Thus, similarly to the rate of dephasing events by AB collisions that lead to collisional broadening (eqn. 10.52), the quenching rate Q_{AB} per molecule of type A by collisions with molecules of type B is described by

$$Q_{AB} = N_B \sigma_{AB}^Q \left(\frac{8\pi kT}{\mu_{AB}} \right)^{1/2}. \tag{11.13}$$

Similarly, the rate of single-component quenching events per molecule is

$$Q_{AA} = N_A \sigma_{AA}^Q \left(\frac{16\pi kT}{m_A} \right)^{1/2} \tag{11.14}$$

(cf. eqn. 10.53). The total quenching rate per molecule of type A in a gas mixture with i different species is thus

$$Q_C = Q_{AA} + Q_{AB} + \cdots = \sum_i Q_{Ai} \tag{11.15}$$

(cf. eqn. 10.54). In reacting gases where the mixture composition varies continuously, correction for quenching effects may require accurate preliminary knowledge of the temperature, pressure, and mixture composition – the same parameters that are to be measured by LIF. When the pressure or temperature are low, these corrections may be neglected. However, for measurements in realistic conditions, other approaches where the effect of quenching is negligible must be used.

The effect of collisional quenching can become negligible at high temperatures and pressures only if the other energy loss effects in the Stern–Volmer coefficient (eqn. 11.6) are significantly larger. Of the four terms in the denominator of (11.6), only stimulated emission can be actively controlled. By focusing the laser beam, the irradiance can be increased until the magnitude of S exceeds the magnitude of all the other rate constants that deplete the excited state (Daily 1977). When this is achieved, the transition is said to be *saturated* – further increase in the incident laser energy will not result in a corresponding increase in the LIF signal. However, at saturation the Stern–Volmer coefficient is

$$\text{SV} \approx \frac{A_{e3}}{S}. \tag{11.16}$$

With the effect of quenching now neglected, the measurement depends only on the population density of the ground state. In an alternative technique for the reduction of the effect of collisional quenching, the molecule is excited to a strongly predissociated state where $P \gg Q_C$ (Andresen et al. 1988). Although the strong predissociation reduces the number of radiating molecules, the Stern–Volmer coefficient can now be simplified as in Example 11.1 to

$$\text{SV} \approx \frac{A_{e3}}{P}. \tag{11.17}$$

Predissociative LIF techniques have been demonstrated for OH and O_2. The severe signal loss due to predissociation is compensated by the use of a brighter light source or by accumulation of signal over a larger number of laser pulses.

The distribution of OH in Figure 11.5 was obtained by an excitation to a predissociation state. The high image quality was obtained by an accumulation of 200 pulses. Owing to the relatively low fluorescence yield, the sensitivity of predissociative LIF is limited and measurements are normally possible only when the concentration of the detected species is high.

The fluorescence signal of an LIF experiment, as predicted by (11.8), includes neither the effects of polarization of the incident radiation on the scattered LIF signal nor the effect of the polarization of the fluorescence on the detection efficiency. Since any single-photon absorption is a dipole interaction, the dipole moment of the absorber must have a component that is parallel to the incident polarization; when the dipole moment is perpendicular to the radiation polarization, excitation may not be possible. Similarly, the polarization of the emission depends on the orientation of the emitting dipole. In the gas phase, molecules rotate at a frequency that is defined by the quantum number J and is of the order of 10^{11} Hz. Thus, even for highly predissociated states, an excited molecule can complete at least one spin around its rotational axis before emission (or before it dissociates). Owing to this fast spin, part of the incident polarization is scrambled; that is, some of the dipoles in the gaseous medium point in random directions, thereby emitting random polarization. Nevertheless, owing to the precession of the J vector around the m_1 axis (Figure 9.3), the rotation around J is not fully scrambled. A component of their orientation relative to a fixed reference frame is still preserved and with it part of the original polarization. Of course, when the orientation of the vector of the magnetic angular momentum is changed by collisions, the polarization of the fluorescence changes with it.

When the emission is polarized, as may be the case when the exciting radiation is polarized, its directional distribution is non-uniform. Thus, if the incident source is linearly polarized, (11.7) and (11.8) must be supplemented with a factor that accounts for the increase, or decrease, in the emission efficiency due to the polarization of the source (Doherty and Crosley 1984). Furthermore, when the detection system inherently discriminates against one polarization (e.g., by the use of diffraction gratings for spectral analysis), additional corrections may be necessary (see Research Problem 11.2). Of course, when the exciting source is unpolarized, or when it is circularly polarized, detection is isotropic and is independent of depolarizing collisions.

The primary application of LIF is for temperature and density measurements of select species. Nevertheless, one of the earlier applications of LIF was to measuring the velocity of a supersonic helium flow (Zimmermann and Miles 1980). In that experiment, the flow was seeded with sodium vapor, which was subsequently excited by a tunable dye laser with a linewidth sufficiently narrow to resolve part of the hyperfine splitting of the sodium D lines. The beam was directed at 67°40′ toward the flow (see Research Problem 9.1) and, after passing the test section, was retro-reflected to cross it once again. Because of the Doppler effect, the excitation frequency of the moving atoms was upshifted

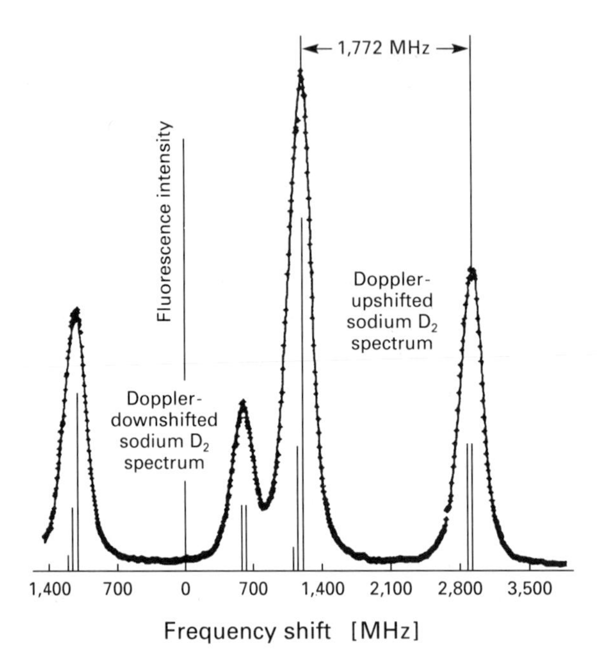

Figure 11.8 Excitation spectrum of the hyperfine structure of Na in a supersonic He flow. [Zimmermann and Miles 1980, © American Institute of Physics, Woodbury, NY]

relative to the forward propagating beam but downshifted relative to the retro-reflected beam. Figure 11.8 illustrates the excitation spectrum of the hyperfine structure obtained by tuning the laser while monitoring the fluorescence emitted by the seed atoms. Owing to the frequency upshift of one beam and the frequency downshift of the other, the scan included two spectra shifted relative to each other by twice the extent of the Doppler shift. The two lines at the high-frequency side of this spectrum are part of the Doppler upshifted structure of the ground-state hyperfine splitting. The difference between the peaks of these lines is 1,772 MHz (Figure 9.5). Similarly, the two lines at the low-frequency end of the spectrum are part of the Doppler downshifted hyperfine structure. By comparison with the known hyperfine splitting of 1,772 MHz, the frequency separation between the two Doppler-shifted excitation spectra was determined to be 2,258 MHz. Using (6.24) for $\Delta\omega_i$, the velocity associated with half of that frequency separation is 1,750 m/s (Zimmermann and Miles 1980).

The need to tune the laser through an entire spectral structure limits this approach to measurements in steady-state flows. However, pulsed PLIF techniques have been developed for instantaneous imaging of the velocity distribution. In one such experiment (Paul, Lee, and Hanson 1989), a relatively broadband laser was used for the excitation of narrowband absorption of NO molecules in a supersonic flow. The beam was introduced at a prescribed angle to the direction of the major component of the flow velocity and then retro-reflected. The wavelength of the incident beam was tuned so that excitation of stationary NO

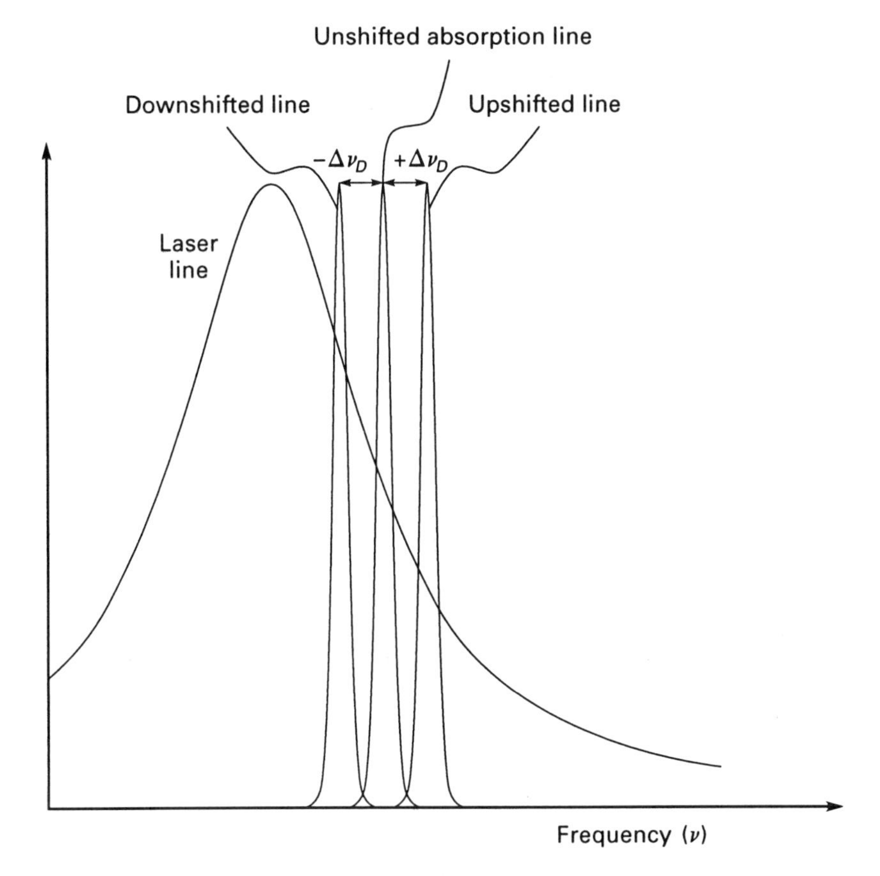

Figure 11.9 Overlap between a broadband laser beam and a Doppler upshifted, unshifted, and downshifted absorption line.

molecules occurred at approximately half the maximum point of the laser lineshape function (Figure 11.9). Therefore, excitation of the Doppler upshifted beam occurred nearer to the peak of the laser spectral irradiance while the Doppler downshifted excitation (by the retro-reflected beam) was induced by a weaker spectral irradiance. With this configuration, the Doppler upshift and the Doppler downshift were translated into higher and lower LIF signals, respectively. Two such fluorescence measurements, together with a reference LIF measurement, can yield the velocity of the seed molecules. Planar distribution of the velocity field may also be obtained by this technique by spreading the laser beam into a light sheet and imaging the fluorescence from that plane.

11.3 Spectroscopic Techniques for Surface Temperature and Pressure Measurements

The LIF techniques discussed in the previous section are primarily used for gas- and liquid-phase measurements. However, potential applications of LIF also include noncontact surface diagnostics, such as: the measurement of turbine blade temperature in jet engines; pressure or temperature of aerodynamic models, either in test facilities or in flight; temperature measurement in furnaces; tests in internal combustion engines; and more. Infrared pyrometry that depends on blackbody emission of test objects is frequently used for noncontact temperature measurements. However, the dependence on surface emissivity and spectral characteristics of the object, as well as the effect of the surrounding temperature on the measurement, limit the use of these techniques. Alternatively, resonance optical techniques, which are similar to LIF, can circumvent many of these difficulties. With these techniques, the test surface is coated with either a temperature-sensitive or pressure-sensitive paint, and is then illuminated by radiation that induces excitation of active components of the paint. The subsequent fluorescence is imaged and interpreted to provide the distribution of the surface temperature or pressure. The techniques can be adapted for measurements over large objects in hostile environments or when access is limited. The principles of both these techniques are discussed here.

For temperature measurements, a group of phosphoric crystals with several unique properties is used. These *thermographic phosphors* include rare-earth ions such as europium or dysprosium doped at low concentration into host crystals such as yttrium aluminum garnet (YAG), lanthanum oxysulfide (La_2O_2S), or yttrium oxide (Y_2O_3). Because of their broadband absorption, thermographic phosphors can be excited by noncoherent broadband ultraviolet sources and with moderate power requirements. The lifetime of their excited states is relatively long (1–5,000 μs), thereby permitting the thermalization of the excited states – that is, redistribution of the population of the excited states until it follows the temperature-dependent Boltzmann distribution (eqn. 10.33). The resultant spontaneous emission, which owing to its long lifetime is called *phosphorescence,* presents characteristics that are temperature-dependent. Depending on the selected material, these characteristics involve an increase in the phosphorescence with temperature (Goss, Smith, and Post 1989), a decrease in the phosphorescence with temperature (Fonger and Struck 1970), or a decrease in the emission lifetime (Alaruri et al. 1993). By grinding these thermographic phosphors into a fine powder which is then incorporated in a binder, test surfaces can be coated with temperature-sensitive paint that emits temperature-dependent phosphorescence when illuminated by the required UV radiation.

One well-tested thermographic phosphor is the europium-doped lanthanum oxysulfide ($La_2O_2S{:}Eu^{3+}$). When illuminated by UV radiation, this complex fluoresces brightly in the visible spectrum. Unlike its absorption, the emission of the Eu^{3+} dopant is not broadband; instead, it consists of several sharp lines. This emission pattern is characteristic of the rare-earth dopants. The outermost shell of these atoms is a closed shell. Therefore, absorption and emission occur by electronic transitions within the partially filled 4f shell that resides inside the completed external shell. The presence of the outer shell shields the emitting shell from effects (of the host crystal) that tend to scramble the absorption and emissions of other phosphors. Nevertheless, there exist some interactions between these states of the dopant ion and its host crystal that render its emission temperature-sensitive.

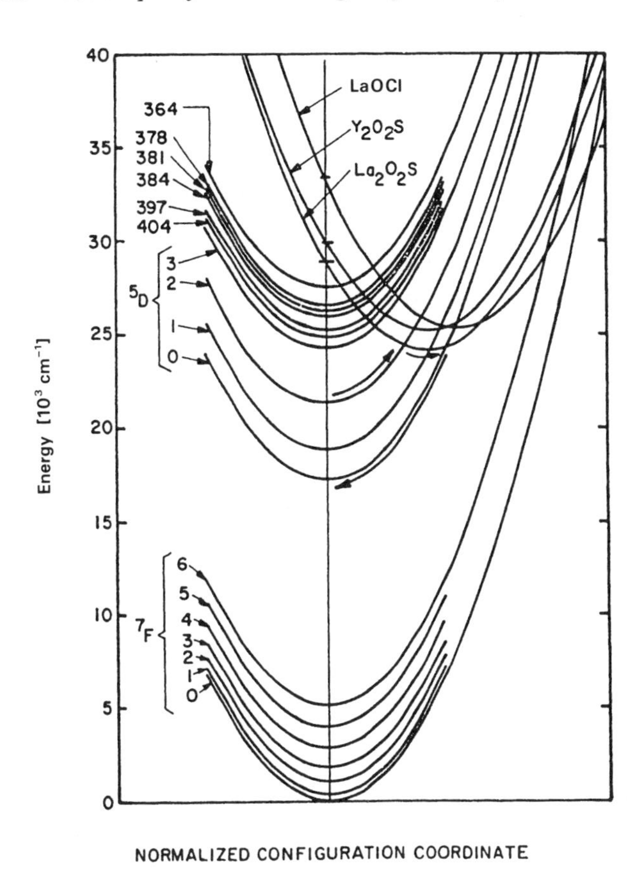

Figure 11.10 Energy-level diagram of the CTS and 4f states of Eu^{3+} in three hosts. The 4f states above $^5D_3$ are indexed by their absorption wavelengths [nm] from $^7F_0$. [Fonger and Struck 1970, © American Institute of Physics, Woodbury, NY]

Figure 11.10 illustrates the energy-level diagram of the 4f levels of Eu^{3+}. The configurational coordinate, similar to the internuclear separation of diatomic molecules, is a measure of the Eu coordinates in the lattice. This diagram includes the potential curves of the ground electronic ^7F states, the ^5D states from which emission is observed after excitation, and a group of states above the ^5D states. The numbers adjacent to these upper states indicate the wavelengths in nanometers required to excite them by absorption from the $^7F_0$ state. To the right of these potential curves are three curves representing the conduction bands, or charge transfer states (CTS), of three potential host lattices: La_2O_2S, Y_2O_2S, and LaOCl. Each of these states can be populated by broadband absorption from the Eu ground state. However, owing to the shift of their vertices relative to the ^7F states, excitation of each of these CTS is controlled by the Franck–Condon principle. Thus, from the $^7F_0$ state, the highest probability for excitation is along the vertical line emerging from its vertex to the horizontal tick-mark at the CTS. Although the ^5D states are observed to be the strongest emitters (Fonger and Struck 1970), the strongest absorption is through the CTS or through those states above ^5D. Therefore, the emitting ^5D states are fed by the excited CTS through the cross-over between these states and the ^5D_j ($j = 0, 1, 2, 3$) states. These cross-overs are similar to those that lead to the dissociation of diatomic molecules such as O_2 (Figure 9.11) or OH. Therefore, while the CTS are thermalized, the various ^5D states are filled up by CTS $\rightarrow$ ^5D cross-over transitions. Similarly, there are cross-over transitions that allow population transfer from certain upper 4f states, which were excited directly by absorption from the ground state, to the CTS. Part of this population is also subject to a CTS $\rightarrow$ ^5D transfer. Although other relaxation processes of the CTS or any of the excited 4f states also exist, the temperature-controlled CTS $\rightarrow$ ^5D transfer is the primary avenue for populating the emitting ^5D states.

Once the ^5D states are populated, they too undergo thermalization. Therefore, as the upward-pointing arrow along the potential curve of the $^5D_2$ state indicates, some excited atoms may (owing to thermalization) cross back to the CTS. When such cross-over occurs, the energy added to the CTS is rethermalized, thereby transferring energy to a lower ^5D_j state as the downward-pointing arrow indicates. The $^5D_j \rightarrow$ CTS transfer is opposed to the CTS $\rightarrow$ ^5D process that initially populated the higher states. As the rate of this latter process increases (i.e., as temperature rises), the phosphorescence is quenched. The higher ^5D_j states, where the bottom of the potential curve is near the cross-over point, require less energy to cross over to the CTS and are therefore quenched at lower temperatures than the lower-lying states.

For illustration, Figure 11.11 presents the dispersion spectrum following excitation of La_2O_2S:1.0%Eu at room temperature by a UV lamp at a nominal

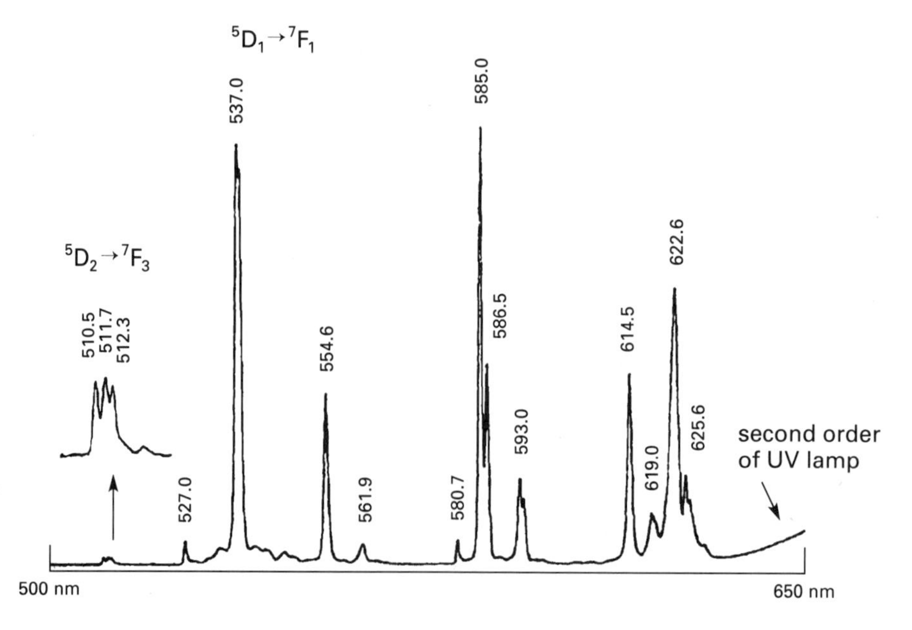

Figure 11.11 Variation of the emission intensity with wavelength for La_2O_2S:Eu at room temperature following 366-nm broadband excitation. [Krauss, Hellier, and McDaniel 1994, © Optical Society of America, Washington, DC]

wavelength of 366 nm (Krauss, Hellier, and McDaniels 1994). Two transitions, $^5D_2 \rightarrow {}^7F_3$ and $^5D_1 \rightarrow {}^7F_1$, are identified in the figure. The first is a weak triplet at nominally 512 nm. At room temperature, quenching by cross-over from the $^5D_2$ state to the CTS was sufficient to extinguish it almost completely. The other line, at 537 nm, from the $^5D_1$ that lies below the $^5D_2$ state, is still bright at this temperature. To reach the cross-over point this state needs higher energy, which can be attained only by raising the temperature. Therefore, the emission from this state has not been significantly quenched. Based on this figure, by spectrally isolating the emission from one of these transitions it is possible to determine the temperature by measuring the phosphorescence.

To quantify the temperature dependence of the phosphorescence, recall that the cross-over quenching depends on the activation energy E_j, which is the energy required to raise the ion from the bottom of the potential curve of the emitting ^5D_j state to the cross-over point with the CTS. At thermodynamic equilibrium (i.e., when the Boltzmann distribution of eqn. 10.33 prevails), the number density of Eu ions that can reach this point increases as $\exp\{-E_j/kT\}$. The frequency of the cross-over events increases proportionally with this density with a rate constant of G_j. Therefore, like the Stern–Volmer coefficient (eqn. 11.6), the phosphorescence efficiency F_j from each of the ^5D_j states is proportional to the ratio between the rate of emission A_j from that state and the sum of all the rates that deplete the emitting state (Fonger and Struck 1970):

$$F_j = \frac{A_j}{\sum A_j + G_j e^{-E_j/kT}},$$

where $\sum A_j$ is the total rate of all the downward emission processes. The coefficients that defined F_j can be determined empirically. Alternatively, the phosphorescence I_j, which is proportional to F_j, can be measured relative to the phosphorescence at a reference point. Therefore, relative to $I_j^{\max}$ (the phosphorescence when $T \rightarrow 0$),

$$I_j = \frac{I_j^{\max}}{1 + Ge^{-E_j/kT}}. \tag{11.18}$$

To determine the variation of I_j with temperature relative to its maximum value, only G and E_j must be known. For La_2O_2S:Eu, the following values were determined experimentally (Fonger and Struck 1970): $G = 10^{7.5}$ for all the ^5D_j transitions, $E_0 = 6{,}300\ \text{cm}^{-1}$, $E_1 = 5{,}000\ \text{cm}^{-1}$, $E_2 = 3{,}000\ \text{cm}^{-1}$, and $E_3 = 1{,}200\ \text{cm}^{-1}$. Although in general the fluorescence from each of the ^5D_j states decreases as temperature increases, this effect is negligible at low temperatures and the phosphorescence remains constant until the exponentially temperature-dependent term in (11.18) increases. The temperature T_j at which the phosphorescence declines to 95% of its maximum is considered as the threshold for the onset of thermal quenching of the phosphorescence from the ^5D_j state. Therefore, the phosphorescence from a ^5D_j state is considered to be temperature-sensitive only while $G\exp\{-E_j/kT\} > 0.053$:

$$T_j > -\frac{E_j}{k \ln(0.053/G)}. \tag{11.19}$$

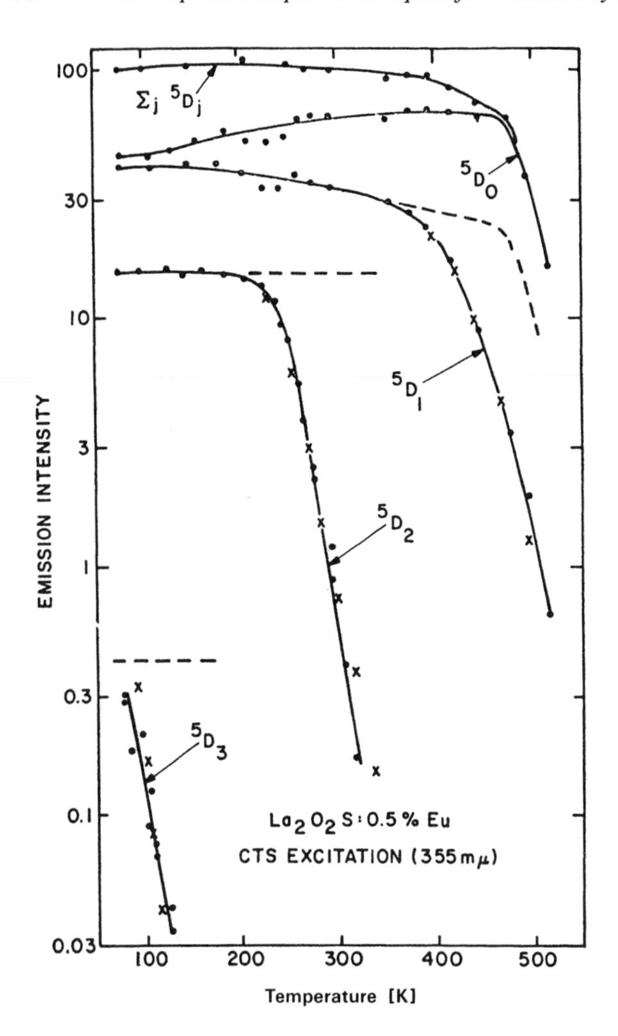

Figure 11.12 Temperature dependence of the ^{5}D emissions of La_2O_2S:0.5%Eu for CTS excitation at 355 nm. [Fonger and Struck 1970, © American Institute of Physics, Woodbury, NY]

Thus, of the four states, the phosphorescence from the state with lowest activation energy, the $^5D_3$ state, becomes temperature-dependent when $T_3 > 86$ K, and the state with the highest activation energy, the $^5D_0$ state, is temperature-sensitive at $T_0 > 449$ K. Figure 11.12 illustrates the temperature variation of the phosphorescence from four of the ^5D_j states. As predicted by (11.19), the phosphorescence from $^5D_3$ appears to be quenched for temperatures above ~90 K, whereas the onset of quenching of the $^5D_0$ state appears at ~470 K. For temperature measurement, the phosphorescence emitted by one of the temperature-sensitive states (following excitation by a UV source) is isolated spectrally from the emission from the other states. The measured irradiance then serves as a temperature indicator. As the temperature increases further, the phosphorescence diminishes and emission from other states may need to be measured.

Clearly, the dynamic range of each state is limited by this declining signal level (see Problem 11.4).

For surface temperature imaging, the test object is coated with a thin layer of a selected thermographic phosphor encased in a binder. The surface is then illuminated and the emitted phosphorescence is imaged by a CCD camera. The output of the camera is digitized and stored. To spectrally separate the emission of the selected transition from background radiation (e.g. room light), the camera is equipped with a bandpass filter. Similarly to (11.8), the signal (in photoelectrons) measured by each camera pixel may be expressed as

$$n_{pe}(x,y) = \left(\frac{\Omega}{4\pi}\eta_F\eta_D\right)t_e\frac{M(x,y)}{h\nu_0}N_j[T(x,y)], \tag{11.20}$$

where t_e is the camera exposure time, $M(x,y)$ is the radiant flux incident on the sample at point (x,y), ν_0 is the phosphorescence emission frequency, and where $N_j[T(x,y)]$ is the temperature-dependent quantum efficiency of the phosphorescence (Krauss et al. 1994). Most of the parameters of (11.20) cannot be accurately determined – in particular, the spatial distribution of the incident irradiance. Therefore, absolute measurements of the phosphorescence signal may not be useful. Alternatively, two separate images, one at a reference temperature T_0 and the other at the actual temperature $T(x,y)$, are recorded. By dividing the measured signal of each pixel of the first image by the measured signal of the corresponding pixels of the other image, the following temperature-dependent distribution is obtained:

$$R_j(x,y) = \frac{n_{pe}(x,y)|_T}{n_{pe}(x,y)|_{T_0}} = \frac{N_j[T(x,y)]}{N_j[T_0]}. \tag{11.21}$$

Here $R_j(x,y)$ is a function of the temperature-dependent parameter $N_j[T]$, which can either be approximated by (11.18) when $La_2O_2S{:}Eu$ is used or determined by direct calibration, using a sample with a prescribed temperature field (Problem 11.3).

Normalization of the temperature-dependent image by an image obtained at a reference temperature requires measurements at two different times. When the source illumination is unstable, or when the duration of the experiments is limited, normalization may be obtained by recording simultaneously the phosphorescence images of two separate transitions (Goss et al. 1989). This is often done by splitting the collected radiation into two separate beams. Each beam is then passed through a separate bandpass filter that was selected to transmit the phosphorescence of the desired transition. Normalization of the phosphorescence of the jth transition with that of the $(j+1)$th transition results in the following distribution:

$$R_{j,j+1}(x,y) = K(x,y)\frac{N_j[T(x,y)]}{N_{j+1}[T(x,y)]}, \tag{11.22}$$

where $K(x,y)$ represents the pixel-to-pixel variation in the camera quantum efficiency and the variation in the filter transmission that did not cancel out through this normalization. For these measurements, independent calibration is necessary.

Although measurements of the phosphorescence are corrected for background radiation and other experimental parameters such as non-uniformity of the detectors and the illumination, absolute irradiance measurements are subject to many incorrigible errors. To avoid these errors and the consequent need for normalization, the temperature can be determined from the decay time of the phosphorecence emanating from a specific state. Because the thermal quenching rate competes effectively with the emission rate, the depletion rate of the emitting state increases as temperature rises. Therefore, the phosphorescence lifetime of a temperature-sensitive transition of Y_2O_3:Eu may decrease from 1,000 μs at 700°C to 0.1 μs at 1,000°C (Alaruri et al. 1993). This is a large variation that permits excellent temperature resolution. For such measurements, the sample is illuminated by a brief laser pulse with typical duration of 10 ns. The subsequent phosphorescence is detected by a fast PMT, and the time-dependent output is then recorded by an oscilloscope or a multichannel analyzer. Unlike the direct recording of phosphorescence, these time-decay techniques are limited to pointwise measurements.

Noncontact fluorescence techniques may also be used for surface pressure measurements of aerodynamic models. Such techniques are used in wind-tunnel applications and may replace pressure taps, which are limited to pointwise measurements. Images of the planar pressure distribution available by such techniques permit detailed analysis of the load on the wings of aircraft models or turbine blades. Similarly, measurements of the pressure distribution over the surfaces of automotive models may be used for contour optimization.

Like thermographic phosphors, the technique depends on a surface coating that incorporates an encapsulated probe molecule. Following excitation by a UV source (Peterson and Fitzgerald 1980) or visible radiation (Morris et al. 1993), the encapsulated molecules emit spontaneously. The radiating states emit brightly in a spectral range that is sufficiently removed from the wavelength of the incident source to allow detection of the fluorescence without interference by radiation scattered elastically from the incident beam. The excited state of these special probe molecules is readily quenched by O_2 molecules that are naturally present in air flows. Because of this quenching, the emission is reduced. However, unlike thermal quenching, this process depends on the rate of collisions of O_2 at the surface. In subsonic flows, where the temperature is approximately uniform throughout the entire flow field, the rate of collisions is simply proportional to P_{O_2}, the partial pressure of O_2. In nonreacting air flows, measurement of P_{O_2} is equivalent to measurement of the air pressure. Therefore, the emission I when $P_{O_2} \neq 0$, relative to the emission I_0 when $P_{O_2} = 0$, is

$$I = \frac{I_0}{1 + K_q P_{O_2}}, \tag{11.23}$$

where K_q is proportional to the rate constant of quenching by O_2 (Peterson and Fitzgerald 1980). This constant is characteristic of the pressure-sensitive paint, as well as of the experimental conditions such as illumination and emission wavelengths and temperature. Like (11.18), (11.23) is a variation of the Stern–Volmer coefficient (eqn. 11.6) that determines the fluorescence yield of excited states.

The coefficient K_q determines the sensitivity of this pressure measurement technique. Thus, when K_q is large, the paint is sensitive to pressure. However, when K_q is large the emission may fall below the detection limit even at moderate pressures. Therefore, selection of a pressure-sensitive paint requires prior consideration of the experimental conditions as well as of the expected dynamic range of the measurement.

The experimental procedure for imaging of pressure distribution is similar to the procedure required for temperature imaging. In a typical experiment, the painted surface is illuminated by the excitation source. The subsequent emission is detected by a CCD camera through a bandpass filter that discriminates against background radiation. To correct for illumination and detection nonuniformities, each image is normalized by a reference image that was obtained at uniform and known pressure conditions. In addition, a calibration may be required to correlate the normalized emission with the actual pressure. Unfortunately, K_q is temperature-dependent and so a single calibration may suffice only when the flow is isothermal. When temperature variation is expected, separate calibration for each temperature may be needed. Furthermore, the temperature distribution at the surface of the object may need to be measured independently to allow point-by-point application of the calibration parameters.

11.4 Rayleigh Scattering

The primary advantage of resonant spectroscopic techniques is their high signal yield, which is accompanied by high sensitivities to the measured thermodynamic properties. Even with moderate illuminating power, the emission of a resonance excitation may be sufficient for two-dimensional imaging of various thermodynamic properties. Furthermore, some of these resonant techniques, particularly when the emission lifetime is short, can be used for time-resolved imaging of fast transient phenomena such as the development of the shock structure in a shock tunnel or the propagation of detonation waves. However, these techniques cannot be applied indiscriminately to any problem. The excitation depends on an exact match between the wavelength of the source selected for the illumination and the molecule selected for the test. Therefore, LIF techniques have been developed for measurements of relatively few molecules. Development of new LIF techniques to test new species or for new applications is costly, and even when accomplished cannot be globally implemented. Furthermore, some molecules (e.g. N_2) do not have any excitation that is resonant with commercially available laser sources and are thus not amenable to LIF diagnostics. In some cases, nonresonant spectroscopic techniques can be used to overcome these limitations. Unlike resonant techniques, the wavelength of the illuminating source for nonresonant spectroscopic measurements does not need to match any absorption wavelength. Nevertheless, the radiation scattered by the irradiated species can convey many of the thermodynamic characteristics of the scatterers. Depending on the technique, measurements of the concentration of major species, their temperature, or even their velocity are possible with nonresonant illumination. However, unlike resonant techniques, the signal level is typically low, thereby limiting many of these measurements to points or to lines.

Nonresonant spectroscopic techniques require illumination of the sample by a bright monochromatic radiation. The sample – in either the gaseous, liquid, or solid phase – scatters the incident radiation. The scattered radiation includes photons that were scattered *elastically* – without any significant change in their wavelength – as well as photons that were scattered *inelastically* – with a significant frequency change due to energy exchange with the scatterer. Elastic scattering is far more efficient than inelastic scattering and is therefore easier to detect. On the other hand, since elastic scattering does not include any energy exchange with the scatterer, very little information is conveyed regarding its identity. Nevertheless there exist numerous engineering applications (e.g. LDV; see Section 6.7) that depend solely on elastic scattering. Elastic scattering can be further divided into scattering by isolated molecules (Rayleigh scattering) and scattering by clusters of molecules and small particles (Mie scattering). A short discussion of these two elastic scattering processes was presented in Section 4.10. Here the discussion of Rayleigh scattering is extended to include some of its applications and its relation to Raman scattering, one of the inelastic nonresonant scattering processes.

On the macroscopic scale, elastic Rayleigh scattering is viewed as the cumulative emission by dipoles that were forced to oscillate by the electric field of an incident radiation. The combined dipole moment per unit volume was previously defined by the polarization vector $\mathbf{P}$ (eqn. 3.18). For a medium with a molecular density N that is exposed to radiation with an electric field $\mathbf{E}$, the polarization vector is

$$\mathbf{P} = \epsilon_0 \chi \mathbf{E} \tag{3.19}$$

(see Section 3.6). This macroscopic observable represents the response of the medium to the incident field. The *electric susceptibility* χ in this equation is the constant that determines the response of the medium to the incident field; it is also the constant that determines the index of refraction of the medium. Thus χ can be replaced in (3.19) with the index of refraction using (3.20) and (4.19). Accordingly, the polarization vector may be written as

$$\mathbf{P} = \epsilon_0 (n^2 - 1)\mathbf{E}. \tag{11.24}$$

On a microscopic scale, the equivalent of the macroscopic polarization vector $\mathbf{P}$ is the induced dipole moment $\mathbf{p}$ of a single molecule, which in the presence of an incident field is defined by

$$\mathbf{p} = \epsilon_0 \alpha \mathbf{E}, \tag{11.25}$$

where α denotes the molecular polarizability. With the exception of material constants, (11.25) and (3.19) are identical. Hence α, like the susceptibility χ,

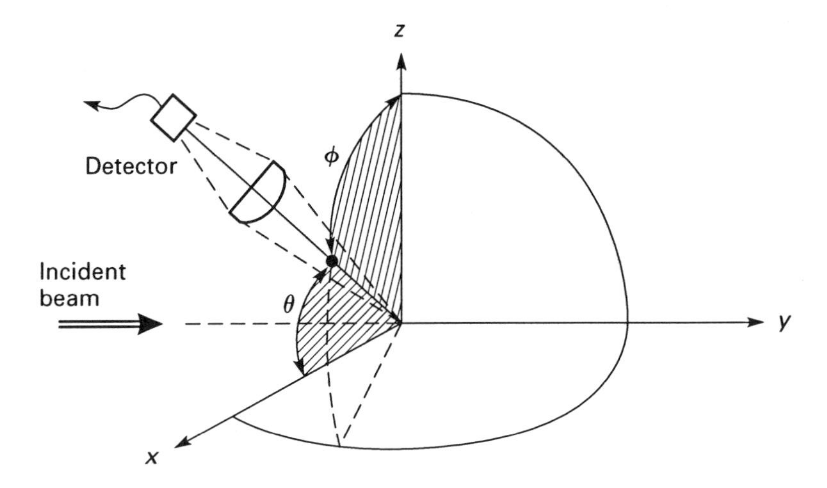

Figure 11.13 Detection geometry of Rayleigh scattering of a collimated beam.

must scale with n^2-1; thus, when the scattering characteristics of two gases (or of the same gas at two different wavelengths) are compared, their polarizabilities, α_1 and α_2, should be related as follows:

$$\left(\frac{\alpha_1}{\alpha_2}\right)^2 = \left(\frac{n_1-1}{n_2-1}\right)^2, \tag{11.26}$$

where the approximation of $n^2-1 \approx 2(n-1)$ was used.

In a typical Rayleigh scattering experiment, a collimated laser beam with an irradiance of I_0 illuminates the sample. Owing to the nonresonant nature of this scattering process, the spectral distribution of the beam does not need to be described. The subsequent scattering, as with LIF, is no longer collimated. Therefore, the scattering from a point is measured by the intensity $I_R(\theta, \phi)$ [W/sr] (eqn. 1.3). For an incident laser beam propagating in the y direction (Figure 11.13) with electric field polarizations of E_x and E_z, the scattering intensity varies with the angles θ and ϕ.

In accordance with (4.42), the incident irradiance must scale quadratically with the polarization vector. To compute the scattering intensity from (11.24), the coefficients that relate I_0 to $\mathbf{E}$ and $I_R(\theta, \phi)$ to $\mathbf{P}$ must be determined. By lumping these coefficients together, the dependence between $I_R(\theta, \phi)$ and I_0 can be reduced to

$$I_R(\theta, \phi) = K(\theta, \phi)(n-1)^2 I_0, \tag{11.27}$$

where $K(\theta, \phi)$ is a coefficient that depends on the polarization of the incident beam, the direction of the detection, the density of the tested gas, the sample volume, and the efficiency of the collection system.

Based on an extended microscopic analysis, a more complete representation of the intensity of Rayleigh scattering has been determined theoretically (Samson 1969) and tested experimentally for several gases (Shardanand and Prasad Rao 1977). These results account for the variation of the scattering intensity with the polarization properties of the scattering molecules, the polarization of the incident radiation, and the direction of the line of sight of the collection system. However, in most applications the collection is at $\theta = 0°$ and $\phi = 90°$, and the radiative power scattered by a volume element V into a small solid angle $\Delta\Omega$ centered along this detection line (Samson 1969) is

$$P(0°, 90°) = I_R(0°, 90°)\Delta\Omega = I_0 NV \frac{1}{1+g} \frac{d\sigma_R}{d\Omega} \Delta\Omega, \tag{11.28}$$

where $g = (E_x/E_z)^2$ is a measure of the incident polarization and $d\sigma_R/d\Omega$ [cm^2/sr] is the differential scattering cross section. This differential cross section is analogous to the absorption cross section σ_{ge} (eqn. 11.1). However, unlike that cross section, $d\sigma_R/d\Omega$ describes the efficiency of the entire scattering process and not just the first step of absorption. Furthermore, to account for the directionally non-uniform scattering, this process cannot be specified by a single value. Instead, the term must include an angular variation and so requires the differential representation $d\sigma_R/d\Omega$. Thus, the differential cross section in (11.28) is typical of this illumination-scattering configuration.

Although $I_R(\theta, \phi)$ depends on the incident polarization and may also vary with θ and ϕ, the polarization of Rayleigh scattering detected along the x axis remains linearly polarized irrespective of the incident polarization. This feature is unique to detection that is orthogonal to the line of illumination. It can be plausibly justified by considering that the E_x component of the incident beam can induce polarization along that axis only. Dipoles that oscillate parallel to the x axis can in turn radiate only along the axes that are perpendicular to their own polarization axes. Thus, E_x cannot induce scattering along the x axis; the only incident polarization that can induce scattering along the x axis is the E_z component. Therefore, $I_R(0°, 90°)$ can reach its maximum only when the incident polarization is coincident with the z axis (E_z), that is, when $g = 0$ (eqn. 11.28). The differential cross section for this unique scattering geometry is labeled by the subscript zz. After including the efficiency of the detection system components η_F and η_D, the signal in terms of photoelectrons induced by a section of length L along a laser beam with a pulse energy E_L polarized along the z direction (cf. eqn. 11.8) is

$$n_{pe} = (\eta_F \eta_D) \frac{E_L}{h\nu} NL \left(\frac{d\sigma_R}{d\Omega}\right)_{zz} \Delta\Omega, \tag{11.29}$$

where the differential cross section $d\sigma_R/d\Omega$ (Eckbreth 1988, p. 176) is

$$\left(\frac{d\sigma_R}{d\Omega}\right)_{zz} = \frac{4\pi^2(n-1)^2}{N_0^2 \lambda^4} \tag{11.30}$$

and N_0 is the Loschmidt number. Differential cross sections for various molecules and at various wavelengths have been reported by Shardanand and Prasad Rao (1977). As predicted by (11.26), the differential cross section increases as $(n-1)^2$, where n for gases is evaluated at standard conditions. However, $n-1$ increases linearly with the gas density (eqn. 6.14), which at standard atmospheric conditions is N_0. Therefore, dividing (11.30) by N_0^2 renders $d\sigma_R/d\Omega$ a molecular property that is independent of the gas density. Using (11.30), (6.14), and the data therein, we obtain: for air at $\lambda = 488$ nm, $(d\sigma_r/d\Omega)_{zz} = 8.30 \times 10^{-28}$ cm^2/sr; for N_2 at $\lambda = 488$ nm, $(d\sigma_R/d\Omega)_{zz} = 8.67 \times 10^{-28}$ cm^2/sr (Shardanand and Prasad Rao 1977).

Most notable in (11.30) is the strong dependence of $(d\sigma_R/d\Omega)_{zz}$ on the incident wavelength. This strong wavelength dependence enhances the scattering efficiency at short wavelengths. The $1/\lambda^4$ variation is valid when the incident wavelength is far from any resonance with an allowed transition. Therefore, for most atmospheric scatterers that do not present such resonance in either the visible or the near UV, any measurement of $d\sigma_R/d\Omega$ at one wavelength may be readily extended to another wavelength. Near resonance, owing to the strong response of the scatterer to the incident radiation, the scattering cross section may increase dramatically within a narrow spectral range. For UV illumination, the scattering intensity may increase sufficiently – by either the $1/\lambda^4$ dependence or the resonance enhancement – to allow imaging of Rayleigh scattering by atmospheric air even at a moderate laser energy. In a recent application, an image of the Rayleigh scattering in the boundary layer of a supersonic air flow was obtained by planar illumination by a pulsed ArF laser beam at nominally 193 nm (Smith, Smits, and Miles 1988; Grünefeld, Beushausen, and Andresen 1994). This choice of laser represents an optimization between the desired short wavelength and high pulse energy. The following example illustrates this possible application of Rayleigh scattering.

Example 11.2 A design of a system for imaging the Rayleigh scattering from a 30×30-cm section of a supersonic air flow calls for illumination by a pulsed laser beam at $\lambda = 388.9$ nm with a pulse energy of 30 mJ. The flow conditions are expected to reach a temperature of 1,500 K and a pressure of 0.2 atm. The scattered radiation can be detected through a port that limits the collection to a NA = 0.1, and the scattering is to be separated from background radiation by a bandpass filter with a transmission $\eta_F = 0.3$. The image is recorded by a CCD camera with a 300×300-pixel array and a quantum efficiency of 0.3. Estimate the signal per pulse (expressed in photoelectrons) to be detected by each pixel.

Solution The configuration for this imaging experiment is similar to the configuration used for PLIF imaging (Figure 11.4). Therefore, the incident laser beam can be viewed as divided into $n_p = 300$ rows with each row divided into $m_p = 300$ pixels. The signal at each pixel is obtained by modifying (11.29) as follows:

$$n_{pe} = (\eta_F \eta_D) \frac{E_L}{n_p h\nu} N \frac{L}{m_p} \left(\frac{d\sigma_R}{d\Omega} \right)_{zz} \Delta\Omega.$$

For the specified NA, the collection angle Ω is

$$\Delta\Omega = \frac{\pi}{4} (2\text{NA})^2 = 3.14 \times 10^{-2}$$

(cf. eqn. 2.14). The density of the scatterers at the specified flow conditions is determined from the ideal gas law and using the known density at standard atmospheric conditions:

$$N = N_0 \frac{P}{P_0} \frac{T_0}{T} = (2.69 \times 10^{19}) \frac{0.2}{1} \frac{273}{1{,}500} = 9.8 \times 10^{17} \text{ cm}^{-3}.$$

Rayleigh scattering of air consists of scattering by its major constituents N_2 and O_2. The effective cross section of air is obtained by combining the contributions of these molecules weighted by their molar fraction. In order to obtain the cross section at $\lambda_2 = 388.9$ nm from the cross section at $\lambda_1 = 488$ nm listed in Table 11.1, the wavelength dependence of (11.30) must be used. Thus, from the data for O_2 and N_2 (Table 11.1), the effective cross section of air at 388.9 nm is:

$$\left(\frac{d\sigma_R}{d\Omega}\right)_{zz} = \left(\frac{\lambda_1}{\lambda_2}\right)^4 \left(0.21 \frac{d\sigma_R}{d\Omega}\bigg|_{O_2}^{488} + 0.79 \frac{d\sigma_R}{d\Omega}\bigg|_{N_2}^{488}\right)$$

$$= \left(\frac{488}{388.9}\right)^4 (0.21 \cdot 0.9 + 0.79 \cdot 1)(8.66 \times 10^{-28}) = 2.1 \times 10^{-27} \text{ cm}^2/\text{sr}.$$

With these parameters, the Rayleigh signal detected by each pixel is:

$$n_{pe} = (0.3 \times 0.3) \frac{0.03}{300(6.63 \times 10^{-34})(7.71 \times 10^{14})} (9.89 \times 10^{17})$$

$$\times \frac{30}{300} (2.1 \times 10^{-27})(3.14 \times 10^{-2}) = 114 \text{ photoelectrons/pixel}.$$

Table 11.1 *Differential zz cross sections for Rayleigh[a] and vibrational Raman[b] scattering for illumination at 488 nm*

Molecule	Vibrational Raman shift [cm^{-1}]	Raman $(d\sigma_S/d\Omega)_{zz}$	Rayleigh $(d\sigma_R/d\Omega)_{zz}$
N_2	2,331	1.0	1.0
O_2	1,556	1.3	0.9
H_2	4,160	2.4	0.22
CO	2,145	1.0	—
CO_2 (ν_1)	1,388	1.4	3.17
CO_2 ($2\nu_2$)	1,286	0.89	3.17
H_2O	3,657	2.5[c]	—
NO	1,877	0.27	—
CH_4 (ν_1)	2,914	6.0	2.29
C_2H_6 (ν_1)	3,062	7.0	—
C_2H_6 (ν_2)	992	9.1	—

Note: Values are relative to the cross section of N_2. The absolute differential Raman cross section of N_2 is $d\sigma_S/d\Omega = 5.61 \times 10^{-31}$ cm^2/sr (Penney and Lapp 1976); for Rayleigh scattering, $d\sigma_R/d\Omega = 8.67 \times 10^{-28}$ cm^2/sr.
[a] Shardanand and Prasad Rao (1977).
[b] Fenner et al. (1973).
[c] Penney and Lapp (1976).

For comparison, the signal projected for imaging the OH distribution in a PLIF experiment (Example 11.1) was 1,004 photoelectrons/pixel. Although the resolution and the signals are comparable, recall that the density of the minor species emitting the LIF signal was approximately 100 times lower than the density of the Rayleigh scatterers. Evidently, LIF is a significantly more sensitive technique for imaging of trace amounts. On the other hand, Rayleigh scattering is not limited to the imaging of particular flows or species. ■

Rayleigh scattering was described here as a nominally elastic interaction between a scatterer and an incident photon – that is, the wavelength of the scattered radiation was considered to be nominally the same as the incident radiation. For scattering to occur, each scatterer forms a dipole moment that oscillates at the frequency sensed by an observer moving with the dipole. Although the scattering itself does not involve transitions between quantum mechanical states, the linear motion of the scatterer relative to the incident radiation induces a Doppler shift. Similarly, the frequency of the scattered radiation when detected by a stationary observer is also Doppler-shifted. The resultant Doppler shift (eqn. 6.24) depends on the velocity vector of the scatterer and the propagation vectors of the incident and scattered radiation. In the absence of a directed flow, the scatterers move randomly and the scattering is Doppler-broadened. The resultant linewidth is the convolution of the incident spectral distribution and the Doppler broadening induced by the scatterers. However, when the entire medium is moving relative to a stationary observer, the Rayleigh line is not only broadened – its center is also shifted owing to this relative velocity. In fast gas flows, this shift is large relative to the broadening linewidth and can be measured when excited by a narrowband laser.

A technique proposed by Shimizu, Lee, and She (1983) and subsequently demonstrated in supersonic flows (Miles and Lempert 1990) can be used to determine the flow velocity from the Doppler shift of Rayleigh scattering. Thanks to its characteristically large signal, the technique may even be used to image major flow structures (e.g. shockwaves) by using a laser light sheet for illumination and a camera for detection. With this technique, the Rayleigh signal is collected through a sharp cutoff filter consisting of a transparent cell containing vapors of molecular or atomic absorbing species. At least one of the absorption lines of the species inside the cell coincides with the laser wavelength. Thus, when the laser is used to induce Rayleigh scattering in a stationary gas, all or most of the Rayleigh signal is absorbed by the cell's content before reaching the camera. However, when the tested gas is flowing, radiation scattered by the moving molecules is Doppler-shifted and its wavelength no longer coincides with the absorption line center. Depending on the spectroscopic characteristics of this atomic (or molecular) filter and the flow velocity, the Rayleigh signal may be partially or fully transmitted. Thus, regions of the flow where the velocity-induced Doppler shift exceeds the absorption linewidth of the filter will have their Rayleigh signal transmitted by the filter, while scattering from other sections will be blocked. Although the gas density affects the level of the Rayleigh signal, images obtained through this filter are influenced primarily by the velocity field (Miles and Lempert 1990). Note that this technique is fundamentally different from LIF. Instead of tuning the laser to an absorption line of a molecule in the test flow, it is tuned to an absorption line of the species in the atomic filter. Thus, unlike LIF, this measurement is possible for any flowing gas species while using a single laser system for excitation.

The irradiance I_ν [W/cm^2] transmitted through an atomic filter of length L [cm] irradiated by an incident narrowband irradiance $I_{0\nu}$ at frequency ν (cf. eqns. 10.14 and 10.24) is

$$\frac{I_\nu}{I_{0\nu}} = \exp\left\{-\kappa_0 g(\nu) L\right\}, \tag{11.31}$$

where $\kappa_\nu = \kappa_0 g(\nu)$ and where $\kappa_0 = \int \kappa_\nu \, d\nu$. This integrated absorption coefficient can be readily related to the Einstein B coefficient (eqn. 10.21′) or the oscillator strength f_{12} by

$$\kappa_0 = (2.64 \times 10^{-11}) N_a f_{12}, \tag{11.32}$$

where N_a is the density [cm^{-3}] of the atomic vapor and the coefficient is evaluated for use with $g(\nu)$ [GHz^{-1}] to produce an absorption coefficient [cm^{-1}]. When the pressure in the atomic cell is a few millitorrs, the absorption linewidth is primarily Doppler-broadened and the lineshape factor (cf. eqn. 10.60) is

$$\begin{aligned} g(\nu) &= \frac{2}{\Delta\nu_D}\sqrt{\frac{\ln 2}{\pi}} \exp\left\{-4 \ln 2 \left(\frac{\nu - \nu_0}{\Delta\nu_D}\right)^2\right\}, \quad \text{where} \\ \Delta\nu_D &= \frac{2.1472 \times 10^{-5}}{\lambda_0 \text{ [cm]}} \sqrt{\frac{T_0}{M}} \text{ GHz}, \end{aligned} \tag{11.33}$$

where M is the atomic (or molecular) weight of the absorbing species. The resolution of this velocity measurement technique is ultimately limited by the width of the absorption line. Therefore, the resolution can be enhanced by selecting an absorber with a large atomic (or molecular) weight. In addition, to minimize the length of the cell, the absorber is expected to have a large oscillator strength at a wavelength that coincides with available laser sources. A possible absorber for such a cell is Cesium. For the Cs $6s \rightarrow 8p^2P_{3/2}$ transition (Shimizu et al. 1983) at 388.865 nm, the oscillator strength is $f_{12} = 2.8 \times 10^{-3}$ while for the $6s \rightarrow 7p^2P_{3/2}$ transition at 455.53 nm the oscillator strength is $f_{12} = 1.23 \times 10^{-2}$ (Marling et al. 1979; Stone 1969). At a temperature of 370 K, the number density of Cs vapor is $N_a = 1.2 \times 10^{13}$ cm^{-3} and the linewidth of that absorbing transition is 0.92 GHz. The following example illustrates the use of such an atomic filter for imaging the velocity field of air flow ahead of a shock wave and behind it.

Example 11.3 A supersonic flow is illuminated by a laser at either $\lambda_1 = 388.865$ nm or $\lambda_2 = 455.528$ nm. The beam propagates opposite to the flow direction. The resultant Rayleigh scattering is collected perpendicularly to the incident laser beam and passed through a 5-cm–long cell, equipped with windows at its opposite ends and containing Cs vapor at 370 K. Draw the variation with flow velocity of the transmission of the Rayleigh-scattered intensity.

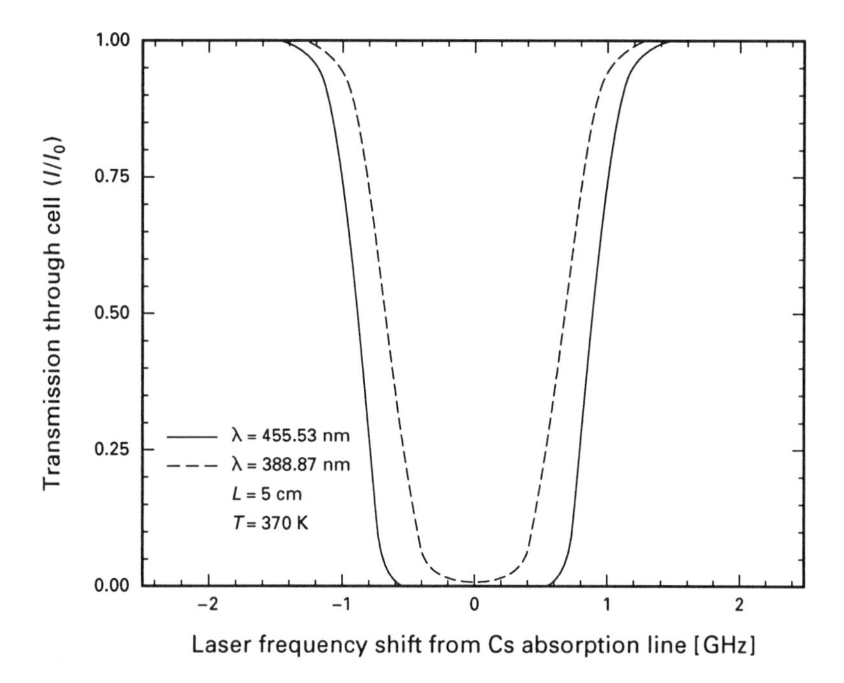

Figure 11.14 Variation of the transmission through a 5-cm-long Cs vapor cell at 370 K with frequency shift $\nu - \nu_0$ relative to the nominal absorption frequency ν_0. The transmission was computed for absorption around 455.528 nm (solid line) and 388.865 nm (dashed line).

Solution The transmission $I_\nu / I_{0\nu}$ through the 5-cm-long Cs vapor cell is calculated using the oscillator strengths for the $6s \rightarrow 8p\,^2P_{3/2}$ transition when the incident wavelength is 388.865 nm and the $6s \rightarrow 7p\,^2P_{3/2}$ transition when the incident wavelength is 455.528 nm.

Figure 11.14 illustrates the variation of the transmission with frequency shift $\nu - \nu_0$ relative to the nominal absorption frequency ν_0 of each of these transitions. The solid line represents the transmission around 456 nm while the dashed line represents the transmission around 389 nm. Thus, when the scattered frequency is $\nu = \nu_0$ (i.e., when the scattering is by motionless gas), the Rayleigh signal is blocked almost entirely by the filter. Even when the laser is tuned to within ± 0.6 GHz from 455.528 nm or within ± 0.4 GHz from 388.865 nm, the transmission by this filter is negligible (the range of frequencies blocked by the 455.528-nm line is wider than the range blocked by the 388.865-nm line owing to its larger oscillator strength). Thus, unless the scattering gas density is increased by many orders of magnitude above the atmospheric density, Rayleigh scattering by motionless gas cannot be detected through the filter while the laser is at either wavelength. However, if the incident wavelength of 455.528 nm is tuned by *more* than 0.6 GHz (or 0.4 GHz for the 388.865-nm line) away from the line center, or (alternatively) if the Doppler shift of the Rayleigh scattering exceeds that range, then the filter sharply becomes transparent. For shifts in excess of 1 GHz from the center of each line, the filter is almost perfectly transparent. Since the primary use of this filter is for imaging the velocity field of gas flows, the laser is usually tuned to the center of one of these Cs absorption lines. Thus, sections of the flow where the Doppler shift $\Delta\nu_D < 0.6$ GHz (when $\lambda = 455.528$ nm) or $\Delta\nu_D < 0.4$ GHz (when $\lambda = 388.865$ nm) away from the line center appear dark when viewed through this filter. Similarly, scattering by the wall, the model, or the windows is blocked. On the other hand, sections where the Doppler shift exceeds half of the absorption bandwidth appear bright. By tuning the laser frequency, selectivity with respect to velocity range can be achieved.

The velocity range that can be resolved by this Cs vapor filter is obtained by correlating the Doppler shift $\Delta\nu_D = \nu - \nu_0$ with the flow velocity V in the direction of the incident beam as determined by

$$\Delta\nu_D = \pm\nu_0 V/c$$

(cf. eqn. 6.24).

The solid line in Figure 11.15 illustrates the variation with flow velocity of the Rayleigh scattering of a laser tuned to 455.528 nm when detected through a Cs vapor filter. This is the velocity of the flow in the direction of the laser beam when the detection is normal to that velocity. Clearly, regions of the flow where the velocity is less than 300 m/s are expected to appear dark upon imaging through this filter, while regions with velocity in excess of 400 m/s are bright. By tuning the laser away from the absorption line center, other sections (either

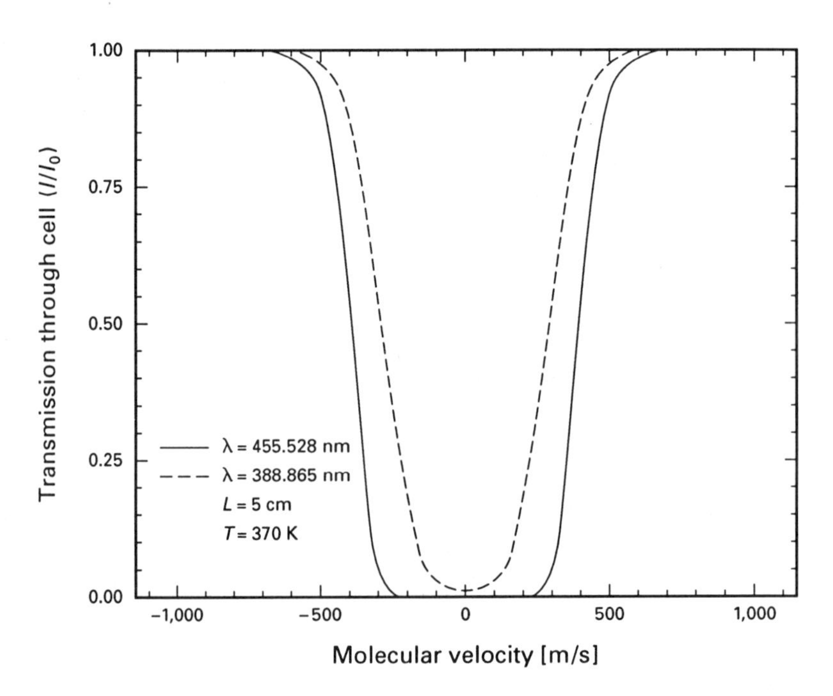

Figure 11.15 Variation with flow velocity of the Rayleigh scattering of a laser tuned to 455.528 nm (solid line) and 388.865 nm (dashed line) when detected through a 5-cm-long cell of Cs vapor held at 370 K.

faster or slower) can be marked. Owing to the relatively sharp cutoff of the filter, this method is used mostly for demarcation of velocity ranges and major flow structures. Alternatively, when the laser is tuned to the 388.865 line, the velocity discrimination occurs at approximately 250 m/s (dashed line in Figure 11.15). The 388-nm line can offer a cutoff similar to the 455.528 line either by increasing the length of the cell or by raising its temperature. Thus, by minor adjustments, the filtration performance of both lines can be made comparable. ■

Rayleigh scattering is an excellent nonresonant spectroscopic technique for imaging gross flow structures. In conjunction with an atomic filter, it can be used to distinguish between the Rayleigh scattering of gas flows and scattering by stationary objects such as the windows or walls of a wind tunnel. Alternatively, in the presence of large density gradients such as a shockwave or boundary layers, Rayleigh scattering can be used to locate these gradients and to image their loci in a plane. Rayleigh images of the shock waves in supersonic flows can reveal their structure in the plane of the illuminating laser sheet. This is unlike refractory techniques such as schlieren and shadowgraphs, where images of flow structures are obtained by integration along the entire propagation path of the illuminating beam.

Despite their potential, the use of Rayleigh techniques is limited to specific applications. Lacking selectivity, Rayleigh techniques cannot be used to identify species in chemical reactions or to determine their concentration. Furthermore, the temperature sensitivity of Rayleigh scattering is poor, which prevents the use of such techniques for measurement of thermodynamic properties. Therefore, for species concentration and temperature measurements by nonresonant illumination, an alternative technique must be used.

11.5 Raman Scattering

Raman scattering is the result of an inelastic, nonresonant interaction between an incident photon and a scatterer. These properties of the interaction imply that the frequency of the incident radiation does not need to match a transition frequency of the tested species and that the frequency of the scattered photon is different from the frequency of the incident photon. The energy lost (or gained) by this frequency shift must be matched by an energy gain (or loss) by the scattering species, which in turn must equal the energy difference between quantum mechanical states. Therefore, Raman frequency shifts are discrete. However, owing to the complexity of the quantum mechanical structure of most scatterers, Raman spectra usually contain more than just one spectral component, some of them overlapping. The information that can be derived from Raman spectra is similar to the information contained in absorption or emission spectra. The extent of the scattering corresponds to certain spectroscopic characteristics of the scattering species, their density, and their temperature. Thus, when the scattering characteristics of the tested species are known, its thermodynamic properties can be measured from Raman spectra.

Because of the nonresonant excitation, Raman spectra can be induced in any desired spectral range and not just where single-photon absorptions or emissions normally occur. With Raman scattering, vibrational or rotational spectra of molecules can be studied using visible or ultraviolet radiation; one is not limited to the infrared illumination required for many absorption or emission studies. For combustion diagnostics or chemical reaction analysis, where significant infrared background emission interferes with infrared spectroscopy, Raman scattering can be obtained using a short wavelength source, thereby shifting the signal away from the background luminosity. On the other hand, to overcome the low probability of nonresonant inelastic interaction, Raman scattering experiments require a bright source for illumination. Moreover, in order to preserve the discrete structure of the Raman spectrum, the illuminating source must be nominally monochromatic.

The experimental set-up of a typical Raman scattering experiment is similar to that for Rayleigh scattering (Figure 11.13). A monochromatic laser beam propagating in the y direction irradiates the sample, which may be in either the gas, liquid, or solid phase. The incident polarization may include E_x and E_z components. The scattered radiation is collected by a lens with a collection solid angle Ω at an angle ϕ measured from the positive z axis and an angle θ measured relative to the positive x axis. The collected radiation is then projected on the slits of a spectrometer, where the spectrum of the scattered radiation is analyzed. At or near the incident frequency ν_0, the scattered radiation contains mainly the elastic Rayleigh scattering component. The linewidth of this component is determined primarily by the laser linewidth and Doppler broadening by the scatterers. When the spectrometer is tuned away from this elastic scattering component, the detected scattering may include much weaker (typically $\sim 10^{-3}$) radiation at a frequency of ν_s. The difference $\nu_0 - \nu_s$ between these frequencies equals the transition frequency ν_v between adjacent vibrational or ν_r between neighboring rotational levels of the scattering molecule.

The results of this typical experiment represent two of the three quantum mechanically possible outcomes of any nonresonant scattering process. Figure 11.16 is an energy-level diagram that illustrates these three possible scattering interactions with a molecule. The incident photon in this diagram is represented by an upward-pointing arrow that terminates at an intermediate state. Owing to the nonresonant interaction, this is not an existing energy level; instead, for modeling purposes it is considered as a *virtual state*. Virtual states can be viewed as the "wings" of the probability function of nearby states. Therefore, the molecule can in principle exist there, albeit at extremely low probabilities and with an extremely short lifetime (see e.g. Williams and Imre 1988; Williams, Rousseau, and Dworetsky 1974). From this virtual state, the molecule decays almost instantaneously back into an existing quantum state which may be either the ground state or a new vibrational state. When originated at the ground state, the decay back to the ground state is Rayleigh scattering. On the other hand, the decay to the $v = 1$ state raises the energy of the scatterer by $h\nu_v$. To conserve

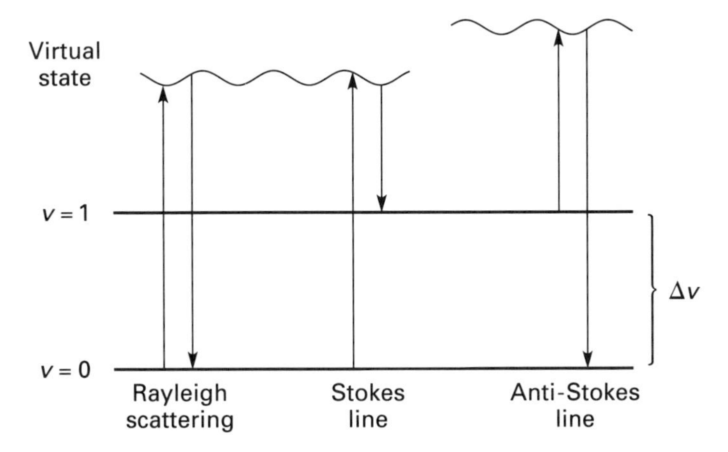

Figure 11.16 Energy-level diagram for Rayleigh and Stokes and anti-Stokes Raman scattering.

energy, the frequency of the scattered photon must be reduced relative to the frequency of the incident photon, so that $\nu_s = \nu_0 - \nu_v$. An interaction resulting in a reduction in the scattered photon energy is called *Stokes scattering*. Similar interactions occur when the molecule is initially at an upper vibrational state. From there, the scatterer may return to its original state by Rayleigh scattering or to yet a higher vibrational state by Stokes scattering. However, it may also end at a lower state. With this, the scattered photon gains energy relative to the incident photon. This is the *anti-Stokes scattering*. Of course, for anti-Stokes scattering the tested species must first be prepared in an excited state. Therefore, anti-Stokes Raman scattering is observed only when the tested sample is hot or when it is not in thermodynamic equilibrium.

Comparison of this schematic energy-level diagram with the energy-level diagram for LIF shows certain similarities between the various inelastic scattering processes. Indeed, when the incident source is near resonance, the distinction may be difficult (see e.g. Williams and Imre 1988; Williams et al. 1974). However, since LIF occurs when the frequency of the incident radiation coincides with an absorbing transition of the tested species, the cross section for LIF is significantly higher than the cross section for Raman scattering, thereby rendering LIF a more efficient process.

The association between the three nonresonant scattering processes can be derived from semiclassical considerations, that is, by using a combination of classical and quantum mechanical arguments. Assuming that the scatterer can be modeled as a harmonic oscillator (eqn. 9.22) oscillating at a frequency ν_v, its time-dependent displacement r from its equilibrium separation r_e is:

$$r = r_e \cos 2\pi\nu_v t. \tag{11.34}$$

The interaction between such an oscillator and the incident field is described by the dipole moment $\mathbf{p}$, which in turn depends on the polarizability α (eqn. 11.25).

However, since the dipole moment and hence polarizability depend on the momentary displacement r of the oscillator, α must be a time-dependent function that can be approximated by the following linear expansion:

$$\alpha = \alpha_0 + \left(\frac{\partial \alpha}{\partial r}\right)_0 r \tag{11.35}$$

(Herzberg 1989, p. 87). With this expression for the polarizability, the dipole moment (eqn. 11.25) induced by an electric field $\mathbf{E} = \mathbf{E}_0 \cos 2\pi\nu_0 t$ is

$$\begin{aligned}\mathbf{p} &= \left[\alpha_0 + \left(\frac{\partial \alpha}{\partial r}\right)_0 r_e \cos 2\pi\nu_v t\right] \epsilon_0 \mathbf{E}_0 \cos 2\pi\nu_0 t \\ &= \alpha_0 \epsilon_0 \mathbf{E}_0 \cos 2\pi\nu_0 t + \left(\frac{\partial \alpha}{\partial r}\right)_0 \epsilon_0 \frac{r_e \mathbf{E}_0}{2} [\cos 2\pi(\nu_0 - \nu_v)t + \cos 2\pi(\nu_0 + \nu_v)t].\end{aligned} \tag{11.36}$$

The first term in (11.36) describes an oscillation of the dipole at the frequency of the incident field and must therefore be the term that represents Rayleigh scattering. The frequency of the second term corresponds to the frequency of Stokes scattering, while the third term is the anti-Stokes scattering term. Similar terms appear also in the analysis of classical coupled oscillators, where they represent the "beat" frequencies. Beat frequencies, which are observed in acoustics or in mechanical vibrations, are therefore the result of a superposition of two or more fundamental frequencies. By analogy, Raman scattering may be interpreted as the beat between the incident field frequency and the frequency of the oscillation or rotation of the scatterer. But in quantum mechanics, these beat frequencies also imply a transition of the scatterer from one level to another. This cannot be inferred from the semiclassical model of (11.36) and so must be derived from purely quantum mechanical concepts. Nevertheless, (11.36) still predicts that the probability for a Stokes scattering event is identical to the probability for an anti-Stokes scattering event. On the other hand, Stokes scattering results in the promotion of the scatterer to an upper state, whereas anti-Stokes emission can occur only if the scatterer is already at an upper vibrational state from which it decays to a lower state. Thus, the rate of Stokes scattering, which depends both on the transition probability and on the population of the ground state, normally exceeds the rate for anti-Stokes scattering. At high temperature, when the population density of the upper state is high, the signal associated with anti-Stokes scattering may become significant.

As (11.36) shows, the three nonresonant scattering processes – Rayleigh, Raman Stokes, and Raman anti-Stokes – are but different aspects of the same interaction. Accordingly, many observations related to Rayleigh scattering pertain also to Raman scattering. When a collimated laser beam with an irradiance of I_0 illuminates the sample, the subsequent scattering is no longer collimated and, as with Rayleigh scattering, the distribution of the Raman intensity $I_S(\theta, \phi)$ [W/sr] varies with the angles θ and ϕ. By analogy to (11.28), the Raman Stokes signal varies with the density of the scatterers and the volume of the scattering sample. But, unlike Rayleigh scattering, the Raman signal associated with a selected shift is induced by molecules that occupy a specific vibrational level (e.g. $v=0$). Therefore, the density of the scatterers varies both with the total density N and with the temperature-dependent Boltzmann fraction $\beta_g(T)$ (eqn. 10.38). The total Raman signal detected through a solid angle $\Delta\Omega$ at an angle $\theta = 0°$ and $\phi = 0°$ when the incident polarization is E_z is therefore

$$I_S(0°, 90°)\Delta\Omega = I_0 N\beta_g(T)V\left(\frac{d\sigma_S}{d\Omega}\right)_{zz}\Delta\Omega, \tag{11.37}$$

where $I_S(0°, 90°)$ is the Raman (Stokes or anti-Stokes) scattering intensity observed at $\theta = 0°$ and $\phi = 90°$ (Figure 11.13). Owing to its dependence on N and $\beta_g(T)$, the Raman signal is sensitive to variations in both temperature and species density, and can therefore serve to measure both thermodynamic properties (see Problem 11.8).

To evaluate the signal obtained by electronic detection, parameters of the detection efficiency must also be included. By analogy to (11.29), the Raman signal (in photoelectrons) following a pulsed illumination with a pulse energy of E_L is:

$$n_{pe} = (\eta_F\eta_D)\frac{E_L}{h\nu_L}N\beta_g(T)L\left(\frac{d\sigma_S}{d\Omega}\right)_{zz}\Delta\Omega, \tag{11.38}$$

where ν_L is the incident frequency. When illumination is continuous, E_L is the total energy falling on the sample during the detector exposure time. Calculations of the expected Raman signal, or the relation between the signal and the density of the scattering species, are similar to the calculation of the Rayleigh signal (see Example 11.2 of Section 11.4). With the exception of the differential cross section, both calculations depend on the same parameters.

The angular distribution of the Raman scattering intensity and its polarization strongly depend on the polarization of the incident beam and on the depolarization characteristics of the scatterers. Since the scattering is nearly instantaneous, effects of collisions or other interactions with the surrounding medium on the scattered intensity or polarization can be neglected. Therefore, the angular dependence must be included entirely in the Raman cross section. For the special case of detection along the x axis (Figure 11.13) of scattering of the E_z polarization, the Raman differential cross section is labeled $(d\sigma_S/d\Omega)_{zz}$. This definition is similar to the zz Rayleigh cross section.

Evidently, the differential cross section determines most of the properties of Raman scattering. Many of these properties, particularly its angular and polarization dependence, must be derived quantum mechanically. However, some can be inferred from (11.36), where the scattering is seen to depend on the polarizability and on its rate of change with r. Therefore, like the Rayleigh cross section, $(d\sigma_S/d\Omega)$ varies with wavelength as $1/\lambda^4$. For many molecules of interest, visible radiation is not resonant with any transition and therefore this $1/\lambda^4$ wavelength dependence can be used to calculate the Raman cross section for illumination by any wavelength λ_2 when the differential $(d\sigma_S/d\Omega)_1$ at another wavelength λ_1 is known. Thus, the Raman cross section for scattering of radiaation at λ_2 is

$$\left(\frac{d\sigma_S}{d\Omega}\right)_2 = \left(\frac{\lambda_1}{\lambda_2}\right)^4 \left(\frac{d\sigma_S}{d\Omega}\right)_1. \tag{11.39}$$

The differential vibrational Raman *zz* cross sections for several molecules of interest have been evaluated and are reported in Table 11.1 along with their vibrational Stokes shifts and Rayleigh cross sections.

The data of Table 11.1 show clearly that, for all molecules, the cross sections for Raman scattering are significantly smaller than for Rayleigh scattering. Therefore, whereas Rayleigh scattering (e.g. blue skies) can be seen with the naked eye, observation of the ~1,000-times weaker Raman scattering requires illumination by bright source, efficient signal collection, and sensitive detectors. The data of Table 11.1 also show that polyatomic molecules may have several Raman active vibraional modes, each with its own frequency and Raman cross section. Therefore, the Raman spectrum of any of these species has more than one line. The relative height of these lines must depend therefore on their differential cross sections as well as on their relative population density. As a result, the Raman spectra of hydrocarbon flames are typically complex and may involve numerous superimposed lines.

Away from resonance, $(\partial\alpha/\partial r)_0$ and $(d\sigma_S/d\Omega)$ depend on general molecular properties. But just as with mechanical structures, when the incident field frequency is resonant with a characteristic frequency of the scatterer, the amplitude of the oscillation and with it $(\partial\alpha/\partial r)_0$ and $(d\sigma_S/d\Omega)$ increase rapidly. Resonance occurs when the incident frequency ν_L approaches one or more allowed transitions including vibrational, electronic, or ro-vibronic transitions. Therefore, as the incident wavelength is nearing resonance, the differential Raman cross sections no longer increase with wavelength as $1/\lambda^4$. Resonance effects, particularly when UV radiation is used for illumination, may enhance the Raman cross section well beyond the predictions of (11.39), in which case new values must be obtained. The development of powerful UV excimer lasers generated new opportunities for combustion diagnostics. Not only can the Raman signal be enhanced, but the emission is in the UV where interference by flame luminosity is minimal. The following equation is an acceptable approximation of the resonant cross section of several molecules of interest in combustion diagnostics (Bischel and Black 1983):

$$\left(\frac{\partial\sigma_S}{\partial\Omega}\right)_{zz} = A\frac{\nu_S^4}{(\nu_i^2-\nu_L^2)^2}, \tag{11.40}$$

where A and ν_i are coefficients defined in Table 11.2 for several molecules, with the Stokes frequency ν_S and the incident laser frequency ν_L expressed in units of cm^{-1}.

Although Raman scattering does not depend on a resonant excitation, it does involve transitions between quantum energy levels. Therefore, just as with

Table 11.2 *Coefficients for the evaluation of resonantly enhanced differential Raman cross section (eqn. 11.40)*[a]

Molecule	A [$\times 10^{-28}$ cm^2/sr]	ν_i [$\times 10^4$ cm^{-1}]
H_2	8.74	8.48
D_2	3.90	7.81
N_2	3.02	8.95
O_2	0.459	5.69
CH_4 (ν_1)	10.4	7.23

[a] Bischel and Black (1983).

resonance excitations, the probability for such transitions must depend on selection rules. Accordingly, the Raman cross sections listed in Table 11.1 or calculated from (11.40) pertain to allowed transitions. For linear molecular species such as the diatomic molecules or CO_2, the selection rules for rotational transitions are

$$\Delta J = 0, \pm 2 \tag{11.41}$$

(Herzberg 1945, p. 20). With this rule, Raman spectra may include O, Q, and S branches corresponding to $\Delta J = -2$, 0, and 2, respectively. For nonlinear molecules, other selection rules apply. Of particular interest are the symmetric-top molecules. In these molecules (e.g. NH_3), two out of the three possible moments of inertia are equal to each other. The selection rules for rotational Raman transitions in these molecules (Herzberg 1945, p. 34) are

$$\Delta J = 0, \pm 1, \pm 2; \qquad \Delta K = 0, \pm 1, \pm 2, \tag{11.41$'$}$$

where K – similar to Λ of the diatomic molecule (p. 265) – is the rotational quantum number pointing along the molecular axis of symmetry. These selection rules indicate that pure rotational Raman spectra are allowed (with the exception of the Q branch, which for pure rotational spectra does not involve a net transition). However, the frequency shift associated with pure rotational transitions is small, and without an extremely effective rejection, Rayleigh scattering may obscure the spectrum. Therefore, most rotational Raman spectra are obtained from mixed vibrational and rotational transitions.

For vibrational transitions, the selection rule is

$$\Delta v = \pm 1 \tag{11.42}$$

(Herzberg 1945, p. 249). Of course, from the ground state, only Stokes excitation to $v = 1$ is possible. This selection rule is identical to the selection rule for dipole-allowed infrared absorption or emission. But unlike dipole-allowed (or single-photon) transitions, Raman scattering is allowed even in the absence of a permanent dipole moment. Therefore, homonuclear molecules such as O_2 and N_2, which ordinarily do not have infrared spectra (Section 9.5), do have vibrational Raman spectra. Overtone Raman transitions (i.e., transitions involving $\Delta v = \pm 2, 3, \ldots$) are unallowed and therefore usually very weak. Nevertheless, near resonance, when the cross section for allowed transitions increases (eqn. 11.40), the probability of overtone transitions likewise increases and Raman bands with $\Delta v > 1$ can be detected (Williams and Imre 1988). The cross sections and the Raman shifts listed by Table 11.1 or calculated by (11.40) are for the Q-branch vibrationally allowed Raman transitions from the ground vibrational state.

When used to analyze combustion or chemical reactions, Raman spectra can contain the spectral details of all the major species. From these details, the concentration distribution of these species may be determined. At high temperature, when the population density of the upper vibrational states increases significantly, the spectrum of each species includes detectable Raman scattering from each of the thermally populated states. This scattering may be either Stokes or anti-Stokes shifted. For a Stokes shift, the Raman line of an upper vibrational state is often spectrally separated from the Stokes line of the lower state. The relative position – that is, the difference between the frequency shifts of two adjacent vibrational lines – is

$$\Delta\nu_S = \Delta G(v+1) - \Delta G(v) = -2\omega_e x_e \tag{11.43}$$

(cf. eqn. 9.27). As expected, owing to the anharmonicity of molecular vibrations, the Raman shift decreases progressively with increasing vibrational quantum number.

For anti-Stokes shifts, the first Raman line appears at the "blue" side of the exciting frequency only when the population of the $v = 1$ level is sufficiently high to raise the signal above background levels. At higher temperatures, anti-Stokes lines of even higher vibrational states can be observed. When either the Stokes or the anti-Stokes lines of an upper vibrational state are resolved, the temperature of the scatterer is calculated from their intensities relative to the intensity of the Stokes line originating at the ground vibrational state.

Figure 11.17 illustrates the Raman spectrum of N_2 in an H_2–air flame (Collins 1990). This spectrum was obtained using a frequency-doubled Nd:YAG laser at 532 nm for illumination and a double monochromator to resolve the spectrum. Although the $v = 0$ and the $v = 1$ Raman lines are resolved, the fine ro-vibrational structure of each of these vibrational lines is not. The envelope of these vibrational lines is determined primarily by the relative location of the rotational lines within the Q branch, by the laser bandwidth, and by the slit function of the spectrometer. Therefore, part of the scattering from $v = 0$ is superimposed on the lines from $v = 1$ and vice versa, so accurate temperature determination from this spectrum is possible only by fitting the experimental spectrum with theoretically calculated spectra (Eckbreth 1988). However, as in Example 10.4 of Section 10.5, the approximate temperature can be obtained from (10.31), where the ratio of the population of the vibrational states is obtained by dividing the integrated area under the $v = 1$ line by the integrated area under the $v = 0$ line (see Problem 11.8).

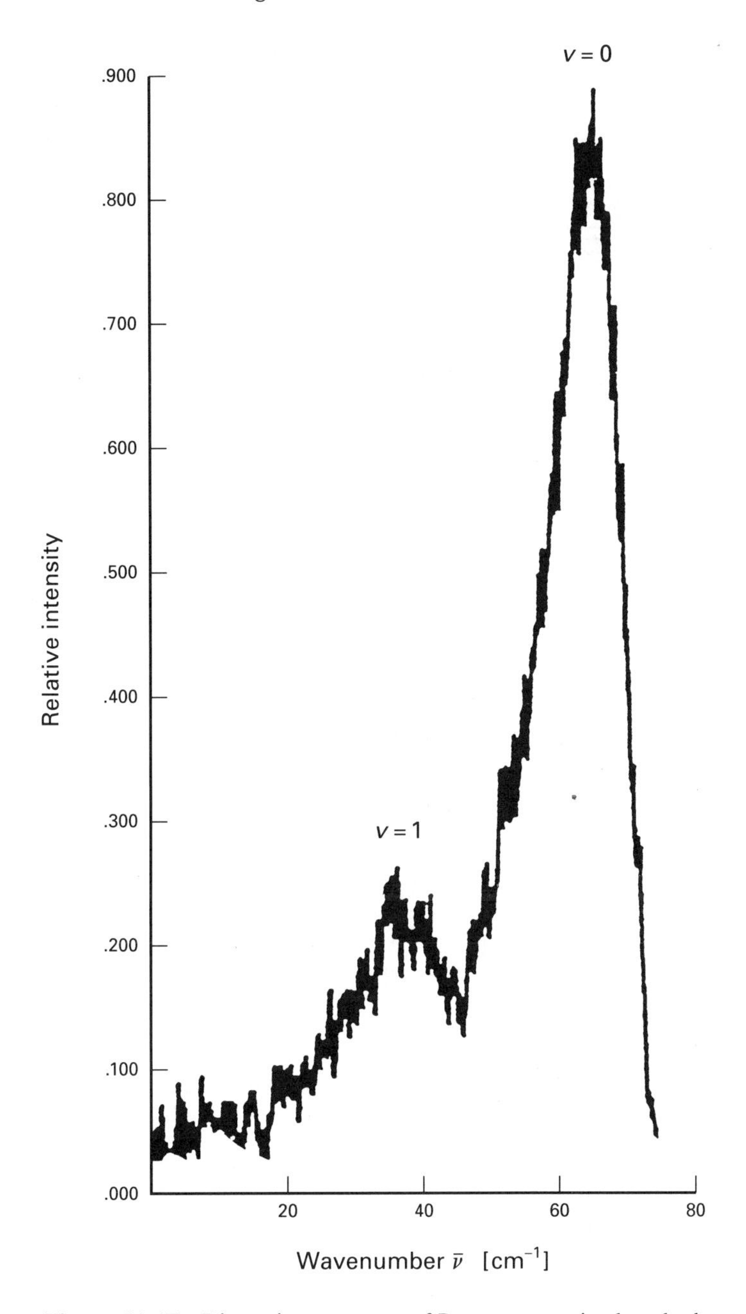

Figure 11.17 Dispersion spectrum of Raman scattering by a hydrogen–air flame, showing the relative positions of the N_2 $v = 0$ and $v = 1$ lines.

There exist numerous applications of Raman techniques for point measurements of the temperature and density of chemical species in chemical reactions, combustion flows, surface diagnostics, and more. However, with the development of sensitive CCD cameras, new Raman techniques have been introduced that allow simultaneous acquisition of Raman spectra from several points along

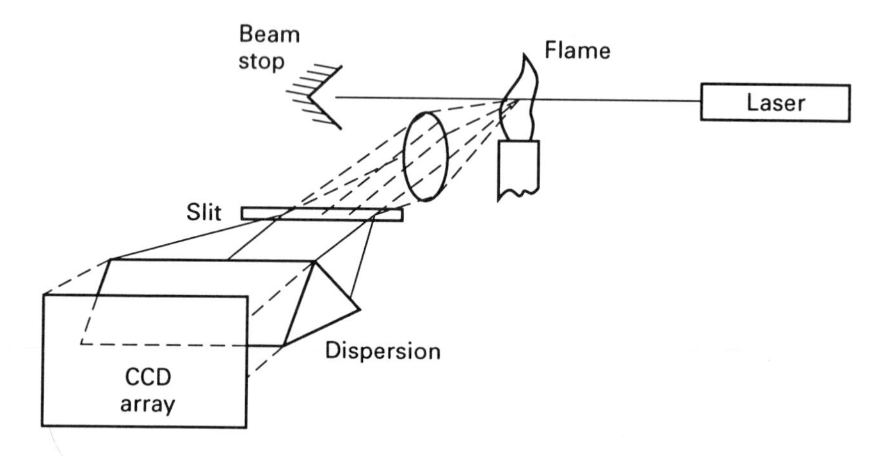

Figure 11.18 Experimental layout for linear Raman imaging of a flame.

the laser beam (see the review by Campion and Perry 1990). In these techniques, a segment of the incident laser beam is imaged onto the slit of a spectrometer, which is represented schematically in Figure 11.18 by a prism. The radiation passing through the slit is dispersed and the portion of the spectrum that contains the Raman signal is projected on the two-dimensional CCD array placed at the back plane of the spectrometer. In Figure 11.18, the dispersion is along the columns of the array, while each row of the array is a monochromatic image of the laser beam. Thus, each column represents the entire Raman spectrum of one point along the laser beam. The number of such sample points is limited only by the number of pixels in a row. However, when the illumination of each pixel is below a desirable level, the signal of several columns may need to be electronically combined, thereby reducing the spatial resolution.

The spectral resolution is likewise limited by the available number of pixels in a column, but it is also limited by the physical size of the array. When the spectrum is resolved by a highly dispersive element, the frequency range of the Raman spectrum covers a large area. Therefore, if the size of the CCD array is insufficient to accommodate the entire Raman spectrum then a less dispersive element must be selected, thereby reducing the spectral resolution. The following example illustrates these considerations.

Example 11.4 A 0.5-m spectrometer is to be used for multipoint imaging of the Raman scattering of a KrF laser operating at 248 nm by the major species of a flame. When equipped with a 1,200-line/mm grating, the dispersion of the spectrometer is $D = 1.7$ nm/mm. The length of the column of the CCD array used for these measurements is $L = 25$ mm and it is divided into $N = 576$ pixels. The system is designed for detection of the following species: CO_2 ($2\nu_2$), CO, O_2, NO, N_2, H_2O, and H_2. Assuming that the Raman signal of CO_2 ($2\nu_2$) is to be imaged on the first pixel of the column, find the pixel separation between the first and second vibrational Raman lines of N_2 and determine the pixel number where each of the other species is detected.

Solution For these measurements, the camera is placed at the back port of the spectrometer where radiation is fully dispersed. Therefore, if the camera is mounted with columns of the detector array parallel to the direction of the dispersion axis, the number of pixels N_p needed to cover 1 nm is

$$N_p = \frac{N}{L \cdot D} = \frac{576}{25 \cdot 1.7} = 13.56 \text{ pixels/nm}.$$

The Stokes shifts of the species to be detected are listed in Table 11.1 in units of cm^{-1}, but the problem, including the terms D and N_p, is specified by wavelength units [nm]. Therefore, using the known laser wavelength and the Raman shift, the wavelength of each of the detected Raman lines must be calculated. For the first Raman line of N_2:

$$\lambda_{N_2} = \frac{10^7}{\dfrac{10^7}{\lambda_L \text{ [nm]}} - \nu_v} = \frac{10^7}{\dfrac{10^7}{248} - 2{,}331} = 263.22 \text{ nm}.$$

The Stokes line of N_2 when at $v = 1$ is shifted relative to the first Stokes line by

$$2\omega_e x_e = 28.64 \text{ cm}^{-1}$$

(cf. eqn. 11.43 and Table 9.1) and its wavelength is accordingly

$$\lambda_{N_2} = \frac{10^7}{\dfrac{10^7}{248} - (2{,}331 - 28.64)} = 263.02 \text{ nm}.$$

The wavelength difference between the two lines corresponds to three pixels of the camera, which implies that the system resolution is insufficient to reliably distinguish between the two vibrational lines.

A similar calculation is repeated to determine if the various species can be individually identified by the system. Table 11.3 summarizes these results. Clearly, the pixel separation between the Raman lines of these species is much larger than the separation between the two Raman lines of N_2 and so it should

Table 11.3 *Correspondence between the Raman shift of major combustion species, the wavelengths of these lines when excited by a KrF laser, and the pixel location when detected by an imaging spectrometer*

Species	Raman shift [cm^{-1}]	Wavelength [nm]	Pixel number
CO_2 ($2\nu_2$)	1,286	256.17	1
O_2	1,556	257.95	24
CO	2,145	261.93	78
N_2	2,331	263.22	96
H_2O	3,657	272.73	225
H_2	4,160	276.53	276

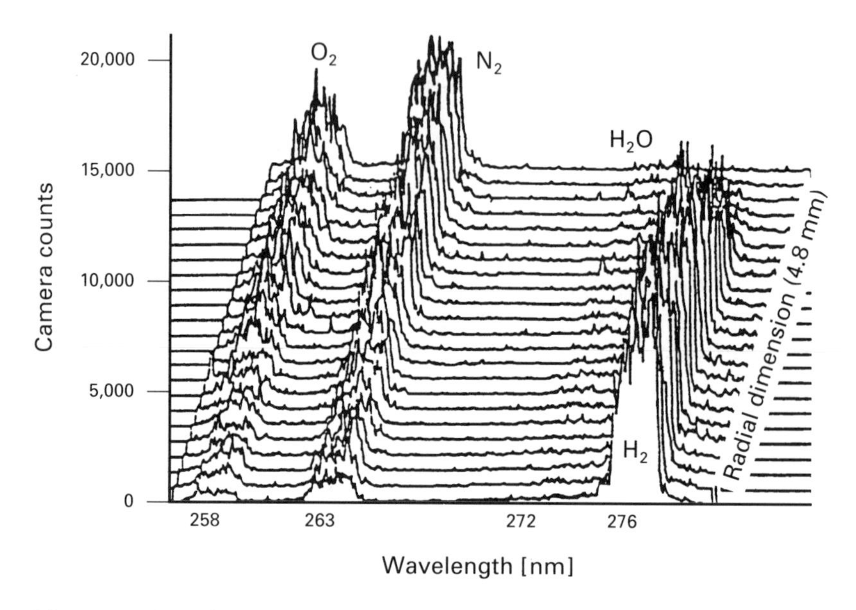

Figure 11.19 Single-pulse, simultaneous multipoint Raman spectrum in turbulent nonpremixed hydrogen–air flame. [Nandula, Brown, Pitz, and De Barber 1994, © Optical Society of America, Washington, DC]

be sufficient to resolve them. With the use of a 2,400-line grating, the dispersion of the spectrometer can be doubled; this can certainly improve the resolution. However, with larger dispersion, the spectral coverage by the detector may be limited and hence Raman lines such as that of H_2 may fall outside the array. ■

Figure 11.19 presents a single-pulse, simultaneous multipoint Raman spectrum in turbulent nonpremixed hydrogen–air flame (Nandula et al. 1994). The spectrum was obtained by passing the beam of a tunable KrF laser through the flame, with its wavelength tuned away from the OH and O_2 absorption bands to avoid interference by LIF of these flame species. The figure is a three-dimensional rendering of the spectrum of the Raman scattering at several points along the radius of the flame. Using the available Raman cross section for each of the species, their mole fraction could be determined. Alternatively, with calibration the absolute density of each species could be determined.

This result illustrates the potential of Raman spectroscopy. Other applications of this and other Raman and Rayleigh techniques are continuously being introduced. Development of new laser sources as well as faster and better-resolved recording devices may extend the present limits even further.

References

Alaruri, S. D., Brewington, A. J., Thomas, M. A., and Miller, J. A. (1993), High-temperature remote thermometry using laser-induced fluorescence decay lifetime measurements of Y_2O_3:Eu and YAG:Tb thermographic phosphors, *IEEE Transactions on Instrumentation and Measurement* 42: 735–9.

Andresen, P., Bath, A., Gröger, W., Lülf, H. W., Meijer, G., and ter Meulen, J. J. (1988), Laser-induced fluorescence with tunable excimer lasers as a possible method for instantaneous temperature field measurements at high pressures: checks with an atmospheric flame, *Applied Optics* 27: 365–78.

Bischel, W. K., and Black, G. (1983), Wavelength dependence of Raman scattering cross sections from 200–600 nm. In *Excimer Lasers - 1983* (C. K. Rhodes, H. Egger, and H. Pummer, eds.) [AIP Conference Proceedings, ser. no. 100(3)], New York: American Institute of Physics.

Campion, A., and Perry, S. S. (1990), CCDs shine for Raman spectroscopy, *Laser Focus World* 26(August): 113–24.

Collins, D. J. (1990), Multi-point Raman spectroscopy in a laminar, pre-mixed H_2–air flat flame, M.Sc. thesis, University of Virginia, Charlottesville.

Crosley, D. R. (1981), Collisional effects on laser-induced fluorescence flame measurements, *Optical Engineering* 20: 511–21.

Daily, J. W. (1977), Saturation effects in laser induced fluorescence spectroscopy, *Applied Optics* 16: 568–71.

Dimpfl, W. L., and Kinsey, J. L. (1979), Radiative lifetimes of OH ($A^2\Sigma^+$) and Einstein coefficients for the A–X system of OH and OD, *Journal of Quantitative Spectroscopy and Radiative Transfer* 21: 233–41.

Doherty, P. M., and Crosley, D. R. (1984), Polarization of laser-induced fluorescence in OH in an atmospheric pressure flame, *Applied Optics* 23: 713–21.

Eckbreth, A. C. (1988), *Laser Diagnostics for Combustion Temperature and Species,* Cambridge, MA: Abacus Press.

Fenner, W. R., Hyatt, H. A., Kellam, J. M., and Porto, S. P. S. (1973), Raman cross-section of some simple gases, *Journal of the Optical Society of America* 63: 73–7.

Fonger, W. H., and Struck, C. W. (1970), Eu^{+3} ^5D resonance quenching to the charge-transfer states in Y_2O_2S, La_2O_2S, LaOCl, *Journal of Chemical Physics* 52: 6364–72.

Goss, L. P., Smith, A. A., and Post, M. E. (1989), Surface thermometry by laser-induced fluorescence, *Review of Scientific Instruments* 60: 3702–6.

Grünefeld, G., Beushausen, V., and Andresen, P. (1994), Planar air density measurements near model surfaces by ultraviolet Rayleigh/Raman scattering, *AIAA Journal* 32: 1457–63.

Heard, D. E., Crosley, D. R., Jeffries, J. B., Smith, G. P., and Hirano, A. (1992), Rotational level dependence of predissociation in the $v'=3$ level of OH $A^2\Sigma^+$, *Journal of Chemical Physics* 96: 4366–71.

Herzberg, G. (1945), *Infrared and Raman Spectra of Polyatomic Molecules,* New York: Van Nostrand.

Herzberg, G. (1989), *Molecular Spectra and Molecular Structure I. Spectra of Diatomic Molecules,* Malabar, FL: Krieger.

Krauss, R. H., Hellier, R. G., and McDaniel, J. C. (1994), Surface temperature imaging below 300 K using La_2O_2S:Eu, *Applied Optics* 33: 3901–4.

Kychakoff, G., Howe, R. D., Hanson, R. K., and McDaniel, J. C. (1982), Quantitative visualization of combustion species in a plane, *Applied Optics* 21: 3225–7.

Laufer, G., McKenzie, R. L., and Fletcher, D. G. (1990) Method for measuring temperature and densities in hypersonic wind tunnel air flows using laser-induced fluorescence, *Applied Optics* 29: 4873–83.

Lee, M. P., McMillin, B. K., Palmer, J. L., and Hanson, R. K. (1992), Planar fluorescence imaging of a transverse jet in a supersonic crossflow, *Journal of Propulsion and Power* 8: 729–35.

Marling, J. B., Nilsen, J., West, L. C., and Wood, L. L. (1979), An ultrahigh-*Q* isotropically sensitive optical filter employing atomic resonance transitions, *Journal of Applied Physics* 50: 610–14.

Massey, G. A., and Lemon, C. J. (1984), Feasibility of measuring temperature and density fluctuations in air using laser-induced O_2 fluorescence, *IEEE Journal of Quantum Electronics* 20: 454–7.

Miles, R. B., and Lempert, W. (1990), Flow diagnostics in unseeded air, AIAA Paper no. 90–0624, American Institute of Aeronautics and Astronautics, Washington, DC.

Morris, M. J., Donovan, J. F., Kegelman, J. T., Schwab, S. D., Levy, R. L., and Crites, R. C. (1993), Aerodynamic applications of pressure sensitive paint, *AIAA Journal* 31: 419–25.

Nandula, S. P., Brown, T. M., Pitz, R. W., and De Barber, P. A. (1994), Single-pulse, simultaneous multiphoton multispecies Raman measurements in turbulent nonpremixed jet flame, *Optics Letters* 19: 414–16.

Paul, P. H., Lee, M. P., and Hanson, R. K. (1989), Molecular velocity imaging of supersonic flows using pulsed planar laser-induced fluorescence of NO, *Optics Letters* 14: 417–19.

Penney, C. M., and Lapp, M. (1976), Raman scattering cross sections for water vapor, *Journal of the Optical Society of America* 66: 422–5.

Peterson, J. I., and Fitzgerald, R. V. (1980), New technique of surface flow visualization based on oxygen quenching of fluorescence, *Review of Scientific Instruments* 51: 670–1.

Quagliaroli, T. M., Laufer, G., Kruass, R. H., and McDaniel, J. C., Jr. (1993), Laser selection criteria for OH fluorescence measurements in supersonic combustion test facilities, *AIAA Journal* 31: 520–6.

Quagliaroli, T. M., Laufer, G., Krauss, R. H., and McDaniel, J. C., Jr. (1995), Planar imaging of absolute OH concentration distributions in a supersonic combustion tunnel, *Journal of Propulsion and Power* 11: 1083–6.

Samson, J. A. R. (1969), On the measurement of Rayleigh scattering, *Journal of Quantitative Spectroscopy and Radiative Transfer* 9: 875–9.

Shardanand, and Prasad Rao, A. D. (1977), Absolute Rayleigh scattering cross sections of gases and freons of stratospheric interest in the visible and ultraviolet regions, NASA Technical Note no. D-8842, Washington, DC.

Shimizu, H., Lee, S. A., and She, C. Y. (1983), High spectral resolution lidar system with atomic blocking filters for measuring atmospheric parameters, *Applied Optics* 22: 1373–81.

Smith, M., Smits, A., and Miles, R. B. (1989), Compressible boundary-layer density cross sections by UV Rayleigh scattering, *Optics Letters* 14: 916–19.

Stern, O., and Volmer, M. (1919), Über die Abklingungszeit der Fluoreszenz, *Physikalische Zeitschrift* 20: 183–8.

Stone, P. M. (1962), Cesium oscillator strengths, *Physical Review* 127: 1151–6.

Williams, P. F., Rousseau, D. L., and Dworetsky, S. H. (1974), Resonance fluorescence and resonance Raman scattering: lifetimes in molecular iodine, *Physical Review Letters* 32: 196–9.

Williams, S. O., and Imre, D. G. (1988), Raman spectroscopy: time-dependent picture, *Journal of Physical Chemistry* 92: 3363–74.

Zimmermann, M., and Miles, R. B. (1980), Hypersonic-helium-flow-field measurements with the resonant Doppler velocimeter, *Applied Physics Letters* 37: 885–7.

Homework Problems

Problem 11.1

A certain laser-induced fluorescence technique is designed to measure the temperature of N_2 in atmospheric air. The laser is focused into the gas sample and, following a multi-photon absorption, N_2 molecules are ionized and also excited to the $v' = 8$ level of the upper $B^2\Sigma$ state of the N_2^+ ion. From this vibrational level, the ions cascade to $v' = 0$, from which they radiate to a lower electronic state. The temperature of the ions is determined from the rotational dispersion spectrum of the emitted fluorescence. Most of the ions reach the $v' = 0$ level by cascading down the vibrational ladder through collisional energy exchange with other molecules. For every collision with any neutral molecule, an excited molecule drops one step in the vibrational level. Assume that the neutral molecules are at 300 K and that the apparent rotational temperature represents a temporary equilibrium between the excited ions and their collisional partners. What will be the rotational temperature of the ions just before they radiate spontaneously from $v' = 0$? For N_2^+ in the $B^2\Sigma$ state, $B_e = 2.083$ cm^{-1}, $\omega_e = 2{,}419.84$ cm^{-1}, and $\omega_e x_e = 23.19$ cm^{-1}. (*Hint:* The effective size of the heat bath is defined in this problem by the total number of collisions, not by the density of the surrounding molecules.)

Problem 11.2

The temperature of O_2 is to be measured by a laser-induced fluorescence technique, where the $J'' = 15$ state is to be excited to an upper state and the subsequent fluorescence is to be measured. Since the upper state is strongly predissociated, most excited molecules dissociate and never return to the ground state. Therefore, if the energy density [J/cm^2] of the incident laser is high and the laser pulse duration is short, the population of the original state may be severely depleted and the temperature measurement may be distorted. Assume that there are no mechanisms to repopulate the $J'' = 15$ state during the excitation. Find the maximum laser energy density that can be used without depleting the ground-state population by more than 1%, where $B_{12} = 661.66$ cm^2/s-erg, $\nu_{12} = 51{,}783$ cm^{-1}, and $\Delta\nu = 3.2$ cm^{-1}.

Problem 11.3

A thin rectangular plate is heated uniformly on one side while cooled on the opposite (Figure 11.20). The two other sides are insulated. The temperature on the $y = 0$ side is $T_0 = 120$°C; on the $y = 15$-cm side, the temperature is $T_1 = 25$°C. The plate is coated with a thermographic phosphor consisting of La_2O_2S:1.0%Eu powder encased in a binder. The phosphor is illuminated by a UV lamp at a nominal wavelength of 366 nm.

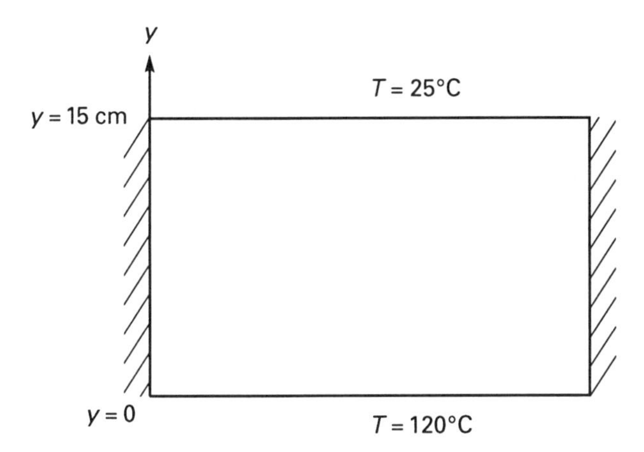

Figure 11.20 Rectangular plate heated uniformly on one side and cooled on the other.

(a) Calculate the temperature distribution of the plate. (*Answer:* $(T-T_1)/(T_0-T_1)=y/15$.)

(b) Determine which of the ^5D_j phosphorescence transitions is likely to have the best sensitivity for the temperature range expected for this plate.

(c) Calculate the normalized variation of the phosphorescence emitted by the coating relative to the phosphorescence at $y=15$ (in other words, calculate $I(y)/I(15)$).

(d) Plot the variation with y of the dimensionless temperature, $(T-T_1)/(T_0-T_1)$, and the dimensionless phosphorescence variation $I(y)/I(15)$.

Problem 11.4

Determine the dynamic range (i.e., the upper and lower temperatures) for temperature measurements made using a La_2O_2S:Eu thermographic phosphor. Assume that each of the ^5D_j ($j=0,1,2,3$) transitions becomes temperature-sensitive when the phosphorescence is quenched to 95% of its maximum, and that the error in the phosphorescence measurement becomes unacceptable when the signal declines to 5% of its maximum.

Problem 11.5

The sky appears blue due to Rayleigh scattering of the sunlight. Using the blackbody representation of the solar spectrum and the wavelength dependence of the cross section for Rayleigh scattering, derive a mathematical description of the spectral distribution of skylight in the visible range. Plot your result and determine graphically the wavelength where this spectrum peaks.

Problem 11.6

Determine, either experimentally or by heuristic arguments, the polarization of skylight during sunset when viewed with the sun on your right and then with the sun behind you. (Do not view the sun directly! Direct viewing of the sun is dangerous, and in any case Rayleigh scattering cannot be resolved from directly transmitted radiation.)

Problem 11.7

Derive an expression similar to (11.38) that describes the variation in the number of photoelectrons per pixel that can be detected in a Raman or Rayleigh experiment as a function of the wavelength of the illuminating source. Assume that the signal n_{pe}^0 for illumination at λ_0 is known. (*Hint:* The detected signal does *not* vary as $1/\lambda^4$.)

Problem 11.8

The trace in Figure 11.17 describes the vibrational Raman spectrum of N_2 recorded in a hydrogen–air flame. The tall line corresponds to a Raman transition from $v=0$, and the second line corresponds to a transition from $v=1$. Owing to the limited resolution of the detection system, the rotational structure could not be resolved. Estimate the vibrational temperature of the gas.

Research Problems

Research Problem 11.1

Read the paper by Massey and Lemon (1984).

(a) Derive equations (3) and (6) of the paper.

(b) Calculate the total fluorescence detected from a "slice" of 1 mm along a laser beam tuned to the $v'=4 \leftarrow v''=0$ R(19) line in the Schumann–Runge band of atmospheric O_2 ($T = 300$ K, $P = 1$ atm). The fluorescence is detected in the R(19) $v'=4 \rightarrow v''=7$ band. The pulse energy is 250 mJ, and the wavelength is 193.27 nm. The radiation is collected by a lens with NA = 0.25. For O_2, $A_{4\rightarrow 0} = 516$ s^{-1} and $A_{4\rightarrow 7} = 3.964 \times 10^6$ s^{-1}. The rotational constants are given in Table 9.1. In addition, the excited state undergoes rapid predissociation. The rate of this process can be estimated from the apparent linewidth, $\Delta\nu = 3.2$ cm^{-1}, of the transition from the excited state. (*Hint:* See Example 10.2 of Section 10.3.)

Research Problem 11.2

Read the paper by Doherty and Crosley (1984).

(a) A horizontally polarized laser beam is used for the excitation of the $P_1(8)$ line of OH ($J = N + \frac{1}{2}$). The fluorescence is collected along a line that is parallel to the polarization axis of the laser. The collected fluorescence is resolved by a diffraction grating that can transmit either the fluorescence of the $Q_1(7)$ line or the fluorescence of the $R_1(6)$ line. The grating transmits the p polarization at an efficiency that is only 50% of the transmission of the s polarization (see Figure 7.8). Find the polarization correction factor for each polarization of the detected emissions.

(b) Assume that the experiment is repeated with a circularly polarized laser beam for the excitation. Find the polarization correction factor for each transition when detected through the spectrometer grating, assuming that the incident laser energy is the same as in the experiment of part (a). Calculate the ratio between the signals of the Q line when excited by the linearly polarized laser to the Q line when excited by the circularly polarized laser. Calculate this ratio also for the R line (*Hint:* Determine first the energy of each of the polarization components of the circularly polarized beam. *Answer:* $Q_{lin}/Q_{cir} = 1.078$, $R_{lin}/R_{cir} = 0.95$.)

Research Problem 11.3

Read the paper by Nandula et al. (1994).

(a) For the conditions specified by the paper, calculate what should be the collection *f*/# required to obtain a signal similar to that reported by the authors when using a pulsed illumination of 400 mJ/pulse by a frequency-doubled Nd:YAG laser at 532 nm.

(b) Repeat the calculations for an illumination of 200 mJ/pulse at 193 nm by an ArF laser.

(c) One of the concerns when using illumination at 248 nm is OH fluorescence induced by its inadvertent excitation. Emission of excited OH in the $3 \rightarrow 1$ band at nominally 272 nm overlaps the Raman line of H_2O and may not be spectrally separated. Therefore, unless negligible, this emission may interfere with the measurements. Assume that the OH density in the flame is approximately uniform at 1×10^{15} cm^{-3} and its temperature is 2,000 K. Use the parameters of Example 11.1 of Section 11.2 for LIF of the $(3 \leftarrow 0)$ $P_1(8)$ line, let $A_{31} = 1.14 \times 10^5$ s^{-1} for the strongest emission line, and assume that approximately 2% of the laser energy is coupled into the inadvertent OH excitation. Compare the expected OH LIF signal with the measured H_2O Raman signal.

12 Optical Gain and Lasers

12.1 Introduction

Previous discussions (see Section 10.4) suggested that stimulated emission can be used to generate optical gain, that is, to amplify radiation. The reader certainly has experience in the amplification of electronic signal. For example, radio receivers capture faint radio waves and turn them into a signal that is powerful enough to drive large speakers. This electronic amplification increases the amplitude of the signal while faithfully preserving its acoustic frequencies and modulation characteristics. Similarly, optical amplification is expected to increase the amplitude of an optical signal while preserving its frequency, its modulation characteristics, and its coherence. The latter requirement is of particular significance for optical radiation, where the coherence of naturally occurring radiation rarely exceeds one micrometer. Lasers are the primary source for coherent radiation. They depend on stimulated emission for amplification and for the generation of coherent radiation (the word *laser* is the acronym of Light Amplification by Stimulated Emission of Radiation). For amplification, an atomic system that is part of the laser medium must be prepared with a sufficiently large number of particles in the excited state. (The term *atomic system* is used here to describe all microscopic systems including molecules and free electrons.) Radiation passing through that excited medium encounters multiple events of stimulated emission, each event contributing one photon that is added coherently to the propagating beam. When the number of events of stimulated emission exceed all losses by absorption or scattering, the incident radiation is amplified. As we will show in more detail, this requirement can be met only when the density of the particles in the upper (excited) state exceeds the density of the particles in the lower (ground) state.

Although the need for a large number of atomic particles in the excited state is evident, achieving the necessary excitation level was an impediment to the early development of lasers. Unfortunately, direct excitation of the upper state from the ground state cannot force the population of the upper state to exceed that of the ground state. Thus, although two-level systems can be used to estimate the gain itself, the description of how the gain medium is formed requires consideration of the interaction and coupling between three or more energy levels. Consequently, the gain term – the mathematical expression that describes the population density requirements for net optical gain – will first be derived using the interaction between two levels. However, the conditions for maintaining stable gain must be derived using a multi-level system. Finally, the gain required for steady emission of a laser beam must overcome not only the losses by absorption between the interacting levels, but also absorption losses of the laser beam energy by impurities, scattering, reflection losses at interfaces, and (of course) losses by output coupling. Therefore, the energy requirements for stably maintaining the gain condition must include parameters that describe the operation of the entire laser system.

12.2 Optical Gain and Loss

Optical gain and optical loss are counterparts of the same process. *Loss,* or absorption, occurs when an optical beam at a certain frequency passes through a medium having an absorbing transition that is resonant with the incident frequency. Similarly, gain occurs when an optical beam passes through a medium that has been conditioned to present gain at the incident frequency. Thus, when absorption occurs, the energy of the incident beam diminishes while the energy of the absorbing medium increases. By contrast, *gain* is an emission process that transfers energy from the gain medium to the transmitted beam. Unlike spontaneous emission, where photons may be scattered uniformly in all directions with a random phase and finite range of frequencies, photons that contribute to optical gain are emitted only when a beam is traveling through the medium. These photons are an identical replica of the incident photons and so propagate in their direction, with the same phase, frequency, and polarization. The gain is considered positive when stimulated emission exceeds absorption. Of course, gain extracts energy from the medium and therefore cannot exceed the extent of stored and added energy.

Preparation of an excited medium that provides a positive gain is a complex task that will be discussed in the next section using a multi-level system modeling. However, once established, gain can be evaluated by simply using a two-level system. Consider a collimated narrowband beam, with an irradiance of $I_\nu = I(\nu)\,d\nu$ (Section 10.4), traveling through a gain medium; see Figure 12.1. The frequency ν of the beam is at or near the transition frequency between the two levels 1 and 2. The population densities of these levels are N_1 and N_2 and their degeneracies are g_1 and g_2, respectively. As the beam propagates through the medium, absorption by the lower-state population depletes the beam energy while stimulated emission from level 2 amplifies it. Spontaneous emission from level 2 may also contribute to the beam irradiance. However, since the beam is

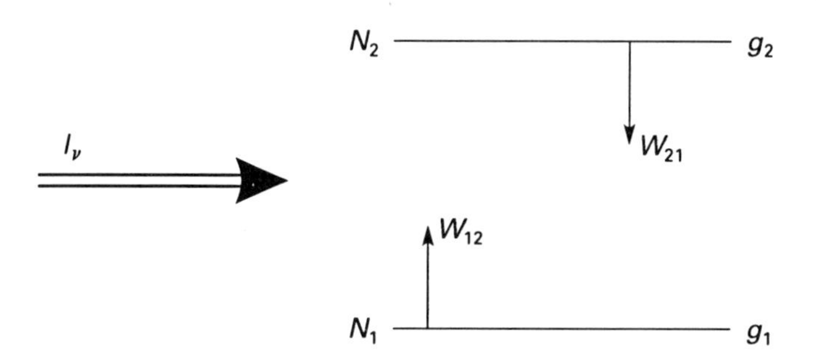

Figure 12.1 Absorption and stimulated emission induced by narrowband radiation passing through a two-level system.

nearly collimated, the solid angle it encompasses is $\Omega \to 0$ and the fraction $\Omega/4\pi$ of the total fluorescence emission that propagates with the beam is negligible. Therefore, the net change in the beam irradiance when propagating a distance dx through this medium must match the net change in the energy density, excluding spontaneous emission, stored in the upper state. Using (10.15), (10.27), and (10.43), this net change may be written as

$$\frac{dI_\nu}{dx} = (N_2 W_{21} - N_1 W_{12})h\nu = \left(N_2 - \frac{g_2}{g_1}N_1\right)W_{21}h\nu. \tag{12.1}$$

Of course, W_{21} increases linearly with the incident irradiance, and – similarly to the absorption rate coefficient – this dependence can be expressed in terms of the following cross section for stimulated emission $\sigma_{SE}(\nu)$:

$$W_{21} = \frac{I_\nu}{h\nu}\sigma_{SE}(\nu) \tag{12.2}$$

(cf. eqn. 10.23). Including this cross section in (12.1) reduces it to the following single-variable, first-order differential equation:

$$\frac{dI_\nu}{dx} = \Delta N \sigma_{SE}(\nu) I_\nu = \gamma(\nu) I_\nu, \tag{12.3}$$

where

$$\Delta N = N_2 - \frac{g_2}{g_1}N_1 \tag{12.4}$$

is the population difference term and

$$\gamma(\nu) = \Delta N \sigma_{SE}(\nu) \tag{12.5}$$

is the gain constant.

Although (12.3) is a simple differential equation that describes the net gain or loss of irradiance, it depends on an unspecified coefficient, σ_{SE}, which needs to be related to tabulated spectroscopic constants such as the Einstein coefficients or the oscillator strength. By analogy to (10.25), $\sigma_{SE}(\nu)$ can be represented in terms of the Einstein A coefficient by introducing the relation between

A_{21} and B_{21} (eqn. 10.44). Therefore, in terms of tabulated coefficients, the cross section for stimulated emission is

$$\sigma_{SE}(\nu) = \frac{A_{21}c_0^2}{8\pi n^2 \nu^2} g(\nu). \tag{12.6}$$

The term $g(\nu)$ is the lineshape factor of the two-level system, which for a homogeneously broadened system is mostly Lorentzian and for an inhomogeneously broadened system is Gaussian. This cross section may also be expressed by the oscillator strength, using (10.48) to replace A_{21} with f_{21}.

With all of its coefficients defined, (12.3) can now be solved to obtain the variation of the irradiance of a beam traveling through a gain medium. If the incident irradiance $I_\nu(0)$ is weak then the effect of the incident beam on the population density of the gain medium is negligible, and the population difference term (12.4) can be assumed to be independent of I_ν. Accordingly, the irradiance $I_\nu(x)$ at any point x along the beam path is

$$I_\nu(x) = I_\nu(0)e^{\gamma(\nu)x} \tag{12.7}$$

(Verdeyen 1989, p. 176). This can be used to determine the amplification G_0 of a weak beam after propagating a distance L through the medium:

$$G_0 = \frac{I_\nu(L)}{I_\nu(0)} = e^{\gamma(\nu)L}. \tag{12.8}$$

Thus, when the gain constant $\gamma(\nu) > 0$, the medium acts as an amplifier and $G_0 > 1$. By contrast, if $\gamma(\nu) < 0$ then the medium acts as an attenuator. In the special case where $\gamma(\nu) = 0$, the beam can pass through the medium without any gain or loss and the medium is considered transparent. The gain constant depends on two parameters, ΔN and σ_{SE}. Because $\sigma_{SE} > 0$, the sign of $\gamma(\nu)$ and with it the gain characteristics of the medium are controlled only by the sign of ΔN (eqn. 12.5), which can be positive only when

$$\frac{N_2}{g_2} > \frac{N_1}{g_1}; \tag{12.9}$$

that is, when the population in each of the degenerate states at the upper energy level exceeds the population of each degenerate state of the lower energy level, the medium becomes an amplifier. Accordingly, ΔN is a measure of the inversion density. However, at thermodynamic equilibrium, the population of each of the upper degenerate states must be lower than the population of any of the lower degenerate states (eqn. 10.31). Thus, at thermodynamic equilibrium the medium is always lossy. To produce gain, the population distribution must depart from Boltzmann's distribution by creating a *population inversion* (eqn. 12.9). Such inversion can be produced only by an energy input that selectively excites one or more energy levels. This selective excitation is called *pumping*. Although population inversion is a condition of thermodynamic nonequilibrium, it can be maintained indefinitely (i.e., at steady state) as long as pumping is maintained. At steady state, the effect of pumping is balanced by stimulated emission and other losses of the excited population. The extent of inversion that needs to be achieved depends on the desired level of gain and on various system losses. The following example illustrates the inversion requirements for an excimer laser.

Example 12.1 A KrF laser radiating at 248 nm is designed to provide a gain of 5% per centimeter. Find the minimum inversion density ΔN required to maintain this gain. The lifetime of the KrF molecule is approximately 9 ns and the emission bandwidth is 180 cm^{-1}. Also find the total amplification experienced by a beam traveling a distance of $L = 1$ m through the excited gas.

Solution To obtain the population difference ΔN from (12.5), the cross section for stimulated emission $\sigma_{SE}(\nu)$ must first be calculated using (12.6). The parameters $g(\nu)$, A_{21}, and ν of that equation can be obtained by the following calculations:

$$g(\nu) \approx \frac{1}{\Delta\nu} = \frac{1}{180(3\times 10^{10})} = 1.85\times 10^{-13}\ \text{s};$$

$$A_{21} = \frac{1}{9\times 10^{-9}} = 1.11\times 10^{8}\ \text{s}^{-1};$$

$$\nu = c/\lambda = 1.21\times 10^{15}\ \text{Hz}.$$

Note that, although the lineshape of this excimer transition is natural, it is not controlled by A_{21}. Instead, as illustrated by Example 9.2 of Section 9.4, the shape and width of this line are determined by the structure of the predissociated ground state. The cross section is obtained by direct substitution of these results into (12.6) as follows:

$$\sigma_{SE}(\nu) = 5.03\times 10^{-16}\ \text{cm}^2;$$

from (12.5), the required inversion density is

$$\Delta N = \frac{\gamma(\nu)}{\sigma_{SE}(\nu)} = \frac{0.05}{5.03\times 10^{-16}} \approx 10^{14}\ \text{cm}^{-3}.$$

Because only 1% of the excimer molecules can survive collisional and other losses, the actual inversion density required to maintain this level of gain is approximately $\Delta N = 10^{16}\ \text{cm}^{-3}$. When compared with the Loschmidt number, this result implies that if the gas is at near-atmospheric conditions then approximately 1/1,000 of all molecules must be in the excited state. Owing to the instability of the ground state (Example 9.2 of Section 9.4), the population of the excited state represents the inversion almost entirely.

The amplification of the irradiance after propagating a distance of 1 m through this excimer medium is obtained by directly substituting into (12.8) the available parameters L and $\gamma(\nu)$:

$$G_0 = e^{0.05\cdot 100} = 148.4.$$

The length of $L = 1$ m used in this problem is comparable to the length of commercially available KrF lasers. Therefore, the irradiance of a weak beam passing through the entire length of the excited-gas medium may increase by a factor of 148. ■

The results of this example show that even an apparently modest gain of 5% per centimeter can result in an appreciable enhancement of a weak beam. If the amplified beam is reflected back on itself and allowed to pass through the medium once again, the gain predicted by (12.7) will be $148^2 = 21{,}904$. With such a large gain, even if the incident or signal beam is initially weak, it can become sufficiently powerful after traveling through the gain medium to start depleting the population of the excited state. Thus, the assumption that the population density of the interacting states is independent of the passing irradiance is true only within a limited propagation distance, and (12.7) may not be generally valid. To determine the general conditions for gain, the effects of the passing radiation on the inversion and the results of the ongoing pumping must be considered simultaneously. Thus, before (12.7) can be modified, excitation methods that create inversions must be identified.

Pumping of a laser gain medium may be achieved by any technique that can induce thermodynamic nonequilibrium: electrical discharges, rapid chemical reactions, shock waves of a supersonic expansion into the active medium, and so forth. Of course, the most selective (and hence the most efficient) pumping method would be by absorption of radiation tuned to the excitation frequency between levels 1 and 2 of the gain medium (Figure 12.1). However, it can be shown (see Problem 12.1) that when the frequency of the pumping radiation coincides with the frequency of the amplified beam, the population of the upper state cannot be forced optically to exceed the population of the lower state. Stated differently, inversion cannot be achieved by the interaction of radiation with a two-level system. When the incident irradiance approaches infinity, the population density of the excited state can only approach the density of population in the ground state. With such a high optical pumping rate, $W_{21} \gg A_{21}$ and the entire radiative process is dominated by absorption and stimulated emission that simply offset each other. Thus, if at low levels of irradiance the medium is absorbing, at higher levels of irradiance the medium becomes transparent and is said to be *bleached* (Problem 12.2).

Since the population of a two-level system cannot be inverted, the simplest system that permits inversion by optical pumping must include three levels. Figure 12.2 illustrates such a system. Although the illustration suggests optical pumping at a rate of W_{02}, other pumping schemes may be used as well. The inversion in this system is required between level 1 and the ground state 0. Pumping of level 1 is achieved by initial excitation of level 2, followed by rapid relaxation to level 1. The objective is to increase the population of level 1 at a rate that exceeds the rate of losses from that state. Therefore, the rate of spontaneous emission from level 2 to level 1 must exceed the spontaneous emission rate from level 1 back to the ground state:

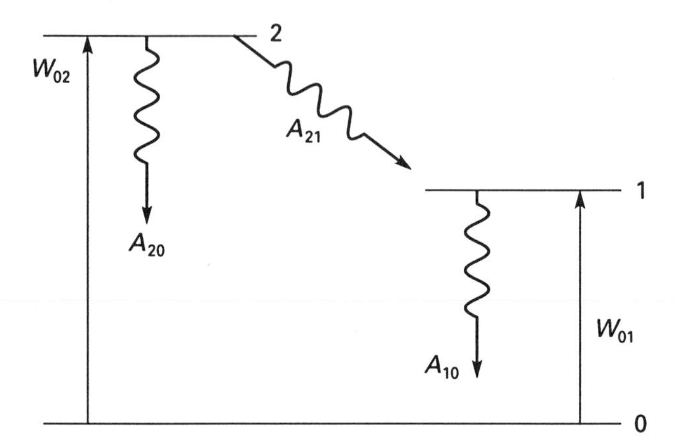

Figure 12.2 A three-level system and the transitions associated with the inversion between levels 1 and 0.

$$A_{21} \gg A_{10}. \tag{12.10}$$

If, in addition, the spontaneous emission rate from level 2 back to the ground state is maintained at a sufficiently low rate (see Problem 12.3), the population of level 1 can be increased until inversion between level 1 and the ground state is achieved.

Although the first laser to be demonstrated, the ruby laser (Maiman 1960b), was a three-level system, this method is by far the least effective pumping system. Because of the large energy difference between the ground state and levels 1 and 2, the thermally induced population of both levels 1 and 2 is negligible and almost all the participating particles are in the ground state. Thus, if $g_1 = g_0$, at least half of all the participating atoms or molecules must be raised to level 1 before inversion can be established. Furthermore, even after inversion is established, each decay from level 1 to the ground state not only decreases the population of the upper laser state but also increases the population of the lower state, thereby imperiling the inversion at both ends.

An alternative and more effective laser pumping scheme involves four levels (Figure 12.3). In this scheme, the upper laser level 2 is populated by first pumping level 3. The representation of level 3 by a band suggests that a group of states may be pumped simultaneously, thereby relaxing the need for selective pumping. If the lifetime τ_{32} of spontaneous decay from the directly pumped states to level 2 is short relative to the time τ_{31} of decay to level 1, then level 2 can be filled faster than level 1. And if the spontaneous decay τ_{21} from level 2 to level 1 is long relative to τ_{32}, the population in level 2 can increase until an inversion between levels 2 and 1 is created. Unlike the three-level system, where at least half of the ground-state population had to be excited, inversion can be achieved here merely by raising the population of level 2 above the thermally induced population of level 1, which is often negligible. Furthermore, if the spontaneous decay rate from level 1 is fast, inversion can be sustained not only by pumping level 2 but also by effectively draining level 1. Therefore, the power

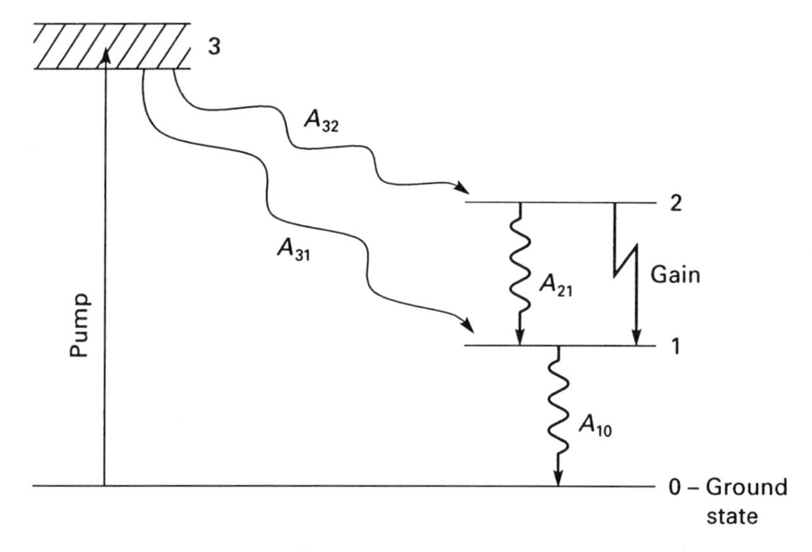

Figure 12.3 A four-level system and the transitions associated with the inversion between levels 2 and 1.

required both to form and sustain the inversion is reduced relative to the requirements of a three-level system (Yariv 1985, p. 155). To derive the conditions for inversion in a realistic laser system, pumping of the participating states must be considered simultaneously with the depletion of the inverted states by a propagating laser beam. This analysis will be presented next.

12.3 Gain in Homogeneously Broadened Laser Systems

The four-level system presented in Figure 12.3 is an ideal configuration. Although it indicates that level 3 may consist of a group of states, it still represents a highly selective excitation where only the upper level of the amplifying transition is pumped. Such highly discriminating pumping of only one group of states is rarely achieved. Typical pumping techniques, including excitation by an electrical discharge or an illumination by the bright flash from a flashlamp, tend to fill also the lower level of the gain system (level 1), thereby reducing the inversion density that could be achieved if only levels 3 and 2 were pumped. Therefore, a realistic model of the four-level system must also include the effect of inadvertent pumping of level 1. Figure 12.4 illustrates such an energy-level system for optical gain generation. The gain is generated by inversion between levels 2 and 1. Although an intermediate level, like level 3 in Figure 12.3, may be part of the pumping scheme, it is not presented here. Instead, the net effect of pumping level 2 at a rate of P_2 is marked by an arrow from the ground state, while a second arrow marks the inadvertent pumping of level 1 at a rate of P_1. Nevertheless, pumping is designed to favor level 2, $P_2 > P_1$, and its rate exceeds the rate of all losses from level 2 and so creates an inversion between levels 2 and 1.

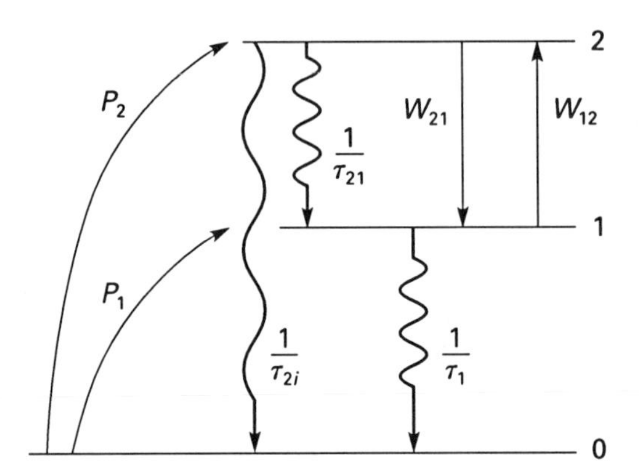

Figure 12.4 Energy-level diagram of an optical gain medium. The inversion is between levels 2 and 1. The details of the pumping are not shown and may involve additional intermediate states.

The density of the inversion depends on the pumping rates and also on the effects of spontaneous transitions from levels 1 and 2. All spontaneous transitions from level 2 decrease the inversion density. However, transitions from level 2 to level 1 not only decrease the upper-state population but also increase the lower-state population, thereby doubling their adverse effect on the inversion. Thus, from level 2 the rate of spontaneous transitions $1/\tau_{21}$ should be considered separately from the combined rate $1/\tau_{2i}$ of spontaneous transitions to all other levels. On the other hand, all spontaneous transitions from level 1 benefit the inversion; hence only the total rate $1/\tau_1$ needs to be considered. Because of the possibly different degeneracies g_1 and g_2 of the participating levels 1 and 2, a distinction must be made between the rate of absorption and the rate of spontaneous emission, which are related by $W_{12} = (g_2/g_1)W_{21}$ (eqn. 10.43).

To establish the interaction between a propagating beam and the gain medium, assume that the narrowband irradiance I_ν is resonant with the transition frequency between levels 2 and 1, and that its bandwidth is narrow relative to the transition linewidth. Although most laser systems are partly homogeneously broadened and partly inhomogeneously broadened, our simplified analysis will consider only the two extremes, where broadening is dominated exclusively by one or the other mechanism. When broadening is mostly homogeneous – that is, when the lineshape of each excited particle is identical to the lineshape of its neighbor – all particles present the same cross section for induced transition, be it absorption or stimulated emission. This cross section is determined by $g(\nu)$ (eqn. 12.6) and by the frequency ν of the propagating beam. Thus, even when the propagating beam is off the line center and so $g(\nu)$ and consequently $\sigma_{SE}(\nu)$ are below their peaks, any stimulated transition removes a particle from the pool of inverted population. Still, all the particles are available for interaction with the beam, albeit at a reduced probability. By contrast, when the broadening is inhomogeneous and hence dominated by the random motion of excited particles, only those particles whose frequency is Doppler-shifted to coincide with the amplified beam can undergo stimulated emission or absorption. Thus, when stimulated emission in an inhomogeneously broadened medium occurs off the line center, only a small pool of particles is available for interaction while all the others do not contribute to the gain. Of course, at a later time, when their velocity is changed by collisions, new particles may contribute to the gain or the loss. Thus, at any given time, the pool of particles that participate in the generation of the gain is only a fraction of all the excited particles. Clearly, when the gain medium is homogeneously broadened, the analysis includes all the particles that occupy levels 1 and 2 and does not require the identification of a select group. While it may not always be justified physically, the assumption of homogeneous broadening greatly simplifies the analysis of optical gain and will therefore be considered first.

The optical gain of a homogeneously broadened medium depends ultimately on the inversion density ΔN between levels 2 and 1 (eqn. 12.5). To establish this inversion we must solve the following rate equations for a continuously pumped medium irradiated by a narrowband irradiance I_ν:

$$\frac{dN_1}{dt} = P_1 - \frac{N_1}{\tau_1} + \frac{N_2}{\tau_{21}} + \left(N_2 - \frac{g_2}{g_1}N_1\right)W_{21} \tag{12.11a}$$

and

$$\frac{dN_2}{dt} = P_2 - \frac{N_2}{\tau_2} - \left(N_2 - \frac{g_2}{g_1}N_1\right)W_{21}, \tag{12.11b}$$

where

$$\frac{1}{\tau_2} = \frac{1}{\tau_{2i}} + \frac{1}{\tau_{21}}$$

(Yariv 1985, p. 142). For steady-state gain, each of these equations can be considered to be independently at steady state, $dN_1/dt = dN_2/dt = 0$, thereby reducing (12.11) to a pair of algebraic equations that can be readily solved for N_1 and N_2. The inversion density ΔN is obtained as follows:

$$\begin{aligned}\Delta N &= N_2 - \frac{g_2}{g_1}N_1 \\ &= \frac{P_2\tau_2 - [P_2(g_2/g_1)\delta + P_1(g_2/g_1)]\tau_1}{1 + [\tau_2 + (g_2/g_1)(1-\delta)\tau_1]W_{21}},\end{aligned} \tag{12.12}$$

where

$$\delta = \frac{1/\tau_{21}}{1/\tau_2}$$

is the relative rate of decay from level 2 to level 1.

In an operating laser system, there exist numerous loss mechanisms that were not included in the rate equations for levels 1 and 2 (eqns. 12.11). Therefore, for steady-state operation, the gain must overcome these yet-unspecified losses and the inversion must accordingly exceed a threshold level ΔN_t beyond which radiation can be amplified despite the losses. The only losses included in (12.12) are by absorption from level 1. Therefore, (12.12) can be used to determine only the lowest threshold for gain, which merely requires that $\Delta N > 0$. This condition is met when the numerator exceeds zero, that is, when

$$\frac{1/\tau_1}{1/\tau_2} > \frac{P_1/g_1}{P_2/g_2} + \frac{g_2}{g_1}\delta. \tag{12.13}$$

The left side of (12.13) is the ratio between the total rate of spontaneous decay from level 1 and the total rate of decay from level 2. While decay from level 1 promotes the inversion, decay from level 2 depletes it. Therefore, for inversion, the relative rate of decay from level 1 must exceed the relative rate of additions to that level caused either by inadvertent pumping (the first term on the right side of eqn. 12.13) or by spontaneous decay from level 2 to level 1 (the last term).

The condition for inversion stated by (12.13) is sufficient to initiate optical gain when all other loss mechanisms are absent, but it does not state the projected gain, which depends not only on an initial inversion but also on the irradiance of the beam traveling through the medium. When that irradiance becomes appreciable, the depletion of the inversion by stimulated emission is no longer negligible. This is apparent in the denominator of (12.12), where the irradiance-dependent (eqn. 12.2) stimulated emission rate W_{21} is included. Thus, as I_ν increases, W_{21} becomes the dominant term in the denominator and $\Delta N \to 0$. Conversely, when $I_\nu \to 0$ the inversion is

$$\Delta N_0 = P_2\tau_2 - \left(P_2\frac{g_2}{g_1}\delta + P_1\frac{g_2}{g_1}\right)\tau_1. \tag{12.14}$$

For given pumping rates, ΔN_0 is the largest possible inversion. Associated with ΔN_0 is the largest available gain - the *small-signal gain*. This gain, which is experienced when the incident irradiance approaches zero, is obtained by combining (12.14) with (12.5):

$$\gamma_0(\nu) = \Delta N_0 \sigma_{SE}(\nu). \tag{12.15}$$

As the incident irradiance I_ν increases, the denominator of (12.12) increases and the available gain decreases. The actual gain of a homogeneously broadened medium is obtained by combining (12.5) and (12.12). To account for the effect of the propagating irradiance on ΔN, the stimulated emission rate W_{21} is replaced with $I(\nu)$ (eqn. 12.2). The equation for the gain can now be reduced to the following form (Rigrod 1963):

$$\gamma(\nu) = \frac{\gamma_0(\nu)}{1 + I_\nu/I_{\text{sat}}}, \tag{12.16}$$

where

$$I_{\text{sat}} = \frac{h\nu}{[\tau_2 + (g_2/g_1)(1-\delta)\tau_1]\sigma_{SE}(\nu)} \tag{12.17}$$

defines the *saturation irradiance.* Equation (12.16) describes the diminishing of the gain with increasing irradiance. Thus, whereas for $I(\nu) \to 0$ the gain approaches $\gamma_0(\nu)$, when $I(\nu) \to \infty$ the gain approaches zero and the medium becomes transparent. When $I_\nu = I_{sat}$ the gain is reduced to one half of its peak level. Although an arbitrary choice, I_{sat} is considered to be a convenient measure for the saturation of the gain of a homogeneously broadened medium.

12.4 Gain in Inhomogeneously Broadened Laser Systems

When the gain medium is inhomogeneously broadened, the lineshape for stimulated emission is determined primarily by the motion of the atomic particles. As before, the lineshape of each particle is defined by its own natural linewidth and collisional broadening. However, the Doppler shift induced by the particles' motion is much larger than their individual linewidths. The resultant lineshape of randomly moving particles is therefore defined by numerous overlapping Lorentzian distributions. When an incident narrowband beam is transmitted through such a medium, it interacts most favorably with the group of particles whose absorption frequency is shifted to coincide with the incident wavelength. Other groups may also interact, but since their center frequency is not coincident with the radiation frequency, their cross section for stimulated emission is smaller than the cross section of the resonant group. The net gain is obtained by the summation of the contributions of all the participating groups. Thus, for an inhomogeneously broadened medium the gain (Rigrod 1963; see also Verdeyen 1989, p. 206) is

$$\gamma(\nu) = \frac{\gamma_0(\nu)}{(1+I_\nu/I_{sat})^{1/2}}, \tag{12.18}$$

where I_{sat} is defined as before (eqn. 12.16).

Note that, although the broadening is primarily inhomogeneous, I_{sat} is defined by the lineshape factor of the Lorentzian of the homogeneously broadened transition. This can be understood by recalling that each group, including the group that is resonant with the incident radiation, is homogeneously broadened. Hence I_{sat} represents the saturation of that group, and when $I_\nu = I_{sat}$ the gain provided by this group decreases to half of its peak. Although the gain by this group is saturated, other groups in the wings of the spectral distribution of the irradiance can still contribute to the gain. Therefore, saturation of the gain of an inhomogeneously broadened medium is realized more slowly than in a homogeneously broadened medium. Furthermore, even after saturation is realized, it eliminates only one group of particles. Other particles that do not coincide with the passing beam can still produce gain when a beam at a different frequency is transmitted. This is illustrated in Figure 12.5, where the gain curves of homogeneously and inhomogeneously broadened media are presented. These gain curves illustrate the spectral variation of the available gain. Thus, for homogeneously broadened media (Figure 12.5(a)) the gain curve approximates

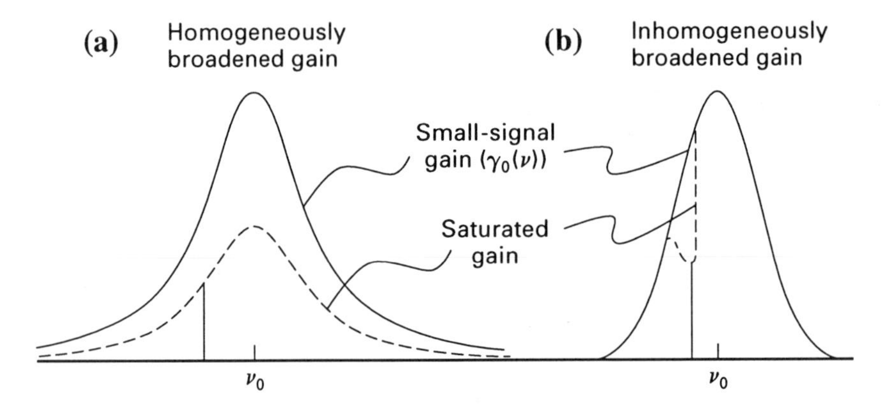

Figure 12.5 Gain curves of (**a**) homogeneously and (**b**) inhomogeneously broadened media, when I_ν is low (solid line) and when the gain is saturated (dashed line). The vertical solid lines at frequency ν mark the saturated gain available to a beam at that frequency.

the homogeneously broadened lineshape of the transition, whereas for inhomogeneously broadened media (Figure 12.5(b)) it approximates the convolution between the homogeneous and Doppler broadening lineshape factors. The solid lines in Figure 12.5 represent the unsaturated gain curves $\gamma_0(\nu)$ while the dashed line represents the gain curve when $I_\nu = I_{\text{sat}}$. Note that the gain curve of the homogeneously broadened medium is reduced uniformly until the gain is saturated at the incident beam frequency (marked by the vertical solid line at ν.) This saturation is achieved even though the frequency of the incident beam does not coincide with the transition line center. By contrast, the gain curve of the inhomogeneously broadened medium shows a "hole" at the frequency of the incident radiation while the remainder of the gain curve is unaffected. The hole represents the depletion of the group that is resonant with the incident beam. This selective depletion effect is known as *hole burning* (Bennett 1962). Since other groups may still provide gain, inhomogeneously broadened media can support several laser frequencies simultaneously.

12.5 The Laser Oscillator

The application of stimulated emission for optical gain and the development of lasers had to await the solution of two technical problems. The first obstacle was the identification of optical gain media and methods of pumping them. The second problem was the development of a device in which radiation could be successfully amplified by the gain medium and grow until emitted at a selected moment or power level. Although in hindsight the first hurdle seems more difficult, gain media along with compatible pumping techniques were identified long before the demonstration of the first laser device. Gain by stimulated emission was used for more than a decade in *masers* (microwave amplification by stimulated emission of radiation) before the first practical device

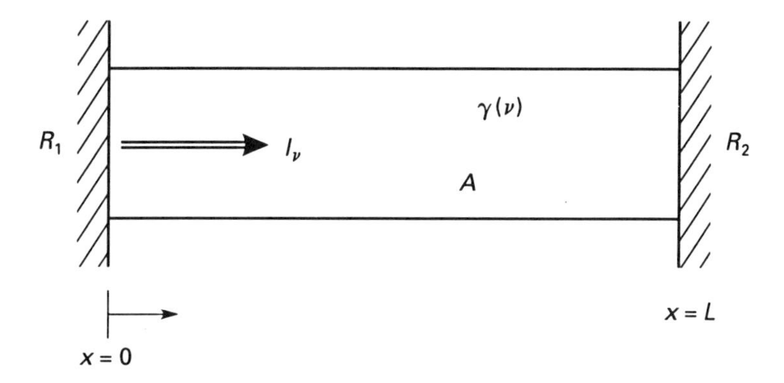

Figure 12.6 Illustration of a simplified laser cavity showing only the gain medium and two plane mirrors.

for an "optical maser" was proposed (Schawlow and Townes 1958). The breakthrough in the newly proposed device was a suggestion that gain media can be shaped as columns and enclosed between two parallel mirrors – facing each other – thereby forming a linear cavity (Figure 12.6). When an enclosed gain medium is excited and its inversion density exceeds a threshold level ΔN_t, randomly emitted photons inside the cavity begin to travel through the medium and experience amplification. However, those that propagate axially through the column have a longer propagation distance, and consequently larger amplification, than photons that select alternative paths. Therefore, these axially propagating photons act as the seed of a laser beam. When the nascent beam is incident upon the first mirror it is fed back into the gain medium for an additional gain and then fed back once again by the opposite mirror. If both mirrors are perfect reflectors, the irradiance can increase by repeatedly oscillating between the mirrors until the gain is reduced to its threshold (eqn. 12.16 or 12.18). From there on, oscillation between the mirrors can continue at steady state if pumping is maintained at a level sufficient to overcome internal losses. Of course, with a perfect reflector at each end there can be no output and such a device would be of little practical use. Therefore, one of the cavity end mirrors is replaced with a partially reflecting mirror. This mirror serves as the output coupler. It allows a predetermined portion of the incident energy to escape as a beam while reflecting back into the cavity the remainder, which must be sufficient to sustain steady oscillation. The other cavity end mirror continues to serve as the feedback mirror.

A critical parameter of the laser cavity that needs to be specified is the reflectivity of the output coupling mirror. This is the fraction of the incident energy that is reflected back toward the gain medium. On the one hand, with low reflectivity, a large portion of the intracavity energy can be emitted by the laser. On the other hand, the reflectivity must remain sufficiently high to permit the reflected energy to overcome internal losses such as scattering by impurities or reflections at interfaces. Optimally, the reflectivity of each mirror, R_1 and

R_2, must depend on the available gain $\gamma(\nu)$ and on the internal cavity losses. Assume that, in addition to the losses by absorption between levels 1 and 2 of the gain transition (Figure 12.4), there exist internal cavity losses that can be represented by an average or distributed loss factor A. After combining the cavity gain with the distributed loss, the net variation of the incident irradiance as it propagates through the gain medium is

$$\frac{dI_\nu}{dx} = I_\nu[\gamma(\nu) - A] \tag{12.19}$$

(Rigrod 1963); or, for a homogeneously broadened medium (eqn. 12.16),

$$\frac{dI_\nu}{dx} = I_\nu\left[\frac{\Delta N_0 \sigma_{SE}(\nu)}{1 + I_\nu/I_{\text{sat}}} - A\right]. \tag{12.20}$$

With the exception of the loss factor, (12.19) is an explicit presentation of (12.3). Thus, if propagation is arbitrarily set to begin at $x = 0$, gain for one intracavity round trip can be obtained by integrating (12.20) from $x = 0$ to $x = L$, multiplying the result by the reflectivity R_2 of the mirror at $x = L$, repeating the integration for the return path using the new irradiance, and multiplying the result by the reflectivity R_1 of the first mirror.

Equation (12.20) can be solved using the approximation $I_\nu \ll I_{\text{sat}}$, which simulates the onset of oscillation (Rigrod 1963). However, an alternative analysis of the intracavity electric field can provide not only the conditions for the onset of oscillation but also the requirements for stable oscillation. The two parallel cavity end mirrors form an etalon (Section 6.5), thereby restricting oscillation inside the cavity to stable etalon modes. To identify these modes, instead of the irradiance it is the intracavity propagation of the electric field that must be analyzed, starting at $x = 0$ where the amplitude is $\mathbf{E}_0$. The equation for the propagation of that electric field wave through a gain medium is identical to (4.36), where propagation through a lossy medium was described. The only exception is that the negative loss term is replaced by a positive gain term. Thus, after one round trip the electric field is

$$\mathbf{E}(2L) = \text{Re}[r_1 r_2 \mathbf{E}_0 e^{2ikL}] \exp\left\{\frac{\gamma_0(\nu) - A}{2} 2L\right\},$$

where $r_i = \sqrt{R_i}$ is the reflection constant of mirror i. Since the irradiance associated with this field (for $x = 0$) is $I_\nu \ll I_{\text{sat}}$, $\gamma(\nu)$ was replaced by the small-signal gain $\gamma_0(\nu)$ (eqn. 12.15), which is independent of the irradiance. This approximation may be acceptable even after few passes through the gain medium. Originally, $\gamma_0(\nu)$ and A were defined as the gain and loss factors of the irradiance. To describe the gain and loss of a propagating electric wave they must be divided by 2 (eqn. 4.42) before introducing them in the exponential term of the electric field. Owing to the high oscillation frequency of optical fields, the oscillatory term in (4.36) was omitted.

When $A > \gamma_0(\nu)$ (i.e., when the loss exceeds the gain), the initial field is attenuated after one round trip and oscillation is terminated. Conversely, when the gain exceeds all losses, the amplitude increases after each round trip and a laser beam can be formed. Therefore, for a steady oscillation, the electric field after one round trip must be restored to its initial value – both in amplitude and phase. Thus, the threshold condition for the onset of steady oscillation is:

$$r_1 r_2 e^{2ikL} e^{[\gamma_0(\nu)-A]L} = 1. \tag{12.21}$$

The first exponential term of (12.21) defines the condition that the phase of the electric field must meet for any stable oscillation:

$$e^{2ikL} = 1.$$

The second exponential term defines the minimum gain requirement for the onset of steady oscillation:

$$r_1 r_2 e^{[\gamma_0(\nu)-A]L} = 1.$$

Although both conditions must be met simultaneously, they can be analyzed independently. Thus, from the second condition we derive the small-signal gain – the gain required to start an oscillation in a cavity with end mirrors having reflectivity R_1 and R_2 – as

$$\gamma_0(\nu) = A - \frac{1}{2L} \ln R_1 R_2. \tag{12.22}$$

Alternatively, the threshold inversion density ΔN_t can be defined by combining (12.15) and (12.6) with (12.22):

$$\Delta N_t = \frac{8\pi n^2}{A_{21}\lambda^2 g(\nu)}\left(A - \frac{1}{2L} \ln R_1 R_2\right). \tag{12.23}$$

Although all losses in the laser gain medium – as well as reflection losses and output coupling requirements – are met by this inversion density, this threshold is sufficient only for starting the oscillation. As the intracavity irradiance grows, depletion of the excited state by stimulated emission increases and the inversion may fall below its threshold level. Therefore, to produce net gain in the presence of a nonvanishing irradiance, the inversion must be increased by raising the pumping rate so that $\Delta N_0 > \Delta N_t$. With larger inversion, the irradiance circulating inside the cavity can increase until the net gain as described by (12.16) or (12.18) offsets the internal cavity losses. If pumping continues at the same rate, the circulating irradiance approaches a steady state. This steady-state irradiance I_{SS} can be obtained from (12.20) for a cavity with both R_1 and $R_2 \approx 1$ and an inversion ΔN_0 by assigning $dI_\nu/dx = 0$:

$$I_{SS} = I_{\text{sat}}\left(\frac{\Delta N_0 \sigma_{SEw}(\nu)}{A} - 1\right) = I_{\text{sat}}\left(\frac{\gamma_0(\nu)}{A} - 1\right). \tag{12.24}$$

Clearly, I_{SS} increases linearly with the small-signal gain but only when $\gamma_0(\nu) > A$. Otherwise, the cavity is lossy and cannot support a laser beam. Although $\gamma_0(\nu)$ must exceed the distributed losses for oscillation to begin, as $I_\nu \to I_{SS}$ the gain declines owing to depletion of the inversion; at steady state, $\gamma(\nu) = A$ (eqn. 12.19); that is, the saturated gain exactly matches the loss.

The other condition for stable oscillation requires that the initial phase of the electric field be restored after each round trip between the cavity mirrors. This is implied by the first exponential term of (12.21), which must be unity. This condition is met if

$$\begin{aligned} 2kL &= 2m\pi \quad \text{for } m = 1, 2, 3, \ldots \quad \text{or} \\ 2nL &= m\lambda_m. \end{aligned} \tag{12.25}$$

Equation (12.25) simply states that for stable oscillation, the integral multiple of a sustainable wavelength λ_m must equal the optical path of an intracavity round trip. Therefore, it also suggests that the phase of the electric wave after the mth round trip is restored to its initial value. Incidentally, (12.22) is also the condition that reinforces transmission by an etalon of a normally incident beam (cf. eqns. 6.18 and 6.21). However, unlike an etalon, most optical laser cavities are long relative to the wavelength they support. (A new technology of quantum-well lasers, where subwavelength cavities are used, is being developed but will not be discussed here.) Therefore, unless the bandwidth of the gain curve is exceptionally narrow, most laser cavities can support stable oscillation of more than just one discrete wavelength. Each of these wavelengths λ_m belongs to a separate wavetrain, or a *longitudinal mode,* that complies with (12.25). The free spectral range of a laser cavity – that is, the frequency difference between two adjacent longitudinal modes λ_m and λ_{m+1} – is identical to the free spectral range of an etalon at normal incidence (eqn. 6.21). This free spectral range is

$$\Delta\bar{\nu} = \frac{1}{\lambda_{m+1}} - \frac{1}{\lambda_m} = \frac{1}{2nL}\ \text{cm}^{-1}. \tag{12.26}$$

Example 12.2 The length of a He–Ne laser in a supermarket checkout scanner is 24.1 cm. The bandwidth of its gain curve is approximately 1.5 GHz. Find the free spectral range of the laser and the number of longitudinal modes it can possibly support.

Solution He–Ne lasers consist of an electric discharge tube containing a gas mixture of He and Ne. The end mirrors of the laser are fused directly to the ends of that tube. Therefore, the intracavity index of refraction depends only on the gas mixture and can be approximated as $n \approx 1$. The free spectral range of the cavity is obtained using (12.26) as follows:

$$\Delta\nu = \frac{c}{2L} = \frac{3 \times 10^{10}}{2 \times 24.1} = 622\ \text{MHz},$$

where multiplication by c was required to convert the frequency units to hertz. This free spectral range is approximately one third of the width of the gain profile, suggesting that the number of longitudinal modes that can be supported is three or four. The actual number depends on the position of the modes relative to the center of the gain curve. If one of the modes is near the edge of the curve, the gain there may not be sufficient to overcome the losses and that mode may not be sustained. ■

Although the maximum number of longitudinal modes in a laser cavity depends on its length, the actual number of realized modes depends on the nature of the broadening mechanism. When the broadening is primarily homogeneous, each member of the excited population can interact with any sustainable longitudinal mode. Therefore, the allowed longitudinal mode nearest to the peak of the gain curve experiences the largest gain and is hence more likely than the others to stimulate emission. Each stimulated emission event that contributes energy to that preferred mode deprives the other modes of their gain and so weakens them even further until eventually they fall below the loss level, leaving only the preferred mode to oscillate. This is illustrated in Figure 12.7(a), where a homogeneously broadened small-signal gain profile is represented by the solid line and the distributed loss A by a horizontal line. Four longitudinal modes that could potentially be sustained by the cavity are marked by vertical lines. The small-signal gain of at least two of these modes exceeds the distributed loss. When oscillation begins and the irradiance of any sustainable mode increases, the total inversion density declines. It is typical of homogeneous broadening that this decline affects the entire gain curve; stimulated emission by one excited particle – at any wavelength – reduces the gain everywhere on the curve. When the irradiance of the preferred mode (marked by an arrow) approaches I_{SS}, the gain curve at that frequency exactly matches the loss, thereby offering no net gain. Meanwhile, the gain of all other modes falls below that line. Therefore, as all the other modes are extinguished, the preferred mode can continue to oscillate (although without further increase in its irradiance) while emitting radiation through the output coupler. Such processes, where the mode with the highest gain is naturally selected for oscillation, is by evolutionary analog often dubbed "survival of the fittest" (Herbert Spencer).

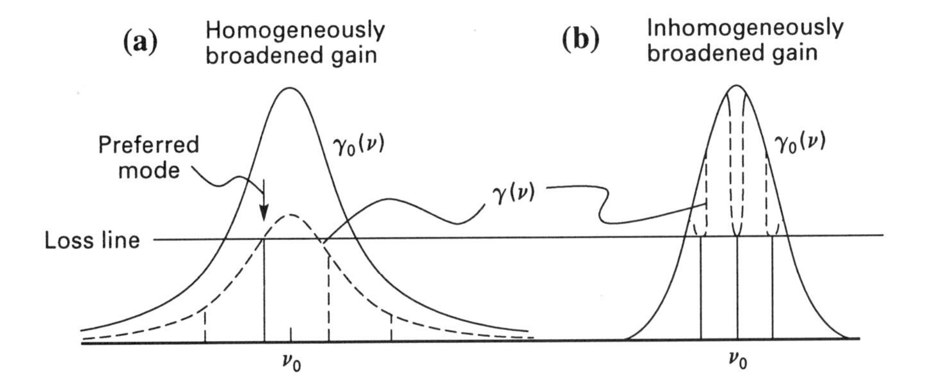

Figure 12.7 Gain curves of (**a**) homogeneously and (**b**) inhomogeneously broadened media in a cavity that can support several longitudinal modes, when I is low (solid line) and when the gain is saturated (dashed line).

By contrast, inhomogeneously broadened media include numerous independent groups of homogeneously broadened particles, each group moving at a different velocity. Because of the Doppler shift, the frequency of some of the allowed modes appears to observers moving with these particles to be resonant with their own emission frequency. The gain of each of these groups is set initially by their individual inversion density. The small-signal gain curve, marked by the solid line in Figure 12.7(b), follows the statistical velocity distribution of the particles. Modes where the small-signal gain is large are simply coincident with a group having a large membership of inverted population. As the irradiance of any mode with positive net gain increases, it depletes the inversion density only of the group that coincides with its frequency. Therefore, unlike homogeneously broadened gain media, inhomogeneously broadened media sustain numerous modes. Each longitudinal mode can reach its own steady-state irradiance specifid by $\gamma_0(\nu)$ (eqn. 12.24) while the gain curve experiences hole burning. Note that each longitudinal mode consists of two counterpropagating beams - each resonant with a different velocity group in the gain medium. If the beam propagating to the left in the cavity of Figure 12.6 is resonant with particles moving in its direction, then the counterpropagating beam must be resonant with particles moving to the right. Therefore, in a linear laser cavity holes are "burned" in pairs. This is illustrated in Figure 12.7(b) by the pair of holes burned symmetrically relative to the centerline frequency, while the mode at ν_0 burns only one hole at ν_0.

Throughout the preceding analysis of longitudinal modes it was assumed that the optical length of the laser cavity is fixed. However, owing to random changes in the index of refraction that may be induced by temperature fluctuations inside the cavity or due to mechanical vibrations, the effective length of the cavity changes randomly, thereby allowing the development of other longitudinal modes. Thus, from time to time the wavelengths supported by the cavity can hop from one mode to another. Owing to such *mode hopping,* the single mode of a homogeneously broadened laser or the multitude of modes of inhomogeneously broadened lasers can rarely be resolved, and the time-averaged spectral distribution of most laser beams includes a significantly larger number of modes. These modes, even if unresolved, are nevertheless essential for the formation of subnanosecond laser pulses (Section 12.8).

12.6 *Q*-Switching

One of the unique advantages of lasers, when compared to incoherent light sources, is the potential of controlling the temporal characteristics of the beam. Laser beams can be compressed into pulses of selected duration, and the time of emission can be synchronized with various events. Intracavity devices permit reduction of the pulse duration of certain laser systems to 10^{-8} s while forcing most of the available energy into the pulse. Other devices allow even further reduction of the pulse duration, to the picosecond [ps] or even the femtosecond [fs] range. When combined with their well-defined spatial and spectral distribution, energy delivery by lasers can be extremely precise, thereby allowing unique applications where resolution or synchronization with transient events is necessary. By contrast, the temporal control of incoherent radiation is limited to the pulsation of the source or to intermittent blocking, or "chopping," of the beam. Pulsed incoherent sources usually consist of flashlamps (e.g., the lamps used for photography). These lamps are excited by an electric discharge and the pulse duration and its energy are limited by the parameters of the driving circuitry. Normally, the discharge energy is stored in capacitors, which limit the response time of the entire system, particularly when the capacitance is as large as is required for a large discharge energy. Fast stroboscopic sources may be pulsed at a rate of 100,000 Hz with pulse durations well in excess of 1 μs. Furthermore, the uniformly scattered illumination precludes the possibility of beam shaping or tight focusing. Temporal control of incoherent emission can also be achieved by chopping the beam with a fast-spinning slotted wheel placed between the source and the target. The pulse width is limited by the spinning rate and by the geometrical parameters of the chopping wheel and the beam. Typical chopping rates do not exceed 4,000 Hz. Furthermore, synchronization of the illumination with desired events may not be possible. Two techniques for laser pulse shaping and their technical potential are discussed in this and the following sections.

In one method for the compression of laser beam energy into brief pulses, the onset of stimulated emission is delayed by an intracavity shutter. The shutter consists of an optical medium that can be made temporarily lossy. While the loss exceeds the gain, a laser beam cannot develop in the cavity and the depletion by stimulated emission of the inversion density is negligible. Therefore, as long as the shutter is lossy, $W_{21} \to 0$ and the population of the excited laser state is controlled only by the pumping, at a rate P_2, of the upper laser emission state and by spontaneous emission with a time constant τ_2 from that state (eqn. 12.11b). When the shutter is suddenly opened, the threshold for laser oscillation is reduced and the net gain increases abruptly. That massive gain is the result of the large inversion that was allowed to develop without competition by stimulated emission. Therefore, the irradiance can grow rapidly to a level that is not otherwise possible. However, owing to the rapidly growing signal, saturation is achieved after few oscillations in the cavity, the gain is reduced below the cavity losses, and the pulsed emission is terminated. The cavity quality factor Q is defined (Boyd and Gordon 1961) as

$$Q = \omega \frac{\text{energy stored}}{\text{energy lost per second}}.$$

We can see that the shutter increases the cavity Q factor, and this laser pulsing technique is accordingly called *Q-switching*.

Two parameters are critical for successful Q-switching: rapid switching of the shutter and relatively long lifetime of the excited state. Rapid switching can be obtained electro-optically. Figure 12.8 illustrates a Q-switched laser cavity. The shutter consists of the combination of a Pockels cell and a polarizer

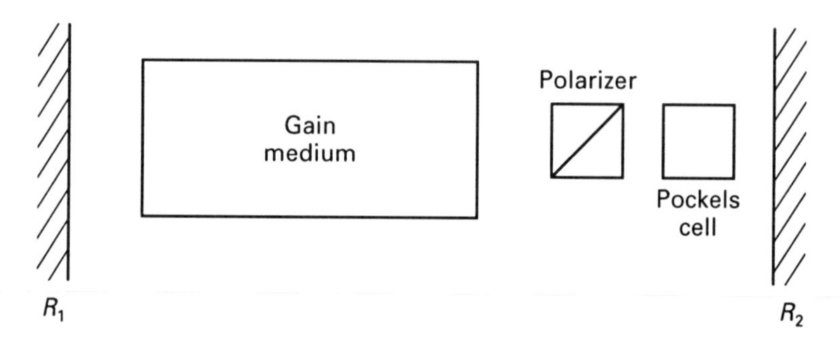

Figure 12.8 Simplified Q-switched laser cavity, showing the cavity end mirrors, gain medium, and optical switch consisting of a polarizer and Pockels cell.

(Section 5.4). When the Pockels cell acts as a quarter-wave plate, the polarization of the beam returning from the feedback mirror at the right is rotated by 90° relative to the polarization of the beam entering the cell from the left. Thus, if the polarizer is set to transmit the beam entering it from the left, it blocks the beam that returns from the right and so prevents the oscillation. When the cell is switched to a half-wave mode, or alternatively to a nonbirefringent mode, the polarizations of both the incident and returning beams are identical, and oscillation in the cavity is unimpeded. Rapid switching (of ~1 ns) is possible with this cell. Using such opto-electric switching permits synchronization of the emission with desired events.

The other parameter that is critical for successful Q-switching is the lifetime of the excited state. Ideally, if that lifetime is infinitely long, the entire population of the excited state can be preserved and increased indefinitely as long as cavity oscillation is blocked. Of course, the lifetime in existing lasers is finite, so when oscillation is blocked by the shutter for a duration exceeding τ_2, losses by spontaneous emission begin to compete effectively with the pumping of the excited state (see Problem 12.8). Therefore, although any laser system can be electro-optically switched, only gain media with a long lifetime (typically $\tau_2 > 100\ \mu s$) can produce high-energy Q-switched pulses. In those systems, the cavity can remain blocked for durations that are comparable to τ_2 while pumping continues, thereby allowing the build-up of a large inversion.

To determine the duration of a Q-switched laser pulse, consider the laser cavity as a container of optical energy with an initial energy density of ρ_0. After the shutter is opened, a laser beam develops rapidly and the density of the stored energy decreases as the beam is emitted through the output coupling mirror. Even if pumping continues, its effect on the stored energy is negligible relative to the high rate of laser energy output. Therefore, the density of the stored energy after the switch is opened decays at a rate that is both independent of external effects and proportional to the momentary density ρ. With such simplifying assumptions, the differential equation describing this process is

$$\frac{d\rho}{dt} = -\frac{\rho}{\tau_m}, \tag{12.27}$$

where τ_m is the *cavity time constant* – a property of the laser cavity. Therefore, if the energy density at the time the shutter is opened is ρ_0, the stored energy is subsequently

$$\rho = \rho_0 e^{-t/\tau_m}. \tag{12.28}$$

To evaluate τ_m, assume that the only energy loss mechanism is the output by the cavity end mirrors. Therefore, at a time $T_C = 2L/c$ (i.e., after one round trip in a cavity of length L), the energy density is

$$\frac{\rho}{\rho_0} = \exp\left\{\frac{-T_C}{\tau_m}\right\} = R_1 R_2. \tag{12.29}$$

Using the approximation of $-\ln x \approx 1-x$, the cavity time constant can be readily obtained for (12.29) as

$$\tau_m = -\frac{T_C}{2\ln\sqrt{R_1 R_2}} \approx \frac{T_C}{2(1-R)}, \tag{12.30}$$

where $R = \sqrt{R_1 R_2}$ is the geometric mean of the reflectivities of the cavity end mirror. The cavity time constant can be regarded as the lifetime of a photon in the cavity. By analogy to the spontaneous lifetime (eqn. 10.4), it can be used to estimate the duration of a Q-switched pulse (Wagner and Lengyel 1963). For a 30-cm–long cavity with $R = 0.9$, the pulse duration is approximately 10 ns. This is compatible with the duration of Q-switched pulses in commercially available Nd:YAG lasers.

To estimate the energy of a Q-switched pulse, assume again that – after the shutter is opened – the entire energy stored in the gain medium is dumped into the beam with no losses, but also without energy addition that may accrue by continued pumping. Although the entire inversion density $\Delta N = N_2 - N_1$ at the time the switch is opened is available for gain, only half of its stored energy can be added to the beam. During the short build-up time of the laser pulse, N_1 is controlled only by the stimulated emission, where every transition from level 2 reduces N_2 while increasing N_1. When half of the initial inversion density is removed from the upper state and added to the lower state, the inversion is terminated and the gain disappears. The energy added to the beam may be further limited by the mismatch between the cross section of the laser beam and the cross section of the gain volume; only the inversion within the volume V that is swept by the laser beam can contribute to its energy. With these restrictions, the energy available for the laser pulse is

$$E = \tfrac{1}{2}\Delta N V h\nu \tag{12.31}$$

(Wagner and Lengyel 1963). The average power of the laser pulse is readily obtained by dividing the pulse energy (eqn. 12.31) by its approximate duration (eqn. 12.30):

$$P = \frac{\Delta N V h\nu}{2\tau_m}. \tag{12.32}$$

Q-switching is normally used in those lasers where the excitation is already pulsed. In "free run" the duration of the laser pulse may extend to 100 μs, whereas Q-switching can compress it to 10 ns. The so-called insertion losses of the Pockels cell combined with the losses of the switching process are less than 50%. Therefore, the peak power of the Q-switched pulse is approximately 10^4 times higher than the power of the free-run laser. Of course Q-switching is also possible in continuously run lasers, where the shutter must remain closed between successive Q-switched pulses for a duration that is sufficient for the recovery of the gain to its unsaturated level.

The need for a long spontaneous lifetime in the excited state of the gain medium limits the selection of gain media that are amenable to Q-switching. Two solid-state lasers, the Nd:YAG and the ruby, can be Q-switched by commercially available devices. The first laser, emitting at $\lambda = 1{,}064$ nm, consists of Nd^{3+} ions trapped in a host crystal of yttrium aluminum garnet (YAG). The second laser, emitting at $\lambda = 694.3$ nm, consists of Cr^{3+} ions in Al_2O_3. Both are typically pumped by a flashlamp. Pulses of 1 J and durations of 10 ns can be obtained from commercially available Q-switched Nd:YAG lasers. Certain gas-phase lasers such as the CO_2 laser, which emits at $\lambda = 10.6$ μm, can also be Q-switched. For more discussion of these and other Q-switched lasers, see Hecht (1986). The following example illustrates a few parameters of a Q-switched laser.

Example 12.3 The density of the ions in the crystal of a flashlamp-pumped ruby laser is 1.62×10^{19} cm^{-3}. The crystal is shaped into a rod with a diameter of 5 mm and an optical length of 30 cm. The flashlamp that is used to pump the crystal induces an inversion that constitutes 15% of the ion density. The geometric mean of the reflectivity of the cavity end mirror is 0.9. The nominal diameter of the beam propagating through the gain medium is 1 mm and its wavelength is 694.3 nm. Find the average power of the Q-switched pulse.

Solution The pulse duration of the Q-switched laser can be estimated using (12.30) as follows:

$$\tau_m = \frac{L/c}{1-R} = \frac{30/(3 \times 10^{10})}{0.1} = 10^{-8} \text{ s}.$$

The transit time between the end mirrors, $T_C = 2L = c$, was calculated using the speed of light in free space because the optical length of the crystal, as specified in the problem, already includes the effect of its index of refraction. The average pulse power is obtained from (12.32) as

$$P = \frac{\Delta N V h\nu}{2\tau_m}$$

$$= \frac{0.15(1.62 \times 10^{19})(0.1)(30)(6.63 \times 10^{-34})(4.32 \times 10^{14})}{2 \times 10^{-8}} = 105 \text{ MW}.$$

The energy of this pulse is approximately 1 J. Note that the power of the beam is determined by its diameter and not by the volume of the gain medium. Thus, energy stored outside the beam is lost. Unique designs of laser cavities (e.g., unstable oscillators; see Section 13.4) force the beam to fill the entire gain medium and so capture most of this energy. The increase in available power varies quadratically with the beam diameter (or linearly with its cross-section area). For the parameters of this example, such design could increase the available energy and power by a factor of 25. ■

12.7 *Q*-Switched Lasers for Material Processing

One of the earliest applications realized by lasers was the processing of materials, semiconductors, plastics and even human tissue, where the laser beam is used as a well-controlled heat source. Because of its defined shape, heating by a laser beam can be localized, thus confining its effects to a small region. Furthermore, by controlling the exposure duration (e.g. by *Q*-switching), relatively slow effects such as conduction or convection of heat may become negligible during the exposure time. With sufficient heating, a thin surface layer (<0.1 mm) may be melted or its crystalline structure modified while the material underneath it remains at its initial temperature. When illumination is completed, heat from the surface layer is rapidly convected by the steep temperature gradient induced by the laser, thereby quenching the surface and freezing its modified structure. With such excellent control, the surface of metals can be hardened (Duley 1983) or annealed. Numerous techniques are available for metal surface processing. Such parameters as the incident wavelength, beam power, exposure duration, pre- and post-exposure cooling, the chemistry of the atmosphere enclosing the processed medium, and so on can be defined to achieve various desirable effects (see e.g. Bass 1983). Alternatively, with sufficient laser energy, material can be removed for cutting or engraving (Bass 1983, chap. 2). At lower incident energy only melting is possible, whereby welding is achieved (Bass 1983, chap. 3).

Detailed discussion of these techniques is beyond the scope of this book. However, to illustrate the effect of rapid heating by a laser pulse on the temperature distribution below the surface of a metal, consider the problem of one-dimensional (1-D) transient conduction. Figure 12.9 illustrates the surface of a metal piece processed by a pulsed laser beam. The entire medium is initially at temperature T_0. However, immediately after the impact of a laser beam, the temperature at the surface rises to T_s. Detailed solution of the effect of the laser beam on the surface must include the spatial and temporal energy distribution of the beam, heat losses at the surface by conduction and radiation, variation of these parameters with temperature, and more (Armon et al. 1989). However, at the center of the illuminated spot, transverse conduction is negligible and the conduction of heat from the surface can be approximated by the following 1-D heat conduction equation (Mills 1992, p. 127):

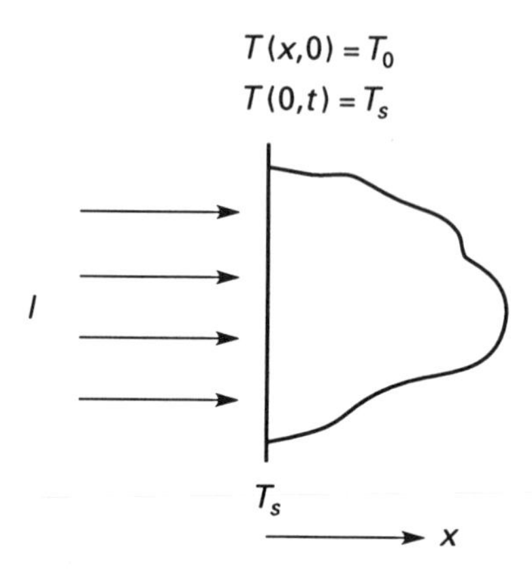

Figure 12.9 Surface of a workpiece processed by a pulsed laser beam showing the heat-affected zone beneath the surface.

$$\frac{\partial T}{\partial t} = \alpha \frac{\partial^2 T}{\partial x^2}, \tag{12.33}$$

where α denotes the thermal diffusivity. The initial and boundary conditions are assumed to be:

$$T(x, 0) = T_0,$$
$$T(0, t) = T_s,$$
$$T(\infty, t) = T_0.$$

The second condition (i.e., that the surface temperature T_s remain unchanged) may not always be valid. However, during melting or evaporation phases of the process, this significant simplification may be justified. With the solution of this 1-D problem, fundamental aspects of this laser–metal interaction – such as the characteristic time for conduction and the requirements for exposure duration – can be easily identified. The solution of (12.33) is readily obtained (Incropera and DeWitt 1990, p. 260) as

$$\frac{T - T_0}{T_s - T_0} = \operatorname{erfc} \frac{x}{(4\alpha t)^{1/2}}, \tag{12.34}$$

where

$$\operatorname{erfc}(\eta) = 1 - \frac{2}{\pi^{1/2}} \int_0^{\eta} e^{-u^2} du$$

is the *conjugate error function*. Numerical values of the conjugate error function are tabulated in texts on heat transfer (see e.g. Incropera and DeWitt 1990, p. B3). Analysis of this solution can be simplified by introducing a characteristic time constant

$$\tau_c = \frac{L^2}{\alpha}, \tag{12.35}$$

where L is a characteristic dimension, for example, the depth of the layer that is affected by the heat conducted from the surface. With this characteristic time for conduction, (12.34) can be represented in a dimensionless form:

$$\frac{T-T_0}{T_s-T_0} = \operatorname{erfc}\left(\frac{\tau_c}{4t}\right)^{1/2}. \tag{12.34'}$$

At $t = \tau_c$, the dimensionless temperature at $x = L$ is approximately $(T-T_0)/(T_s-T_0) = 0.5$. At twice the distance from the surface but still at $t = \tau_c$, the dimensionless temperature is only 0.16. Thus, if significant temperature rise is to be limited to a layer of thickness L, the exposure duration must be kept at $t < \tau_c$. Even after the heating at the surface is terminated, temperature beneath the surface may increase. Nevertheless, this result can be useful for estimating the upper limit of the exposure duration.

Example 12.4 Find the pulse duration required for surface treatment of 2024-T6 aluminum alloy when the thickness of the treated layer is to be limited to 0.01 mm.

Solution The thermal diffusivity of aluminum is $\alpha = 7.3 \times 10^{-5}$ m^2/s (Incropera and DeWitt 1990, p. A3). Therefore, the characteristic time for conduction (eqn. 12.35) of heat to a depth of $L = 10^{-5}$ is:

$$\tau_c = \frac{L^2}{\alpha} = \frac{10^{-10}}{7.3 \times 10^{-5}} = 1.4\ \mu\text{s}.$$

This duration is 100 times longer than the duration of a typical Q-switched laser pulse. However, application of a Q-switched pulse may be useful. Most metals, including aluminum, are highly reflective when cold. As their temperature rises, their reflectivity declines (Duley 1983, pp. 75, 79) and the coupling of energy between the laser and the workpiece improves. By application of Q-switched illumination, energy dissipation is initially limited to a layer that is much thinner than the projected depth of 0.01 mm. Therefore, even though aluminum is highly reflective, the little energy initially deposited in the upper layer of the aluminum workpiece can raise its temperature to the point where its reflectivity declines. Consequently, the temperature of surfaces illuminated by Q-switched pulses is expected to be higher than the temperature achieved by the same pulse energy spread over 1 μs. When the irradiation is complete, the energy deposited at the thin surface layer is dissipated by conduction and the material underneath is heated. Depending on the extent of deposited energy, metallurgical transformation is possible in subsurface layers. By carefully monitoring the energy dosage, the processed depth and its width can be controlled. ■

Q-switched lasers are ideally suited for high-power delivery of energy in a time that is shorter than τ_c. Applications of Q-switched lasers for material processing include ablation of thin layers, drilling of narrow holes, surface glazing, annealing, and more (see Example 13.1). Q-switched lasers are also required for processing certain metals (e.g. aluminum) that are highly reflecting at low temperatures but are strong absorbers when hot. Tightly focused Q-switched pulses can avoid competition by heat conduction that may otherwise combine with the high reflectivity to further depress energy deposition. Each short pulse can establish a highly absorbing melted point or *keyhole,* which can then absorb the energy of subsequent pulses - even if non-Q-switched - thereby permitting the cutting or treatment process to continue. Therefore, such metals are processed either by a series of Q-switched pulses or by alternating sequences of a keyhole-forming Q-switched pulse and a longer (~1-μs) highly energetic pulse.

12.8 Mode-Locking

Q-switching of lasers is an excellent method of generating high-power pulses with a duration of approximately 10 ns. The duration of the pulse is ultimately determined by the cavity time constant τ_m (eqn. 12.30), and unless the cavity is made unpractically short this duration cannot be reduced significantly. For applications where even shorter transients are to be probed, pulses that are much less than 1 ns can be formed by forcing the longitudinal modes inside a laser cavity to interfere with each other. Since each longitudinal mode represents a standing wave oscillating at its own frequency (Problem 12.9), constructive interference can occur only at prescribed times when all the modes are at the same phase, thereby resulting in a laser pulse with an amplitude that exceeds the amplitude of any individual mode. At other times, the modes interfere with each other destructively, thereby resulting in almost no emission. For each of the longitudinal modes, the time for one round trip in the cavity is $T_C = 2L/c$. Consequently, any event that requires simultaneous participation of the phases of these modes must be periodic with a period of T_C; once conditions for constructive interference are established, new pulses are emitted at regular intervals of T_C. These pulses are much shorter than Q-switched pulses, with durations that can fall below 100 fs and repetition rates ranging from 0.1 GHz to tens of gigahertz. The technique of coupling the longitudinal modes together with the objective of generating a train of short pulses is called *mode-locking*. Before discussing the methods of mode-locking we should consider the interaction between longitudinal modes and the requirements for their interference.

Figure 12.10 illustrates the gain curve of an inhomogeneously broadened laser with a bandwidth of $\Delta\omega$ centered at ω_0 rad/s. The longitudinal modes supported by the laser are marked by vertical lines in the figure and are evenly separated by a frequency (cf. eqn. 12.26) of

$$\omega_c = 2\pi/T_C, \tag{12.36}$$

Assuming that the laser operation has already reached steady state, the temporal component of the electric field of the nth longitudinal mode is:

$$\mathbf{E}_n = \mathbf{E}_0 \exp\{-i(\omega_0 + n\omega_c)t + \phi_n\} \quad \text{for } n = 0, \pm 1, \pm 2, \ldots, \tag{12.37}$$

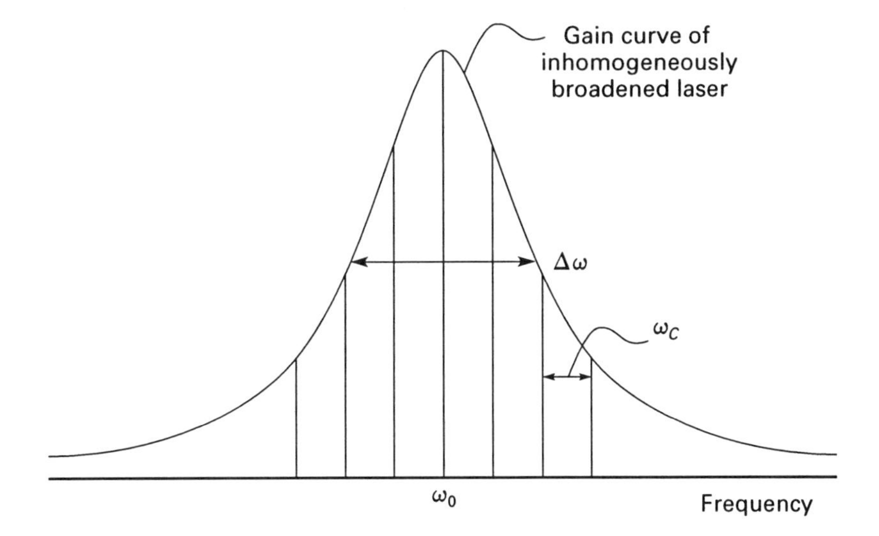

Figure 12.10 Gain curve of an inhomogeneously broadened laser with a bandwidth of $\Delta\omega$ centered at ω_0 rad/s. The longitudinal modes supported by the laser are marked by vertical lines.

where $n = 0$ corresponds to the mode at the line center of the gain curve. To simplify the analysis, all modes were assumed to have the same amplitude $\mathbf{E}_0$. When the laser oscillates naturally, the phase ϕ_n of each mode is a random parameter; the phases of the various modes are unrelated, and when the phase of one mode changes it does so without coordination with the other modes of the cavity. Thus, even if every mode is coherent (i.e., its own phase changes infrequently), it may interfere only with itself to form a standing wave and cannot interfere with other modes. To induce such multimode interference, the phases of all the modes must be locked together. That is, at a set time $t = 0$, the phases of all the modes are forced to the same value, for example $\phi_n = 0$. (The choice of $t = 0$ and $\phi_n = 0$ is arbitrary but simplifies the subsequent analysis.)

Even before introducing a technique of "clamping" the phases of the longitudinal modes together, it is possible to contemplate its results using phasor diagrams (DeMaria 1968). In a phasor diagram (Figure 12.11), each cavity mode is represented by an arrow with a length of unity - a *phasor* - attached to an origin. As the phase of the mode evolves, the phasor representing it spins in the plane of the diagram with the origin as its axis. The angle θ_n measured from a reference line is the phase angle at time t:

$$\theta_n = \omega_n t + \phi_n. \tag{12.38}$$

These phasors can be viewed as vectors in the time domain. Assuming that at $t = 0$ all initial phases are indeed $\phi_n = 0$, the phasors of all N cavity modes point at that time in the same direction (Figure 12.11(a)) and can therefore be added algebraically to form a resultant with an amplitude that is N times the amplitude of an individual mode. Because the frequency of the nth mode increases

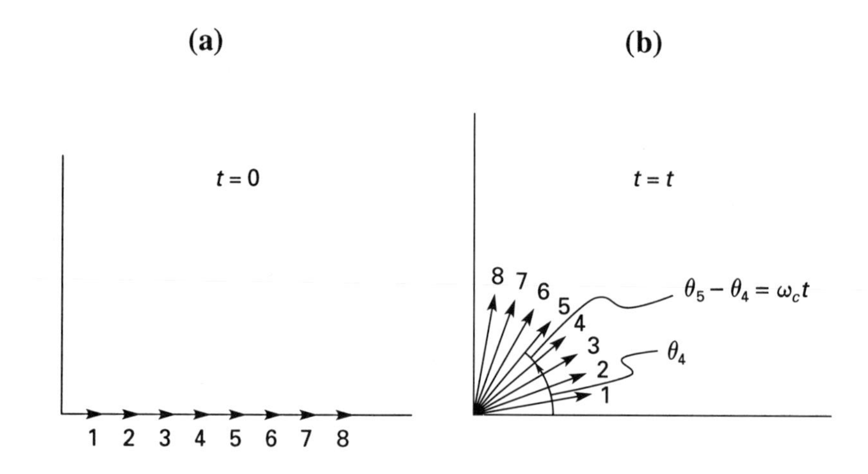

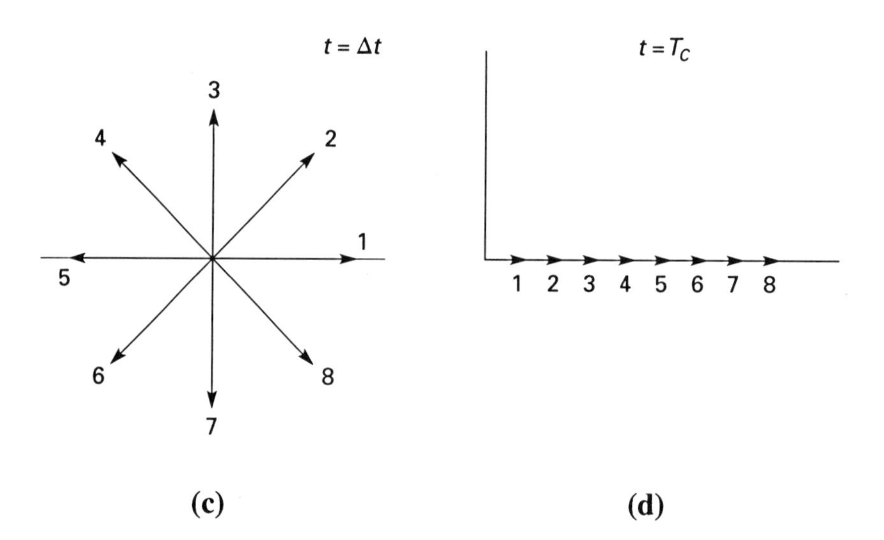

Figure 12.11 Phases of the cavity-supported longitudinal modes at the following times: (**a**) $t=0$; (**b**) t; (**c**) $t=2\pi/N\omega_c$; (**d**) $t=T_C$. The four parts of this phasor diagram are not drawn to the same scale (each arrow has length of unity).

by an increment of ω_c relative to the $(n-1)$th mode, at a time t, shortly after the modes start their spin, the phasors fan out as shown in Figure 12.11(b) with an angle between any two adjacent modes of:

$$\theta_n-\theta_{n-1}=\omega_c t. \tag{12.39}$$

As the angle between the phasors increases, the amplitude of their resultant decreases until a time Δt (Figure 12.11(c)) when each phasor is aligned with another phasor that points in exactly the opposite direction, thereby cancelling each other out. The total electric field at that moment vanishes by destructive interference. From simple geometrical observation it is evident that, at that time, the phase angle between two adjacent modes is

$$\theta_n - \theta_{n-1} = 2\pi/N. \tag{12.40}$$

Therefore, with the help of (12.39), it can be shown that

$$\Delta t = \frac{2\pi}{N\omega_c}. \tag{12.41}$$

As the angle between each phasor continues to increase, their combined electric field may increase slightly and then decline again several times until the angle between two adjacent modes is $\theta_n - \theta_{n-1} = 2\pi$. At that time, once again, they all line up along the reference line of the diagram and their amplitudes can be added algebraically. Again with the help of (12.39), the time between two successive such events is found to be

$$\omega_c t = 2\pi$$

or (cf. eqn. 12.36)

$$t = \frac{2\pi}{\omega_c} = T_C. \tag{12.42}$$

Thus, as anticipated by the earlier heuristic arguments, the time between pulses (the period) is T_C.

The peak power of each of the pulses increases quadratically with the field. Since the field of N modes – when constructively interfering with each other – is N times larger than the field of an individual mode, the power of each pulse is N^2 higher than the power of a single mode and is N times higher than the power of all modes when incoherently combined.

The primary objective of the mode-locking technique is to achieve a pulse with the shortest possible duration. To identify the parameters that determine the pulse length, assume that the duration is measured from the time of its peak to the time Δt when the field vanishes the first time. Inspecting (12.41), it is evident that Δt decreases as the number N of longitudinal modes increases. On the other hand, the number of modes of an inhomogeneously broadened laser is limited by the bandwidth of the gain curve:

$$N = \Delta\omega/\omega_c.$$

By replacing N in (12.41) with this ratio, it can be seen that the pulse duration obtained by mode-locking is limited only by the bandwidth of the gain curve:

$$\Delta t \Delta \nu \approx 1. \tag{12.43}$$

This result, not surprisingly, is compatible with the uncertainty principle and with the Fourier transform limit (Example 8.3).

Similar results can be obtained somewhat more rigorously by combining the fields of all the modes (Yariv 1985, p. 167). For N modes arranged symmetrically around the center frequency ω_0 (Figure 12.10), the total field is

$$\mathbf{E} = \mathbf{E}_0 e^{-i\omega_0 t} \sum_{-(N-1)/2}^{(N-1)/2} e^{-in\omega_c t}. \tag{12.44}$$

Implicit in (12.44) is the assumption that the initial phases of all the modes are zero and their polarizations are identical. Equation (12.44) is a geometrical series that can readily be summed to yield the following expression for the total field:

$$\mathbf{E}(t) = \mathbf{E}_0 e^{-i\omega_0 t} \frac{\sin(N\omega_c t/2)}{\sin(\omega_c t/2)}. \tag{12.45}$$

This result represents a pulse that peaks at $t = 0$ and then again at periods of $T_C = 2\pi/\omega_c$. This is consistent with (12.42), which was derived using a phasor diagram. As before (eqn. 12.41), the pulse duration Δt is defined as the time from the peak to the time of the first node, that is, when:

$$\frac{N\omega_c \Delta t}{2} = \pi. \tag{12.46}$$

Finally, with the implementation of (4.42), the pulse power is:

$$P(t) = I_0 \frac{\sin^2(N\omega_c t)/2}{\sin^2(\omega_c t)/2}, \tag{12.47}$$

where I_0 is the average power of a single mode. The peak power of the pulse, reached at $t = 0$, is N^2 times larger than the average power of a single mode or N times larger than the average power of the unlocked laser.

These results demonstrate that once mode-locking is achieved, the emission consists of a train of pulses separated by T_C, the time for one round trip between the cavity end mirrors (Figure 12.12). This train can also be visualized as the result of a single pulse bouncing back and forth between the mirrors; each time this oscillating pulse strikes the output coupling mirror, a portion of its energy is passed through the mirror and is detected as a newly emitted pulse. The remainder of the oscillating pulse is reflected back into the cavity for an additional gain. At steady state, each such trip through the gain medium compensates for the energy output and for cavity losses. Controlling the oscillation of this pulse and preventing the development of other parasitic modes can be achieved by an intracavity fast optical modulator, placed next to one of the

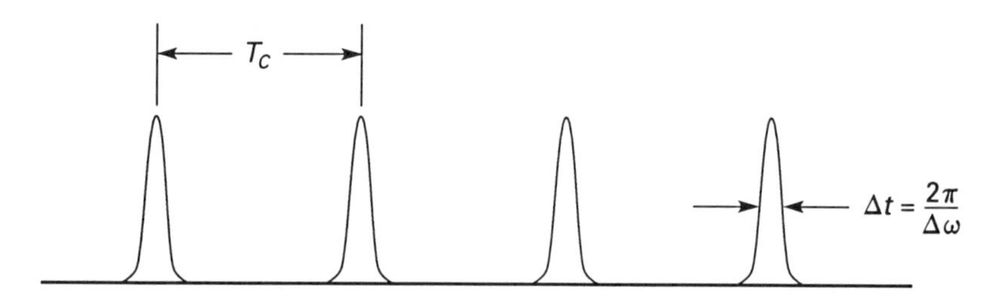

Figure 12.12 Train of mode-locked pulses.

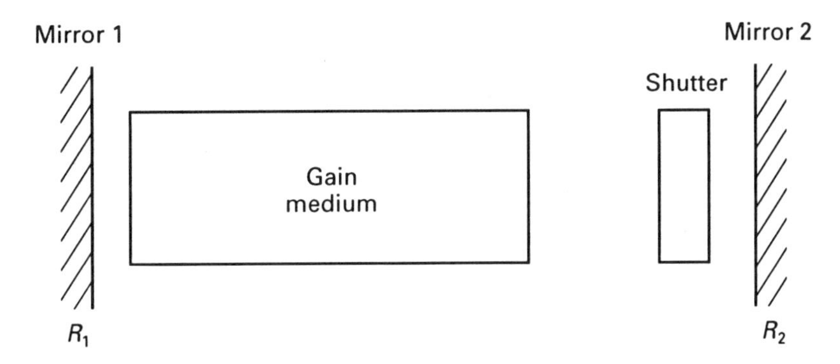

Figure 12.13 Simplified mode-locked laser cavity.

cavity end mirrors (Figure 12.13). The modulator, which acts as a shutter, is opened periodically at a rate of $1/T_C$ for the duration needed for a mode-locked pulse to pass through the modulator to the mirror and back to the cavity. Immediately after the pulse has cleared it, the modulator is closed; it is reopened only when the oscillating pulse returns after the completion of the next round trip. With this shutter, laser oscillation is prevented at any time other than a brief period every cycle. All parasitic modes, or combinations of modes, that would otherwise develop are blocked by the shutter, and their energy remains at the minimum level corresponding to a single or double pass through the cavity. However, out of the infinite number of combinations, the one group of modes with identical polarization and identical phases is preserved and is allowed by the modulator to oscillate. This is the group that can produce the pulse that travels in the cavity synchronously with the modulator cycle; it can pass through the shutter uninterrupted, return to the cavity for further gain and thence back to the shutter, where it can pass once again uninterrupted. The energy coupled into this group can increase unimpeded by the shutter at the expense of the parasitic modes, which are depressed by the shutter. Thus, while the shutter does not by itself select the phases of the various modes, it enforces the survival of one select group while extinguishing all competitors.

Fast modulation for mode-locking can be achieved by active modulators (see e.g. DiDomenico et al. 1966) or passive modulators (Garmire and Yariv 1967). Active modulators may consist of a Bragg cell (an acousto-optic modulator) or a Pockels cell that can be forced by an external source to open and close at prescribed times and for the desired duration. Passive modulation, on the other hand, cannot be controlled externally. It can depend either on optical phenomena (such as self-focusing) within the laser gain medium or on passive devices, such as a cell containing a saturable absorber. For either choice, passive modulators are activated by the level of the intracavity irradiance. Therefore, when a thin cell of a saturable dye is placed next to one of the cavity end mirrors, it can introduce sufficient loss to keep laser oscillation below a threshold when a stray mode is passing. However, a group of locked modes can have sufficient energy to force nearly half of the absorbers at the ground level into an excited state, thereby turning the dye transparent. Consequently, a mode-locked pulse can pass through the thin cell with only a minimal loss, for which the laser gain medium can easily compensate. If the dye in the cell is selected to have a short excited-state lifetime, the population of its ground state can be restored briefly after the pulse has cleared the cell, thereby blocking other modes. Starting oscillation in a cavity that is equipped with passive devices is difficult and may require additional devices. However, once a stable mode-locked oscillation is established, passive devices can support it indefinitely without outside intervention, or at least until a parasitic mode acquires sufficient irradiance to bleach the passive device and compete with the mode-locked oscillation for available laser energy. By contrast, oscillation of such parasitic modes is effectively blocked by active modulators.

Example 12.5 Determine the response time and frequency required by a passive and an active modulator to successfully mode-lock a Nd:YAG laser having a bandwidth of 4 cm^{-1} and a cavity length of 40 cm. The length of the proposed active modulator is 3 cm while the layer of the proposed saturable absorber of the passive modulator is 100 μm thick. Assume that the indices of refraction of both modulators are $n = 1.5$.

Solution The design parameters to be determined are the period T_C of the modulation cycle and the total time T_0 that each modulator needs to remain open in order to allow the mode-locked pulse to pass toward the adjacent cavity end mirror and then bounce back toward the gain medium. The cavity round-trip time is

$$T_C = \frac{2 \times 40}{3 \times 10^{10}} = 2.67 \text{ ns}.$$

This is the period for the cycle of both modulators.

The time that a mode-locking modulator needs to remain open is the longest of either the pulse duration or the transit time through the modulator to the adjacent mirror and back (Figure 12.13). For this laser, the pulse duration is approximately

$$\Delta t \approx \frac{1}{\Delta \nu} = \frac{1}{4(3 \times 10^{10})} = 8.33 \text{ ps}$$

(cf. eqn. 12.43). For comparison, the transit time through the active modulator toward the mirror and back is

$$t_t = \frac{2nL_m}{c} = \frac{2 \times 1.5 \times 3}{3 \times 10^{10}} = 300 \text{ ps},$$

whereas the transit time through the much thinner layer of saturable dye is

$$t_t = \frac{2 \times 1.5 \times 0.01}{3 \times 10^{10}} = 1 \text{ ps}.$$

The transit time through the active modulator $t_t \gg \Delta t$ is significantly longer than the pulse duration, thereby requiring the modulator to remain open for a duration of $T_o \approx 300$ ps. To achieve this, the rise time of the modulator must be fast relative to T_o, and the recovery time (i.e., the time required by the electronic circuitry to reactivate the modulator) must be short relative to the cavity transit time $T_C = 2.67$ ns.

The transit time through the saturable dye film is significantly shorter than the expected pulse duration. Therefore, it needs to remain transparent for a duration of approximately $T_o \approx \Delta t = 8.33$ ps. In addition, once the pulse has left the dye film, the population density at the ground state needs to be restored rapidly, which means that the lifetime of the excited state must be short relative to the pulse duration. Therefore, the Einstein A coefficient of the excited state of that dye must be

$$A_{21} > 1.2 \times 10^{11}\ \mathrm{s}^{-1}$$

(cf. eqn. 10.1). ■

Although the first mode-locked laser operation was successfully demonstrated using a He–Ne laser (Hargrove, Fork, and Pollack 1964), in which inhomogeneous broadening permits several modes to coexist, mode-locking is also possible when the medium is homogeneously broadened. In free oscillation, one mode can deplete the gain available for other modes, but a modulated cavity oscillation can force the coexistence of several longitudinal modes even in homogeneously broadened gain media. Therefore, such pulse parameters as its duration, its power relative to the average power of one mode, and so forth are identical to those expected when the medium is inhomogeneously broadened. More detailed discussion of mode-locking of homogeneously broadened gain media is available in Verdeyen (1989, p. 280).

The primary impetus for the development of mode-locked lasers was their potential application for communications. With a 1-cm laser cavity and a bandwidth in excess of 100 cm^{-1} (both are realistic parameters), 0.3-ps pulses can be generated at a rate of 30 GHz. This is an exceptionally fast rate, which is attractive for transmission of large files of digital data such as electronically recorded high-resolution images.

The development of mode-locked lasers yielded other applications, such as the measurement of fast transients. With subpicosecond pulses, the lifetime of short-lived excited states can be measured directly. In a typical experiment, an excited state is prepared by one laser pulse emitting at the wavelength requirement for its excitation. After a short and controllable delay, a second pulse is used to probe the excited state – for example, by measuring the attenuation by absorption from that state to another even higher state. The variation of this absorption with the delay is used to determine the lifetime of the probed state.

Similarly, fast surface phenomena such as the sequence of events leading to melting or annealing of metals can be readily studied using picosecond laser pulses. In such applications, one laser pulse is used to raise the surface temperature before a second pulse probes the surface characteristics (such as reflection) as they vary with time after the initial exposure.

Short laser pulses are also used for measuring the distance to a target (e.g., a moving automobile) or even the velocity of that target. In such applications, a brief pulse from the measuring station illuminates the target. The scattered light is then collected by a detector adjacent to the laser and the time of travel (or time of flight) between the laser device and the target is measured. The distance to the target can be determined using known parameters of dispersion by air. In this measurement, the resolution and accuracy are limited by the detector time response and the laser pulse duration. Application of ultrashort laser pulses improves the performance of such measurements. An accuracy of several centimeters is achievable for distances ranging from a few meters to a few kilometers. For measurements of larger distances, retro-reflectors are used.

A device using the time-of-flight approach for distance measurements has been developed to measure the rate of change of the distance to the target – that is, the target velocity. Measurements of up to 300 km/hr have been demonstrated with an accuracy of ±1.5 km/hr using a train of short pulses and a train duration of 0.3 s. Present applications include automobile speed enforcement, space measurements, and topography.

References

Armon, E., Zvirin, Y., Laufer, G., and Solan, A. (1989), Metal drilling with a CO_2 laser beam. I. Theory, *Journal of Applied Physics* 65: 4995–5002.

Bass, M. (1983), *Laser Materials Processing,* Amsterdam: North-Holland.

Bennett, W. R., Jr. (1962), Hole burning effects in a He–Ne optical maser, *Physical Review* 126: 580–93.

Boyd, G. D., and Gordon, J. P. (1961), Confocal multimode resonator for millimeter through optical wavelength masers, *The Bell System Technical Journal* 40: 489–508.

DeMaria, A. J. (1968), Mode locking opens door to picosecond pulses, *Electronics* 41: 112–22.

DiDomenico, M., Jr., Geusic, J. E., Marcos, H. M., and Smith, R. G. (1966), Generation of ultrashort optical pulses by mode locking the YAlG:Nd laser, *Applied Physics Letters* 8: 180–3.

Duley, W. W. (1983), *Laser Processing and Analysis of Materials,* New York: Plenum.

Forsyth, J. M. (1967), Single-frequency operation of the argon–ion laser at 5145 Å, *Applied Physics Letters* 11: 391–4.

Garmire, E. M., and Yariv, A. (1967), Laser mode-locking with saturable absorbers, *IEEE Journal of Quantum Electronics* 3: 222–6.

Goela, J. S., and Thareja, R. K. (1982), Cooling or heating of gases through energy transfer using lasers, *Optics Communications* 42: 417–18.

Hargrove, L. E., Fork, R. L., and Pollack, M. A. (1964), Locking of He–Ne laser modes induced by synchronous intracavity modulation, *Applied Physics Letters* 5: 4–5.

Hecht, J. (1986), *The Laser Guidebook,* New York: McGraw-Hill.

Incropera, F. P., and DeWitt, D. P. (1990), *Fundamentals of Heat and Mass Transfer,* 3rd ed., New York: Wiley.

Maiman, T. H. (1960a), Optical and microwave-optical experiments in ruby, *Physical Review Letters* 4: 564–6.
Maiman, T. H. (1960b), Stimulated optical radiation in ruby, *Nature* 187: 493–4.
Rigrod, W. W. (1963), Gain saturation and output power of optical masers, *Journal of Applied Physics* 34: 2602–9.
Schawlow, A. L., and Townes, C. H. (1958), Infrared and optical masers, *Physical Review* 112: 1940–9.
Verdeyen, J. T. (1989), *Laser Electronics,* 2nd ed., Englewood Cliffs, NJ: Prentice-Hall.
Wagner, W. G., and Lengyel, B. A. (1963), Evolution of the giant pulse in a laser, *Journal of Applied Physics* 34: 2040–6.
Yariv, A. (1985), *Optical Electronics,* 3rd ed., New York: Holt, Rinehart & Winston.

Homework Problems

Problem 12.1

Write a rate equation for a two-level system irradiated by a narrowband beam with an irradiance of I_ν at frequency ν. The population and degeneracy of the lower state are N_1 and g_1 (respectively) and of the upper state N_2 and g_2. The total number of particles in these two states is constant, $N_0 = N_1 + N_2$. The spontaneous emission lifetime is τ_{21}, the rate of absorption is W_{12}, and the rate of stimulated emission is W_{21}. Find the inversion density ΔN at steady state and show that, as $I_\nu \to \infty$, the population density of the upper state approaches the population density of the lower state.

Answer:

$$\Delta N = N_2 - \frac{g_2}{g_1} N_1 = -\frac{(g_2/g_1)N_0}{1 + W_{21}(1 + g_2/g_1)\tau_{21}}.$$

Problem 12.2

Saturable absorbers are used as passive optical switches. When a thin cell containing such an absorber with a density of N_A is placed along the path of a beam, the cell can attenuate the beam almost completely when the irradiance is low. However, as the incident irradiance increases, the population of the absorbing state is depleted and the absorber becomes almost perfectly transparent. Assume that such an absorber can be modeled as a two-level system with $g_1 = g_2 = 1$. Show that when saturated (i.e., when $W_{12} \gg A_{21}$) the attenuation of a narrowband beam with an irradiance I_ν is

$$\frac{d(I_\nu/h\nu)}{dx} \approx \frac{N_A A_{21}}{2}.$$

(*Hint:* Write the rate equation for N_2 and solve it. Use eqn. 12.1 to find the attenuation of I_ν.)

Problem 12.3

The three-level system in Figure 12.2 is used to generate optical gain for transitions between the ground state (level 0) and level 1. The system is pumped optically from level 0 to level 2 at a rate of W_{02}, and the rate of absorption to level 1 is W_{01}. Assume that the lifetimes for spontaneous transitions from level 2 are τ_{20} and τ_{21}, and from level 1, τ_{10}.

(a) Write the rate equations that describe the time-dependent population density of levels 0 and 1, assuming that the degeneracies of all states are unity.

(b) Using the condition that $N = N_0 + N_1 + N_2$, show that at steady state the population differential $N_1 - N_0$ between the gain-producing states is:

$$N_1 - N_0 = N\left[\frac{W_{02}\left(\frac{1}{\tau_{21}} - \frac{1}{\tau_{10}}\right) - \frac{1}{\tau_{10}}\left(\frac{1}{\tau_{21}} + \frac{1}{\tau_{20}}\right)}{\left(W_{02} + \frac{1}{\tau_{21}} + \frac{1}{\tau_{20}}\right)\left(2W_{01} + \frac{1}{\tau_{10}}\right) + W_{02}\left(W_{01} + \frac{1}{\tau_{21}} + \frac{1}{\tau_{10}}\right)}\right].$$

(c) Determine the conditions that must be met by τ_{10} and τ_{21} to assure inversion and gain.

(d) If the *pumping efficiency* is defined by

$$\eta_p = \frac{1/\tau_{21}}{1/\tau_{21} + 1/\tau_{20}},$$

show that for inversion $W_{02}\eta_p > 1/\tau_{10}$ and explain this condition.

Problem 12.4

The gain curve of an Ar^+ laser is broadened homogeneously as well as inhomogeneously. The homogeneous broadening is with a typical linewidth of ~500 MHz and the inhomogeneous broadening is with a typical linewidth of 5 GHz (Forsyth 1967).

(a) Find the total number of modes that can be accommodated by a 0.5-m-long cavity.

(b) How many modes can be observed in an instantaneous image of the dispersion spectrum of that laser?

(*Answer:* (a) 16–17; (b) 10.)

Problem 12.5

Mode hopping occurs when the optical path of a laser cavity changes either mechanically or by variations in the index of refraction.

(a) Show that the variation in the intracavity index of refraction that is required to shift the wavelength of one longitudinal mode to the wavelength of its neighbor is

$$\Delta n = \lambda/2L.$$

(b) The index of refraction in gas lasers is approximately $n = 1 + \alpha$, where $\alpha \approx 10^{-4}$ is proportional to the gas density. The Ar^+ laser is broadened homogeneously as well as inhomogeneously. The linewidth of the homogeneous broadening component is 500 MHz (Forsyth 1967). Thus, mode hopping occurs when the wavelength is shifted by an entire homogeneous bandwidth. Estimate the gas density fluctuation $\Delta N/N$ required for the hopping of a longitudinal mode of a 1-m-long Ar^+ laser operating at $\lambda = 514.5$ nm.

Problem 12.6

To characterize the gain properties of a laser medium, a laser beam with an incident irradiance of $I_{in} = 1$ W/cm^2 was transmitted through it. After a single pass through this medium, the measured irradiance was $I_{out} = 7$ W/cm^2. The measurement was repeated using a $I_{in} = 2$-W/cm^2 beam, where the amplified beam was measured to have an irradiance of $I_{out} = 12$ W/cm^2. Assume homogeneous broadening, that the distributed losses are negligible, and that the gain $\gamma(\nu)$ does not change along the length of the medium (this is a poor assumption, but it certainly simplifies matters).

(a) What is the amplification when $I_{in} \to 0$ W/cm^2 (small-signal amplification)?
(b) What is I_{sat}?
(c) Assume that this gain medium is enclosed between two mirrors with a reflectivity of 95% and 99%. What is the steady-state irradiance that can be developed in the cavity?
(d) What will be the irradiance emitted through the 95% reflecting mirror?

Problem 12.7

The upper level of a three-level gain medium (Figure 12.14) is pumped at a rate of $P = 5 \times 10^{18}$ cm^{-3}-s^{-1}. The lifetimes of the two excited states are $\tau_{10} = 1$ μs and $\tau_{21} = 3$ μs. The energy of level 2 and the energy of level 1 are represented by the frequency [cm^{-1}] of the emission from these levels to the ground state. The medium is enclosed in a laser cavity that is blocked, thereby suppressing stimulated emission and absorption. Assume that initially the population of both excited states is negligible.

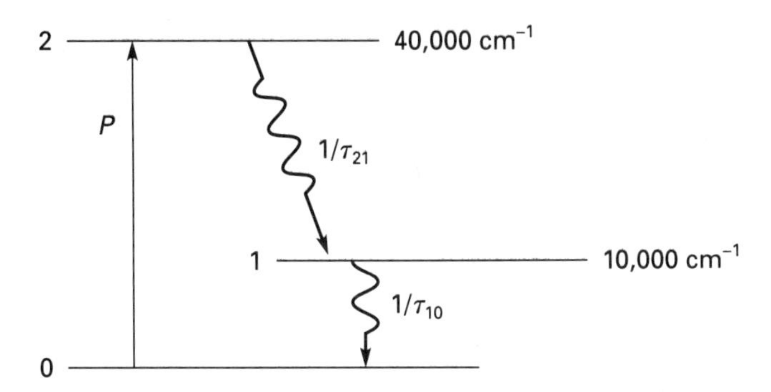

Figure 12.14 Three-level gain medium.

(a) Write the rate equations for levels 1 and 2.
(b) Determine the time-dependent variation of the population of levels 1 and 2 (N_1 and N_2, respectively).
(c) What are the steady-state values of N_2 and N_1?
(d) What is the ideal energy efficiency that can be expected from a laser using this gain medium? (Assume that all the cavity losses are negligibly small.)

Problem 12.8

Write the rate equation for the excited state of a laser gain medium in a *Q*-switched cavity while the shutter is closed. Solve the equation and find the maximum population of the excited state, N_2^{max}, that can be achieved by a pump rate of P_2 when the lifetime is τ_2. Determine how long the shutter should remain closed after the pumping begins in order to reach a population of $N_2 = 0.63 N_2^{max}$.

Problem 12.9

Show that each longitudinal mode inside a laser cavity forms a standing wave (i.e., that the nodes of the oscillating electric field are fixed in space and time).

Problem 12.10

In order to mode-lock a dye laser, it is pumped by a mode-locked Ar^+ laser. Since the lifetime of the excited dye molecules is short, the gain medium of the dye laser can provide net gain only while being illuminated by the exciting laser pulse. At all other times the dye medium is lossy. This mode-locking technique is called *synchronous pumping*.

(a) What must the length of the dye laser cavity be (relative to the length of the exciting laser cavity) for successful mode-locking?

(b) If the linewidth of the dye is 300 GHz and the linewidth of the Ar^+ laser is 5 GHz, what are the expected pulse durations of each of the mode-locked lasers?

(c) What is the largest error in the adjustment of the dye laser cavity length that can be allowed before mode-locking is terminated?

Research Problems

Research Problem 12.1

Read the paper by Goela and Thareja (1982).

(a) Consider an alternative cooling method in which level 3 is "drained" by a stimulated emission process (see Figure 12.15). For that assumption to work, level 4 must be significantly above the ground level. When that occurs, even a slight increase in the population of level 3 will generate an inversion. Assume that all the parameters in the paper remain the same (e.g. $E_3 - E_4$, $E_3 - E_2$, or k_{34}) *except* that the energy of level 2 is 10,000 cm^{-1} and accordingly E_3 and E_4 are higher. Assume also that $A_{40} = 10^8\ s^{-1}$.

(b) Find the energy of level 4. (*Answer:* $E_4 = 7{,}857\ cm^{-1}$.)

(c) Find the value of W_{34} assuming that $\Delta\nu_{34} = 0.1\ cm^{-1}$ and $I_\nu = 1\ W/cm^2$ at $\nu_{34} = 2{,}349\ cm^{-1}$.

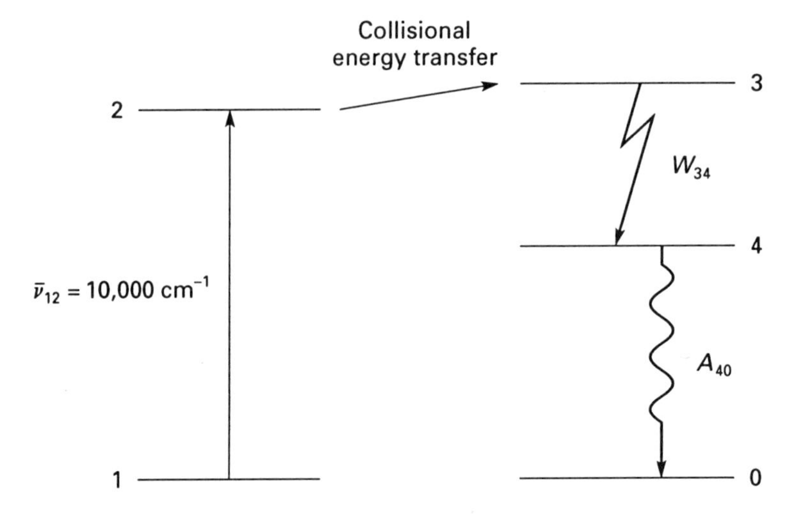

Figure 12.15 Energy-level diagram for molecular cooling by stimulated emission.

(d) Find the steady-state population of level 3 in terms of n_2.

(e) Find the critical pressure for cooling at the conditions described by part (c).

(f) Can you obtain cooling at any pressure using this approach? Explain.

Research Problem 12.2

Read the papers by Maiman (1960a,b) describing the first operating laser – a ruby laser.

(a) Retrieve the values for A_{21}, A_{31}, S_{31}, and S_{32} in units of s^{-1} and for ν_{31} and ν_{21} in units of cm^{-1}.

(b) Find the pumping efficiency ζ. (*Note:* You must modify slightly the definition given in this book.)

(c) Find the minimum value of the absorption rate coefficient W_{13} required to obtain inversion.

(d) Find the irradiance [W/cm^2] of the incident radiation required to maintain this absorption rate coefficient when the incident linewidth is 10 cm^{-1}.

13 Propagation of Laser Beams

13.1 Introduction

Many of the characteristics of laser beams are determined by properties of their gain medium and by the loss and gain characteristics of the laser cavity. The previous chapter discussed factors that determine the wavelength and spectral bandwidth of laser beams, the characteristics of their longitudinal modes, gain requirements for steady-state oscillation, the ultimate power (or energy) of laser beams, the duration of a laser pulse when Q-switched or mode-locked, and so on. However, this wealth of information is insufficient for design applications where the spatial pattern of the energy delivery must be well defined. To illustrate this, recall that when a laser is used for illumination (such as in PLIF), a relatively uniform distribution of the energy may be required; for material processing, the beam energy may need to be concentrated into a narrow well-defined spot; and for holography or interferometry, the shape of the incident wavefronts may need to be geometrically simple. Furthermore, in all applications, the distribution of the energy passing through any optical element must be carefully controlled to prevent laser-induced damage by localized high-energy concentration. Popular belief has it that laser beams are always collimated and that their wavefronts are planar. But this is true only in the limit, when the beam diameter approaches infinity. Because of diffraction, the beam cannot remain collimated indefinitely when the diameter is finite; with the exception of a narrow range where the beam may be considered as nearly collimated, it must either converge or diverge. Accordingly, the transverse distribution of the electromagnetic field, the diameter of the beam, and the shape of its wavefronts must vary along the propagation path, and their values at one location influence their values elsewhere. The analysis of these propagation characteristics provides the tools for the design of many laser beam delivery systems.

Ultimately, propagation characteristics of a laser beam are controlled by only two parameters: the transverse field distribution at any cross section of the beam, and the shape of its wavefronts. Both parameters are determined initially by properties of the laser cavity. However, outside the cavity, interaction with optical elements and other transmitting media (or simply diffraction) can alter the beam's initial energy distribution, the shape of its wavefronts, or both. For certain simple transverse distributions and for some common optical elements, mathematical description of these variations is possible and the propagation characteristics of the beam can be calculated throughout its entire path. But when the field distribution is complex, or when the optical elements are irregularly shaped - either by design or owing to imperfections and impurities in optical elements - propagation characteristics may need to be determined experimentally by measuring the energy distribution profile at several cross sections of the beam. Nevertheless, mathematical description of even the simplest mode can serve as a yardstick for estimating the characteristics of the more complex modes.

Although many beam structures, or *modes,* can be described mathematically, the description here of the propagation of laser beams includes the details of only the most fundamental mode: the *Gaussian* mode. This description includes the necessary conditions for a laser cavity to produce such a beam, as well as the propagation characteristics outside the cavity, through optical media and through lenses. These results can then be used to determine, for example, the distribution of the beam irradiance at the focal point of a lens or the conditions for the formation of a nominally collimated beam for illumination.

13.2 The Fundamental Mode - The Gaussian Beam

In the absence of gain and loss, the propagation of a laser beam can be described by the wave equation. Although laser beams are almost never collimated, their divergence is usually so small that the propagation can be approximated by a single component (e.g., the z component k) of the propagation vector $\mathbf{k}$. For a forward-propagating and linearly polarized beam, the time-independent term of the electric field is

$$\mathbf{E}(x, y, z) = \mathrm{Re}[\mathbf{E}_0(x, y, z)e^{ikz}] \tag{13.1}$$

(cf. eqn. 4.33). Although this solution of the wave equation lacks detail, it does suggest that even for a paraxial propagation the distribution of the field amplitude, $\mathbf{E}_0(x, y, z)$, may vary both longitudinally along the z axis and transversely along the x and y axes. Experience shows that while most laser beams propagate nominally in a uniaxial direction, their irradiance distribution is not uniform and can vary significantly. Therefore, (13.1) may represent a number of solutions of the wave equation, with each solution describing a different field distribution. Since laser beams are coherent, $\mathbf{E}_0(x, y, z)$ consists of well-defined wavefronts that are described by a complex function: the real part represents the amplitude, while the imaginary part represents the transverse and longitudinal variations of the phase. Possible solutions for $\mathbf{E}_0(x, y, z)$ can be obtained by introducing (13.1) into the three-dimensional (3-D) wave equation (eqn. 4.8). However, for paraxial propagation (i.e., when propagation is mostly along the z axis), the 3-D wave equation can be simplified considerably by recognizing that

$$\left|\frac{\partial^2 \mathbf{E}_0}{\partial z^2}\right| \ll \left|2k\frac{\partial \mathbf{E}_0}{\partial z}\right|, \left|\frac{\partial^2 \mathbf{E}_0}{\partial x^2}\right|, \left|\frac{\partial^2 \mathbf{E}_0}{\partial y^2}\right|. \tag{13.2}$$

Thus, in addition to the previous assumption that $\mathbf{k}$ is described by a single component, the problem can be further simplified by neglecting the small term of $\partial^2 \mathbf{E}_0/\partial z^2$ relative to the larger components of the wave equation. With these approximations, the 3-D equation of the paraxial propagation of a coherent laser beam reduces to

$$\frac{\partial^2 \mathbf{E}_0}{\partial x^2} + \frac{\partial^2 \mathbf{E}_0}{\partial y^2} + 2ik\frac{\partial \mathbf{E}_0}{\partial z} = 0. \tag{13.3}$$

Together with its boundary conditions, solution of this equation provides the 3-D distribution of the electric field of a coherent beam propagating nominally along the z axis.

Although the boundary conditions of (13.3) were not specified here, one of its possible solutions can be shown (Boyd and Gordon 1961; see also the review article by Kogelnik and Li 1966) to be

$$\mathbf{E}_0 = \epsilon_0 \exp\{-iP\} \exp\left\{i\frac{k}{2q}r^2\right\}, \tag{13.4}$$

where ϵ_0 is an amplitude term. The second term of the solution includes a complex parameter $P = P(z)$ that describes the variation of the field along the beam. This term will be further discussed in what follows. However, the solution is characterized by its last term, which includes the variation of the field transverse to the beam, that is, in the $r^2 = x^2 + y^2$ direction. This term is similar to the statistical Gaussian distribution and accordingly this transverse distribution – or the transverse mode – is called Gaussian (cf. eqn. 10.59 describing the distribution of molecular speed in gases). The term $q = q(z)$ is the beam parameter; it is a yet-unspecified complex function that varies with the radius of the beam (*spot size*) and with the curvature of the wavefront at cross sections along the beam path. Although this is not the only solution of (13.3), it is an important one: it represents the simplest and most fundamental transverse mode that can be sustained by a circularly symmetric cavity having a pair of infinitely large spherical (or plane) mirrors parallel to each other. Consequently, it is also often called the *fundamental mode*. Furthermore, the diffraction losses associated with the propagation of the Gaussian mode are smaller than the losses of alternative modes. Therefore, when a randomly distributed field occurs anywhere inside an ideal and well-aligned laser cavity soon after the gain medium is excited, the dominating field distribution that develops after a finite number of cavity transits is Gaussian (Fox and Li 1961). Owing to its lower diffraction

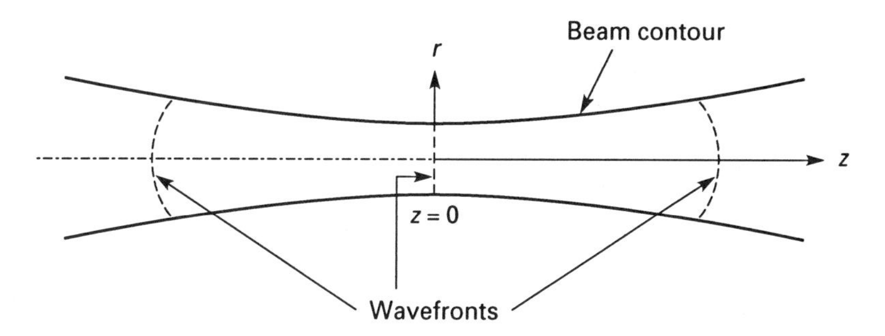

Figure 13.1 Contour of a laser beam focused by a lens as it converges toward its narrowest point and then diverges. Note the wavefronts ahead of the focal point, at the focus, and past it.

losses, this mode suppresses other, higher-order, lossier modes that were contained in the initial random distribution.

Substituting (13.4) into (13.3) and comparing terms with equal powers of r provides the following conditions, which need to be satisfied by both P and q everywhere along the beam:

$$\frac{dq(z)}{dz} = 1 \quad \text{and} \quad \frac{dP(z)}{dz} = -\frac{i}{q(z)}. \tag{13.5}$$

Of course, the solution of the first of these equations is simply

$$q = q_0 + z, \tag{13.6}$$

where q_0 (which is yet to be determined) is the beam parameter at the origin $z = 0$.

To specify the location of the origin of the coordinate system, consider the contour of a laser beam focused by a lens as it converges toward its narrowest point and then diverges (see Figure 13.1). That narrowest point is the *beam waist,* where convergence ends and divergence begins. The wavefronts, which are also shown in the figure, are perpendicular to the propagation vector at every point and are concave when the beam converges and convex when the beam diverges. Thus, at the waist the wavefronts must be planar. By placing the origin of the coordinate system at the waist, the value of the term $P(z)$ (eqn. 13.4) can be conveniently set to

$$P(z=0) = 0. \tag{13.7}$$

Although this choice may seem arbitrary, one should note that $P(z)$ describes phase and amplitude variations of the electric field along the beam. At the waist, where the wavefronts are planar, the initial phase value can thus be set to zero uniformly for the entire cross section. The amplitude at the origin is set by this choice to ϵ_0. Therefore, the field distribution at the waist is simply

$$\mathbf{E}_0(z=0) = \epsilon_0 \exp\left\{\frac{ikr^2}{2q}\right\}.$$

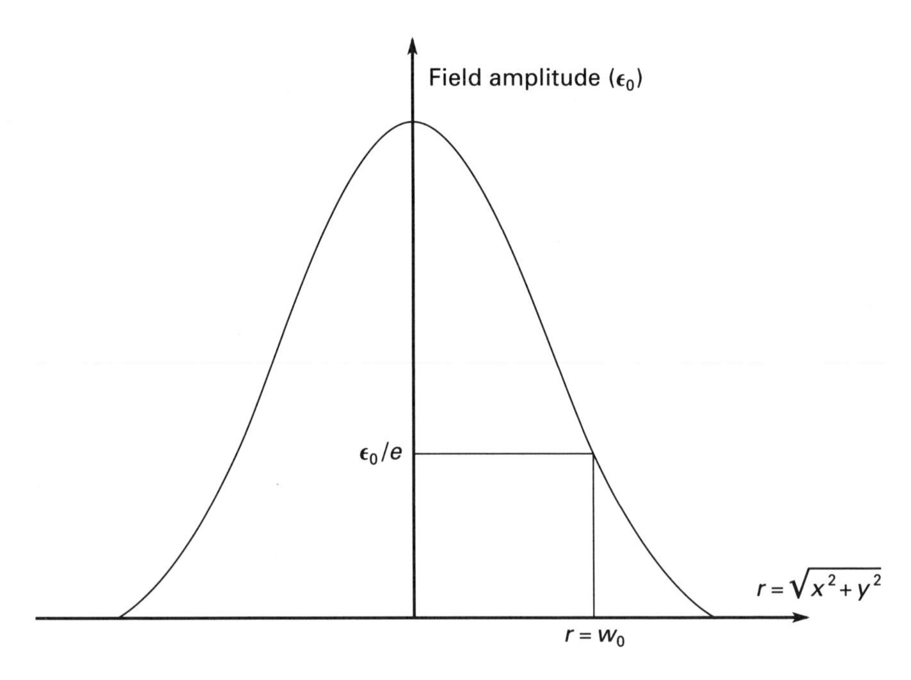

Figure 13.2 Distribution of the electric field at the waist of a Gaussian laser beam.

Although the phase at the waist is uniform and the beam at that point is collimated, the transverse distribution of the field amplitude varies. Figure 13.2 illustrates this distribution, which can be obtained experimentally by measuring the variation of the beam irradiance along a radial line and correcting for the functional dependence between the field and irradiance terms (eqn. 4.42). This is a circularly symmetric profile decaying radially from a maximum reached at the centerline where $r = 0$. Although the field extends indefinitely, it is customary to measure the beam radius of the wasit to the point $r = w_0$, where the field amplitude declines to $1/e$ of its peak value. Thus, the field distribution at the waist should be described by

$$\mathbf{E}(x, y, z=0) = \epsilon_0 \exp\left\{-\frac{x^2+y^2}{w_0^2}\right\}. \tag{13.8}$$

Equation (13.8) together with (13.4) establishes the condition for q_0; that is, the value of the beam parameter $q(z)$ at $z = 0$ is $q_0 = -i\pi w_0^2/\lambda$. By introducing q_0 into (13.6) we obtain the following solution for the beam parameter:

$$q(z) = q_0 + z = -i\frac{\pi w_0^2}{\lambda} + z. \tag{13.9}$$

This solution can be introduced into (13.5) to obtain the following solution for $P(z)$:

$$iP(z) = \ln\left[1 + \frac{i\lambda z}{\pi w_0^2}\right] \tag{13.10}$$

(Kogelnik and Lee 1966). These explicit terms for $P(z)$ qnd $q(z)$ can now be inserted into (13.4) and then into (13.1) to obtain the following equation for the propagation of a Gaussian beam:

$$\mathbf{E}(x,y,z)=\epsilon_0\frac{e^{ikz}}{1+i\lambda z/\pi w_0^2}\exp\left\{\frac{i\pi(x^2+y^2)}{\lambda(-i\pi w_0^2/\lambda+z)}\right\}. \tag{13.11}$$

Since the index of refraction was assumed here to be $n=1$, the solution must be corrected for the change in optical path when $n\neq 1$ by dividing everywhere λ with n.

Although this result is complete, the physical meaning of its various components is obscured by complex mathematical terms. To clarify (13.11), both $P(z)$ and $q(z)$ may be rewritten. The beam parameter $q(z)$ is expressed by:

$$\frac{1}{q(z)}=\frac{1}{R(z)}+i\frac{\lambda}{\pi w^2(z)}, \tag{13.12}$$

where

$$R(z)=z\left[1+\left(\frac{\pi w_0^2}{\lambda z}\right)^2\right] \tag{13.13}$$

and

$$w^2(z)=w_0^2\left[1+\left(\frac{\lambda z}{\pi w_0^2}\right)^2\right]. \tag{13.14}$$

As shown in what follows, $R(z)$ is the radius of curvature of the wavefront of the beam while $w(z)$ is the spot size, or the effective beam radius, both determined at any cross section perpendicular to the z axis.

The other parameter of the beam, the *phase-shift parameter* $P(z)$, is modified by using simple operations of complex variables to read

$$\begin{aligned} iP(z)&=\ln\sqrt{1+\left(\frac{\lambda z}{\pi w_0^2}\right)^2}+i\tan^{-1}\left(\frac{\lambda z}{\pi w_0^2}\right)\\ &=A+i\Phi. \end{aligned} \tag{13.15}$$

In spite of its name, $P(z)$ includes an amplitude term A that can be reduced with the help of (13.14) to

$$A=\ln\left(\frac{w(z)}{w_0}\right). \tag{13.16}$$

In addition, $P(z)$ includes a phase shift term Φ. Both terms describe variations of the field that are associated with the geometry of propagation along the z axis. Thus, A describes an increasing amplitude when the beam converges and a diminishing amplitude when it diverges. Similarly, Φ is associated with the shape of the wavefronts of the converging/diverging beam. Introducing these transformed parameters into (13.4), we obtain a physically more meaningful expression for the propagation of a Gaussian laser beam:

$$\mathbf{E} = \left(\epsilon_0 \frac{w_0}{w(z)}\right) \exp\left\{-\frac{r^2}{w(z)^2}\right\} \exp\{i(kz - \Phi)\} \exp\left\{i \frac{kr^2}{2R(z)}\right\}. \tag{13.17}$$

The four terms of (13.17) were arranged so that the first two terms describe variations of the field amplitude, longitudinally and transversely, while the last two terms describe the phases - also longitudinally and transversely.

The variation of the beam amplitude along its centerline (i.e., when $r = 0$) is described by the first term. It is controlled by the ratio $w_0/w(z) \leq 1$ (eqn. 13.14); only at the waist, where $z = 0$ and $w_0/w(z) = 1$, can the amplitude along the centerline reach its full value of ϵ_0. Since the irradiance scales as $|\mathbf{E}|^2$ (eqn. 4.42), it must scale along the centerline of a Gaussian beam as $w_0^2/w(z)^2$. But $w(z)$ is the effective radius of the beam and accordingly this dependence implies that the irradiance varies inversely with the beam cross-section area. Thus, in the absence of absorption, this dependence suggests that the energy flux through any cross section remains constant; that is, energy is conserved as the beam propagates - hardly a surprise.

The second term of (13.17) describes the variation of the field in the transverse direction. At $z = 0$, where $w(z) = w_0$, the remaining three terms of (13.17) reduce to ϵ_0, and thus (13.17) becomes equivalent to the distribution at the waist presented by (13.8). Away from the waist, the amplitude of the field at $r = w(z)$ declines according to the second term to $1/e$ of its value at $r = 0$. Thus, similarly to w_0 at the waist, $w(z)$ is the spot size of the beam when $z \neq 0$. Although the transverse field distribution remains Gaussian, the effective radius of the beam increases to $w(z)$. This is illustrated in Figure 13.3, where the field distributions at the waist and at cross sections ahead and past the waist are depicted. The spread of these distributions is narrowest at the waist, while the peak amplitude is highest there. The latter is compatible with the ramifications of the first term of (13.17), where the peak amplitude was projected to reach its maximum at the waist and to decline as the beam expands. The irradiance at $r = w(z)$ is of course $1/e^2 = 0.135$ of its peak at $r = 0$. Thus, effectively, most of the beam energy is transported through a spot with a cross section that can be measured by $w(z)$ - the spot size of the beam.

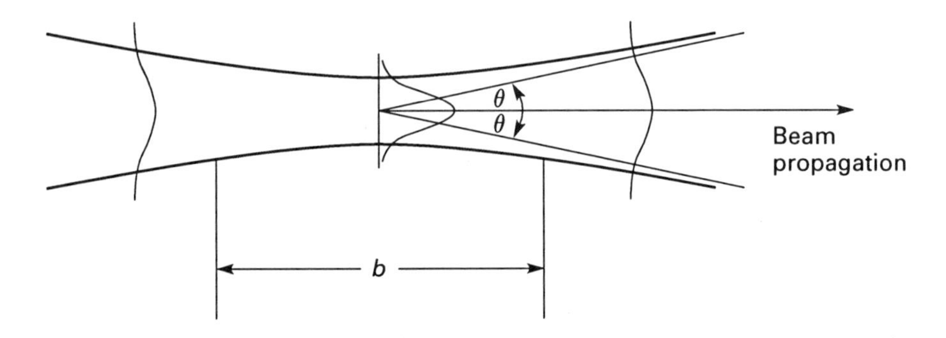

Figure 13.3 Electric field distributions at the waist and at cross sections ahead of and past the waist of a Gaussian laser beam.

As Figure 13.1 illustrates, laser beams are never collimated; they either converge or diverge. However, in the vicinity of the waist, divergence is relatively small and a section of the beam there can be considered as nearly collimated. The length of this section is an important parameter of the beam. It is used to determine the depth of the cut that can be obtained with a focused beam, the distance that a laser beam can travel without significant reduction in its irradiance, and the dimensions of the field that can be illuminated uniformly. The decision of where to locate the boundaries of this nearly collimated section depends on the application. In applications where large divergence is acceptable, the boundaries may be set farther apart than in applications where a well-defined collimated beam is required. However, a standard definition of this range sets its limits at both sides of the waist at

$$|z| < \frac{\pi w_0^2}{\lambda} \tag{13.18}$$

(Siegman 1986, p. 669). Within this range, the second bracketed term of the effective beam radius (eqn. 13.14) is less than unity and by this approximation can be neglected relative to the first term, thereby eliminating the only term that describes variation of the beam radius with z. Consequently, the approximate beam radius throughout the entire range remains at w_0. The length of this range, measured from the left side of the waist in Figure 13.3 to its right side, is

$$b = \frac{2\pi w_0^2}{\lambda}. \tag{13.19}$$

This is the *Rayleigh range*. For reasons to be explained shortly, b is also called the *confocal parameter*. Clearly, the beam is never perfectly collimated. In fact, the area of its cross section at the edges of the Rayleigh range is twice its area at the waist. However, when $w_0 \gg \lambda$ the Rayleigh range is long enough to be measured in meters; the slow variation in the beam cross section along this range is almost unnoticeable and the beam appears as collimated. Conversely, when the beam is focused tightly (e.g., in metal processing or for cutting), the Rayleigh range is short and the beam converges and diverges over a short distance that can be measured in millimeters.

Outside the Rayleigh range, Gaussian beams can no longer be considered collimated. Although variations in the beam radius anywhere along its path can be readily calculated by (13.14), the expression can be simplified for calculations at the far field where the beam diverges rapidly. Thus, at a distance from the waist that is large relative to $b/2$ (i.e., when $|z| \gg \pi w_0^2/\lambda$), the second bracketed term of (13.14) is dominant relative to unity and the radius of the beam can be approximated by

$$w(z) \approx \frac{\lambda z}{\pi w_0}. \tag{13.20}$$

This simple result shows that at the far field, the radius of the beam expands linearly with z and the beam envelope asymptotically approaches a cone. This is illustrated in Figure 13.3, where the two asymptotes of the beam contour are drawn. The angle θ between each of these asymptotes and the z axis is

$$\theta \approx \frac{w(z)}{z} \approx \frac{\lambda}{\pi w_0} \tag{13.21}$$

(Kogelnik and Lee 1966). For comparison, the divergence of a beam emerging from a uniformly illuminated aperture with a radius of a is

$$\theta \approx \frac{0.61\lambda}{a} \tag{13.22}$$

(cf. eqn. 7.10). Recall that this divergence angle was obtained with the assumption that the incident field is coherent. Thus, although both beams in this comparison are coherent, the transverse distributions of their irradiance are substantially different and as a result the divergence through a uniformly illuminated aperture is nearly twice as large as the divergence of a Gaussian beam. This example illustrates one of the most important properties of the fundamental mode: slow diffraction-related divergence. Inside the laser cavity, this slow divergence assures that diffraction losses remain low. In other words, the low divergence makes this mode the least lossy and thus the dominant mode in the cavity. Outside the cavity, the same property guarantees long and relatively collimated propagation.

Although (13.21) describes the divergence of a beam propagating away from its waist, symmetry suggests that the same cone angle should correspond to the far-field convergence of that beam. Such conical convergence can be induced by a focusing lens. If the beam fills the entire lens aperture, θ is the angle of a cone with the lens as its base and with its apex at the waist of the focused beam. Comparison of (13.21) and (7.11) suggests also that the spot size at the focus of a Gaussian beam is smaller than the spot size obtained by focusing a uniformly illuminating beam by the same lens. This observation can be generalized to show that the smallest spot size is obtained by focusing a coherent beam with Gaussian distribution. Thus, devices where spatial resolution is critical – such as laser printers, scanners, or laser machining systems – must use lasers that operate in the fundamental mode.

The remaining two terms of the Gaussian beam equation (eqn. 13.17) determine the phase and shape of the beam wavefronts. Almost all applications that depend on the interference between two overlapping coherent beams require regularly shaped wavefronts. Thus, certain configurations of speckle pattern interferometry (Section 6.6) or laser Doppler velocimetry (Section 6.7) require a structure of planar and parallel interference fringes that can be obtained only when the wavefronts of the interacting beams are planar as well. Similarly, recording and reconstruction of certain holograms requires the use of well-defined wavefronts, preferably planar or spherical. Finding a region along a laser beam with the needed wavefronts, or (alternatively) shaping the laser beam to form such wavefronts, can be accomplished only with the guidance of the remaining two terms of (13.17).

aussian distribution the irradiance of the beam were kept uniformly at I_0, same power could be transmitted through a cross section with an area that only half the area of the Gaussian beam. Although this is only a hypotheti-al distribution, it is used to define the *equivalent area,* which for a Gaussian beam is

$$A_{eq} = \frac{\pi w_0^2}{2} = \frac{b\lambda}{4}. \qquad (13.25)$$

To illustrate the use of these last results for the design of laser systems for engineering applications, consider the following example.

Example 13.1 A 1-kW CO_2 laser emitting at $\lambda = 10.6\ \mu m$ is used to cut a 2-mm–thick steel plate. To obtain a narrow cut, the beam must be focused tightly. On the other hand, to obtain a cut with edges that are perpendicular to the faces of the sample, the Rayleigh range must be long relative to the depth of the cut.

(a) Find what the beam diameter must be in order for b to exceed 3 mm.
(b) Find the peak irradiance of that beam and its equivalent area.
(c) If the diameter of the focusing lens is 1 cm and its aperture is filled by the laser beam, estimate its distance from the sample.

Solution (a) Figure 13.4 illustrates the contour of a Gaussian beam penetrating the steel plate. By adjusting the position of the plate (or the focusing lens), the waist of the beam can be located halfway through the sample. A good-quality cut is obtained when the Rayleigh range b is long relative to the plate thickness d. On the other hand, when b is too long, the diameter of the beam may exceed the cut resolution limits and the irradiance may not be sufficient to induce the necessary heating. Thus, as an optimization between conflicting requirements, the Rayleigh range was selected to approximately match the plate thickness; that is, $b = 3$ mm. For the parameters of this problem, the diameter of a Gaussian beam for which $b > 3$ mm is

$$2w_0 \geq 2\sqrt{\frac{b\lambda}{2\pi}} = 2\sqrt{\frac{3(10.6\times 10^{-3})}{2\pi}} = 0.14 \text{ mm}$$

(cf. eqn. 13.9). Neglecting effects of heating by conduction, the cut obtained by this laser can be expected to be approximately 0.14 mm wide. Other lasers, operating at shorter wavelengths, may produce thinner cuts while still having the same Rayleigh range. For example, in the absence of conduction effects and for the same energy absorption, the width of a cut obtained by a Nd:YAG laser at $\lambda = 1.064\ \mu m$ can be one third of the width obtained by a CO_2 laser. Thus, although selection of a laser for metal cutting must be motivated primarily by the absorption characteristics of the treated metal, the effect of the laser wavelength on the desired cut quality and the thickness of the workpiece must also be considered.

Like the amplitude terms, the phase terms ar
propagation along the beam axis where $r = 0$ (th
transverse to the beam (the fourth term of e
scope of this book, it is possible to show (Brors
Gaussian beams are shaped as ellipsoids. The foci o
in the $z = 0$ plane at a distance of $\pm b/2$ from the axis
parameter, or the Rayleigh range (eqn. 13.19). With this
13.13) is the radius of curvature of the wavefronts of the Ga
at $z = 0$, the curvature is $R \to \infty$ and the wavefronts are plan
radius of curvature is $R \to z$ and the wavefronts appear as conc
centered at the waist. Having spherical wavefronts at the far field is
with the conical divergence projected by (13.20). The shapes of the wav
both at $z = 0$ and $z \to \infty$, are consistent with the ellipsoid presentation of
son (1988). Clearly, for applications that require planar wavefronts, the las
beam must be used at or near its waist. Although away from the waist the wavefronts are slightly curved, their curvature is considered to be negligible within the Rayleigh range. Thus, intersection of the two beams that are used for laser Doppler velocimetry must occur within their Rayleigh ranges. If one or both of these beams is intersected outside that range, their interference pattern consists of curved surfaces that are no longer parallel to each other (see Figure 6.15), resulting in velocity measurements that may be in error.

Although the distributions of the field amplitude and its phase are fully characterized by (13.17), most laser applications require instead a prediction of the distribution of the beam irradiance. Such applications include use of high-energy beams where the location of optical elements must be selected to avoid irradiance that exceeds their damage threshold, or where the targets must be placed to obtain desired effects such as evaporation, melting, or controlled heating – all with predictable resolution. The distribution of the irradiance can be determined directly from (13.17) by introducing it into (4.42). Thus, with a peak irradiance of I_0 at the waist, the irradiance elsewhere in the beam is:

$$I(z) = I_0 \left(\frac{w_0}{w(z)} \right)^2 \exp\left\{ -\frac{2r^2}{w(z)^2} \right\}. \tag{13.23}$$

Because the irradiance at $r = w(z)$ declines to $1/e^2$ of its peak at $r = 0$, the spot size of the beam can be defined by a disc with a radius of $w(z)$. The area of that disc scales as $w(z)^2$, and accordingly the irradiance along the centerline varies inversely with that area.

The power P_0 transmitted by the beam can now be calculated from (13.23) by integrating the irradiance over the entire cross section. For a circularly symmetric beam, this integral is simply

$$P_0 = 2\pi \int I_0 \left(\frac{w_0}{w(z)} \right)^2 \exp\left\{ -\frac{2r^2}{w(z)^2} \right\} r\,dr = I_0 \frac{\pi w_0^2}{2}. \tag{13.24}$$

Thus, in the absence of absorption, P_0 is constant along the beam and depends only on the peak irradiance and the spot size at the waist. If instead of

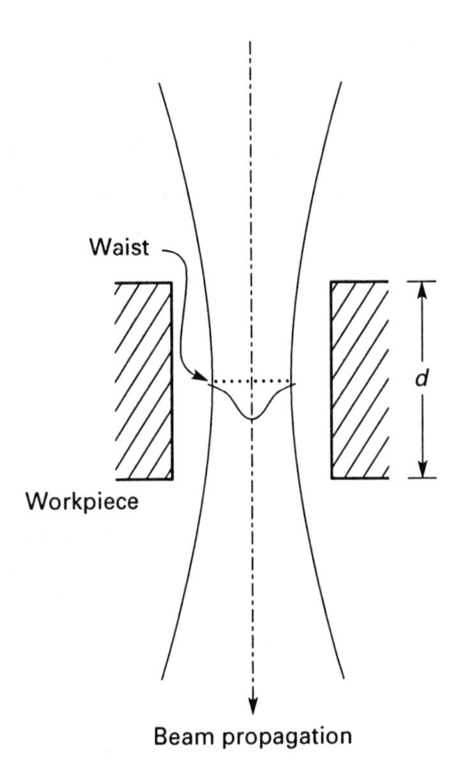

Figure 13.4 Penetration of a metal plate by a cutting laser beam.

(b) The peak irradiance of this laser beam can be determined from (13.24). For a 1-kW beam focused into a diameter of 0.14 mm, the peak power is

$$I_0 = \frac{2 \times 1{,}000}{\pi \times (0.07)^2} = 1.3 \times 10^5 \text{ W/mm}^2 = 13 \text{ MW/cm}^2.$$

Note that this irradiance is delivered continuously. A significantly higher irradiance can be delivered when a laser is Q-switched. However, owing to the short duration of such pulses, the energy delivered by a typical Q-switched laser is much lower than the energy of the laser of this example.

(c) To estimate the distance f between the focusing lens and the plate, assume that the lens aperture D is filled entirely by the yet-unfocused beam. In addition, assume that the distance between the lens and the workpiece is large relative to the length of the Rayleigh range, and that the converging beam may be viewed as a cone with a cone angle θ (eqn. 13.21). Thus:

$$\theta \approx \frac{\lambda}{\pi w_0} \approx \frac{D}{2f}.$$

With the available parameters expressed in millimeters, the distance to the plate is

$$f \approx \frac{\pi \times 10 \times 0.07}{2(10.6 \times 10^{-3})} = 104 \text{ mm}.$$

Note that f is approximately the focal length of the lens. The exact parameters of a Gaussian beam transformed by a lens will be discussed in the next section. However, the present result already illustrates one of the difficulties associated with metal processing by tightly focused laser beams: to achieve the small diameter needed for thin cuts, the focusing lens must be placed at a perilously short distance from the workpiece; debris and smoke may sputter from the melting surface and deposit on the lens, thereby blocking the clear path of the beam and reducing the damage threshold of the lens. To prevent such contamination, the lens in most laser machining systems is enclosed in a protective sleeve, and inert gas flows coaxially with the beam at a rate that is sufficient to blow away most sputtering material. ■

The fundamental mode described here can be delivered by carefully designed laser cavities, or by low-gain lasers where only the least lossy mode survives. Consequently, the modes of many high-power or gas-phase lasers deviate from this ideal distribution. Some of these more complex modes can be described mathematically (Fox and Li 1961, Kogelnik and Li 1966). However, most commonly observed modes are too complex for mathematical description. Nevertheless, they can still be viewed as a mixture of several fundamental modes – not necessarily collinear – summed together. A mixed-mode beam may deliver more power than a single-mode beam emitted by a comparable laser, but it also has a larger divergence angle. Thus, the propagation of a single-mode beam is viewed as a yardstick against which other modes can be measured. A comparison between mixed-mode laser beams and single- (or Gaussian) mode beams can be made using the propagation factor of mode quality M, or the M^2 *factor.* This is the ratio between [the product of the diameter D_w of the complex mode beam at its waist and its divergence angle] and [the product of the same parameters of the fundamental mode]. With this definition, the M^2 factor may be calculated as

$$M^2 = \frac{D_w \times \theta}{2w_0(\lambda/\pi w_0)} = \frac{\pi D_w \times \theta}{2\lambda}, \tag{13.26}$$

where (13.21) was used to specify the divergence angle of the fundamental mode. Owing to diffraction, the beam diameter at the waist and its divergence angle are inversely proportional. Thus, the M^2 factor is independent of the actual beam diameter and is thus a dimensionless measure of the mode quality. For the fundamental mode, $M = 1$; otherwise, $M > 1$. For systems (such as scanners or laser printers) that depend on the mode quality, measurement of D_w and θ that produce $M \approx 1$ may be a sufficient test. Other applications (such as surface heat treatment) may require uniform or "top-hat" illumination, which can be obtained only by complex modes where $M \gg 1$.

The discussion here of Gaussian beams simply described their properties without offering any means to control them. However, selection of a focusing lens for laser-assisted cutting, or of a cylindrical lens to shape a light sheet for illumination of a test section for LIF imaging, requires equations that describe the transformation of Gaussian beams by lens. Such equations are expected to describe the size of the waist behind a lens or its location relative to the lens. Similarly, equations that describe the oscillation of Gaussian beams inside laser cavities are needed for cavity design. Although the formation of Gaussian beams in a resonator logically should be considered before examining beam transformation by lenses, the latter information is needed to analyze the cavities and will therefore be presented next.

13.3 Transformation of a Gaussian Beam by a Lens

The beam parameter $q(z)$ (eqn. 13.12) and the phase-shift parameter $P(z)$ (eqn. 13.15) fully characterize Gaussian laser beams. Therefore, the propagation of a laser beam through optical media, or its transformation by lenses or mirrors, can be generally modeled by evaluating $q(z)$ and $P(z)$ at the input and output planes of an optical element. However, in the absence of absorption losses (or gain), the real and imaginary components of the complex beam parameter $q(z)$ can be used to fully specify $P(z)$ (see Problem 13.2). Therefore, the propagation through nonlossy media can be described by simply evaluating $q(z)$ at each point. Similarly, transformation by an optical element, such as a lens, can be described by evaluating q_1 and q_2, the beam parameters at the input and output planes of that element (Kogelnik 1965a). The relation between q_1 and q_2 for such transformation must then be specific to that element. Beam propagation through a train of optical elements and through the separating media between them can be described by applying consecutively the transformation function of the various elements.

The transformation of radiation by optical elements can also be analyzed by geometrical optics, where the paths of geometrical rays is drawn. Therefore, it can be expected that some of the techniques of geometrical optics will also be applicable to the analysis of propagating laser beams. In particular, the method of ray transfer matrices (Section 2.8) is adopted here to describe the transformation of Gaussian beams by lenses and mirrors. This method is extremely useful for describing the transformation by several consecutive elements or an oscillation between the mirrors of a laser cavity. Similarly to geometrical optics, each element will be described by a 2×2 matrix (Figure 13.5). However, unlike geometrical optics, where the matrix is used to compute r and r' (eqn. 2.17), here it will be used to calculate the real and imaginary components of q_2 in terms of the two components of q_1.

To develop the equation for the transformation of a Gaussian beam by a lens, consider first as an analogy the transformation by a lens of a spherical wave with a radius of curvature R_1. The curvature is defined as positive if it is concave when viewed from the left side of the front. For radiation emitted by an ideal point source, this curvature for propagation to the right is simply the distance to the source $R_1 = s_1$. After being focused by a lens of focal length f, the wave is transformed into a new spherical wave with a curvature $-R_2$ propagating toward a point at a distance $s_2 = -R_2$ away. Therefore, by replacing s_1 and

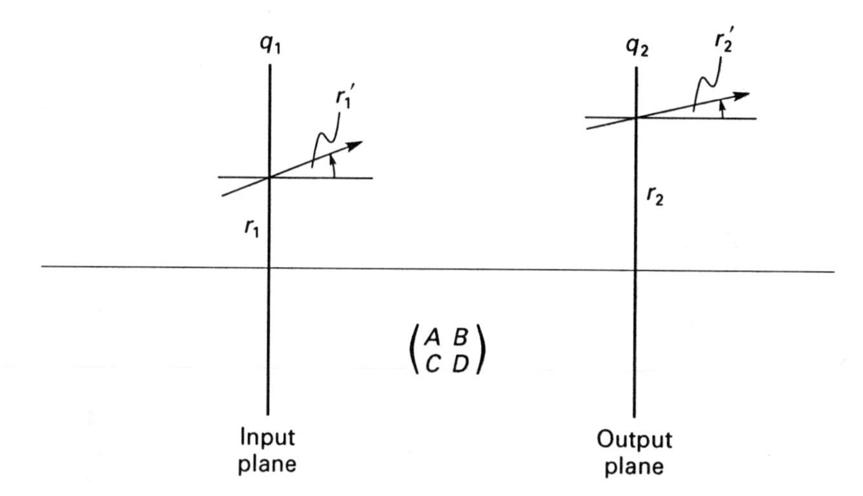

Figure 13.5 Schematic presentation of a transforming optical element, showing the ray and beam parameters at the input and output planes and the ray transfer matrix of that element.

s_2 with R_1 and $-R_2$, the lens equation (eqn. 2.10) can also be used to describe the curvatures of the incident and transmitted beams:

$$\frac{1}{R_2} = \frac{1}{R_1} - \frac{1}{f}$$

(see Problem 13.3).

A more general expression that describes the transformation of a spherical wave by any optical element can be obtained by assuming that, for paraxial propagation, $R \approx r/r'$. Since r_2 and r_2' at the output plane of an optical element can be calculated by the ray transfer matrix when r_1 and r_1' are known, the curvature of the spherical wave at the output plane can also be determined by that matrix from the curvature in the input plane. It can be shown (Problem 13.3) that, for paraxial propagation through an optical component with ray transfer matrix elements A, B, C, and D, the transformation of a spherical wave is

$$R_2 = \frac{AR_1 + B}{CR_1 + D} \tag{13.27}$$

(Kogelnik 1965a).

To extend this result for the analysis of Gaussian laser beams, consider the transmission of a Gaussian beam by a thin, nonabsorbing lens. The beam parameter at the input plane of the lens (eqn. 13.12) is

$$\frac{1}{q_1} = \frac{1}{R_1} + i\frac{\lambda}{\pi w_1^2}.$$

For a thin lens, the radius of the beam immediately past the lens is $w_2 = w_1$, and the only effect of that lens on the beam is to transform its radius of curvature from R_1 to R_2:

$$\frac{1}{q_2} - \frac{1}{q_1} = \frac{1}{R_2} - \frac{1}{R_1}.$$

Thus, similarly to the transformation of spherical waves by a lens, the transformation of the beam parameter by a lens is

$$\frac{1}{q_2} = \frac{1}{q_1} - \frac{1}{f}. \tag{13.28}$$

Although the analogy between the beam parameter q and the radius of curvature of the wavefront R was shown here only for a transformation by a thin lens, it was generalized (Kogelnik 1965b) to describe transformation by arbitrary elements. For an optical component with ray transfer matrix elements A, B, C, and D, the transformation of the beam parameter similar to (13.27) is:

$$q_2 = \frac{Aq_1 + B}{Cq_1 + D}. \tag{13.29}$$

This is the *ABCD law*. It has been demonstrated to accurately describe paraxial propagation of Gaussian beams through optical elements of interest, including thick lenses, spherical mirrors, and media with varying refraction indices.

To illustrate the utility of the *ABCD* law for deriving the transformation by elements other than a thin lens, consider the propagation through a distance z in free space. The matrix elements for this propagation are $A = 1$, $B = z$, $C = 0$, and $D = 1$ (Table 2.1). After inserting these elements into (13.29), the following transformation of q_1 is obtained:

$$q_2 = q_1 + z. \tag{13.30}$$

Although this is not a new result (compare it to eqn. 13.6), it demonstrates that the *ABCD* law is not restricted to transformations by thin elements.

The *ABCD* law can also be used to describe the propagation of a Gaussian beam from its waist where its radius is w_1 to a lens located a distance d_1 away, through that lens and then to the new waist a distance d_2 away, where the radius is w_2 (Figure 13.6). When modeled by ray transfer matrices, this transformation

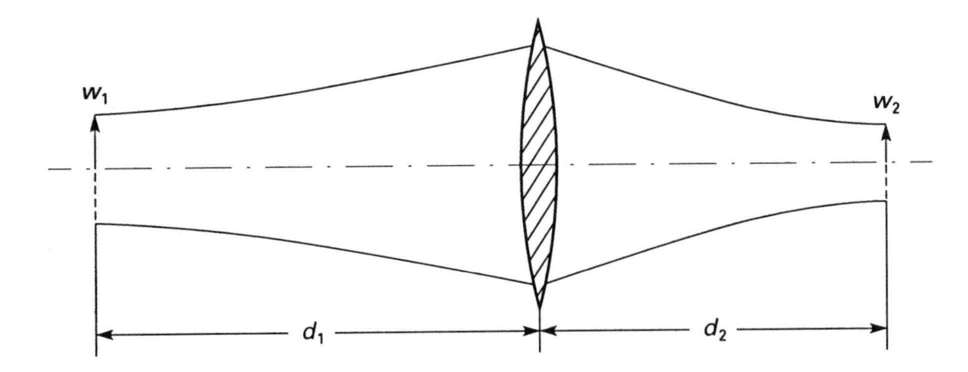

Figure 13.6 Transformation of a Gaussian beam by a lens.

includes three elements: two of these elements are the flat media with $n=1$ (Table 2.1) that represent the propagation to and from the lens; the third element is the lens with a focal length of f. The combined ray transfer matrix of these elements was previously calculated (eqn. 2.18). Therefore, using the *ABCD* law, the transformation of the beam may be written as

$$q_2 = \frac{\left(1-\frac{d_2}{f}\right)q_1 + \left(d_1 - \frac{d_1 d_2}{f} + d_2\right)}{\left(-\frac{q_1}{f}\right) + \left(1 - \frac{d_1}{f}\right)}. \tag{13.31}$$

Owing to the selection of the input and output planes at the waists, where $1/R_1 = 1/R_2 = 0$, q_1 and q_2 can be reduced to

$$q_1 = -i\frac{\pi w_1^2}{\lambda} \quad \text{and} \quad q_2 = -i\frac{\pi w_2^2}{\lambda}$$

(cf. eqn. 13.12). Although q_1 and q_2 are now purely imaginary, (13.31) still includes real and imaginary components. By separating and equating the imaginary and real parts at both sides of this equation, one obtains:

$$\frac{d_1 - f}{d_2 - f} = \frac{w_1^2}{w_2^2} \quad \text{and} \quad (d_1 - f)(d_2 - f) = f^2 - f_0^2, \tag{13.32}$$

where

$$f_0 = \frac{\pi w_1 w_2}{\lambda}$$

(Kogelnik and Li 1966).

With the two equations of (13.32), the spot size of the focused beam w_2 and the distance d_2 to its waist from a lens with a focal length f can be calculated when λ, w_1, and d_1 are known. Although this result can be viewed as the counterpart to the lens equation (eqn. 2.10), one should note that focusing a Gaussian laser beam may be significantly different from the results of geometrical optics. This is illustrated in Problem 13.4, where the location of the waist of a focused beam is seen to fall short of the geometrical focus of that lens. The difference between the two results may be attributed to diffraction effects that are neglected in geometrical optics. Thus, the location of the waist coincides with the geometrical focus of the lens only when $b_1 \gg f$, that is, when diffraction is indeed negligible.

The two parameters solved by (13.32) – the location of the waist of the focused beam and its radius – determine other important beam parameters such as the peak irradiance and the Rayleigh range. Alternatively, these equations may be used to select a focusing lens for particular beam-shaping needs. For applications such as laser-assisted cutting or surface heat treatment, where high irradiance is required, the lens is selected primarily to minimize w_2. By contrast, for illumination, the beam is expected to remain collimated over a prescribed length and the lens is selected to extend the Rayleigh range by increasing w_2.

Although most system designs involve an optimization between the Rayleigh range and the spot size, the beam radius is usually the first parameter to be calculated. Other parameters, such as d_2, are only secondary and can be determined after selecting an optimal lens. Therefore, the following more useful equation for w_2 was obtained by eliminating d_2 between the two equations (13.32):

$$\frac{1}{w_2^2} = \frac{1}{w_1^2}\left(1 - \frac{d_1}{f}\right)^2 + \frac{1}{f^2}\left(\frac{\pi w_1}{\lambda}\right)^2 \tag{13.33}$$

(Kogelnik 1965b). With this equation, w_2 is expressed explicitly in terms of known parameters (λ, w_1, d_1) of the incoming laser beam and in terms of the focal length of the selected focusing lens.

It is evident from (13.33) that to tightly focus a beam, the lens must have a short focal length. This is hardly surprising. However, a less expected result is that for Gaussian beams, the spot size at the focus depends on the spot size of the incident beam. With the same focusing lens, the spot size of the focused beam may increase or decrease when the radius of the incident beam increases (Problem 13.5). Usually, to obtain small w_2, the spot size at the waist of the incoming beam must be made as large as possible; as w_1 increases, the first term of (13.33) diminishes while the second term increases until it dominates (13.33), which can now be reduced to:

$$w_2 \approx \frac{f\lambda}{\pi w_1}. \tag{13.34}$$

Therefore, for high irradiance or for fine spatial resolution, a Gaussian beam must be focused by a lens with a short focal length and the spot size of the incoming beam at its waist must be expanded until the beam fills the clear aperture of the lens.

An alternative presentation of this result can be obtained by introducing the divergence angle θ_1 of the incoming beam. Using (13.21) to replace w_1 in (13.34), we have

$$w_2 \approx f\theta_1. \tag{13.35}$$

Thus, for tight focusing, the far-field divergence angle of the income beam must be reduced. Hypothetically, with $\theta \to 0$ (i.e., when $w_1 \to \infty$), the spot size at the focus can be reduced to a point.

When the diameter of the incoming beam is comparable to the diameter of the focusing lens, (13.34) may also be expressed in terms of the lens $f/\#$ (Verdeyen 1989, p. 94) as follows:

$$w_2 \approx \frac{2\lambda}{\pi}(f/\#), \tag{13.36}$$

where $f/\# \approx f/2w_1$. This result shows a linear dependence between the lens $f/\#$ and the spot size of the focused beam.

Equations (13.34) and (13.35) can be inverted to describe the cone angle θ_2 obtained by focusing an incident beam with a spot size of w_1. The result is identical to the cone angle projected by Example 13.1 of the previous section, where focusing was considered using concepts of geometrical optics.

Although the analysis here was limited to transformations of axially symmetric beams by lenses with spherical symmetry, a beam can have independent parameters in two perpendicular planes. Thus, with the use of a cylindrical lens, the beam cross section can be shaped into an ellipse with an effective width along one axis that is different from its width along the other axis. The axes of the ellipse and the beam propagation axis define the two orthogonal planes. As the beam propagates, it may have two divergence angles; the shallower divergence is in the plane that includes the wider effective width and the fast divergence is in the plane that includes the narrower width. Such disparity in the beam properties is necessary to shape laser beams into light sheets. With cylindrical focusing, the beam parameters q_1 and q_2 can no longer be defined by a single application of the *ABCD* law (eqn. 13.29). Instead, two beam parameters at the output plane of the cylindrical lens must be evaluated separately, using its two focal lengths f_1 and f_2 (see Research Problem 13.1).

13.4 Stable and Unstable Laser Cavities

The equations of the transformation of laser beams by a lens, together with the *ABCD* law, can now be used to identify criteria for stable oscillation of the fundamental mode in a laser cavity. When a cavity is properly designed and aligned, random field distributions anywhere within it must gradually evolve until they give way to the least lossy distribution – the Gaussian mode (Fox and Li 1961). Thus, during an initial transition period, the field inside the cavity can vary from transit to transit. Such variations may include an increase in the amplitude until I_{SS} is reached (eqn. 12.24), or evolution of the transverse distribution until the mode that has the lowest diffraction losses is fully developed. During this transition period, the irradiance is typically low, and most loss mechanisms such as absorption, scattering, or reflections at interfaces do not discriminate among the various transverse modes and are therefore extraneous to the selection of a stable field distribution. (Note that, in high-gain lasers, the field may develop rapidly and such nonlinear effects as saturation or bleaching may participate in shaping the transverse mode.) Thus, initially the primary loss mechanism that discriminates among transverse modes is diffraction, and the mode that presents the lowest diffraction losses survives to the detriment of other potentially competing modes. One such stable distribution is the Gaussian mode. After steady state is established, the distribution of the field, as well as its gain, remains unchanged during consecutive cavity transits. This condition was used to find the amplitude of the electric field and the irradiance (eqn. 12.24) and will also be used here to establish the condition for steady-state oscillation of the Gaussian mode.

Figure 13.7 illustrates a laser cavity designed to support steady oscillation of a Gaussian beam. This cavity is shown empty – the gain medium and other intracavity devices were omitted for simplicity. In the absence of nonlinear

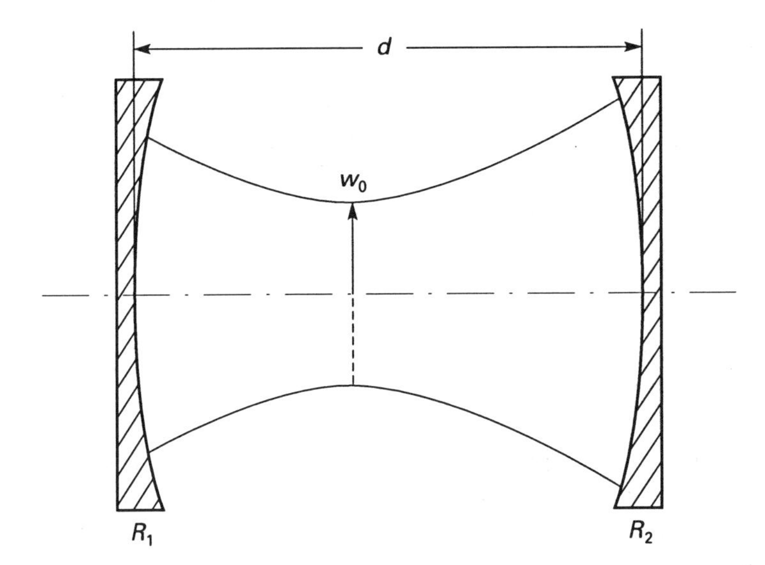

Figure 13.7 Schematic presentation of an empty laser cavity.

effects, some of these devices merely extend the effective length of the cavity and their influence on the mode can be easily included subsequent to the empty-cavity analysis. Effects of non-uniform distributions of the gain or non-uniform mirror reflectivity have been analyzed (see Casperson and Yariv 1968) but will not be discussed here. Other intracavity devices such as lenses, tilted etalons, or diffraction gratings may also have a significant effect on the transverse mode, but they too will not be discussed here.

The two cavity end mirrors shown in Figure 13.7 are curved. To determine their shape, we use the condition that for steady-state oscillation the curvatures of the wavefronts of the stable mode must match the curvature of the mirror reflecting it. Since the wavefrons of Gaussian beams are shaped as ellipsoids (Brorson 1988), the cavity end mirrors should be expected to be shaped as ellipsoids as well. Fortunately, at sufficiently large distances from the waist, the shape of the wavefronts of the Gaussian mode can be approximated by concentric spherical sectors centered at the waist (eqn. 13.13). Thus, if the cavity is sufficiently long, both mirrors may be spherical (or planar). Assuming that the diameters of the mirrors are large and that diffraction induced by their finite aperture is negligible, the location of the waist must depend only on the curvatures R_1 and R_2 of these mirrors and the separation d between them. The only diffraction effects to be included in the analysis here are induced by the finite width of the beam and the transverse distribution of the field.

A relation between the beam and cavity parameters can be obtained from the *ABCD* law (eqn. 13.29) using the ray transfer matrix elements of the entire cavity – including cavity end mirrors, the space between them, and (when appropriate) intracavity devices. To evaluate these matrix elements, the cavity may be replaced with an equivalent periodic sequence consisting of transmitting elements such as lenses, spacers, and intracavity components that match

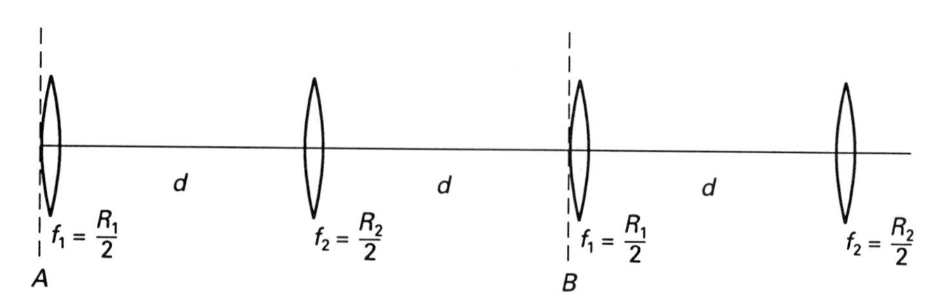

Figure 13.8 Periodic sequence of an empty cavity (equivalent to the empty cavity in Figure 13.7).

the cavity components. Each lens in the sequence represents one reflection by a cavity end mirror, and each spacer in the sequence represents the distance between two consecutive components. The focal lengths of the lenses of the sequence must be identical to the focal lengths of the cavity end mirrors. Thus, for spherical cavity end mirrors with curvatures of R_1 and R_2, the focal lengths of the equivalent lenses are

$$f_1 = \frac{R_1}{2} \quad \text{and} \quad f_2 = \frac{R_2}{2}$$

(cf. eqn. 2.11).

Figure 13.8 illustrates a periodic sequence, representing an unfolded empty cavity. Thus, rays passing through this periodic sequence travel through trajectories that are equivalent to trajectories in a cavity consisting of two spherical mirrors separated by a distance of d. One cell in this periodic sequence consists of the transit from plane A to plane B in the figure, and is equivalent to one cavity transit. Thus, the matrix elements of one such cell are identical to the matrix elements of one cavity transit.

Once the matrix elements of a cell of the periodic sequence are determined, the conditions for cavity oscillation of a Gaussian beam can be found. For stable oscillation, the beam parameter q_2 after one transit, or after passage through one cell of the periodic sequence, must equal the beam parameter of the previous transit q_1. This self-consistency postulate $q_1 = q_2 = q$ can be used with the $ABCD$ law (eqn. 13.29) to establish the following condition for stable oscillation (Kogelnik 1965b):

$$q = \frac{Aq+B}{Cq+D}, \tag{13.37}$$

where A, B, C, D are the terms of the ray transfer matrix for one cavity transit and must be determined separately for each laser cavity. Equation (13.37), which is quadratic in q, has the following solutions:

$$\frac{1}{q} = \frac{D-A}{2B} + \frac{i}{2B}\sqrt{4-(A+D)^2}. \tag{13.38}$$

This result was structured to resemble (13.12), including the positive sign of the root. Thus, its first term must correspond to $1/R$, which may be either positive, negative, or zero, while the second term corresponds to the second term in (13.12), which must be purely imaginary when w is real. Therefore, for the second term in (13.38) to be purely imaginary, the argument of the square-root term must be positive. Stated explicitly, this condition is

$$-1 < \frac{A+D}{2} < 1. \tag{13.39}$$

When $(A+D)/2 = \pm 1$, the second term in (13.38) disappears, thereby implying that $w \to \infty$ which practically is an unacceptable result. When $(A+D)/2$ exceeds the limits of (13.39), w becomes imaginary. Thus, to obtain a Gaussian laser beam with a real finite radius, the A and D terms of the laser cavity must meet the condition stated by (13.39) (Kogelnik and Li 1966). This is the condition for stable cavity oscillation of the fundamental mode.

To derive explicitly the stability condition for oscillation in an empty cavity consisting of two spherical mirrors, consider the transition through a single cell of the equivalent periodic sequence (Figure 13.8). The matrix must represent the transformation of a ray by lens f_1 followed at a distance d away by lens f_2 and by a transit through a distance d in free space to plane B just ahead of the next lens f_1. Using the ray transfer matrices of the individual components (Table 2.1) and the procedure outlined in Section 2.8, we obtain the following terms for the entire cavity:

$$\begin{aligned} A &= 1 - \frac{2d}{f_1} + \frac{d^2}{f_1 f_2} - \frac{d}{f_2}, \\ B &= 2d - \frac{d^2}{f_2}, \\ C &= -\frac{1}{f_1} + \frac{d}{f_1 f_2} - \frac{1}{f_2}, \\ D &= -\frac{d}{f_2} + 1. \end{aligned} \tag{13.40}$$

The stability condition for the empty cavity is obtained by introducing the A and D matrix elements into the general stability condition of (13.39):

$$0 < \left(1 - \frac{d}{2f_1}\right)\left(1 - \frac{d}{2f_2}\right) < 1,$$

or, after reintroducing the curvature of the cavity end mirrors:

$$0 < \left(1 - \frac{d}{R_1}\right)\left(1 - \frac{d}{R_2}\right) < 1. \tag{13.41}$$

Although other terms for A, B, C, and D may be obtained by selecting reference planes elsewhere in the cavity (Problem 13.10), the stability condition of (13.41) is independent of that choice.

Equation (13.41) was derived for empty cavities that include only two cavity end mirrors. Most lasers, however, include intracavity components. Fortunately, most of these components (such as the gain medium, polarizers, or retarders) are flat and their only effect on the ray transfer matrix is to extend the effective cavity length d by their higher index of refraction. Thus, with the exception of special systems that include intracavity lenses or focusing elements, (13.41) may be applied to evaluate the stability of most laser cavities by replacing d with the effective intracavity optical paths.

The stability condition (eqn. 13.41) can be used to evaluate laser cavities with convex, concave, or flat mirrors. For a broad overview of the conditions for stable oscillation see the graphical presentation by Kogelnik and Li (1966) and also Problem 13.9. However, simple combinations of cavity end mirrors can be readily evaluated by inspection. Thus, when one of the cavity end mirrors is flat (i.e., $R_1 \to \infty$), stable oscillation is possible only when the other mirror is concave and its radius of curvature is $R_2 > d$ (i.e., the center of curvature of that mirror lies outside the cavity). When both mirrors are flat, the cavity is marginally stable and while oscillation is possible, it may persist only when both mirrors are perfectly parallel. Of course, when the gain is high, steady-state emission may develop between two flat mirrors after a few cavity transits even when the cavity is slightly misaligned. When both mirrors are convex (R_1 and $R_2 < 0$), the cavity is unstable.

Confirming the stability of a cavity is only one step in its design. Complete design of a laser cavity involves selection of mirrors that not only support stable oscillation but also form a beam with desired parameters. Whereas stability is determined by (13.41), the relation between beam parameters and cavity geometry is described by (13.38). One of the beam parameters, w_0 (or b), can be expressed in terms of R_1, R_2, and the cavity length d by locating the reference plane for the oscillation at the waist of the beam and using q_0 and the matrix terms for a single-cavity transit starting at the waist (Problem 13.10). Alternatively, since the curvatures of the cavity end mirrors match the curvatures of the reflected wavefronts, (13.13) can be used to express the beam parameters in terms of R_1, R_2, and d.

To illustrate this last procedure, assume that the distance from the mirror with curvature R_1 to the waist is z_1 and is z_2 from the other mirror (Figure 13.7). Normally the waist is inside the cavity, that is, $d = z_1 + z_2$. Therefore, the curvatures of the wavefronts at the cavity end mirrors are

$$R_1 = z_1\left[1+\left(\frac{b}{2z_1}\right)^2\right] \quad \text{and} \quad R_2 = z_2\left[1+\left(\frac{b}{2z_2}\right)^2\right]$$

(cf. eqns. 13.13 and 13.19). Combining these equations and solving for b, one obtains

$$\frac{b}{2} = \frac{\sqrt{d(R_1-d)(R_2-d)(R_1+R_2-d)}}{R_1+R_2-2d} \tag{13.42}$$

(Kogelnik and Li 1966). Using (13.19), this result can readily be transformed to show the beam radius w_0 in terms of the cavity parameters R_1, R_2, and d. These equations can also be solved to determine the location of the waist z_1 relative to the R_1 mirror:

$$z_1 = \frac{d(R_2 - d)}{R_1 + R_2 - 2d}. \tag{13.43}$$

In cavities consisting of two identical convex end mirrors, by symmetry the waist is in the middle. This is readily confirmed by substituting $R = R_1 = R_2$ in (13.43). The length of the Rayleigh range of the beam in such a cavity is:

$$b = \sqrt{d(2R - d)}. \tag{13.44}$$

In addition, when the foci of both mirrors coincide (i.e., when the cavity length is $d = R$), the Rayleigh range is

$$b = d. \tag{13.45}$$

This unique cavity is the *confocal* cavity. As determined by the stability condition (eqn. 13.41), this cavity is only marginally stable. Nevertheless its length is used to define the Rayleigh range, or the confocal parameter.

Although cavities with two concave ($R > 0$) mirrors are usually stable, practical considerations require that one of these mirrors be flat. Otherwise, the beam will have to exit through one spherical mirror, which upon transmission will act as a lens and alter the beam parameters. With one flat mirror, the cavity can be stable only when the center of curvature of the spherical mirror is outside the cavity (i.e., when $R > d$). To serve as an output coupler, the inner surface of the flat mirror's substrate is coated with a partially reflecting film (Figure 13.9); the radiation transmitted by the mirror is the cavity output while the reflected part returns to the gain medium to sustain steady oscillation (eqn. 12.21). Meanwhile, the spherical mirror is coated with a highly reflecting film. If the substrate of the flat mirror consists of two parallel surfaces, the back surface may also reflect a fraction of the emerging beam back into the cavity. Depending on the geometry of the cavity, this parasitic reflection – although not as bright as the reflection by the coated surface of the mirror – may pass through the gain medium, experience gain, compete with the oscillating mode

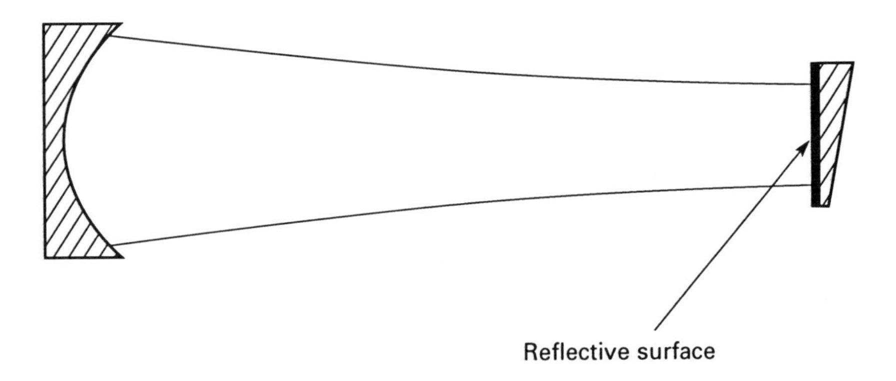

Figure 13.9 An empty cavity with a wedged output coupling mirror.

for the available energy, and eventually deform its structure. To avoid such competition between the oscillating mode and parasitic reflections, the back surface may be coated with an antireflection coating. Alternatively, the substrate may be shaped as a wedge (Figure 13.9) with an angle that is sufficiently large to tip the back-reflection away from the gain medium. In the absence of such competing modes, a stable fundamental mode can develop with its waist at the inner face of the output coupler. The Rayleigh range of that mode, obtained by setting $R_1 = R$ and $R_2 \to \infty$ in (13.42), is

$$b = 2\sqrt{d(R-d)}. \tag{13.46}$$

The same Rayleigh range is obtained by cavities twice as long and consisting of two mirrors with the same curvature R (cf. eqn. 13.44). Thus, cavities with one flat mirror can be viewed as one half of symmetrical cavities with two spherical mirrors (Problem 13.11).

Example 13.2 Find the Rayleigh range and the beam radius at the waist of a Nd:YAG laser with a 30-cm-long cavity consisting of a flat output coupler and a total reflector with a 3-m radius of curvature.

Solution The Rayleigh range is readily obtained by substitution into (13.46) as follows:

$$b = 2\sqrt{0.3(3-0.3)} = 1.8 \text{ m}.$$

Since the waist is located at the face of the output coupling mirror, the beam has the smallest radius when leaving the laser. For a wavelength of 1.064 μm this radius is

$$w_0 = \sqrt{\frac{b\lambda}{2\pi}} = \sqrt{\frac{1.8(1.064\times 10^{-6})}{2\pi}} = 5.5\times 10^{-4} \text{ m}$$

(cf. eqn. 13.19). The diameter of the beam as it exits the cavity is approximately 1 mm. For comparison, the diameter of a typical Nd:YAG rod is 3–5 mm. Thus, the cross section of the 1-mm beam overlaps less than 10% of the cross section of the gain medium. Since laser energy is derived by stimulated emission, the part of the gain medium that is not overlapped by the beam cannot contribute to the output energy, and thus more than 90% of the available energy is lost owing to this cavity design. ■

As illustrated by this example, the diameter of the Gaussian mode of a laser beam can be small relative to the diameter of the gain medium. By not filling the gain medium, the energy of the Gaussian mode is limited by its own size, and energy stored outside the beam is either lost or – what may be even worse – feeds competing non-Gaussian modes that develop in the unused portion of the gain medium. When such parasitic modes form, they combine with the fundamental mode to form a complex mode that limits the possibilities for tight focusing of the laser beam or for uniform illumination. Although such undesirable parasitic modes can be prevented by installing intracavity apertures and baffles, the energy of the fundamental mode still remains limited by the beam spot size. To improve the match between the diameters of the gain medium and the beam, the curvature R of the spherical mirror and/or the cavity length d may be increased. However, as R increases relative to d, the stability of the cavity becomes marginal (eqn. 13.41). Increasing the cavity length may thus be limited either by stability considerations or by practical considerations such as available space. An alternative option – reduction of the diameter of the gain medium – is equivalent to the insertion of intracavity apertures; although it chokes parasitic modes, it also limits the available laser energy. An unexpected solution to this problem is to force a more uniform oscillation through the entire volume of the gain medium by designing the cavity as an unstable oscillator (Siegman 1965). Depending on the design, a large gain medium can be successfully filled by a laser beam with a simple mode structure, thereby permitting higher output energy together with an acceptable mode quality.

Figure 13.10 illustrates one possible configuration of an unstable oscillator. It consists of two diverging mirrors (R_1 and $R_2 < 0$). With this configuration, the diameter of one of the mirrors (e.g., the mirror on the right) is much smaller than the diameters of the gain medium and of the other mirror. A typical ray starts at the smaller mirror, passes through the gain medium while diverging, bounces back from the feedback mirror, returns for a second path through the yet-unused part of the gain medium, and emerges from the cavity by "spilling over" the edges of the small mirror. By diverging continuously as it oscillates between these mirrors, the beam uniformly fills the volume of the gain medium. The highly reflecting spot at the output coupling side projects a shadow into the laser beam, which now has a hollow cross section; owing to its shape, this

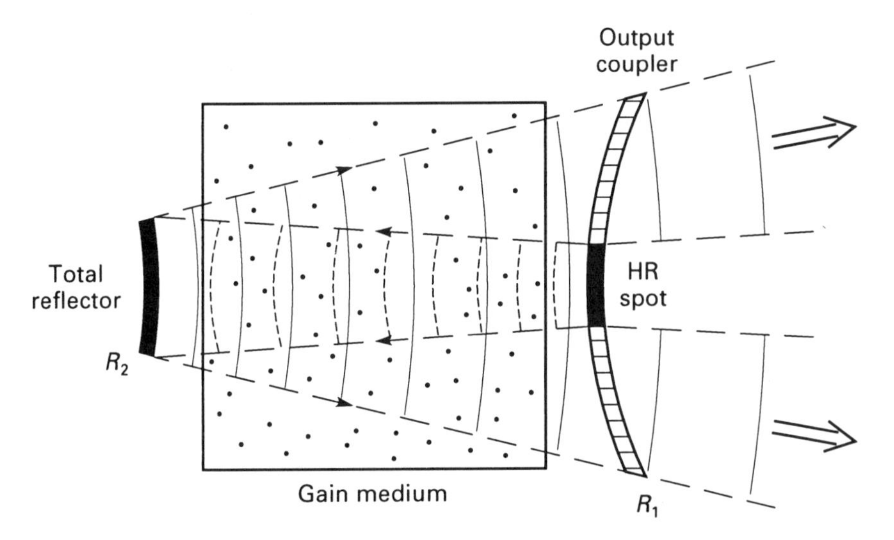

Figure 13.10 An unstable oscillator consisting of two convex mirrors. Output coupling is through the clear annulus around the small mirror on the right. HR denotes "highly reflecting."

is known as the *donut mode*. This cavity design is characterized by diffraction losses that are significantly higher than those of a stable oscillator. However, in media with exceptionally high gain (such as Q-switched lasers), a single or double pass through the medium may suffice to saturate the gain medium (eqn. 12.16). Laser energy can now be derived from the entire volume of the gain medium, thereby more than offsetting the larger diffraction losses. Other designs of unstable resonators and their analysis are presented by Siegman (1974).

Although stable resonators are the most common design of laser cavities, other requirements may force designs that are either unstable or marginally stable. Certain tunable lasers operate with a diffraction grating as the feedback mirror or as an intracavity device. When placed in a Littrow configuration (Figure 7.9), one grating mode - typically the blazed mode - can be forced back to the cavity, thereby assuring the oscillation of a narrowband beam. However, with this configuration, one cavity end mirror is no longer parallel to the other. While stable oscillation may still be possible, the mode is no longer Gaussian.

Diode lasers represent another group of nonstandard cavities. These lasers are usually manufactured by cleaving a semiconductor crystal. Thus, the shape of the cavity end mirrors and their orientations are dictated by the crystalline faces, which are typically flat and parallel to each other. Although the stability of these cavities is theoretically marginal, stable oscillation can still be supported owing to their short cavity length and high gain. Despite this variety of options, most laser beam analyses assume that the beam is Gaussian and so establish a reference against which the actual beam can be compared.

References

Boyd, G. D., and Gordon, J. P. (1961), Confocal multimode resonator for millimeter through optical wavelength masers, *The Bell System Technical Journal* 40: 489-508.

Brorson, S. D. (1988), What is the confocal parameter? *IEEE Journal of Quantum Electronics* 24: 512-15.

Casperson, L., and Yariv, A. (1968), The Gaussian mode in optical resonators with a radial gain profile, *Applied Physics Letters* 12: 355-7.

Fox, A. G., and Li, T. (1961), Resonant modes in a maser interferometer, *The Bell System Technical Journal* 40: 453-88.

Herbst, R. L., Komine, H., and Byer, R. L. (1977), A 200 mJ unstable resonator Nd:YAG oscillator, *Optics Communications* 21: 5-7.

Kogelnik, H. (1965a), On the propagation of Gaussian beams of light through lenslike media including those with a loss or gain variation, *Applied Optics* 4: 1562-9.

Kogelnik, H. (1965b), Imaging of optical modes - resonators with internal lenses, *The Bell System Technical Journal* 44: 455-94.

Kogelnik, H., and Li, T. (1966), Laser beams and resonators, *Proceedings of the IEEE* 54: 1312-29; reprinted (1966) in *Applied Optics* 5: 1550-67.

Siegman, A. E. (1965), Unstable optical resonators for laser applications, *Proceedings of the IEEE* 53: 277-87.

Siegman, A. E. (1974), Unstable optical resonators, *Applied Optics* 13: 353-67.

Siegman, A. E. (1986), *Lasers,* Mill Valley, CA: University Science.

Verdeyen, J. T. (1989), *Laser Electronics,* Englewood Cliffs, NJ: Prentice-Hall.

Homework Problems

Problem 13.1

The effective radius $w(z)$ of a Gaussian beam is considered as the spot size. Find what fraction of the total beam power is transmitted through the area of the disk defined by the spot size. (*Answer:* $1-1/e^2$.)

Problem 13.2

All the parameters of a Gaussian beam are fully characterized at any point where $P(z)$ and $q(z)$ are known (eqn. 13.4). However, since $P(z)$ depends on $q(z)$, it may be possible to determine the beam parameters simply by specifying the real and imaginary parts of the complex beam parameter $q(z)$. Show that if (at a given cross section of the beam) $q=A+iB$, where A and B are real numbers, then $P=P(A,B)$. In other words, show that by determining the numerical value of q, all other beam parameters can be determined.

Problem 13.3

A beam having a spherical wavefront with radius of curvature R_1 is passed through a medium with matrix elements A, B, C, and D (eqn. 2.17) and with an effective focal length f (see Figure 13.11).

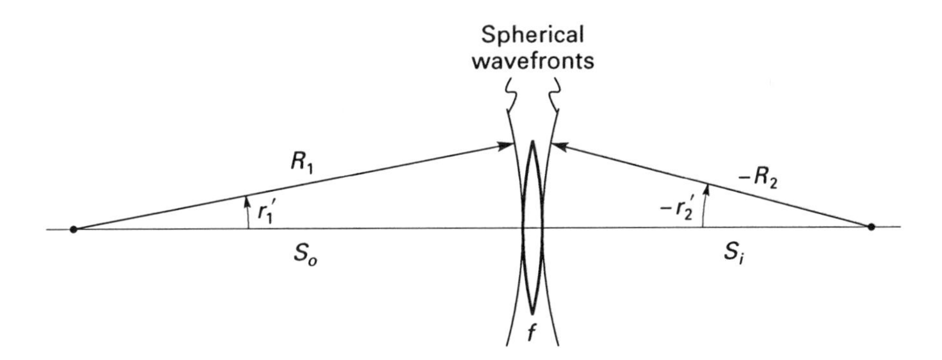

Figure 13.11 Transformation of a spherical wave by an optical element with an effective focal length f.

(a) Show that the radius of curvature of the output beam is

$$\frac{1}{R_2}=\frac{1}{R_1}-\frac{1}{f},$$

where R is defined as positive if it is concave when viewed from the left side of the front.

(b) Show that, for paraxial propagation where $R_1 \approx r_1/r_1'$, the radius of curvature of the transmitted beam is

$$R_2=\frac{AR_1+B}{CR_1+D}.$$

Problem 13.4

A lens is used to focus a Gaussian laser beam with a long Rayleigh range b_1. Assume that the lens is placed at the waist of the incident beam. Show that the waist of the focused beam is at

$$d_2 = \frac{f}{1+[f/(b_1/2)]^2}.$$

Compare this to the result of geometrical optics (e.g. eqn. 2.10).

Problem 13.5

For laser-assisted cutting or surface processing, the spot size of the focused beam w_2 must be small. To achieve such tight focusing, the spot size of the incident beam w_1 is expanded and the tightly focused beam is obtained by using a low $f/\#$ lens. Thus, typically $dw_2/dw_1 < 0$. However, by placing the focusing lens at a certain location along the incident beam, dw_2/dw_1 becomes positive and the spot size of the focused beam increases when the spot size of the incident beam is increased. Find where the focusing lens should be placed to obtain $dw_2/dw_1 < 0$.

Problem 13.6

Read Problem 7.1. Assume that the laser beam used for these measurements is Gaussian, with $\lambda = 1.064\ \mu$m. The spot size of the beam at the moon is required to be $w_2 < 10$ km. Determine where the waist of the beam should be located to minimize the diameter of the launching mirror. Find the radius w_1 of that mirror and the radius w_0 of the beam at the waist.

Problem 13.7

A telescope – consisting of two lenses with focal lengths f_1 and f_2 placed parallel to each other at a distance of f_1+f_2 apart – is used to expand a laser beam. Assume that the input and output planes of the telescope are at the focal planes of the two lenses. Find the relation between the beam parameters q_1 and q_2 at the two sides of the telescope and the magnification of the beam if the waist is at the input plane. (*Answer:* $q_2/q_1 = (f_2/f_1)^2$.)

Problem 13.8

Show that, for a cavity with two spherical end mirrors, the condition for stable oscillation is

$$0 < g_1 g_2 < 1,$$

where $g_1 = 1 - d/R_1$ and $g_2 = 1 - d/R_2$.

Problem 13.9

For a better overview of the stability of empty cavities, the stability condition (eqn. 13.41) can be presented graphically (Kogelnik and Li 1966). Obtain a stability diagram by defining the functions $g_1 = 1 - d/R_1$ and $g_2 = 1 - d/R_2$ and drawing the lines of $g_1 g_2 = 1$ on a (g_1, g_2) plane, where g_1 is the ordinate and g_2 the abscissa. Shade the regions of the diagram that correspond to an unstable cavity, while leaving unmarked the regions that represent stable oscillation. Also divide the diagram into regions where stable oscillation is possible with $R_1 > 0$, or with $R_1 < 0$, or with $R_2 > 0$, or with $R_2 < 0$.

Problem 13.10

An empty cavity consists of two mirrors with curvatures of R_1 and R_2 separated by a distance d. Assume that the beam waist is located at a distance z_1 from mirror R_1. Find the ray transfer matrix for one cavity transit starting at the waist, and express b and z_1 in terms of the cavity parameters R_1, R_2 and d.

Problem 13.11

A laser cavity consisting of one flat mirror and one spherical mirror with curvature R can be viewed as one half of a cavity with two identical spherical mirrors with curvature R. Show that the output characteristics (b or w_0) of a cavity formed by two identical spherical mirrors are identical to those of a cavity formed by placing a flat mirror at the center between the spherical mirrors.

Research Problems

Research Problem 13.1

Laser beams are often used for two-dimensional planar illumination of gas flows. In these applications, an axially symmetric Gaussian laser beam is shaped into a light sheet. The thickness of the sheet must be thin relative to the desired spatial resolution, its width must exceed the dimensions of the imaged plane, and the distribution of the irradiance within the sheet may not vary by more than a factor of 2 from maximum to minimum.

Design a beam-shaping optical system, consisting of two spherical lenses and one cylindrical lens, to transform a $\lambda = 532$-nm beam with a radius at the waist of $w_0 = 1$ mm into a light sheet to be used for the imaging of a 5×5-cm test section with a resolution that is better than 1 mm and with illumination uniformity as described above. Considerations for the optimization of your design must include the length of your optical system and the size of your lenses. Assume, arbitrarily, that the cost of your lenses scales quadratically with their diameter and that both cost and space must be minimized. Can the design be improved by the addition of another lens (either spherical or cylindrical)?

Research Problem 13.2

A design for an unstable Nd:YAG oscillator is presented by Herbst, Komine, and Byer (1977). It is intended for use in Q-switched lasers where the single-pass gain is known to be high. The design consists of one convex cavity end mirror with a radius of curvature $R_1 < 0$ and a concave mirror with a radius of curvature $R_2 > 0$. Radiation is emitted around the convex mirror R_1. Since the cavity is assembled as a confocal cavity, the output beam is collimated. Assume that the diameters of the concave mirror and of the laser rod are equal and that diffraction effects are negligible.

(a) Trace the outermost rays of a laser beam, starting at the convex mirror, passing through the gain medium, reflecting back by the concave mirror R_2, and passing outside the edges of the convex mirror.

(b) Derive the equations presented in the paper for R_1 and R_2 in terms of the magnification M, and prove that the geometric output coupling is:

$$\delta = 1 - 1/M^2.$$

(c) Assuming that diffraction effects are negligible, find R_2 and L when $R_1 = -50$ cm and $M = 3.53$; compare your results to those of the paper.

(d) Assuming that scattering and absorption losses are negligible, find an expression for the amplification G that must be experienced by the beam after passing once through the gain medium if the laser is to operate at steady state. Find the numerical value of the amplification necessary to support the cavity described in the paper.

(e) New configurations of Nd:YAG lasers use diode lasers for pumping. In these designs, the Nd:YAG rod is replaced with a slab, which can be pumped by an array of diodes that illuminate it uniformly. To fill the volume of this slab with the Nd:YAG laser beam, the cavity is designed as an unstable resonator using cylindrical cavity end mirrors. These strip-shaped mirrors are curved in only one plane - the plane that contains the wide dimension of the slab. Output, similar to the design presented by Herbst et al., is outside the edges of the convex mirror. Thus, absent diffraction effects, the laser beam appears as a rectangle with a dark strip at its center. Find the radii of curvature of the cylindrical mirror in terms of the cavity length L and the magnification M.

Appendix A

To convert a quantity given in the Gaussian system to the MKSA system, multiply by the appropriate constant listed below under "Conversion coefficient."

Table A.1 *Frequently used physical quantities and their conversion from the Gaussian system of units to MKSA*

Description of quantity	Symbol	Conversion coefficient
Electric field, voltage	$\mathbf{E}$, V	$\sqrt{4\pi\epsilon_0}$
Electric displacement	$\mathbf{D}$	$\sqrt{4\pi/\epsilon_0}$
Charge density	ρ	$1/\sqrt{4\pi\epsilon_0}$
Magnetic field	$\mathbf{H}$	$\sqrt{4\pi\mu_0}$
Magnetic induction	$\mathbf{B}$	$\sqrt{4\pi/\mu_0}$
Specific conductivity	σ	$1/4\pi\epsilon_0$
Dielectric constant	ϵ	$1/\epsilon_0$
Magnetic permeability	μ	$1/\mu_0$
Resistance	R	$4\pi\epsilon_0$
Capacitance	C	$1/4\pi\epsilon_0$
Inductance	L	$4\pi\epsilon_0$
Speed of light	c	$1/\sqrt{\mu_0\epsilon_0}$
Length	l	1
Mass	m	1
Time	t	1
Force	F	1

$\epsilon_0 = 8.854 \times 10^{-12}$ C^2/N-m^2 $= 8.854 \times 10^{-12}$ C/V-m
$\mu_0 = 4\pi \times 10^{-7}$ Wb/A-m
$c_0 = 1/\sqrt{\epsilon_0\mu_0} = 2.998 \times 10^8$ m/s
$\eta_0 = \sqrt{\mu_0/\epsilon_0} = 376.73$

Appendix B

Table B.1 *Frequently used physical constants*

Coefficient	Symbol	Value
Avogadro's number	N_{Av}	6.023×10^{26} kmoles^{-1}
Boltzmann's constant	k	1.38×10^{-23} J/K
Charge of electron	e	1.60×10^{-19} C
Electron rest mass	m_e	9.11×10^{-31} kg
Loschmidt number	N_0	2.69×10^{19} cm^{-3}
Neutron rest mass	m_n	1.67×10^{-27} kg
Proton rest mass	m_p	1.67×10^{-27} kg
Speed of light in free space	c_0	2.998×10^{8} m/s
Electron volt	eV	8,054.3 cm^{-1}
Energy of one wavenumber	cm^{-1}	1.99×10^{-23} J
Planck's constant	h	6.63×10^{-34} J-s
Stefan–Boltzmann constant	σ	5.6697×10^{-12} W/cm^2-K^4
Atmospheric pressure	P_0	1.01325×10^{5} Pa
Universal gas constant	R_u	8.314×10^{3} J/kmoles-K

Index